T0327335

Pearl Millet

A Resilient Cereal Crop for Food, Nutrition, and Climate Security

Editors

Ramasamy Perumal, PhD, is the Professor of Sorghum and Pearl Millet breeding at Kansas State University. His research focuses on the release of several seed and pollinator parents with heat/chill and drought tolerance and mapping populations. He completed his PhD in Plant Breeding and Genetics in 1993 from the Tamil Nadu Agricultural University, India. Dr. Perumal is the receipient of The Rockefeller Foundation Post Doctoral Fellow Award in Sorghum Biotechnology (1998–2000).

P. V. Vara Prasad, PhD, is the University Distinguished Professor, R.O. Kruse Professor of Agriculture and Director of the Feed the Future Sustainable Intensificaiton Innovation Lab at Kansas State University. Dr. Prasad received his MS from Andhra Pradesh Agricultural University, India, and his PhD from the University of Reading, United Kingdom. His research focuses on understanding responses of crops to changing environments and developing best management strategies to improve and protect yields. He is an elected fellow of the American Society of Agronomy; the Crop Science Society of America; the American Association for the Advancement of Science. He is a former President of the Crop Science Society of America.

C. Tara Satyavathi, PhD, is the Director, Indian Council of Agriculture Research-Indian Institute of Millets Research, Hyderabad, India. She has also steered the Global Centre of Excellence for Millets (Research & Development) since 2023. She received her PhD from the Indian Agricultural Research Institute, New Delhi. Dr. Satyavathi is recipient of many awards including; Dr. Panjabrao Deshmukh Outstanding Woman Scientist (2016), Outstanding Millet Scientist (2018), Outstanding Research Contribution in Pearl Millet Improvement (2008, 2018, 2021), Dr. M. S. Swaminathan Outstanding Woman Scientist (2019) and Eminent Scientist (2023).

Mahalingam Govindaraj, PhD, is currently a Senior Scientist for Crop Development at HarvestPlus and the Alliance of Bioversity International and International Center for Tropical Agriculture. He coordinates the biofortification crop development research network and builds research capacities for biofortification. Dr. Govindaraj received his PhD from Tamil Nadu Agricultural University, India, specializing in plant breeding and genetics Dr. Govindaraj is the recipient of many awards including; International Scholar in 2009, Young Scentsit in 2016and Norman E. Borlaug Award for Field Research and Application in 2022.

Abdou Tenkouano, PhD, is the Director General of the International Centre of Insect Physiology and Ecology *(icipe)*, Kenya. Prior to this he served as the Executive Director of CORAF (the West and Central Africa Council for Agricultural Research and Development), an International non-profit Association of National Agricultural Research and Development institutions from 23 countries, with headquarters in Dakar, Senegal. His academic background is in crop science, with a PhD in Genetics (1993) and a MSc in Plant Breeding (1990) from Texas A&M University.

Contributors

Krishnam Raju Addanki
International Crops Research Institute
for the Semi-Arid
Tropics, (ICRISAT)
Hyderabad, India

Supriya Ambawat
ICAR-AICRP on Pearl Millet,
Agriculture University
Jodhpur, Rajasthan, India

Srikanth Bollam
International Crops Research
Institute for the Semi-Arid Tropics
(ICRISAT)
Hyderabad, India

Ignacio A. Ciampitti
Department of Agronomy,
Kansas State University
Manhattan, Kansas, USA

Omar Diack
Center of Excellence on Dry Cereals and
Associated Crops (CERAAS)
Thies, Senegal

Abdoulaye Dieng
Center of Excellence on Dry Cereals and
Associated Crops (CERAAS)
Thies, Senegal

Antonio DiTommaso
School of Integrative Plant Science, Soil
and Crop Sciences Section, Cornell
University
Ithaca, New York, USA

Mamadou T. Diaw
Center of Excellence on Dry Cereals and
Associated Crops (CERAAS)
Thies, Senegal

Maduraimuthu Djanaguiraman
Department of Crop Physiology
Tamil Nadu Agricultural University
Coimbatore, India

Aliou Faye
Center of Excellence on Dry Cereals and
Associated Crops (CERAAS)
Thies, Senegal

Prakash I. Gangashetty
International Crops Research Institute
for Semi-Arid Tropics (ICRISAT)
Hyderabad, India

Mahalingam Govindaraj
HarvestPlus, Alliance of Bioversity
(CIAT), and International and the
International Centre for Tropical
Agriculture (ICRISAT)
Hyderabad, India

S. K. Gupta
International Crops Research Institute
for the Semi-Arid Tropics (ICRISAT)
Hyderabad, India

Drabo Inoussa
CIMMYT
Senegal, Africa

Jagdish Jaba
International Crops Research Institute
for the Semi-Arid Tropics (ICRISAT)
Hyderabad, India

Prashant Jha
Department, Plant, Environment
Management & Soil Sciences
Louisiana State University
Baton Rouge, Louisiana, USA

Rajkumar P. Juneja
Pearl Millet Research Station, Junagadh
Agricultural University
Jamnagar, Gujarat, India

Ghislain Kanfany
Senegalese Agricultural Research
Institute – ISRA
Thies, Senegal

Vinutha Kanuganahalli
International Crops Research Institute
for the Semi-Arid Tropics, (ICRISAT)
Hyderabad, India

Vikas Khandelwal
ICAR-AICRP on Pearl Millet,
Agriculture University
Jodhpur, Rajasthan, India

Kassim-Al-Khatib
Department of Plant Sciences
University of California
Davis, California, USA

Shreeja Kulla
ICAR-Indian Institute of Millets Research
Hyderabad, India

Vipan Kumar
School of Integrative Plant Science,
Soil and Crop Sciences Section, Cornell
University
Ithaca, New York, USA

Christopher R. Little
Department of Plant Pathology
Kansas State University
Manhattan, Kansas, USA

Talla Lo
Center of Excellence on Dry Cereals
and Associated Crops (CERAAS)
Thies, Senegal

R. S. Mahala
SeedWorks International Pvt. Ltd.
Hyderabad, India

Mahesh Mahendrakar
International Crops Research Institute
for the Semi-Arid Tropics (ICRISAT)
Hyderabad, India

Sandeep Marla
Department of Agronomy
Kansas State University
Manhattan, Kansas, USA

Doohong Min
Department of Agronomy
Kansas State University
Manhattan, Kansas, USA

Rezazadeh Mohammed
International Crops Research Institute
for the Semi-Arid Tropics (ICRISAT)
Bamako, Mali.

Anuradha Narala
ICAR-Indian Institute of Millets
Research
Hyderabad, India

Augustine Obour
Kansas State University, Agricultural
Research Center
Hays, Kansas, USA

Sushil Pandey
ICAR-National Bureau of Plants Genetic
Resources (NBPGR)
New Delhi, India

Sabreena A. Parray
Department of Agronomy
Kansas State University
Manhattan, Kansas, USA

Ramasamy Perumal
Agricultural Research Center, Kansas
State University
Hays, Kansas, USA

Shivaprasad Doddabematti Prakash
Department of Grain Science and
Industry
Kansas State University
Manhattan, Kansas, USA

P. V. Vara Prasad
Department of Agronomy
Kansas State University
Manhattan, Kansas, USA

A. S. Priyanka
Department of Crop Physiology
Tamil Nadu Agricultural University
Tamil Nadu, India

Mahesh Pujar
International Crops Research Institute
for the Semi-Arid Tropics (ICRISAT)
Hyderabad, India

Manoj Kumar Pulivarthi
Department of Grain Science and
Industry
Kansas State University
Manhattan, Kansas, USA

Dayakar Rao
ICAR-Indian Institute of Millets Research
Hyderabad, India

P. Rakshith
International Crops Research Institute
for Semi-Arid Tropics (ICRISAT)
Hyderabad, India

Ajay P. Ramalingam
Department of Agronomy, Kansas State
University
Manhattan, Kansas, USA

P. Sanjana Reddy
ICAR-Indian Institute of Millets Research
Hyderabad, India

Mohammed Riyazaddin
International Crops Research Institute
for Semi-Arid Tropics (ICRISAT)
Hyderabad, India

C. Tara Satyavathi
ICAR-Indian Institute of Millets Research
Hyderabad, India

Desalegn D. Serba
USDA-ARS, Water Management and
Conservation Research: Maricopa
Arizona, USA

Arun K. Shanker
Central Research Institute for Dryland
Agriculture
Hyderabad, India

Rajan Sharma
International Crops Research Institute
for the Semi-Arid Tropics (ICRISAT)
Hyderabad, India

Kaliramesh Siliveru
Department of Grain Science and Industry
Kansas State University
Manhattan, Kansas, USA

Kuldeep Singh
International Crops Research Institute
for the Semi-Arid Tropics (ICRISAT)
Hyderabad, India

Parvaze A. Sofi
Stress Physiology Lab, Sher-e-Kashmir
University of Agriculture Sciences and
Technology-Kashmir
Jammu and Kashmir, India

Rakesh K. Srivastava
International Crops Research Institute
for the Semi-Arid Tropics (ICRISAT)
Hyderabad, India

Abdou Tenkouano
Director General of the International
Center of Insect Physiology and
Ecology (*icipe*)
Kenya

Nepolean Thirunavukkarasu
ICAR-Indian Institute of Millets
Research
Hyderabad, India

Timothy C. Todd
Department of Plant Pathology
Kansas State University
Manhattan, Kansas, USA

Midhat Zulafkar Tugoo
Department of Agronomy
Kansas State University
Manhattan, Kansas, USA

S. J. Vaishnavi
Department of Crop Physiology
School of Agricultural Sciences, Amrita
Vishwa Vidyapeetham
Tamil Nadu, India

Mani Vetriventhan
International Crops Research Institute
for the Semi-Arid Tropics (ICRISAT)
Hyderabad, India

Veeresh S. Wali
ICAR-Indian Institute of Millets Research
Hyderabad, India

O. P. Yadav
ICAR-Central Arid Zone Research
Institute (CAZRI)
Jodhpur, Rajasthan, India

EDITORIAL CORRESPONDENCE

American Society of Agronomy
Crop Science Society of America
Soil Science Society of America
5585 Guilford Road, Madison, WI
53711-58,011, USA

SOCIETY PRESIDENTS

Kristen S. Veum (ASA)
Kimberly A. Garland-Campbell (CSSA)
Michael L. Thompson (SSSA)

SOCIETY EDITORS IN CHIEF

David E. Clay (ASA)
Bingru Huang (CSSA)
Craig Rasmussen (SSSA)

BOOK AND MULTIMEDIA PUBLISHING COMMITTEE

Girisha K. Ganjegunte (Chair)
Sangamesh V. Angadi
Xuejun Dong
Fugen Dou
Limei Liu
Shuyu Liu
Gurpal S. Toor
Sara Eve Vero

PUBLISHING STAFF

Matt Wascavage (Director of
Publications)
Richard J. Easby (Content Strategy
Program Manager)
Robert Gagnon (Copyeditor)

Pearl Millet

A Resilient Cereal Crop for Food, Nutrition, and Climate Security

Edited by
Ramasamy Perumal, P. V. Vara Prasad, C. Tara Satyavathi,
Mahalingam Govindaraj, and Abdou Tenkouano

Editorial Correspondence:
American Society of Agronomy, Inc.
Crop Science Society of America, Inc.
Soil Science Society of America, Inc.
5585 Guilford Road, Madison, WI 53711-58011, USA agronomy.org • crops.org • soils.org

Registered Offices:
John Wiley & Sons, Inc., 111 River Street, Hoboken, NJ 07030, USA

For details of our global editorial offices, customer services, and more information about Wiley products, visit us at www.wiley.com.

Wiley also publishes its books in a variety of electronic formats and by print-on-demand. Some content that appears in standard print versions of this book may not be available in other formats.

Library of Congress Cataloging-in-Publication Data Applied for
Paperback: 9780891184041

Cover Design: Wiley
Cover Image: © V. R. Murralinath/Adobe Stock Photos, Gv Image-1/Shutterstock, DEEP/Adobe Stock Photos

Set 9.5/12.5pt STIXTwoText by Straive, Chennai, India

Contents

Foreword

World-wide climatic changes, including unpredictable and highly variable precipitation, intermittent drought, and high temperatures, are becoming common challenges for crop cultivation throughout the growing season. These challenges are creating a slow-developing crisis that threatens the long-term sustainability of global food production. Overall, we are seeing more irrigated land converting to dryland, reduced crop yields, and lower land values. At the same time, the demand for food is continually increasing as the world's population grows, reaching a projected nine billion by 2050.

Pearl millet (*Pennisetum glaucum* [L.] R. Br.) is a climate resilient, water-use efficient, dryland crop. Globally, it is the sixth most important crop after rice, wheat, corn, barley, and sorghum. Pearl millet is a valuable alternative resource to traditionally grown grains and forages and can increase the profitability of cropping systems in dryland cultivation. Pearl millet can be grown in difficult or harsh conditions, including areas of low soil fertility, high pH, low soil moisture, high temperature, high salinity and limited rainfall, where other cereals like maize, rice, sorghum and wheat would fail. Pearl millet is a staple food for approximately 90 million people in Africa, and Asia, and its stalks are used as green and dry fodder for livestock. This book will explore the potential of pearl millet to play a significant role in ensuring global food, forage and nutrition security, particularly in the face of changing climate scenarios.

Pearl Millet is traditionally grown as a subsistence crop in the Sahel region of West Africa. It is well suited to this arid and hot region. Pearl Millet is more nutritious than maize, rice, wheat, and sorghum because its grain contains higher levels of protein, vitamins and essential micronutrients such as iron and zinc. This nutri-cereal crop has the potential to reduce malnutrition in developing countries. It also has the potential to grow in many new areas such as the United States, where it has great potential for the future for both grain and forage. Besides its grain nutritional value, pearl millet also genetically possesses several forage attributes. For example, a short growing season, high photosynthetic efficiency, thin stem more tillers, low lignin increased digestibility, and no prussic acid (animal can be grazed at any growing stages) can make pearl millet a suitable alternative to drought-sensitive food/feed crops in regions with receding groundwater levels, e.g., the Ogallala Aquifer Region.

To highlight the value and potential importance of millets at the global level, the United Nations declared the year 2023 as the "International Year of Millets." We hope this monograph complements and supports the work of the UN by focusing on this valuable crop. Throughout the book the author team explores the genetic and genomic resources of pearl millet, crop production strategies, stress tolerance, nutritional value of grain and forage, and strategies for further research for overall crop improvement. I would like to take this opportunity to thank the extensive and diverse team of experienced scientists who contributed to this important work. I sincerely believe this monograph will become a significant reference as we battle food security and the challenges faced by climate change.

P. V. Vara Prasad
Director, Sustainable Intensification Innovation Lab
University Distinguished Professor, Crop Ecophysiology
R.O. Kruse Professorship in Agriculture
Former President, Crop Science Society of America (2021)

1

Global Production, Status, and Utilization Pattern of Pearl Millet

C. Tara Satyavathi, Supriya Ambawat, Nepolean Thirunavukkarasu, Vikas Khandelwal, Sanjana Reddy, and Anuradha Narala

Chapter Overview

Pearl millet [*Pennisetum glaucum* (L.) R. Br.] is a major cereal and staple food in the developing world, especially in the arid and semi-arid regions of Asia and Africa. It is cultivated on >30 million ha in the arid and semi-arid tropical regions of Asia and Africa, contributing 40.51% of the global millet production, with 60% of the cultivation area in Africa and 35% in Asia. It is primarily cultivated for grain production, but its stover is also valued as dry fodder. It is resilient to climate change due to its inherent adaptability to drought and high temperatures and is well adapted to marginal lands with low productivity. In addition, it has several nutritional properties and is rightly termed as nutricereal. It has high levels of energy, dietary fiber, and proteins with a balanced amino acid profile, many essential minerals, some vitamins, and antioxidants, which help in combating several diseases. Due to its excellent properties, it can play a vital role in overcoming malnutrition to ensure food and nutritional security. The different efforts being taken up to enhance its production and consumption on a global scale will be highly useful in the present scenario to overcome the challenges of the growing population. Thus, pearl millet, with its hardiness and good prospects of genetic enhancement, has the potential of contributing to sustainable food and nutritional security of farmers in the arid and semi-arid tropics and other parts of the world with similar agroecologies.

Introduction

Pearl millet [*Pennisetum glaucum* (L.) R. Br.] is the sixth major cereal crop globally, followed by maize, rice, wheat, barley, and sorghum. It is cultivated on

Pearl Millet: A Resilient Cereal Crop for Food, Nutrition, and Climate Security, First Edition.
Edited by Ramasamy Perumal, P. V. Vara Prasad, C. Tara Satyavathi, Mahalingam Govindaraj, and Abdou Tenkouano.
© 2025 American Society of Agronomy, Inc., Crop Science Society of America, Inc., Soil Science Society of America, Inc. Published 2025 by John Wiley & Sons, Inc.

>30 million ha in the arid and semi-arid tropical regions of Asia and Africa, contributing 40.51% of the global millet production, with 60% of the cultivation area in Africa and 35% in Asia (FAO, 2020). This crop is widely used as a staple food for human consumption and as fodder and feed for livestock. Additionally, it is used in various industries, such as alcohol fuel, starch, and the processed food sectors. Pearl millet is imperative to mitigate the adverse effects of changing climate and can provide income and food security to the farming communities of arid regions. In India, it is the fourth most commonly grown food crop, after rice, wheat, and maize. It shows rapid growth with minimal inputs, high photosynthetic efficiency, a balanced and good nutritional profile, and tolerance to extreme climatic conditions and biotic stresses.

Pearl millet is a hardy crop that can thrive in adverse agro-climatic conditions. It can withstand low soil fertility, high soil pH, high soil Al^{3+} saturation, low soil moisture, high temperature, high soil salinity, and low rainfall, making it a viable option in areas where other staple cereal crops like rice and wheat struggle to survive. As per the fourth advance estimates during 2021–2022, the pearl millet area in India was 6.70 million ha, with an average production of 9.62 million tons and productivity of 1,436 kg/ha (Directorate of Millets Development, 2024). The major pearl millet growing states in India are Rajasthan, Maharashtra, Uttar Pradesh, Gujrat, and Haryana, contributing to 90% of total production in the country. Rajasthan contributes nearly 45%, followed by Uttar Pradesh (19%), Haryana (9%), Gujarat (9%), Maharashtra (6%), and Tamil Nadu (2%). Most pearl millet in India is grown in the rainy (*kharif*) season (June/July–September/ October). Pearl millet is also cultivated during the summer season (February– May) in parts of Gujarat, Rajasthan, and Uttar Pradesh and during the post-rainy (*rabi*) season (November–February) at a small scale in Maharashtra and Gujarat.

Pearl millet is more nutritious than commonly consumed staple crops such as wheat, rice, maize, and sorghum. Its grain is rich in carbohydrates, proteins, fats, fibers, resistant starch (RS), vitamins, antioxidants, and essential micronutrients like Fe and Zn and has a more balanced essential amino acid profile than maize or sorghum. It also contains omega-3 and omega-6 fatty acids (polyunsaturated fatty acids), which are good for heart and brain health. Despite being less expensive than pearls, pearl millet offers pearl-like quality that is advantageous to the body: 100 g of grain contains 360 calories, 12 g moisture, 12 g protein, 5 g fat, 2 g minerals, 1 g fiber, 67 g carbohydrates, 42 mg Ca, 242 mg P, and 8 mg Fe. It is a gluten-free grain with high-quality protein, making it ideal for those with gluten allergies. It is the only cereal that retains its alkaline nature after cooking and has a low glycemic index (GI), making it a diabetic-friendly food. It acts as a probiotic food for microflora in our inner ecosystem and hydrates our colon preventing constipation. Its consumption has been associated with protection against certain types of cancer and cardiovascular diseases. It has high proportions of slowly digestible starch and

RS, which contribute to its low GI, and is much sought after in the recent era of transforming diets, eating habits, and the food industry. As a result, pearl millet is gaining popularity among health-conscious people worldwide.

Pearl millet can play a vital role in overcoming malnutrition to ensure food and nutritional security. Due to its excellent nutritional properties, pearl millet is designated as a "nutri-cereal" (Gazette of India, No. 133 dtd 13th April, 2018) for production, consumption, and trade and was included in the public distribution system. India's prime minister has given it the name of "Shree Anna" for its superior quality over all other grains. In addition, to include millets in the mainstream and to exploit their nutritionally superior qualities and promote their cultivation, the Government of India declared 2018 as the "Year of Millets," and the FAO Committee on Agriculture forum declared 2023 as the "International Year of Millets."

Pearl millet is the next-generation crop holding the potential of nutritional richness and climate resilience. To fully tap into its potential, efforts must be made to enhance its production and consumption on a global scale. The availability of high-quality whole genome sequences and re-sequencing information will provide a good opportunity for genomic dissection and for exploiting the nutritional and climate-resilient attributes of pearl millet. Additionally, an integrated approach that includes genomics, transcriptomics, proteomics, and metabolomics can provide valuable insights into the genetic improvement and biofortification of pearl millet (Ambawat et al., 2020; Anuradha et al., 2017; Dita et al., 2006; Lata, 2015). There is a need to focus on this very important crop and harness its suitability to adverse conditions and to use its capacity to ensure global food and nutritional security. Therefore, we must continue to make advances in the pearl millet improvement program and prioritize this vital crop in the face of changing climatic conditions to ensure food and nutritional security.

History of Pearl Millet: Origin and Spread

Pearl millet originated in tropical Western Africa around 4,000 years ago. It is a member of the grass family and was originally a wild plant in Africa, where the largest members of both wild and cultivated forms of this species occur. Later, it differentiated into two races: the *globosum* race, which moved to the western side, and the *typhoides* race, which reached Eastern Africa and spread to India and Southern Africa around 2,000–3,000 years ago. Due to its evolution under drought and high temperatures, pearl millet can tolerate harsh conditions in Indian and African deserts better than other cereals, such as wheat and rice (Sheahan, 2014). Based on crossability, cross fertility, and the complexity of gene transfer studies of *Pennisetum* species, three gene pools were classified: primary, secondary, and tertiary (Harlan & De-Wet, 1971). All types of weedy, cultivated, and wild diploids (2n = 2x = 14) were included in primary gene pool. The secondary gene pool was comprised exclusively of tetraploid *P. purpureum* (Shum.) (2n = 4x = 28), and the tertiary gene pool contained widely related *Pennisetum* species of different ploidy levels (Dujardin & Hanna, 1989).

Unlike in the primary gene pool, where hybridization among species results in fertile hybrids, gene transfer from the secondary and tertiary gene pools is not straightforward due to genetic incompatibilities. For that, introgression is carried out via embryo rescue–assisted techniques under artificially controlled conditions. Producers are reluctant to work in the pre-breeding sector due to the long breeding cycles required for isolating desired segregants, the low hybrid seed set, seed sterility, the large advanced population size required for selection, and the maintenance of wild germplasm. Earlier pearl millet researchers concentrated on identifying and utilizing traits donors available in cultivated pearl millet and its primary gene pool. A study of 529 wild pearl millet species, consisting of *P. violaceum*, *P. mollissimum*, *P. purpureum*, *P. pedicellatum*, and *P. polystachyon* under artificial and field conditions, identified 223 accessions with no disease symptoms (Singh & Navi, 2000). Crosses with two wild species (*P. pedicellatum* and *P. polystachion*) attempted to introgress downy mildew resistance, but no seed set was noted in the hybrids formed (Dujardin & Hanna, 1989).

Commercial Uses

Pearl millet is a versatile crop that serves as a staple food for humans and as a source of feed for livestock. Its different uses in different parts of the world are summarized in Table 1.1. It is used in industries such as—alcohol, fuel, starch and processed food sectors. Around 90 million people in the Sahelian region of Africa and northwestern India consume pearl millet grain as a primary source of food (Jukanti et al., 2016; Srivastava et al., 2020). It is used in different forms at the global level, such as unleavened bread (roti or chapatti), porridge, gruel, dessert, and is generally defined as a "poor man's bread." Its flour can substitute (10%–20%) for wheat flour in whole-grain breads, pretzels, crackers, tortillas, and dry and creamed cereals (Dahlberg et al., 2003). Although it is mainly grown for food and forage in India and Africa, it serves as a major component of poultry and livestock feed in the American continent (Serba et al., 2020). Pearl millet can be used as a quality-protein grain for many livestock-feeding purposes. It has higher crude protein, essential amino acids, and ether-extract concentrations in comparison to maize and thus can prove more promising for feeding poultry and cattle. In addition to being grown for grain, pearl millet is also cultivated as a forage crop for livestock grazing, silage preparation, hay production, and the cutting of green

Table 1.1   Utilization Pattern of Pearl Millet in Different Parts of the World.

Site no.	Country/city	Use	Popular names of items
1	India	Industries	Alcohol and fuel
		Forage and grain	Livestock and poultry feed
		Traditional food products	Roti or chapatti, rabri, dalia, gruel, kheer, churma, suhaali, khichri, sakarpare, gulgule, mathi, ladoo, barfi, sev, pakoda, dhokla
		Baked products	Biscuits, bread, cookies, cakes, puffs, muffin, nan-khatai
		Extruded products	Noodles, pasta, vermicelli, macroni
		Health products	Sugar free biscuits, idli
2	West Africa/ Sahelian region	Industries	Non-alcoholic or alcoholic beverages
		Forage and grain	Livestock and poultry feed
		Food products	Stiff or thick porridges (Tuwo or Tô)
3	Senegal	Steam-cooked products	Couscous, thin porridge (*bouillie*)
4	Nigeria and Niger	Food products	Porridge (*Fourra*) and fried cake (*Masa*)
5	Senegal	Food products	Soungouf; "Sankhal" and "Araw"

fodder (Newman et al., 2010). The advantage of using this crop as livestock feed is that it may be fed to animals at any stage of crop development without negative consequences (Arya et al., 2013). This has been practiced in the countries like the United States (Sheahan, 2014), during the summer in Australia (Hanna, 1996), Canada (Brunette et al., 2014), and Mexico (Urrutia et al., 2015). It is also used as triple cropping method in southern Kyushu, Japan (Li et al., 2019); Iran (Aghaalikhani et al., 2008); Central Asia (Nurbekov et al., 2013); and Brazil (Dias-Martins et al., 2018). It has become a significant fodder crop in northwest India during the summer (Amarender Reddy et al., 2013) and recently has been used as a cover crop in Brazil (5 million ha) (De Assis et al., 2018; Dias-Martins et al., 2018). Climate change and global warming have prompted a reconsideration of previously used fuel energy sources to reduce greenhouse gas emissions. This has resulted in an emphasis on the usage of various biofuels. Among biofuels, pearl millet lignocellulosic feed stock with high cellulose (41.6%) and hemicellulose (22.32%) concentration is a significant source of biofuel production, which may contribute to an increase in farmer income (Packiam et al., 2018).

Value-added products are also gaining popularity, promoting the consumption of pearl millet. Pearl millet can be used for making various traditional food products (e.g., khichri, roti, sakarpare, gulgule, and ladoo), while industries in India are also using it for making products like noodles, pasta, vermicelli, biscuits, bread, cookies, cakes, and puffs. Several indigenous foods and drinks are also made from flour/meal and malt of the millet in Africa and are nutritionally superior to other cereals. Further, the millets market can be segmented based on type into pearl millet, finger millet, proso millet, foxtail millet, and others. In 2018, pearl millet held the greatest share in terms of product volume in the millets market and is likely to expand at >3% compound annual growth rate by 2025. The massive demand share of pearl millets is due to its rich nutritional value because it has eight times higher Fe content than rice (agriXchange, 2024). Millets contain high protein (up to 9.5 g/100 g for teff and fonio), ash, calcium (up to 344 mg/100 g for finger millet), P, and K (up to 250 mg/100 g), Fe, and Zn levels (Obilana & Manyasa, 2002). In West Africa, different countries have their unique food items made from pearl millet. The stiff or thick porridges (Tuwo or Tô) are very famous and generally used in all the Sahelian countries, while Francophone countries like Senegal, Mali, Guinea, Burkina Faso, Niger, and Chad prefer steam-cooked couscous. The thin porridge "bouillie" is also popular in these countries. In Nigeria and Niger, the thin porridge "Fourra" and *Masa*, a fried cake, are quite popular, whereas "Soungouf" "Sankhal" and "Araw" are mainly popular in Senegal (Ajeigbe et al., 2020; Kaur et al., 2014). Its grains are also locally brewed to produce non-alcoholic or alcoholic beverages in Asia and Africa (Dwivedi et al., 2012) (Table 1.1).

Different Agro-ecologies for Pearl Millet Production in India

Pearl millet can grow in varying conditions from near-optimum with high external inputs to highly drought-prone environments. Prioritization of research in cognizance of production constraints and different requirements of various pearl millet–growing regions has led to the formation of three production zones (Zones A_1, A, and B) along with summer hybrids (Figure 1.1). Different public- and pricate-sector pearl millet breeding programs in India have their own product profiles, depending on the need of their target environment (Table 1.2).

Zone A_1 (Hyper-arid Zone)

Approximately 3.5–4.0 mha of pearl millet production lies under Zone A_1, which includes parts of Rajasthan, Gujarat, and Haryana, which receive <400 mm annual rainfall. Genotypes suitable for this area are early maturing (70 days), dual purpose, high yielding, high tillering, downy mildew, and blast resistant with

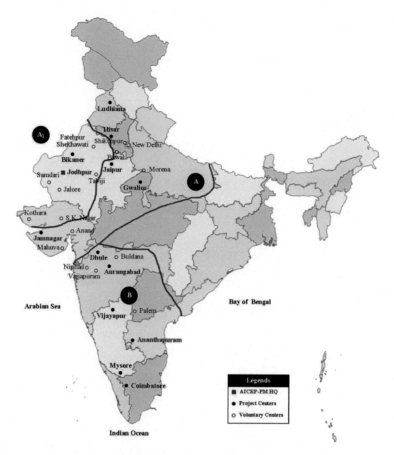

Figure 1.1 Geographical demarcation of Zones A_1, A, and B of pearl millet cultivation in India.

seedling stress tolerance. The seedlings contain benchmark levels of Fe (42 ppm) and Zn (32 ppm).

Zone A (Arid Zone)

Zone A comprises the remaining parts of the states of Rajasthan, Gujarat, and Haryana and the entire pearl millet growing areas of Uttar Pradesh, Madhya Pradesh, Punjab, and Delhi, covering 2.0–2.5 mha. This zone has sandy loam soils and annual rainfall >400 mm. Hybrids and populations suitable for this area are medium maturing (71–80 days), dual purpose, high yielding, downy mildew, and blast resistant with seedling stress tolerance. The seedlings contain benchmark levels of Fe (42 ppm) and Zn (32 ppm).

Table 1.2 Pearl Millet Product Profiles for Different Agro Ecologies of India.

	Zone A_1 (early duration)	Zone A (medium duration)	Zone B (medium duration)	Summer
Name of the commercial line to be replaced	HHB67 (improved)	MPMH 17	Pratap	Proagro 9444
Agro-ecologyzone	Aridzone (Zone A_1)	Zone A	Zone B (medium duration)	Summer pearl millet area
Year of release	2005	2012	2012	2004
Basic traits/unique selling features				
Basic trait 1	Early duration (<45 days)	Medium duration (48 days)	Medium duration (51 days)	Late duration (58 days)
Basic trait 2	Grain yield (~2.0 t/ha)	Grain yield (2.8 t/ha)	Grain yield (2.9 t/ha)	Grain yield (4.0 t/ha)
Basic trait 3	Stover yield (~3.7 t/ha)	–	Stover yield (6.2 t/ha)	Stover yield (7.7 t/ha)
Basic trait 4	Downy mildew (5%)	Downy mildew (5%)	Downy mildew (3.26%) <5%	Downy mildew (2.44%) <5%
Basic trait 5	Blast resistance (3 score)	Blast resistance (3 score)	Blast resistance (4.5 score)	Blast resistance (4.7 score)
Basic trait 6	Rust, smut, and ergot ≤20%	Rust, smut, and ergot ≤20%	Rust, smut, and ergot ≤20%	Rust, smut, and ergot ≤20%
Basic trait 7	Fe ≥42 ppm	Fe 41 ppm	Fe 49 ppm	Fe 48 ppm
Basic trait 8	Zn ≥32 ppm	Zn 34 ppm	Zn 45 ppm	Zn 36 ppm
Value-added trait 1	Grain yield	Grain yield	Grain yield	Grain yield
Future unique selling point	High yield under arid zone	High yield under semi-arid zone	High yield under B zone	High yield under summer cultivation
Trait compared with the benchmark	10% more grain yield	10% more grain yield	10% more grain yield	5% more grain yield
Value added trait 2	Better blast resistance	Better blast resistance	Blast resistance	Downy mildew and blast resistance
Advanced testing for government registration or private sector licensing	5 years	5 years	5 years	5 years

Zone B (Semi-arid Zone)

Zone B covers 1.0–1.5 mha from the states of Maharashtra, Karnataka, Tamil Nadu, and Andhra Pradesh with rainfall >600 mm, heavy soils, and mild temperatures. They are medium- to late-duration hybrids and populations maturing in >80 days or more and are high yielding and downy mildew and blast reistant. The seedlings contain benchmark levels of Fe (42 ppm) and Zn (32 ppm).

Summer Hybrids

The summer pearl millet growing region covers 0.4 mha area, comprising parts of Gujarat and Uttar Pradesh. Suitable genotypes include medium- to late-duration hybrids maturing in >80 days and are high yielding, dual purpose, and downy mildew and blast resistant. The seedlings contain benchmark levels of Fe (42 ppm) and Zn (32 ppm).

Trends in Global Pearl Millet Production

Pearl millet is a descendant of the wild West African grass and was domesticated over 4,000 years ago in the West African Sahel, spreading later to East Africa and India (Sharma et al., 2021). Millet cultivation has expanded globally to >30 million ha, with the largest portion grown in Africa (>18 million ha), followed by Asia (>10 million ha). This crop contributes to half of the world's millet production, with 60% of the cultivation area located in Africa and 35% in Asia (Raheem et al., 2021; Satyavathi et al., 2021). As per the fourth advance estimates during 2021–2022. the pearl millet area in India was 6.70 million ha, with an average production of 9.62 million tons and 1,436 kg/ha productivity (Directorate of Millets Development, 2024). Over 65 years, pearl millet productivity has increased fourfold (or 400%) from 305 kg/ha in 1951–1955 to 1,290 kg/ha in 2016–2020, and it was 1,436 kg/ha during 2021–2022. Large-scale adoption of high-yielding cultivars (of the >60% pearl millet area under hybrids and improved varieties, >90% is under hybrids) by Indian farmers is the main reason for the increase in its production and productivity. The Indian states of Rajasthan, Maharashtra, Uttar Pradesh, Gujarat, and Haryana are the primary producers of pearl millet, accounting for 90% of the total production in the country. Among all the states, Rajasthan contributes a maximum of around 3.75 million tons, followed by Uttar Pradesh (1.94), Haryana (1.12), Gujarat (1.05), Madhya Pradesh (0.86), and Maharashtra (0.47), during 2021–2022 (Directorate of Millets Development, 2024) (Table 1.3). The crop is primarily grown during the rainy season (*Kharif*) from June/July to September/October, although it is cultivated in certain regions of Gujarat, Rajasthan, and Uttar Pradesh during the summer season (February–May). In

Table 1.3 State-wise Area, Production, and Productivity of Pearl Millet in India.

State	2020–2021			2021–2022		
	Area	Production	Yield	Area	Production	Yield
	lakh ha	lakh tons	kg/ha	lakh ha	lakh tons	kg/ha
Andhra Pradesh	0.31	0.71	2,281	0.31	0.55	1,782
Bihar	0.04	0.05	1,134	0.03	0.03	1,134
Chhattisgarh	0.00	0.00	515	0.00	0.00	446
Gujarat	4.60	10.09	2,192	4.46	10.56	2,368
Haryana	5.69	13.50	2,372	4.83	11.20	2,318
Himachal Pradesh	0.01	0.00	557	0.01	0.00	557
Jammu and Kashmir	0.00	0.00	0	0.00	0.00	0
Jharkhand	0.00	0.00	643	0.00	0.00	618
Karnataka	2.22	2.76	1,241	1.47	1.71	1,161
Madhya Pradesh	3.27	7.38	2,256	3.43	8.69	2,533
Maharashtra	6.88	6.57	955	5.26	4.75	903
Odisha	0.01	0.01	622	0.02	0.01	615
Punjab	0.00	0.00	640	0.01	0.00	650
Rajasthan	43.48	45.61	1,049	37.36	37.51	1,004
Tamil Nadu	0.67	1.59	2,357	0.60	1.57	2,616
Telangana	0.10	0.09	930	0.04	0.03	823
Uttar Pradesh	9.07	20.14	2,221	9.04	19.49	2,156
West Bengal	0.00	0.00	425	0.00	0.00	428
Other	0.16	0.13	848	0.16	0.12	751
All India	76.52	108.63	1,420	67.03	96.24	1,436

Note: 2020–2021 APY data: Directorate of Economics and Statistics, GOI and 2021–2022 APY data, fourth advance estimates: Directorate of Millets development, Jaipur.

addition, it is grown on a smaller scale in Maharashtra and Gujarat during the post-rainy (*Rabi*) season from November to February.

Globally, millets are cultivated in 93 countries, and only seven countries (India, Niger, Sudan, Nigeria, Mali, Burkina Faso, and Chad) have >1 M ha harvested area, whereas around 25 countries have >0.1 M ha harvested area. All contribute around 97% of the total world millet harvested area (34.1 M ha). In general, >97%

of millets production and consumption is by developing nations. It has been estimated that from 1961 to 2018 around 25.71% of the area under millets cultivation has declined across the continents. However, global millet productivity has increased by 36% from 1961 (575 kg/ha) to 2018 (900 kg/ha) (Meena et al., 2021). India is the largest pearl millet–producing country in Asia and the world, both in terms of area and production. During 2010–2012 the average pearl millet area in India, with 8.5 million ha cultivated and an average production of 9.4 million tons. Western and Central Africa (WCA) is the largest pearl millet–producing region in Africa and the world. Segregated data on the pearl millet area for WCA are not available, but pearl millet is assumed to account for 95% of the total area in WCA. During 2010–2012, the average pearl millet area was 15.3 million ha, and the average production was 10.3 million tons in WCA. In the Southern, Eastern, and North Africa (SENA) region, there is also no segregated data for pearl millet. Sudan has the largest millet area of 2.18 m ha, of which more than 95% is assumed to be under pearl millet. Overall, the SENA region has ~3.0 m ha under pearl millet cultivation (Jukanti et al., 2016).

In Africa, the WCA region (Nigeria, Niger, Chad, Mali, and Senegal) and east/southern Africa, which includes Sudan, Ethiopia and Tanzania, are the two main areas of pearl millet cultivation. Pearl millet is the third major crop in sub-Saharan Africa, with Nigeria, Senegal, Chad, Mali, Niger, and Burkina Faso as the major producing countries. Pearl millet has socio-economic, food/feed, health, and environmental impacts on the resource-poor people of Africa. The WCA region produces the most pearl millet in the world, with 95% of the crop grown in the region (Jukanti et al., 2016). West Africa is the largest producer, led by Nigeria (41%), Niger (16%), Burkina Faso (7%), Mali (6.4%), Senegal, and Sudan (4.8%). Across Africa, millet is produced in 18.50 million ha by 28 different countries, with a yield of 11.36 million tons. It covers 30% of the SAT areas of the continent and is grown in diverse agro-ecologies. Because the region makes up 49% of the global millet area, it holds great significance (FAO, 2019). The area under millet production in different countries across the world is depicted in Figure 1.2.

According to recent report by the Agricultural and Processed Food Products Export Development Authority (APEDA, 2024), the global production of millets stood at 90.65 million tons in 2022. India was the highest millets–producing country in the year 2022, with production of 17.60 Mn million tons, contributing 19% of the global production. It was followed by Nigeria (9.00 Mn million tons; 10.01%), Sudan (6.50 Mn million tons; 7.23%), the United States (6.21 Mn million tons; 6.91%), and China (5.70 Mn million tons; 6.34%) (Table 1.4).

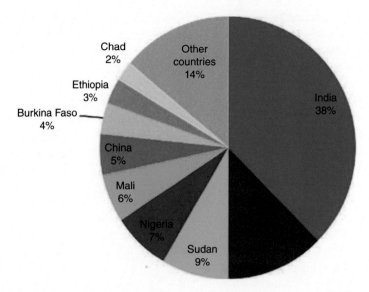

Figure 1.2   Millet production (%) in different countries of the world (FAO, 2018).

Table 1.4   Top Five Countries in Millets Production, 2022.

Rank	Country	Production 000MT	Share of global production %
1	India	17,600	19
2	Nigeria	9,000	10.01
3	Sudan	6,500	7.23
4	United States	6,212	6.91
5	China	5,700	6.34

Note: APEDA (2022).

Trends in Global Millet Consumption

The use of millet is primarily limited to developing nations, particularly following a significant decline in production and consumption within the Commonwealth of Independent States (CIS), which was previously the largest producer in the developed world. According to estimates, ~80% of millet produced worldwide (with a staggering 95% in Asia and Africa) is used for human consumption. The remaining proportion is divided among feed (7%) and other uses, such as seed and beer and waste. The per capita consumption of millet for food varies greatly between countries. It is most prevalent in Africa, where millet serves as a crucial food source in regions with low rainfall. In Niger, millet accounts for ~75% of all cereal food consumption, while in many other Sahelian countries it represents >30%. Additionally, it is a significant crop in Namibia, where it comprises 25% of total cereal food consumption, and Uganda, where it makes up 20%. Outside of Africa, millet is also an important food source in certain regions of India, China, and Myanmar. Millet is a highly nutritious and energizing food and is especially recommended for children, convalescents, and the elderly. Millet is used in various food preparations that vary across countries and regions. These typically include porridge or a flatbread that resembles a pancake. However, because whole meal quickly goes rancid, millet flour (prepared by pounding or milling) can be stored only for short periods.

Worldwide, millet food consumption has grown only marginally over the past 30 years, whereas total food use of all cereals has almost doubled. Millet is nutritionally equivalent or superior to other cereals. However, consumer demand has fallen because of several factors, including changing preferences in favor of wheat and rice, irregular supplies of millet, and cereal-centric government policies and promotions. Particularly in urban environments, the cost of women's time has encouraged the shift from millet to readily available processed foods (milled rice, wheat flour, etc.) that are faster and more convenient to prepare. Different processed food products are prepared from pearl millet flour (Yadav et al., 2012). Traditional foods such as porridge or flat and unfermented bread (chapatti) are commonly prepared from pearl millet flour. Preparations from processed pearl millet flour contain less anti-nutrients and are more easily digested (Singh, 2003). Furthermore, in India, several snacks (sev, laddoo, namkeen, and matari) with good shelf life are prepared from pearl millet. In addition to these traditional foods, pearl millet is also a good raw material for the bakery industry. Pearl millet cookies produced with some supplements have spread characteristics (texture, grittiness, and top grain) at par with those made from wheat flour. Different types of biscuits and cakes with good organoleptic qualities are being produced from pearl millet flour (either blanched or malted) (Singh, 2003). Pearl millet flour can also be used to prepare ready-to-eat (RTE) products. Extrusion (cooking at high temperature for a short time) is being used to prepare RTE products that have better digestibility and probably inactivated anti-nutritional factors. Pearl millet products prepared from blends of different flours (such as gram or soybean) have better protein content and protein efficiency ratios (Sumathi et al., 2007). Popped pearl millet has better nutritional quality. Popped millet is a good source of energy, fiber, and carbohydrates and is used in producing supplementary or weaning foods for children. Malting helps in the production of easily digestible, high-calorie, and low-viscous weaning foods, which are essential during the transition of infants from breast feeding to other type of foods. Pearl millet's high fiber content and gluten-free nature coupled with its relatively good nutritional composition makes it the cereal of choice for preparing various kinds of health foods. Pearl millet flour is also used to prepare different kinds of drinks, such as rab/rabari (Rajasthan), Cumbu Cool (Tamil Nadu), lassi, and buttermilk (Jukanti et al., 2016). The utilization of millet grain as animal feed is not significant. It is estimated that <2 million tons (~7% of total utilization) is fed to animals, compared with about 30 million tons of sorghum (almost half of total output). Animal feed usage of millet grain is largely concentrated in developing countries in Asia , with limited usage in Africa. It is important to mention that the millet fodder and stover are valuable and critical resources in the crop/livestock systems. Based on calculations of feed use in the CIS, it is estimated that ~1.0 million tons per annum are currently used as animal feed in developed countries. Western Europe, North America, and Japan together use slightly over 200,000 tons, almost exclusively as bird seed. The current rise of pearl millet as a more affordable alternative to maize feed in aquaculture, dairy, and poultry farming in India and the southeastern United States is not thoroughly documented. Nonetheless, it only accounts for a minor portion of the overall feed grain usage at present.

Current Global Scenario of Millet Consumption

The global consumption of millets stood at 90.43 Mn MT in the year 2022. The top 10 countries contributed to nearly 80% of the total global millet consumption in the same period. Millets displayed their highest consumption in India with 17.75 Mn MT, followed by China (13.70 Mn MT) and Nigeria (8.80 Mn MT) (APEDA, 2022) (Figure 1.3).

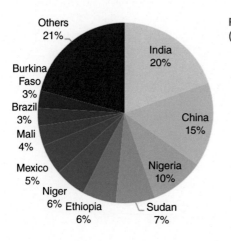

Figure 1.3 Global consumption of millets (APEDA, 2022).

International Trade Market for Pearl Millet

The limited market for pearl millet due to low consumer demand, including among farmers, is a cause for concern. Usually, this crop is being grown for food security instead of large-scale marketing, which is a major obstacle to its growth; hence, the market participation and price are quite low in comparison to other major cereals. The Minimum Support Price system could not be very helpful in this case because pearl millet is mostly sold in the local market and marketing of the produce is usually not in bulk. In addition, it has received inequitable treatment in terms of price in comparison to other major cereals (Deshpande & Rao, 2004). Poor post-harvest processing with no policy support; unorganized markets; the lack of an information system; excessive middlemen; meager transportation associations; inadequate bargaining capacity; lesser vertical coordination among producers, processors, and consumers; and poor processing facilities are some of the other major issues to be addressed to improve its trade and consumption. It is important to establish targeted policies that address the challenges faced by farmers. These individuals often encounter insufficient markets for their crops, particularly in comparison to major staples such as rice, wheat, and oilseeds, as well as pulses. Additionally, pearl millet is sold at a lower price compared with other cereals, causing producers to be economically challenged because they are only able to meet their production costs. The use of value addition strategies along with backward and forward linkages for the value chain is needed in domestic and international markets to gear up the consumption of this nutricereal.

However, millet exports over the past 5 years (2016–2021) show a positive trend in both value (3.10% CAGR) and volume (0.37% CAGR), indicating the recouping of the millet trade. In 2018, pearl millet held highest share in terms of product volume in the millets market and is likely to expand to >3% CAGR by 2025. The massive demand share of pearl millets is due to its rich nutritional value as it has eight times higher Fe content than rice (https://agriexchange.apeda.gov.in). Bajra (pearl millet) and Ragi (finger millet) together contributed 42.12% in terms of value (USD 27.42 Mn) and 54.41% in terms of volume (91,280 MT) in the year 2021 (Figure 1.4).

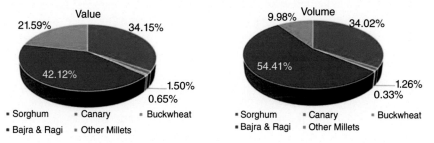

Figure 1.4 Category-wise share of millets in India's export in terms of value and volume (2021) (APEDA, 2022).

Challenges and Opportunities for Enhancing Pearl Millet Production

Pearl millet production has declined in the past several years mainly due to changing consumer preferences and pro-cereal government policies. Its production usually fluctuates widely from year to year due to rainfall variability and drought in the major production areas. In addition, several other factors, such as cultivation on poor soils, use of limited inputs, obsolete farming techniques, inadequate research resources, lack of industrial usage, and non-availability of value-added technologies, are also responsible for this decline. The hybrids and varieties developed so far are mainly for well-endowed production ecologies, whereas most of the breeding programs are not targeted to develop products for arid and hyper-arid ecologies, which is the area of utmost priority. Hence, there is a dire need to give priority to generate breeding programs and augmenting cultivar diversity for the arid and hyper-arid ecologies. It is worth concentrating on developing OPVs specifically for arid regions because it is challenging to produce sustainable hybrids with high yields for this area. It can reduce seed cost, which in turn can lead farmers to retain seed for generations, ultimately reducing the cost of cultivation. Sustainable breeding components (e.g., advanced genomic tools, speed breeding, and pre-breeding) and management components (e.g., natural resource management, better water conservation techniques, high soil fertility, and good production practices) are some of the ways to improve pearl millet productivity (Twomlow et al., 2008). Pearl millet has high nutritional value, which can be exploited for improving nutritional quality and combating malnutrition. Hence, the development of food and value-added products from pearl millet should be focused to make it more acceptable as an alternative crop of the future.

A variety of socio-economic constrictions has also limited the production and consumption of pearl millet, leading to a loss in cultivated diversity. Rural people of western Rajasthan usually prefer local landraces or "Bajri" for consumption. Thus, more efforts should be put into their development and use in the breeding programs. Landraces should be used to develop donor lines and create locally adopted base populations that can be further used as mother stocks to develop target lines in breeding programs of arid ecology. Subsequently, mechanized harvesting and the development of pearl millet hybrids/varieties with high regeneration capability and tolerance to salt and high temperature are among some of the issues which should be focused on for its improvement. Rancidity is another big challenge in pearl millet. Although pearl millet possesses superior nutritional qualities, the poor shelf life of the flour is the main challenge towards promotion and utilization of products from pearl millet. Hence, a major focus should be put on enhancing the shelf life of pearl millet flour. Varieties with reduced phenolic compounds and fat content that can have longer shelf life should be developed to overcome this issue. There is also a need to investigate its nutraceutical value and the associated health benefits for developing new industries. Difficulty in processing is another social issue that needs attention because pearl millet is generally processed at a household level in rural areas. Millet is conventionally pounded in a mortar, but mechanical dehulling and milling are increasingly used because they can eliminate a considerable amount of hard labor and can improve flour quality.

Conclusion

Pearl millet is a climate-resilient crop and an excellent genomic resource for candidate genes responsible for tolerance to climatic and edaphic stresses. It can be exploited for the genetic improvement of other crops. Pearl millet has the potential to become a significant food and feed crop due to its superior tolerance to drought, high temperature, and salinity compared with other cereals. This makes it an excellent alternative for ensuring nutritional security. Extensive collection and characterization of pearl millet germplasm are important for developing a strong breeding program and developing superior varieties. The continuous downfall in the global area under millets may be attributed due to the area shifting to other crops, changing food habits, assured irrigation facilities, and ensured returns from major commercial crops. Thus, it is important to redirect breeding efforts toward identifying specific end-product traits rather than solely focusing on improving productivity to fully unlock the potential of this crop. Further, collaborative efforts of breeders, physiologists, agronomists, policymakers, and donors in both individual and institutional capacities are highly needed. To combat abiotic stresses and develop tolerant varieties, it is important to have specific phenotyping platforms to assess tolerance to high temperatures and drought. To target key traits and speed up cultivar development, modern breeding tools and platforms, along with genomic designing and approaches, are necessary. Specific policies need to be developed to address the issues of farmers because the pearl millet farmers face relatively imperfect markets compared with superior cereals, pulses, and oilseeds. Moreover, the low prices received will make them economically inefficient, merely meeting the production costs. Providing backward and forward linkages for the value chain using innovative value addition in domestic and international markets is also highly needed. Considering its nutraceutical properties, the capacity for social capital formation among farmers and consumers regarding cooperation for millet cultivation and consumption is also essential. Thus, private–public partnership and investment is necessary to breed varieties with high starch content, targeting 100% substitution of pearl millet with maize, rice, etc. in alternative uses. In order to enhance demand and consumption from consumers, there is a need to establish a link between the health and consumption of traditional food grains and to invite initiatives from different stakeholders.

Pearl millet has superior nutritional qualities, but the shelf life of its flour and bioavailability are major challenges in the promotion of pearl millet products. Thus, efforts should focus on enhancement of shelf life of pearl millet flour, bio-fortification for Fe and Zn, and generating authentic data on the nutritional benefits and bioavailability of pearl millet. Varieties must be developed that possess the capacity to enhance the shelf life of flour and reduce the undesirable attributes in the grain (e.g., reducing fat content and phenol compounds), followed by exploring the health benefits and nutriceutical value for pearl millet. Providing backward and forward linkages for the value chain using innovative value addition in domestic and international markets is also needed. High-throughput genome analysis, next-generation sequencing techniques, and genome editing can aid in trait discovery, mapping, and line improvement. Discovering sequence-based molecular markers associated with important agronomic traits will enhance opportunities for pearl millet improvement and cultivar development in the future. There is a great potential for harnessing its positive attributes through genetic improvement, improved crop management, and grain processing and food products technologies. These should help to develop greater global awareness of the importance of this crop for food and nutritional security.

References

Aghaalikhani, M., Ahmadi, M. E., & Modarres Sanavy, A. M. (2008). Forage yield and quality of pearl millet (*Pennisetum americanum*) as influenced by plant density and nitrogen rate. *Pajouhesh Sazandegi, 77*, 19–27.

Agricultural and Processed Food Products Export Development Authority Act, 1985, (2 of 1986), (2024). Agricultural and Processed Food Products Export Development Authority (APEDA), Government of India. https://agriexchange.apeda.gov.in/

agriXchange. (2024). *International data.* Agriexchange.apeda.gov.in.

Ajeigbe, H. A., Angarawai, I. I., Inuwa, A. H., Akinseye, F. M., & Abdul Azeez, T. (2020). *Hand book on improved pearl millet production practices in North Eastern Nigeria.* ICRISAT.

Amarender Reddy, A., Yadav, O. P., Malik, D. P., Singh, I. P., Ardeshna, N. J., Kundu, K. K., Gupta, S. K., Sharma, R., Sawargaonkar, G., Moses Shyam, D., & Sammi Reddy, K. (2013). *Utilization pattern, demand and supply of pearl millet grain and fodder in Western India.* ICRISAT.

Ambawat, S., Singh, S., Shobhit, A., Meena, R. C., & Satyavathi, C. T. (2020). Biotechnological applications for improvement of the pearl millet crop. In S. K. Gahlawat, S. Punia, A. K. Siroha, K. S. Sandhu, & M. Kaur (Eds.), *Pearl millet:*

Properties, functionality and its applications (pp. 115–138). CRC Press. https://doi
.org/10.1201/9780429331732-7

Anuradha, N., Satyavathi, C. T., Meena, M. C., Sankar, S. M., Bharadwaj, C., Bhat, J.,
Singh, O., & Singh, S. P. (2017). Evaluation of pearl millet [*Pennisetum glaucum*
(L.) R. Br.] for grain iron and zinc content in different agro climatic zones of India.
Indian Journal of Genetics and Plant Breeding, 77(1), 65–73.

APEDA. (2022). *Indian superfood millets: A USD 2 billion export opportunity.* https://
apeda.gov.in/milletportal/files/Indian_Superfood_Millet_APEDA_Report.pdf

Arya, R. K., Kumar, S., Yadav, A. K., & Kumar, A. (2013). Grain quality improvement
in pearl millet: A review. *Forage Research, 38*, 189–201.

Brunette, T., Baurhoo, B., & Mustafa, A. F. (2014). Replacing corn silage with
different forage millet cultivars: Effects on milk yield, nutrient digestion, and
ruminal fermentation of lactating dairy cows. *Journal of Dairy Science, 97*,
6440–6449. https://doi.org/10.3168/jds.2014-7998

Dahlberg, J. A., Wilson, J. P., & Snyder, T. (2003). Sorghum and pearl millet: health
foods and industrial products in developed countries. In *Alternative uses of
sorghum and pearl millet in Asia* (pp. 42–59). ICRISAT.

De Assis, R. L., de Freitas, R. S., & Mason, S. C. (2018). Pearl millet production
practices in Brazil: A review. *Experimental Agriculture, 54*, 699–718.

Deshpande, R. S., & Rao, V. M. (2004). *Coarse cereals in a drought-prone region:
A study in Karnataka social and economic change.* Institute for Social and
Economic Change.

Dias-Martins, A. M., Pessanha, K. L., Pacheco, S., Rodrigues, J. A. S., & Carvalho,
C. W. P. (2018). Potential use of pearl millet (*Pennisetum glaucum* (L.) R. Br.) in
Brazil: Food security, processing, health benefits and nutritional products. *Food
Research International, 109*, 175–186. https://doi.org/10.1016/j.foodres
.2018.04.023

Directorate of Millets Development. (2024). *Statewise millet production.* https://
apeda.gov.in/milletportal/files/Statewise_Millet_Production.pdf

Dita, M. A., Rispail, N., Prats, E., Rubiales, D., & Singh, K. B. (2006). Biotechnology
approaches to overcome biotic and abiotic stress constraints in legumes. *Euphytica,
147*(1–2), 1–24.

Dujardin, M., & Hanna, W. W. (1989). Crossability of pearl millet with wild
Pennisetum species. *Crop Science, 29*, 77–80. https://doi.org/10.2135/cropsci198
9.0011183x0029000100

Dwivedi, S., Upadhyaya, H., Senthilvel, S., Hash, C. T., Fukunaga, K., Diao, X.,
Santra, D., Baltensperge, D., & Prasad, M. (2012). Millets: Genetic, and genomic
resources. In J. Janick (Ed.), *Plant breeding reviews* (pp. 247–375). Wiley.

FAO. (2018). FAOSTAT database Rome: Food and Agriculture Organization. http://
www.fao.org/faostat/en/#data/QC Accessed 04-July- 2020.

FAO. (2019). Food and Agriculture Organization of the United Nations. Rome.

FAO. (2020). FAOSTAT. FAO. Food and Agriculture Organization of the United Nations, Rome, Italy. http://www.fao.org/faostat/en/#home.

Hanna, W. (1996). Improvement of millets: Emerging trends. In V. L. Chopra, R. B. Singh, & A. Varma (Eds.), *Proceedings of the 2nd International Crop Science Congress, New Delhi, India* (pp. 139–146). Oxford & IBH Publishing.

Harlan, J. R., & de Wet, J. M. J. (1971). Towards a rational classification of cultivated plants. *Taxon, 20*, 509–517.

Jukanti, A. K., Gowda, C. L. L., Rai, K. N., Manga, V. K., & Bhatt, R. K. (2016). Crops that feed the world 11. Pearl millet (*Pennisetum glaucum* L.): An important source of food security, nutrition and health in the arid and semi-arid tropics. *Food Security, 8*, 307–329. https://doi.org/10.1007/s12571-016-0557-y

Kaur, K. D., Jha, A., Sabikhi, L., & Singh, A. K. (2014). Significance of coarsecereals in health and nutrition: A review. *Journal of Food Science and Technology, 51*, 1429–1441.

Lata, C. (2015). Advances in omics for enhancing abiotic stress tolerance in millets. *Proceedings of Indian National Sciences Academy, 81*(2), 397–417.

Li, B., Ishii, Y., Idota, S., Tobisa, M., Niimi, M., Yang, Y., & Nisihimura, K. (2019). Yield and quality of forages in a triple cropping system in Southern Kyushu, Japan. *Agronomy, 9*, 277. https://doi.org/10.3390/agronomy9060277

Meena, R. P., Joshi, D., Bisht, J. K., & Kant, L. (2021). Global scenario of millets cultivation. In A. Kumar, M. K. Tripathi, D. Joshi, & V. Kumar (Eds.), *Millets and millet technology* (pp. 33–50). Springer.

Ministry of Agriculture and Farmers Welfare, Department of Agriculture, Cooperation and Farmers Welfare, Govt. of India (2018) The Gazette of India. No. 133; F.No. 4-4/2017-NFSM (E)

Newman, Y., Jennings, E., Vendramini, J., & Blount, A. (2010). *Pearl millet (Pennisetum glaucum): Overview and management*. University of Florida.

Nurbekov, A., Jamoliddinov, A., Joldoshev, K., Rischkowskv, B., Nishanov, N., Rai, K. N., Gupta, S. K., & Rao, A. S. (2013). Potential of pearl millet as a forage crop in wheat-based double cropping system in Central Asia. *Journal of SAT Agricultural Research, 11*, 1–5.

Obilana, A. B., & Manyasa, E. (2002). Millets. In P. S. Belton & J. R. N. Taylor (Eds.), *Pseudocereals and less common cereals: Grain properties and utilization potential* (pp. 177–217). Springer-Verlag.

Packiam, M., Subburamu, K., Desikan, R., Uthandi, S., Subramanian, M., & Soundarapandian, K. (2018). Suitability of pearl millet as an alternate lignocellulosic feedstock for biofuel production in India. *Journal of Applied & Environmental Microbiology, 6*, 51–58. https://doi.org/10.12691/jaem-6-2-4

Raheem, D., Dayoub, M., Birech, R., & Nakiyemba, A. (2021). The contribution of cereal grains to food security and sustainability in Africa: Potential application of UAV in Ghana, Nigeria, Uganda, and Namibia. *Urban Science, 5*, 8. https://doi.org/10.3390/urbansci5010008

Satyavathi, C. T., Ambawat, S., Khandelwal, V., & Srivastava, R. K. (2021). Pearl millet: A climate-resilient nutricereal for mitigating hidden hunger and provide nutritional security. *Frontiers in Plant Science, 12,* 659938. https://doi.org/10.3389/fpls.2021.659938

Serba, D. D., Yadav, R. S., Varshney, R. K., Gupta, S. K., Mahalingam, G., Srivastava, R. K., Gupta, R., Perumal, R., & Tesso, T. T. (2020). Genomic designing of pearl millet: A resilient crop for arid and semi-arid environments. In C. Kole (Ed.), *Genomic designing of climate-smart cereal crops* (pp. 221–286). Springer.

Sharma, S., Sharma, R., Govindaraj, M., Mahala, R. S., Satyavathi, C. T., Srivastava, R. K., Gumma, M. K., & Kilian, B. (2021). Harnessing wild relatives of pearl millet for germplasm enhancement: Challenges and opportunities. *Crop Science, 61*(1), 177–200.

Sheahan, C. M. (2014). *Plant guide for pearl millet (Pennisetum glaucum).* USDA-NRCS.

Singh, G. (2003). *Development and nutritional evaluation of value added products from pearl millet (Pennisetum glaucum)* [Doctoral dissertation, CCS Haryana Agricultural University] https://hau.ac.in/.

Singh, S. D., & Navi, S. S. (2000). Genetic resistance to pearl millet downy mildew II. Resistance in wild relatives. *Journal of Mycology and Plant Pathology, 30,* 167–171.

Srivastava, R. K., Singh, R. B., Lakshmi, V., Pujarula, S., Bollam, P. M., Satyavathi, C. T., Yadav, R. S., & Gupta, R. (2020). Genome wide association studies and genomic selection in pearl millet: Advances and prospects. *Frontiers in Genetics, 10,* 1389.

Sumathi, A., Ushakumari, S. R., & Malleshi, N. G. (2007). Physicochemical characteristics, nutritional quality and shelf life of pearl millet based extrusion cooked supplementary foods. *International Journal of Food Science and Nutrition, 58,* 350–362.

Twomlow, S., Shiferaw, B., Cooper, P., & Keatinge, J. D. H. (2008). Integrating genetics and natural resource management for technology targeting and greater impact of agricultural research in the semi-arid tropics. *Experimental Agriculture, 44,* 235–256.

Urrutia, J. M., José, H.,. A. A., Rosales, C. A. N., & Cervantes, B. J. F. (2015). Forage production and nutritional concentration of silage from three varieties of pearl millet (*Pennisetum glaucum*) harvested at two maturity stages. *Journal of Animal and Plant Sciences, 27,* 4161–4169.

Yadav, O. P., Rai, K. N., Rajpurohit, B. S., Hash, C. T., Mahala, R. S., Gupta, S. K., Shetty, H. S., Bishnoi, H. R., Rathore, M. S., Kumar, A., Sehgal, S., & Raghvani, K. L. (2012). *Twenty-five years of pearl millet improvement in India.* In *All India Coordinated Pearl Millet Improvement Project.*

2

Pearl Millet Germplasm Resources

M. Vetriventhan, O. P. Yadav, Rakesh K. Srivastava, Sushil Pandey, and Kuldeep Singh

Chapter Overview

Pearl millet is one of the most nutritious and climate-resilient crops, with the exceptional ability to tolerate high temperatures. It is staple food for ~90 million people of Africa and Asia. Its stalks are used as green and dry fodder for livestock. Thus, it plays a significant role in ensuring food and fodder security and nutrition. Genetic enhancement of pearl millet is governed by various factors, including the availability of genetic resources, inheritance and stability of traits, and the type of varieties cultivated. Globally, a substantial amount of pearl millet germplasm is conserved *ex situ* in genebanks, with the ICRISAT genebank being the largest repository. Characterization and evaluation of these germplasm resources for morpho-agronomic, productivity, quality, and stress tolerance traits have shown significant variability that can be harnessed for crop improvement. Germplasm diversity representative subsets such as core collection and mini core collections have been established in pearl millet. Pearl millet's high outcrossing nature promotes gene flow within and among its wild and cultivated forms, leading to a high level of diversity within its landraces. Therefore, to effectively utilize this genetic variability for crop enhancement, a well-planned strategy is essential. This chapter provides a comprehensive overview of the status of pearl millet germplasm conservation, both ex-situ and in-situ, and attempts to review the progress made in identifying and utilizing trait-specific sources for improving resistance/tolerance to biotic and abiotic stresses in genetic improvement programs.

Pearl Millet: A Resilient Cereal Crop for Food, Nutrition, and Climate Security, First Edition. Edited by Ramasamy Perumal, P. V. Vara Prasad, C. Tara Satyavathi, Mahalingam Govindaraj, and Abdou Tenkouano.

Introduction

Pearl millet [*Pennisetum glaucum* (L.) R. Br., Syn. *Cenchrus americanus* (L.) Morrone] is a nutracereal with wide adaptation to diverse climatic conditions. It is a staple food for 90 million people in Africa and Asia. Its stalks are also used as green and dry fodder for livestock. Thus, it has the potential to play a significant role in ensuring global food, fodder, and nutritional security, particularly in the context of changing climate. Pearl millet is the world's sixth most important cereal crop after rice (*Oryza sativa* L.), wheat (*Triticum aestivum* L.), maize (*Zea mays* L.), barley (*Hordeum vulgare* L.), and sorghum [*Sorghum bicolor* (L.) Moench] and is predominantly grown as a subsistence crop in marginal lands in semi arid and arid regions of Africa and Asia (Satyavathi et al., 2021; Srivastava et al., 2020a). Cultivated pearl millet is a diploid species (2n = 14). The cross-pollinated nature of pearl millet makes it more demanding for collection, conservation, and use of germplasm. Varietal development primarily relies on the genetic variation in the germplasm of the crop species. Over 73,000 germplasm accessions of cultivated, wild and weedy relatives of pearl millet are conserved ex situ in gene banks globally (Bramel et al., 2022). This chapter discusses the status of pearl millet germplasm conservation, both ex situ and in situ, and reviews the progress made in identifying and using trait-specific germplasm for improving resistance to biotic and abiotic stresses in genetic improvement programs.

Pearl Millet Gene Pools and Races

The genus of pearl millet is categorized into five sections with different characteristics, including chromosome numbers, ploidy level (diploid to octoploid), reproductive behaviors (sexual or apomictic), and life cycle. These sections are (a) Penicillaria (tropical Africa and India), (b) Brevivalvula Döll (pan-tropical), (c) Gymnothrix (pan-tropical), (d) Heterostachya (northeast Africa), and (e) Eupennisetum (tropical and subtropical Africa and India) (Sharma et al., 2021). Penicillaria contains the annual diploid cultivated pearl millet *P. glaucum*, its wild species *P. glaucum* ssp. *monodii*, weedy species *P. glaucum* ssp. *stenostachyum*, and the perennial tetraploid *P. purpureum*. Comprehensive reviews on the origin, distribution, taxonomy, and gene pool were published by Pattanashetti et al. (2015) and Sharma et al. (2021). Readers may refer to these reviews for details.

The species belonging to the genus *Pennisetum* are grouped into three gene pools based on their cross-compatibility with cultivated pearl millet: the primary gene pool (GP1), the secondary gene pool (GP2), and the tertiary gene pool (GP3) (Harlan & de Wet, 1971). GP1 includes domesticated diploid species *P. glaucum* ssp. *glaucum* (2n = 2x = 14 with AA genome), its wild progenitor *P. glaucum* ssp.

monodii with its two ecotypes *P. violaceum* and *P. mollissimum* ($2n = 2x = 14$, AA), and the weedy forms shibras (= *P. glaucum* ssp. *stenostachyum* Kloyzesh ex. Müll. Berol. Brunken; $2n = 2x = 14$ with AA genome). The absence of callus formation at maturity is the single and distinct character that separates species *P. glaucum* from *monodii* and *stenostachyum*. The GP1 species are easily crossable under sympatric conditions and form fertile hybrids with normal chromosome pairing (Harlan & de Wet, 1971) and thus have a high possibility of successful introgression of genes among the species in GP1 into cultivated pearl millet. The species in GP2 includes an allotetraploid rhizomatous perennial species, *P. purpureum*, also known as Napier grass or elephant grass ($2n = 4x = 28$ with A′A′BB genome), and the apomictic and octoploid species *P. squamulatum* Fresen. ($2n = 8x = 56$). It can also cross with cultivated pearl millet but produces highly sterile hybrids (Kaushal et al., 2008; Sharma et al., 2021). GP3 includes the remaining species that are cross-incompatible with cultivated pearl millet. *Pennisetum schweinfurthii* Pilg. is the only *Pennisetum* species in GP3 reported to have $2n = 2x = 14$ chromosomes and an annual growth habit, but its chromosomes are nonhomologous to cultivated pearl millet, with different genomic localizations of rDNA probes (Hanna & Dujardin, 1986; Martel et al., 1996, 2004). There are strong reproduction barriers between the members of GP3 and GP1 or GP2, and gene transfer is only possible by radical manipulations involving in vitro techniques or by using complex hybrid bridges (Martel et al., 1996; Sharma et al., 2020). The classification of pearl millet into these gene pools helps plant breeders understand the genetic relationships and compatibility among different species and populations. It also guides breeding efforts to exploit the genetic diversity in the secondary and tertiary gene pools for crop improvement and adaptation to changing environments.

Cultivated pearl millet can also be classified in to four races based on seed shapes (Figure 2.1). The variation in seed morphology among these races is the result of migration events and geographic and ethnographic isolation during early stages of pearl millet domestication (Brunken, 1977).

1) Race typhoides: The seeds of this race are obovate (egg shaped) in both frontal and profile views. They are also obtuse and terete in cross section. The grains are occasionally shorter than an enclosed by the floral bracts. The race typhoides occurs from Senegal to Ethiopia and from North Africa to South Africa. It is the only basis race of pearl millet found outside the continent of Africa and may be the most primitive of the four basic races.
2) Race nigritarum: The seeds of this race are obovate as well, but they are extremely angular in cross section and have three and six facets per grain. The mature grain generally protrudes beyond the floral bracts.
3) Race globosum: The seeds of this race are spherical, with each of its dimensions being approximately equal. It is most common from central Nigeria to Niger.

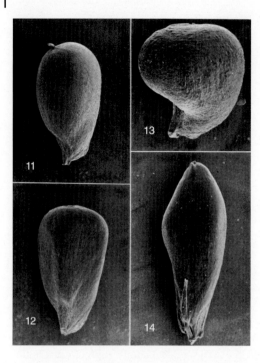

Figure 2.1 Scanning electron micrographs of the grains in the four basic races of pearl millet. 11, typhoides; 12, nigritarum; 13, blobosum; and 14, leonis (Brunken, 1977 / with permission of John Wiley & Sons).

4) Race leonis: The seeds of this race are acute, oblanceolate, and terete. The most distinctive characteristic of the leonis grain is its acute apex, which is terminated by the remnants of the stylar base. At maturity, approximately one-third of the grain protrudes beyond the floral bracts. The inflorescence in leonis is candle-like. Leonis is indigenous to Sierra Leone but occurs frequently as far northern Senegal and southern Mauritania. The greatly elongated character of the grain in leonis may be an adaptation to the relatively high rainfall regimes of Sierra Leone.

Germplasm Conservation

The biodiversity of plant genetic resources for food and agriculture (PGRFA) is preserved through two conservation strategies: in situ conservation and ex situ conservation. In situ conservation involves the conservation of crop wild relatives (CWRs) in their natural habitats or genetic reserves as well as on-farm conservation of traditional landraces or farmer varieties by farmers. Its main objective is to safeguard and manage biological diversity in its natural habitat, which supports the natural evolutionary processes, leading to new variations in the gene pool. Ex situ conservation aims to conserve biological diversity outside of its natural habitat under appropriate storage conditions, such as seed storage, in vitro storage, DNA banks, and field and botanical gardens.

In situ Conservation

In situ conservation is particularly important for maintaining biological diversity in the centers of origin and diversity. This approach helps to protect target species and their genotypes in their natural habitats, where they can continue to evolve and generate new genetic variations. It also promotes the conservation of ecosystem diversity, species diversity, and genetic diversity within species while preserving the critical ecosystem services that support life on Earth. In situ conservation also involves integrating farmers into national PGRFA systems, which can help improve the livelihoods of resource-poor farmers through economic and social development. Many CWRs and landraces are currently at risk of extinction due to various factors, including urbanization, habitat fragmentation, intensification of farming practices, and climate change. In the case of pearl millet, CWRs are conserved in situ by establishing biosphere reserves. There are currently 738 biosphere reserves in 134 countries that belong to the world network of biosphere reserves, promoting the conservation of biodiversity with its sustainable use (UNESCO, 2023). For example, the Arly biosphere in Burkina Faso, which is situated in the West African Savannah, presents a wide variety of natural landscapes, where cereal crops (millet and sorghum), peanut, and cotton are the main crops (UNESCO, 2023). On-farm conservation of landraces is another important strategy for the conservation of PGRFA. Farmers often conserve local landraces of crops like pearl millet, and the establishment of community seed banks supports practices (Vernooy et al., 2015). Many non-governmental organizations, including the M.S. Swaminathan Foundation and Navdanya International, LI-BIRD, promote the conservation of locally important crops and the maintenance of germplasm collections through community seed banks.

Ex situ Conservation

As per the FAO WIEWs, there are 81 genebanks across 55 countries that collectively conserve 60,374 accessions of *Pennisetum* spp (https://www.fao.org/wiews/data/ex-situ-sdg-251/overview/en/). germplasm (Table 2.1). The recent global millets germplasm conservation strategy indicated that a total of 73,578 pearl millet accessions are conserved in 57 genebanks globally (Bramel et al., 2022).

The ICRISAT Genebank holds the largest collection of *Pennisetum* spp. germplasm in the world, which conserves over >25,500 accessions from 52 countries (https://genebank.icrisat.org). Landraces constitute the major portion of this collection (87%), followed by advanced cultivars and/or breeding lines (10%) and wild and weedy relatives (3%). The major portion of pearl millet collection conserved at the ICRISAT genebank is from India (33%), followed by Nigeria (10.1%), Niger (9%), and Zimbabwe (7%); the remaining countries represent <5%. The

Table 2.1   Status of Pearl Millet Germplasm Resources Conserved Ex situ Globally.

Regions	Accessions	Genera	Species	Gene Banks	Countries
Northern Africa	3,366	1	5	4	4
Sub-Saharan Africa	6,338	1	11	25	18
Northern America	5,637	1	19	3	2
Latin America and the Caribbean	1,231	1	6	13	7
Asia	11,254	1	8	14	12
Europe	597	1	18	14	10
Oceania	451	1	6	3	2
Regional	1,789	1	1	1	-
International	29,711	1	18	4	-
TOTAL	60,374	1	21	81	55

Note: https://www.fao.org/wiews/data/ex-situ-sdg-251/overview/en/ (accessed in January 2024)

collection at ICRISAT is conserved in two forms: (a) an active collection stored in aluminum cans with screw caps having rubber gaskets at 4 °C and 20% relative humidity and (b) a base collection stored at −20 °C in vacuum-sealed aluminum foil pouches, where the seed moisture content is between 3% and 7%. The ICRISAT genebank has distributed 85% of the total pearl millet germplasm at least once by supplying over 168,000 seed samples to 86 countries worldwide under the standard material transfer agreement of the International Treaty on Plant Genetic Resources for Food and Agriculture of the FAO. The genebank also serves as a backup for restoring germplasm to source countries when national collections are lost due to natural calamities, civil strife, or any other reason. The ICRISAT genebank has repatriated 8773 pearl millet accessions to India ($n = 7,189$), Cameroon ($n = 922$), Sudan ($n = 594$), Chad (n=42), and Niger (n=26) thus enabling these countries to regain their precious plant germplasm heritage. To ensure further safety, the ICRISAT genebank has submitted over 92% of its total pearl millet collection to Svalbard Global Seed Vault, Norway, as second-level duplication, and ensuring the first-level safety duplication of pearl millet collection in different genebanks (IITA, Nigeria and ICARDA, Morocco). The ICAR-National Bureau of Plant Genetic Resources (ICAR-NBPGR), New Delhi, is conserving the second largest collections of *Pennisetum* spp. (8381 accessions, including 8196 cultivated and 185 wild species accessions; ICAR-NBPGR, 2024) (Table 2.1).

Gap Analysis in The Ex situ Collection

The process of identifying germplasm gaps in ex situ collections helps to locate regions that are poorly represented in a genebank collection of a particular crop.

By analyzing gaps, rare landraces and other germplasm that are at risk of extinction can be collected and conserved. Ideally, a genebank's collection should include accessions collected from the center of diversity of the species as well as representative areas where the crop is extensively grown. The ICRISAT genebank conserves the largest collection of pearl millet, but still considerable gaps exist in the collection that need to be enriched by collecting taxonomical, geographical, and traits gaps. The gap analysis of the ICRISAT collection from the West and Central Africa (WCA) region identified 62 districts in 13 provinces of Nigeria, 50 districts in 16 provinces of Burkina Faso, nine districts in six provinces each of Mali and Mauritania, eight districts in eight provinces of Chad and seven districts in three provinces of Ghana as the major geographical gaps for collection (Upadhyaya et al., 2009b). Similarly, 34 districts located in 18 provinces of four East African countries and 76 districts located in 34 provinces of seven Southern African countries were identified as geographical gaps in the pearl millet collection at the ICRISAT genebank (Upadhyaya et al., 2012b). The gap analysis of germplasm from Asia revealed that 134 districts in 14 provinces in India and 12 districts of Punjab province in Pakistan and southern parts of North Yemen and Lahiz provinces in Yemen were identified as gaps in the collection (Upadhyaya et al., 2010). Crop wild and weedy relatives contribute significantly to trait improvement in almost all crops, including pearl millet. The gap analysis of the CWR collection indicated 194 provinces in 21 countries of Asia and Africa as geographical gaps in the *Pennisetum pedicellatum*, whereas in *P. glaucum* subsp. *monodii,* 354 districts located in 86 provinces of eight countries (158 districts of 35 provinces in Mali, 67 districts of 14 provinces in Burkina Faso, 59 districts of seven provinces in Nigeria, 24 districts of eight provinces in Mauritania, 15 districts of four provinces in Senegal, 14 districts of seven provinces in Sudan, 12 districts of six provinces in Niger, and five districts of five provinces in Chad) in the ICRISAT genebank pear millet CWR collection (Upadhyaya et al., 2014c). ICRISAT is currently enriching its collection by conducting collecting missions and assembling missing diversity to address these geographical and traits gaps.

Factors Shaping Diversity

The plant domestication process involves considerable modification of plant phenotypes through human and natural selection over hundreds to thousands of years to fit human needs. The domestication process includes the evolution of characteristic traits of wild populations early at the beginning of domestication and further diversification caused by local adaptation to the new environments and human needs (Lakis et al., 2012). Plant domestication and local adaption resulted in many landraces displaying tremendous phenotypic variability for various traits, such as flowering time, plant height, panicle length, panicle width, panicle shape, grain characteristics, and grain nutrient content in pearl millet. The abundant diversity of pearl millet is shaped by several factors, including the mating system and gene flow, geographical and environmental factors, farmers' practices, trait preference, and trait selection.

Mating System and Gene Flow

Among the evolutionary factors, gene flow between cultivated and wild and weedy relatives has been considered as a major factor contributing to the evolution of crops (Mariac et al., 2006b). This can lead to the development of new genotypes upon which farmer selection can operate to produce enhanced phenotypes. Pearl millet is a highly outcrossing crop species that favors gene flow over generations within and among wild and cultivated pearl millet forms. Weedy plants with intermediate domesticated/wild phenotypes are present in most pearl millet fields in West Africa (Mariac et al., 2006b). Morphological and AFLP marker data suggests introgression from the wild to the weedy population, but the gene flow between the parapatric wild and domesticated populations is very low (Mariac et al., 2006b). The study also revealed that the level of genetic differentiation between fields from the two different villages was low when considering domesticated and weedy plants. This could be explained by the high gene flow resulting from substantial seed exchanges between farmers (Mariac et al., 2006b). The occurrence of gene flow between early and late landraces was also reported despite differences in flowering time, and this gene flow and subsequent introgression could result in a drastic erosion of diversity for maturity duration in pearl millet. Spatial and socio-ecological factors of gene flow are also particularly key factors shaping the diversity of pearl millet. The genetic diversity of local landrace populations of pearl millet grown by several ethno-linguistic groups in the Lake Chad Basin in Africa was investigated by sampling 69 pearl millet landrace populations characterized using 17 simple sequence repeat (SSR) markers (Naino Jika et al., 2017). This study suggests the existence of a genetic structure of pearl millet associated with ethno-linguistic diversity on the western side of Lake Chad, suggesting limited gene flow between landraces grown by different ethno-linguistic groups. On the eastern side of the Lake, a larger efficient circulation of pearl millet genes between ethno-linguistic groups was reported (Naino Jika et al., 2017). The differently named pearl millet landraces cultivated in the same ethno-linguistic group were not genetically differentiated, and the genetic differentiation between pearl millet populations is better explained by ethno-linguistic differences among farmers than by spatial distances (Naino Jika et al., 2017).

Geographical and Environmental Pattern of Diversity

The diversity in the landraces is attributed to the differences in the prevailing environmental factors such as day length, temperature, rainfall, and elevation of the growing areas. The genebank at ICRISAT conserves over 25,500 accessions originating (collection site) from 52 countries and is the largest collection of *Pennisetum* spp. Among these, a total of 15,979 landraces originating in 34 countries having geographic coordinates of the collection sites were investigated to assess the geographical distribution of pearl millet traits and diversity (Upadhyaya et al., 2017a). The study indicated that most of the landraces originated from the Northern Hemisphere (80.5%), while 19.5% were in the Southern Hemisphere. The 10°–15° on the northern side and 15°–20° on the southern side of the equator are the important source regions from pearl millet germplasm, with 39.6% and 13.1% of accessions, respectively (Upadhyaya et al., 2017a). A higher frequency of early-flowering (33–40 days) landraces is found in Pakistan (18%), followed by Ghana (7.9%). Very late-flowering (121–159 days) was more common in Sierra Leone (96.6%) and in the Central Republic of Africa (88%), while 41% and 48% of landraces from Cameroon and Benin were late flowering. Verytall-growing (>401–490 cm) landraces were found at high frequencies in Chad (43%) (Upadhyaya et al., 2017a) (Figure 2.2).

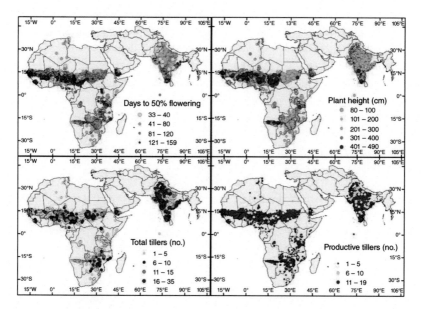

Figure 2.2 Geographical distribution of pearl millet traits: days to 50% flowering and plant height (cm) (Upadhyaya et al., 2017a / with permission of Springer Nature).

Pearl millet is a facultative, short-day species, flowering at all photoperiods but much earlier with short days. However, the critical photoperiod and temperature required to trigger flowering are species and cultivar specific (Upadhyaya et al., 2012a). Days to flowering is the key trait which influences the cropping pattern and adaptation to climate variability. Differences in the maturity duration of pearl millet landraces are partly due to differences in photoperiodism. In general, early-flowering landraces are either insensitive to photoperiod or are preferentially short-day plants, whereas semi-late–flowering or late-flowering landraces are strictly sensitive to day length (Lakis et al., 2011). Among >15,900 georeferenced landraces conserved at the ICRISAT genebank, 52% are categorized as thermo-sensitive, 45.6% are photosensitive, and only 2% of the accessions are insensitive to photoperiod and temperature based on days to flowering in the rainy (long-day condition) and post-rainy (short-day condition) seasons at the Patancheru location, India (Upadhyaya et al., 2012a). Accessions that originated from locations at higher latitudes (>20–35°) on both the northern and southern hemispheres covering countries of the Indian subcontinent and Southern Africa exhibited low sensitivity to both photoperiod and temperature, whereas those collected from lower latitudes (0°–20°) on both sides of the equator covering countries of the Horn of Africa and Western Africa showed a high-sensitivity photoperiod.

The distribution area of pearl millet in WCA harbors a wide range of climate and environmental conditions and diverse farmer preferences and use, which has the potential to lead to local adaptation of pearl millet where a large variability exist for several traits, including flowering time; plant height; panicle exertion, length, thickness, and shape; and seed shape and color (Bhattacharjee et al., 2007; Stich et al., 2010; Upadhyaya et al., 2017b). Similarly, eastern and southern Africa and Asia are the secondary centers of diversity, where large variability exists for seed weight and tillering, respectively (Upadhyaya et al., 2013). A study based on 145 inbred lines derived from 122 different pearl millet landraces from WCA revealed that phenotypic traits such as flowering time, the relative response of photoperiod, and panicle length are significantly associated with population structure but not with environmental factors (Stich et al., 2010). In a study, the geographical pattern of diversity in pearl millet germplasm from Yemen was assessed because the climate conditions in this country differ considerably from other countries in elevation, rainfall, temperature, and soil characteristics (Reddy et al., 2004). The study revealed highly significant differences among three elevational zones (low, <600 m asl; medium, 600–1,600 asl; and high, >1,600 m asl) for flowering, plant height, seed weight, and panicle length and thickness.

Farmer Practices and Trait Preference

Landrace is a dynamic population of a cultivated crop that has a historical origin, has a distinct identity, and lacks formal crop improvement. Landraces are often genetically diverse, locally adapted, and associated with traditional farming systems. Flowering time is one of the key traits selected for adaptation to different climatic conditions. Early flowering landraces ensure minimum production during drought conditions or low-rainfall seasons, whereas late-flowering landraces benefit from late rains. The adaptation of different soils, different culinary uses, and secondary uses (e.g., fodder, fuel, and fence) also explains why farmers conserve both flowering types. In Senegal, farmers prefer late-flowering landraces compared with early flowering landraces for fodder as well as for roof and fence construction because of the longer and more robust stems. This implies a genetic structuring and different selection pressure on these two types (Faye et al., 2022). Pearl millet is an important fodder crop supporting the livestock production system. A study conducted to assess farmers' perceptions of fodder performance of pearl millet indicated that long and wide leaves, abundant tillers, slender stems, large biomass, and regrowth after grazing or mowing are the key traits considered by farmers in Niger as fodder pearl millet (Moussa et al., 2021).

Phenotypic and Genomic Diversity Assessment and Population Studies

Phenotypic Diversity

Immense variability exists in the pearl millet germplasm for morpho-agronomic traits. For example, important quantitative traits in the largest global collection conserved at the ICRISAT genebank for days to 50% flowering varied from 33 to 159 days in rainy conditions and from 31 to 138 days in post-rainy conditions, plant height ranged from 30 to 490 cm in rainy conditions and from 25 to 425 cm in post-rainy conditions, total tillers ranged from 1 to 35, panicle length ranged from 5 to 135 cm in rainy conditions and from 4 to 125 cm in post-rainy conditions, panicle width ranged from 8 to 58 mm in rainy conditions and from 8 to 61 mm in post-rainy conditions, and 1000-seed weight ranged from 1.5 to 21.5 g (Upadhyaya et al., 2009a). Although large and diverse collections of pearl millet germplasm are available, to enhance the use of diversity in crop improvement, the concepts of core (~10% of the entire collection) (Frankel, 1984) and mini-core (~10% of the core collection) (Upadhyaya & Ortiz, 2001) collections were advocated. In the case of pearl millet, a core collection (2084 accessions) (Bhattacharjee et al., 2007; Upadhyaya et al., 2009a) and a mini-core collection (238 accessions) (Upadhyaya et al., 2011) have been developed. A list of accessions in these subsets is available in the Genesys PGR database (https://www.genesys-pgr.org/subsets/v2757WpAPbW).

Genome Sequencing

Genomic characterization and trait discovery in pearl millet are related to the investigation of the genetic framework of the plant species and the detection of specific traits that are regulated by genes. This phenomenon typically comprises various genetic and genomic tools to unravel the genetic and molecular basis of the traits, identify their inheritance patterns, and develop schemes for trait improvement programs. Genome sequencing of the pearl millet reference genome is the breakthrough in the characterization of the pearl millet genome, which enables researchers to mine, map, and align DNA sequences and subsequently identify genes regulating the trait and helps in the investigation of genomic organization and structural framework. A multi-institutional pearl millet genome sequencing consortia led by ICRISAT completed pearl millet genome sequencing (1.76 Gb) by using the bacterial artificial chromosome and whole-genome shotgun methods. Illumina HiSeq 2000 produced 520 Gb of sequencing data, and whole-genome shotgun libraries with varied insert sizes were built using the Tift 23D2B1-P1-P5 genotype (Varshney et al., 2017). Bacterial artificial chromosome libraries from the same genotype were also produced using *Hin*dIII and *Eco*RI 50% scaffolds, and around 90% of the pearl millet genome has been assembled. About 77.2% of the repetitive sequences in the completed genome accounted for an estimated 80% (1.4 Gb) of the genome's total size. Genome sequence information provides important clues about genomic regions associated with agronomically important and nutritional-related traits, along with different biotic and abiotic resistance traits. Pearl millet genome sequencing will help to understand the trait variation and accelerate the genetic improvement of pearl millet (Varshney et al., 2017). Semalaiyappan et al. (2023) used the whole genome sequence of 925 genotypes (Varshney et al., 2017) and developed a new mid-density 4k single-nucleotide polymorphism (SNP) panel. Further, 373 pearl millet inbreds consisting of 195 B lines and 182 R lines were genotyped using the newly developed 4k SNPs, and the SNPs were uniformly distributed across the pearl millet genome and differentiated the accessions into two major groups corresponding to the B lines and R lines.

Genomics-based Germplasm Characterization

In pearl millet, several research groups reported genetic diversity assessment by using molecular markers in diverse germplasm accessions. Genetic diversity analysis of 46 wild and 421 cultivated lines from Niger was performed using 25 microsatellite markers and identified clear differentiation between wild and cultivated lines in Niger (Mariac et al., 2006a). Kapila et al. (2008) reported genetic diversity among 72 pearl millet lines from sub-Saharan Africa and India and assessed using 34 SSR primers. The 72 lines clustered into five groups and were supported by their pedigree and traits. A set of 145 inbreds descended from 122 landraces were characterized for genetic diversity using 20 SSRs, which classified that the lines derived from West and Central African territories clustered into a group (Stich et al., 2010). Sumanth et al. (2013) assessed 42 inbred lines with 17 SSR markers to estimate the genetic diversity of lines for defining the groups. Novel restorer lines ($n = 45$) generated in India and Africa with good combining ability were assessed to study genetic diversity together with 50 SSR markers that grouped lines into eight distinct clusters (Satyavathi et al., 2013). Hu et al. (2015) studied the genetic relatedness of 511 pearl millet genotypes using 83,875 SNPs and reported high genetic diversity in Senegal germplasm compared with other regions of Africa and Asia. In another experiment, a set of 18 pearl millet genotypes were tested with 28 SSR markers and formed three clusters (Kapadia, 2016).

Heterotic Gene Pool

In pearl millet, different research groups used genomic tools to define the heterotic gene pools. ICRISAT partnered with National Agricultural Research Services and the private sector to develop genetic pools by using its germplasm resources from Africa and Asia-based hybrid parental lines to exploit the genetic variability of pearl millet (Ramya et al., 2018; Satyavathi et al., 2013). In pearl millet, heterosis breeding mainly relies on genetically distinct pools of A (Cytoplasmic Male Sterile), B (Maintainer), and R (Restorer) lines to accelerate the genetic gain. Pucher et al. (2016) examined the patterns of combining ability in West African pearl millet landraces, and indicated that hybrid breeding is promising, however there were no clear natural heterotic groups among WA pearl millet landraces. The first study in pearl millet heterotic gene pool identification using genomic markers was reported by the ICRISAT pearl millet molecular breeding research group (Ramya et al., 2018). A set of 342 pearl millet hybrid parental lines, including 160 B and 182 R lines and the control line Tift 23D2B1-P1–P5 (world reference germplasm), were part of the experiment. Expressed sequence tag SSR and genomic SSR markers were used for the characterization and formed 9 and 11 clusters in B and R lines, respectively (Ramya et al., 2018). In another experiment, a set of 190 (95 each of B and R) pearl millet lines was evaluated by using 40 SSR markers and formed two distinct clusters (Anuradha et al., 2018). In another study, different heterotic groups resulted among 150 pearl millet (75 each of B and R) advanced hybrid parental lines by using 56 SSR markers and 75,000 SNPs (Singh et al., 2018). Singh and Gupta (2019) reported eight heterotic gene pools in 150 pearl millet hybrid parental (75 each of the B and R) lines using 56 SSR markers spread across the pearl millet chromosomes. In another experiment, 580 pearl millet hybrid parents (320 R and 260 B lines), developed from different breeding schemes in India, together with genotyping data comprised of RAD-GBS generated ~0.9 million SNPs, defined 12 R and seven B line heterotic groups with minimal outliers (Gupta et al., 2020). To study the magnitude of genetic variability in pearl millet for essential micronutrients a set of 212 core pearl millet germplasm accessions was assessed for genetic diversity, which resulted in eight distinct clusters (Govindaraj et al., 2020). Patil et al. (2021) assessed the diversity of 45 pearl millet populations from Africa and Asia using 29 highly polymorphic SSR markers to define heterotic pools, and these populations were grouped into seven clusters (G1–G7).

Germplasm Evaluation

Productivity improvement and the development of high-yielding varieties and hybrids with desired traits is the major goal of global pearl millet breeding programs. Hybrid breeding technology has proven to be an effective strategy because it has shown a remarkable improvement in pearl millet productivity (Yadav & Rai, 2013). Hybrid breeding programs in cross-pollinated crops like pearl millet demand predefined germplasm pools with significant genetic variability for a specific trait and strategies for improved combining ability for accelerated genetic gain. The accessibility of specific germplasm for a trait of interest is the major basis for exploiting the hybrid vigor in crop plants. Phenotypic and genetic characterization of diverse germplasm for exploring trait-specific genetic variability and grouping them into several pools or heterotic pools is essential for successful hybrid breeding schemes (Gupta et al., 2020; Ramya et al., 2018). Selection of parental material by using trait-specific heterotic pools not only increases the efficiency of breeding programs but also helps reduce crossing time and the cost of field evaluation for the selection of heterotic crosses (Teklewold & Becker, 2005). These trait-specific pools also facilitate generating a greater number of crosses within a short span of time and save the cost of redundant hybrids advancement.

Characterizing and evaluating germplasm is a crucial step in maximizing its use in crop improvement. Without this information, the diversity within germplasm cannot be effectively used. Fortunately, with the emergence of high-throughput phenotypic characterization and evaluation, much germplasm can now be characterized for various traits, enabling the development of trait-specific subsets for use in crop improvement. The literature indicates significant phenotypic and genetic diversity in pearl millet, making it an excellent source for mining genes and alleles of interest for important traits. Researchers have made substantial efforts to characterize pearl millet germplasm, resulting in a pool of trait-specific germplasm lines that show promise for improving grain yield, stover yield, grain nutrition, photoperiod sensitivity, heat tolerance, phonological traits, shelf life, resistance to biotic and abiotic stresses, and specific grain characteristics that meet consumer demands and market trends (Dwivedi et al., 2012; Pattanashetti et al., 2015). Selected germplasm lines with specific traits identified at ICRISAT and elsewhere are given in Table 2.2, and trait-specific genetic stocks registered with ICAR-NBPGR are presented in Table 2.3.

Although large and diverse collections of pearl millet germplasm are available, their use in crop improvement may be limited due to the limited information on traits of economic importance, linkage load with undesirable genes, assumed risks, and restricted access to germplasm collections due to international regulations (Upadhyaya et al., 2014a). Development and evaluation of core/mini-core sets has led to the identification of new sources of variation for grain and fodder traits (Khairwal et al., 2007), resistance to blast and downy mildew (Sharma et al., 2007, 2013), and grain nutritional traits (Govindaraj et al., 2020) (Table 2.2). Pearl millet is an outcrossing crop, and its landraces exhibit a high level of within-accession or within-population diversity. However, limited attempts have been made to investigate the germplasm for trait discovery using genomic methods, such as genome-wide association study and genomic selection. To address this gap, ICRISAT, in association with Aberystwyth University, developed the Pearl Millet inbred Germplasm Association Panel (PMiGAP) comprising 346 lines generated by repeated rounds of selfing (S0–S11) from 1000 accessions representing diverse cultivars, landraces, and mapping population parents of 27 countries. Thus, PMiGAP could be an excellent genetic resource for genomic investigation in pearl millet for traits discovery. This PMiGAP was resequenced, and it has a repository of ~29 million genome-wide SNPs that are being used to map traits related to drought tolerance, grain nutrients, and N use efficiency (Srivastava et al., 2020b).

Crop Wild Relatives as a Source of Important Traits

The *Pennisetum* wild species are the important genetic resources that could be the source of genes of interest for tolerance to abiotic and biotic stresses, forage yield and forage quality, and male sterility and fertility restoration. Pearl millet wild species accessions ($n = 529$) belonging to *P. violaceum*, *P. millisssium*, *P. purpreum*, *P. pedicellatum*, *P. polystachyon*, and *P. schweinfurthii* were screened for downy mildew resistance in the greenhouse and field-disease nurseries (Singh & Navi, 2000). Among these wild species, *P. schweinfurthii* was found to be highly resistant, whereas *P. violaceum* was most susceptible to downy mildew. Screening of 305 wild species accessions belong to *P. violaceum* for resistance to blast under glasshouse conditions against five isolates (Pg 45, Pg 53, Pg 56, Pg 118, and Pg 119) resulted in

Table 2.2 Promising Trait-specific Germplasm Reported in Pearl Millet for Yield, Quality and Stress Tolerance Traits in The Global Collection Conserved at ICRISAT, India.

Trait	Accessions (IP numbers)	References
Early flowering	Early flowering in Kharif season: Bchwaddawana (1417, 1418, 1425, 1427), Chadi local (3155, 3172, 3175, 3220, 3226, 3231, 3235, 3239, 3252, 3253, 3435, 13586), Desert type (3230, 3251), Pitta ganti (1249, 1251, 1252, 11782 to 11784, 11786, 11787, 11789 to 11792, 11794, 11867, 11868, 11873, 11877, 11878, 13460, 13465); Jakhrana (1478–1480, 3123 and 11885), Kala cumbu (1315), Manavari (1304), Peria cumbu (1307), Adchi (9101, 9105, 17867), Arisi (15329 and 15412), Babapuri (1179, 3840, 12556), Gaorani (9107, 9113, 9115, 9121, 9122, 9146, 13554), Kattu cumbu (15335, 15373), Kullan (1306), Pedda sajja (11810, 11811), Sadguru bajra (1266, 1267, 7813), and Vellai (13600).	Upadhyaya et al., (2016, 2017b)
	Early flowering in Rabi season: Gaorani (4144, 4148, 4149), IP 4006, 4149, IP 550, IP 3549, IP 4066, IP 9496, IP 9426, IP 3549	
High grain yield potential	Chadi local (3233, 3234), Desert type (3244, 3245), Peria cumbu (3477), and Gaorani (9119) for grain yield;	Khairwal et al. (2007)
	IP 1763 and IP 1745 promising for long panicles; IP 4394 for thick panicles; and IP 399 for bold grains; IP 128, IP 3416, IP 33, IP 17350, IP 3150, IP 17945, and IP 7095 for high grain yield; IP 13645, IP 15273, IP 17493 for higher productive tillers.	Khairwal et al. (2007), Upadhyaya et al. (2016, 2017b)
	Gero (12188), Guerguera (5311), Haini (10621, 10630, and 10678), Maewa (13129), Mela (11614), Nyali (17681), Sanio (10177 and 10576), and Sounari (10623)	

Fodder yield	Tall plants (11839, 11840, 17287, 20439, 20509, 20538, 20539, 20540, 20544, 20550, 20563, 20571, 20574, 20584, 20585); Higher number of tillers (3080, 3476, 3604, 3613, 3625, 3627, 3628, 3636, 3645, 3663, 3665, 6857, 6892, 8190, 8327, 11838, 13599, 13613, 15257, 15285, 15287, 15288, 15289, 15290, 15301, 15302, 15306, 15307, 15320, 15321, 15322, 15341, 15342, 15343, 15344, 15348, 15351, 15369, 15438, 15556, 20273, 20339, 20344, 20346, 20347, 20348, 20349, 20350, 20379); promising for high fodder yield per plant (3416, 17880 and 104). Green fodder yield: Chadi local (3188 and 3246), Desert type (3173 and 3190), Eravai (3517), Kala cumbu (3586), Peria cumbu (1305, 3610), Amreli (1174), Gaorani (9074 and 9076), Kattu cumbu (3479 and 3596), and Sadguru bajra (16283); Aria (6062), Gaouri (10574), Maewa (17436), Souma (10022 and 10030), Sounari (10014 and 10034), Tiotioni (9997 and 10085), Yadiri (14258), and Zanfaroua (9208). IP 11839 and IP 11840 for tall plant heights and more number of leaves per plant, IP 15710, IP 15735 and IP 15752 for stem thickness and leaf width, and IP 3628, IP 15285, IP 15288, IP 15302, IP 15342, IP 15351, IP 15290, IP 20347 and IP 20350 for total tillers per plant, and IP 3471, IP 3481, IP 3509 and IP 3593 for sweet stalk.	Khairwal et al. (2007), Upadhyaya et al. (2016, 2017b)
High grain nutrients	High Fe: IP# 3329, 7536, 8972, 9301, 10394, 12682, 15817, 17620, 17690, 17707 High Zn: IP# 3329, 3749, 4454, 10471, 11584, 11784, 12181, 13900, 15614, 17217 High Mn: IP# 7536, 8972, 9572, 10394, 13384, 15402, 15614, 17217, 17620, 17707 High Cu: IP# 3329, 3626, 7208, 9351, 10394, 12507, 12682, 13384, 13900, 15614 High Mo: IP# 3329, 13900, 15614, 3749, 10471, 11316, 11353, 11584, 11784, 17217 High Ca: IP# 15402, 9416, 17707, 9351, 15614, 17620, 11584, 14148, 7536, 5316 High Mg: IP# 15614, 9572, 7208, 17217, 7536, 3626, 3329, 4454, 12507, 10471 High K: IP# 9572, 10471, 12682, 11584, 3329, 12507, 14148, 15614, 11316, 11353 High P: IP# 9572, 15614, 3329, 7536, 12507, 14148, 7208, 10471, 12682, 17217 High S: IP# 9572, 17217, 12507, 7536, 7838, 3329, 4454, 15614, 3626, 11784 Low Na: IP# 5316, 13384, 3626, 12181, 17690, 7838, 4454, 11353, 10471, 12682 Low Ni: IP# 11316, 12862, 13384, 7536, 11320, 9407, 11784, 14148, 9416, 10471 Multiple nutrients: IP# 3329, 3749, 4454, 5316, 7208, 7536, 9351, 9407, 9496, 9572, 11316, 11784, 12507, 12939, 14148	Govindaraj et al. (2020)

(continued overleaf)

Table 2.2 *(Continued)*

Trait	Accessions (IP numbers)	References
Downy mildew resistance	IP# 9, 29, 55, 87, 104, 172, 253, 262, 283, 336, 346, 352, 364, 365, 396, 486, 498, 505, 517, 545, 556, 558, 575, 682, 718; IP# 14537 resistant to 7 pathotypes, IP# 9645, 11943, 14542, 14599, 21438, 11930, 12374, 14522, 20715, 21187, 21201, 21244, resistant to 5 pathotypes.	Sharma et al. (2007, 2015)
Blast resistance	*P. violaceum*: IP# 21525, 21531, 21536, 21540, 21594, 21610, 21640, 21706, 21711, 21716, 21719, 21720, 21721, 21724, 21987, 21988, and 22160 (accessions); *P. glaucum*: IP# 7846, 11036 and 21187	Sharma et al. (2013, 2020)
Rust resistance	*P. violaceum*: Single plant selection from IP# 21629, 21645, 21658, 21660, 21662, 21711, 21974, 21975, and 22038	Sharma et al. (2020), Singh et al. (1987)
	P. glaucum: ICML 11, ICML 17, ICML 18, ICML 19, ICML 20, ICML 21, ICML 22	
Smut	ICML 5, ICML 6, ICML 7, ICML 8, ICML 9, ICML 10	Thakur et al. (1992)
Drought tolerance	IP# 3243, 3228, 3424, 3296, 3362, 3180, 3272, 3303, 3252, 3258, 11141, 3318, 3123, 3363, and 3244; IP# 133, 177, 164, 142, 120, 160, 136, 166, 192, 195, IP106, 126, 121, 110, and 117	Choudhary et al. (2021), Yadav et al. (2003)
Salinity tolerance	IP# 22269, 6098, 6105, 3616, 6112, 6104, 3757	Krishnamurthy et al. (2007), Kulkarni et al. (2006)
Heat stress tolerance	IP# 19799, 19877, 19843, 21517, 3175	Sehgal et al. (2015), Upadhyaya et al. (2016), Yadav et al. (2012)

Table 2.3 Pearl Millet Genetic Stock Registered and Conserved at the ICAR-NBPGR, New Delhi, India (Data taken from http://www.nbpgr .ernet.in:8080/registration/Login.aspx).

S. No	Botanical name	National identity	Novel unique features
1	*P. glaucum*	IC296716	Fertility restorer line, resistant to drought and tolerant to high-temperature stress
2	*P. glaucum*	IC296717	Fertility restorer line, excellent combiner for productivity, produces early/extra early hybrids
3	*P. glaucum*	IC296718 (IC471852)	MS line converted in to different background
4	*P. glaucum*	IC296719 (IC471853)	MS line converted into H77/833-2
5	*P. glaucum*	IC296720 (IC471854)	MS line with high tillering (4.2)
6	*P. glaucum*	IC296721 (IC471855)	MS with good tillering and early 50% flowering (46 days)
7	*P. glaucum*	IC296722 (IC471857)	MS line with high grain (1,740 kg/ha) and dry fodder (3.7 t/ha) yield
8	*P. glaucum*	IC296723 (IC471858)	MS line with high combining ability in different genetic background
9	*P. glaucum*	IC296724 (IC471859)	MS line with high combining ability in different genetic background
10	*P. glaucum*	IC296730 (IC471856)	CMS line in different genetic background
11	*P. glaucum*	IC296734	Drought-tolerant good tillering MS line
12	*P. glaucum*	IC567684	Tetraploid male sterile line with A4 cytotype

(continued overleaf)

Table 2.3 *(Continued)*

S. No	Botanical name	National identity	Novel unique features
13	*P. glaucum*	IC568548	Maintainer of Tetra A4 MS line
14	*P. glaucum*	IC0593641	Interspecific hybridbetween pearl millet (*Pennisetum glaucum*) and *P. squamulatum* (novel cytotype, 2n = 56). Obligate sexual in reproduction. Chromosome number, 2n = 42 (14G + 28S), (genomic status GGSSSS).
15	*P. glaucum*	IC0593642	Interspecific hybridBetween pearl millet (*P. glaucum*) and *P. squamulatum* (novel cytotype, 2n = 56). Obligate apomictic in reproduction, chromosome number, 2n = 42 (14G + 28S), (genomic status GGSSSS).
16	*P. glaucum*	IC0593643	Diploid apomeiotic Interspecific hybrid between *P. laucum* and *P. orientale*. Induced apospory (apomixes component). Chromosomes 2n = 16 (7G + 90, genome GS).
17	*P. glaucum*	IC0593644	BC1 interspecific hybrid derived from *P. glaucum* and *P. orientale* cross (a BIII hybrid of F1 with *P. glaucum*). Exhibited inheritable partitioned apomixes components. Chromosomes 2n = 23 (14G + 90, genomes GGO).
18	*P. glaucum*	IC0593645	First tri-specific hybrid between *Pennisetum glaucum*, *P. orientale*, and *P. squamulatum*; chromosomes 2n = 44 (21G + 14S + 90, genomes GGGSSO)
19	*P. squamulatum*	IC0283734	Popping trait
20	*P. glaucum*	IC0617290	High Fe content (91 mg kg^{-1}); high Zn content (78 mg kg^{-1})
21	*P. glaucum*	IC643982	High Fe content (84 ppm); high Zn content (50 ppm)
22	*P. glaucum*	IC650743	High Fe content (95 ppm); High Zn content (53 ppm)
23	*P. glaucum*	IC650744	Resistant to Downy Mildew; High FE (81 ppm); Hight Zn content (49.5 ppm)

the identification of 17 accessions (IP 21525, IP 21531, IP 21536, IP 21540, IP 21594, IP 21610, IP 21640, IP 21706, IP 21711, IP 21716, IP 21719, IP 21720, IP 21721, IP 21724, IP 21987, IP 21988, and IP 22160) that were resistant to all five isolates of the blast (score of 3.0 or less), and 24 accessions were resistant to four pathotypes (Sharma et al., 2020). Rust is another important leaf disease in pearl millet that adversely affects biomass and the quality of forage. The 305 accessions of *P. violaceum* that were screened for blast resistance were also screened for rust resistance (Sharma et al., 2020). Single-plant selections from nine accessions (IP 21629, IP 21645, IP 21658, IP 21660, IP 21662, IP 21711, IP 21974, IP 21975, and IP 22038) were found highly resistant to rust (0% rust severity) after four generations of pedigree selection and subsequent screening (Sharma et al., 2020). Wild species (e.g., *P. hordeoides, P. pedicellatum,* and *P. polystachion*) were reported as promising sources for the blast, downy mildew, and rust resistance (Sharma et al., 2020). The blast-resistant accessions and rust-resistant genetic stocks are being used in a pre-breeding program at ICRISAT to introgress resistance genes from the wild into the parental lines of cultivated and potential pearl millet hybrids and varieties. Wild pearl millet accessions belonging to *P. violaceum* were reported to have resistance to the parasitic weed Striga and identified accessions with consistently low triga emergence (Wilson et al., 2000, 2004). The wild *Pennisetum* species (*P. ciliare, P. orientale, P. setaceum,* and *P. squamulatum* for apomixis; *P. cenchroides, P. hordeoides, P. pedicellatum, P. purpureum,* and *P. setosum* for fodder yield and quality traits; *P. alopecuroides, P. flassidum, P. orientale,* and *P. setaceum* for ornamentals; and *P. sheweinfurthii* for large seed size) are some of the sources for important traits for pearl millet improvement (Sharma et al., 2020). Pennisetum wild species contributed significantly to improving forage yield and quality. Interspecific hybridization between *P. glaucum* and *P. purpureum* has led to the development of forage hybrids with high biomass and better quality.

Pennisetum pedicellatum withstands multi-cuts for several years for green fodder and is generally used as a cut and carry green forage at ear emergence, but it can also be made into silage and hay. When compared with other *Pennisetum* species, *P. pedicellatum* produces more tillers than that of ssp. *monodii* and grows tall, facilitating more green fodder yield in a short period (Upadhyaya et al., 2014b). Most of the *P. pedicellatum* accessions conserved at the ICRISAT genebank produced green stems, the remaining produced purple stems. The wild species accessions belong to *P. pedicellatum,* with stem thickness of >7.5 mm (IP 21850, IP 21861, and IP 21846), longer leaves (>50 cm; IP 21849, IP 22072, and IP 21836), promising for fodder (IP 21850, IP 21853, IP 21853, IP 21861, IP 21850, IP 21849, IP 21862, IP 21845, and IP 21864), and maximum tillers (IP 21821, IP 21866, IP 22220, IP 22071, and IP 22091) have been reported (Upadhyaya et al., 2014b). In comparison to *P. glaucum* ssp. *monodii,* which has a weak stem and is susceptible to lodging, *P. pedicellatum* was found to be early flowering and to have high tillering habit, tallness, and a higher level of resistance to downy mildew, making it important in future pearl millet research and in the development of high-fodder and high-grain-yield cultivars (Upadhyaya et al., 2014b). There are continuous efforts toward introgression of disease resistance genes from CWRs in the development of high-yielding and disease-resistant cultivars (Sharma et al., 2020). The deep understanding of the wild species genetic resources and prioritization of traits for their use, trait discovery using high-throughput phenotyping and molecular tools, the introgression of traits with minimal linkage drag, and the continuous supply of the new and diverse genetic variability derived from CWRs are in the breeding pipeline for further deployment in breeding programs.

Germplasm Use in Crop Improvement

Pearl millet germplasm resources have been extensively used to develop improved composites, open-pollinated varieties (OPVs), and breeding lines with high grain yield; disease and pest resistance; heat, drought, and salinity tolerance; and high Fe and Zn content and have been distributed globally for their use in pearl millet improvement (Rai et al., 2014; Singh et al., 1987; Yadav et al., 2021). One notable example of the direct use of landraces is the development of ICTP 8203 by selection within the *iniadi* landrace from northern Togo (Rai et al., 1990). The ICTP 8203 is a large-seeded and high-yielding, OPV bred at ICRISAT, Patancheru, and this variety was released as MP 124 in Maharashtra and Andhra Pradesh, as PCB 138 in Punjab states of India for commercial cultivation in 1988, as Okashana 1 in Namibia during 1990, and as Nyankhombo (ICMV 88908) in Malawi in 1996. Direct selection within the same landrace from Togo led to the development of a large-seeded and downy mildew–resistant male sterile line ICMA 88004, a seed parent of an early maturing hybrid (ICMH 356) released in India in 1993 (Rai et al., 1995). Further, the intra-population variability within ICTP 8203 was used to develop high-Fe and high-Zn biofortified varieties of pearl millet 'Dhanshakti' and 'Chakti', which were released in India and Africa, respectively (Govindaraj et al., 2019; Rai et al., 2014). Other examples of direct selection from the landraces that were released as varieties include IKMP 3 (IP 11381) and IKMP 5 (IP 11317) in Burkina Faso; OPV ICMV-IS 88102 (IP 6426) in Burkina Faso, and Benkadi Nio in Mali; and IP 6104 and IP 19586 released in Mexico as high forage-yielding varieties (Pattanashetti et al., 2015). CZP 9802, another OPV was developed from early maturing and high-yielding progenies of a population from Rajasthan, India (Yadav, 2004). The varieties JBV 2 and JBV 3 were developed from Early Composite 91 and SRC II, respectively. Two other varieties, HC 10 and ICMV 155, were developed from early maturing and drought-tolerant 'New Elite Composite'.

Initially, in the 1970s and 1980s, the focus was on developing a wide and diverse range of improved trait-specific breeding lines and hybrid parents with higher grain yield and downy mildew resistance as the main selection criteria and with a limited focus toward lines with resistance to smut and ergot. Later, the pearl millet improvement program adopted a country/regional strategy and made a strategic shift toward developing and distributing a diverse range of high-yielding and downy mildew–resistant trait–based breeding lines and hybrid parents (seed parents and restorers) to National Agricultural Research Services and private seed industries for their use in hybrid development and commercialization. In the last two to three decades, about 60–70% of the pearl millet hybrids cultivated on farms in India are directly or indirectly based on ICRISAT-bred hybrid parental lines (Gupta et al., 2022). Germplasm diversity representative subsets such as core and mini--core collection have led to the identification of new sources of variation for use in pearl millet improvement. Wild species accessions have been used in developing pre-beeding lines to expand the cultivated gene pool in pearl millet. Intersubspecific crosses involving cultivated pearl millet with its wild species *P. glaucum* ssp. *monodii* and *P. stenostachyum* (weedy relative) have been successful in transferring desirable traits like rust resistance, male sterility, and alleles for enhancing yield components to cultivated pearl millet (Pattanashetti et al., 2015; Sharma et al., 2021). High-biomass, photoperiod-sensitive accessions are largely used in developing high-biomass, multi-cult forage pearl millet cultivars and released in India and Central Asian Countries, and interspecific hybridization between *P. glaucum* and *P. purpureum* has led to the development of forage hybrids with high biomass and better quality.

Conclusion

Pearl millet is one of the most nutritious and climate-resilient crops. Pear millet has greater potential to tolerate higher stresses compared to other crops. Its productivity, compared with other major cereals like rice, wheat, and maize, is low because it is cultivated primarily under low-input agriculture. Several factors play critical roles in the genetic enhancement of pearl millet, such as the available genetic resources; availability, inheritance, and stability of the traits; variety type, etc. Global genebanks hold over 60,000 accessions of pearl millet genetic resources. This demonstrates the worth of germplasm conservation efforts, especially at ICRISAT. The ICRISAT genebank has the largest repository of pearl millet in the world. Characterization and evaluation of germplasm for various morpho-agronomic, productivity, quality, and stress tolerance traits have shown a significant variability that can be exploited for trait improvement. Pearl millet is a highly outcrossing crop species that favors gene flow over generations within and among wild and cultivated pearl millet species. Its landraces exhibit a high level of within-accession or within-population diversity. Therefore, an appropriate strategy is required to use the existing germplasm variability in crop improvement. Large-scale screening of germplasm for important traits using high-throughput phenotypic and genomic tools, landraces, and CWRs for trait introgression are required for accelerating the genetic improvement of pearl millet. With the avail-ability of a high-quality reference genome as well as SNP chips panel, large-scale characterization, trait discovery, and gene mapping will improve, leading to increased germplasm use. An advanced high-throughput, next-generation sequencing approach can be used to sequence a vast number of pearl millet germ-plasm, enabling the identification and tracking of natural genetic variations within the germplasm collection and the discovery of genes and molecular mark-ers associated with important agronomic traits.

References

Anuradha, N., Satyavathi, C. T., Bharadwaj, C., Sankar, M., Singh, S. P., & Pathy, T. L. (2018). Pearl millet genetic variability for grain yield and micronutrients in the arid zone of India. *Journal of Pharmacognosy and Phytochemistry*, *7*, 875–878.

Bhattacharjee, R., Khairwal, I. S., Bramel, P. J., & Reddy, K. N. (2007). Establishment of a pearl millet [*Pennisetum glaucum* (L.) R. Br.] core collection based on geographical distribution and quantitative traits. *Euphytica*, *155*, 35–45. https://doi.org/10.1007/s10681-006-9298-x

Bramel, P., Giovannini, P., & Dulloo, M. E. (2022). *Global strategy for the conservation and use of genetic resources of selected millets*. Global Crop Diversity Trust.

Brunken, J. N. (1977). A systematic study of Pennisetum Sect. Pennisetum (Gramineae). *American Journal of Botany*, *64*, 161–176.

Choudhary, M. L., Tripathi, M. K., Gupta, N., Tiwari, S., Tripathi, N., Parihar, P., & Pandya, R. K. (2021). Screening of pearl millet [*Pennisetum glaucum* [L.] R. Br.] germplasm lines against drought tolerance based on biochemical traits. *Current Journal of Applied Science and Technology*, *40*, CJAST.73335. https://doi.org/10.9734/CJAST/2021/v40i2331483

Dwivedi, S., Upadhyaya, H., Senthilvel, S., Hash, C., Fukunaga, K., Diao, X., & Santra, D. (2012). Millets: Genetic and genomic resources. *Plant Breeding Review*, *35*, 247–375.

Faye, A., Barnaud, A., Kane, N. A., Cubry, P., Mariac, C., Burgarella, C., Rhoné, B., Faye, A., Olodo, K. F., Cisse, A., Couderc, M., Dequincey, A., Zekraouï, L., Moussa, D., Tidjani, M., Vigouroux, Y., & Berthouly-Salazar, C. (2022). Genomic footprints of selection in early-and late-flowering pearl millet landraces. *Frontiers in Plant Science*, *13*, 880631. https://doi.org/10.3389/fpls.2022.880631

Frankel, O. H. (1984). Genetic perspective of germplasm conservation. In D. W. Arber, K. Illmensee, W. J. Peacock, & P. Starlinger (Eds.), *Genetic manipulations: Impact on man and society* (pp. 161–170). Cambridge University Press.

Govindaraj, M., Rai, K. N., Cherian, B., Pfeiffer, W. H., Kanatti, A., & Shivade, H. (2019). Breeding biofortified pearl millet varieties and hybrids to enhance millet markets for human nutrition. *Agriculture*, *9*, 1–11. https://doi.org/10.3390/agriculture9050106

Govindaraj, M., Rai, K. N., Kanatti, A., Upadhyaya, H. D., Shivade, H., & Rao, A. S. (2020). Exploring the genetic variability and diversity of pearl millet core collection germplasm for grain nutritional traits improvement. *Scientific Reports*, *10*, 21177. https://doi.org/10.1038/s41598-020-77818-0

Gupta, S. K., Patil, K. S., Rathore, A., Yadav, D. V., Sharma, L. D., Mungra, K. D., Patil, H. T., Gupta, S. K., Kumar, R., Chaudhary, V., & Das, R. R. (2020). Identification of heterotic groups in South-Asian-bred hybrid parents of pearl millet. *Theoretical and Applied Genetics, 133*, 873–888.

Gupta, S. K., Patil, S., Kannan, R. L., Pujar, M., & Kumar, P. (2022). *Characterization of ICRISAT bred pearl millet restorer parents (2006–2019)*. International Crops Research Institute for the Semi-Arid Tropics.

Hanna, W. W., & Dujardin, M. (1986). Cytogenetics of *Pennisetum schweinfurthii* Pilger and its hybrids with pearl millet. *Crop Science, 26*, 449–453.

Harlan, J. R., & de Wet, J. M. J. (1971). Toward a rational classification of cultivated plants. *Taxon, 20*, 509–517. https://doi.org/10.2307/1218252

Hu, Z., Mbacké, B., Perumal, R., Guèye, M. C., Sy, O., Bouchet, S., Prasad, P. V., & Morris, G. P. (2015). Population genomics of pearl millet (*Pennisetum glaucum* (L.) R. Br.): Comparative analysis of global accessions and Senegalese landraces. *BMC Genomics, 16*, 1–12.

ICAR-NBPGR. (2024). *National genebank dashboard*. http://genebank.nbpgr.ernet.in

Kapadia, V. N. (2016). Estimation of heterosis for yield and its relevant traits in forage pearl millet [*Pennisetum glaucum* LR Br.]. *International Journal of Agriculture Sciences, 8*(54), 2829–2835.

Kapila, R. K., Yadav, R. S., Plaha, P., Rai, K. N., Yadav, O. P., Hash, C. T., & Howarth, C. J. (2008). Genetic diversity among pearl millet maintainers using microsatellite markers. *Plant Breeding, 127*, 33–37.

Kaushal, P., Khare, A., Zadoo, S. N., Roy, A. K., Malaviya, D. R., Agrawal, A., Siddiqui, S. A., & Choubey, R. (2008). Sequential reduction of *Pennisetum squamulatum* genome complement in *P. glaucum* (2n=28) x *P. squamulatum* (2n=56) hybrids and their progenies revealed its octoploid status. *Cytologia, 73*, 151–158. https://doi.org/10.1508/cytologia.73.151

Khairwal, I. S., Yadav, S. K., Upadhyaya, H. D., Kachhawa, D., Nirwan, B. S., Bhattacharjee, R., & Rajpurohit, B. S. (2007). Evaluation and identification of promising pearl millet germplasm for grain and fodder traits. *Science Agriculture and Technology, 5*, 1–6.

Krishnamurthy, L., Serraj, R., Rai, K. N., Hash, C. T., & Dakheel, A. J. (2007). Identification of pearl millet [*Pennisetum glaucum* (L.) R. Br.] lines tolerant to soil salinity. *Euphytica, 158*, 179–188. https://doi.org/10.1007/s10681-007-9441-3

Kulkarni, V., Rai, K., Dakheel, A., Ibrahim, M., Hebbara, M., & Vadez, M. (2006). Pearl millet germplasm adapted to saline conditions millets research genetic resources. *International Sorghum and Millets Newsletter, 47*, 103–106.

Lakis, G., Navascués, M., Rekima, S., Simon, M., Remigereau, M. S., Leveugle, M., Takvorian, N., Lamy, F., Depaulis, F., & Robert, T. (2012). Evolution of neutral and flowering genes along pearl millet (*Pennisetum glaucum*) domestication. *PloS One*, *7*(5), e36642. https://doi.org/10.1371/journal.pone.0036642

Lakis, G., Ousmane, A. M., Sanoussi, D., Habibou, A., Badamassi, M., Lamy, F., Jika, N., Sidikou, R., Adam, T., Sarr, A., Luxereau, A., & Robert, T. (2011). Evolutionary dynamics of cycle length in pearl millet: The role of farmer's practices and gene flow. *Genetica*, *139*, 1367–1380. https://doi.org/10.1007/s10709-012-9633-1

Mariac, C., Luong, V., Kapran, I., Mamadou, A., Sagnard, F., Deu, M., Chantereau, J., Gerard, B., Ndjeunga, J., Bezançon, G., & Pham, J. L. (2006a). Diversity of wild and cultivated pearl millet accessions (*Pennisetum glaucum* [L.] R. Br.) in Niger assessed by microsatellite markers. *Theoretical and Applied Genetics*, *114*, 49–58.

Mariac, C., Robert, T., Allinne, C., Remigereau, M. S., Luxereau, A., Tidjani, M., Seyni, O., Bezancon, G., Pham, J. L., & Sarr, A. (2006b). Genetic diversity and gene flow among pearl millet crop/weed complex: A case study. *Theoretical and Applied Genetics*, *113*, 1003–1014. https://doi.org/10.1007/s00122-006-0360-9

Martel, E., Poncet, V., Lamy, F., Siljak-Yakovlev, S., Lejeune, B., & Sarr, A. (2004). Chromosome evolution of *Pennisetum* species (Poaceae): Implications of ITS phylogeny. *Plant Systematics and Evolution*, *249*, 139–149. https://doi.org/10.1007/s00606-004-0191-6

Martel, E., Ricroch, A., & Sarr, A. (1996). Assessment of genome organization among diploid species (2n = 2x = 14) belonging to primary and tertiary gene pools of pearl millet using fluorescent in situ hybridization with rDNA probes. *Genome*, *39*, 680–687. https://doi.org/10.1139/g96-086

Moussa, H., Kindomihou, V., Houehanou, T. D., Chaibou, M., Souleymane, O., Soumana, I., Dossou, J., & Sinsin, B. (2021). Farmers' perceptions of fodder performances of pearl millet (*Pennisetum glaucum* (L.) R. Br) accessions in Niger. *Heliyon*, *7*(9), e07965. https://doi.org/10.1016/j.heliyon.2021.e07965

Naino Jika, A. K., Dussert, Y., Raimond, C., Garine, E., Luxereau, A., Takvorian, N., Djermakoye, R. S., Adam, T., & Robert, T. (2017). Unexpected pattern of pearl millet genetic diversity among ethno-linguistic groups in the Lake Chad Basin. *Heredity*, *118*, 491–502. https://doi.org/10.1038/hdy.2016.128

Patil, K. S., Mungra, K. D., Danam, S., Vemula, A. K., Das, R. R., Rathore, A., & Gupta, S. K. (2021). Heterotic pools in African and Asian origin populations of pearl millet [*Pennisetum glaucum* (L.) R. Br.]. *Scientific Reports*, *11*(1), 1–13.

Pattanashetti, S. K., Upadhyaya, H. D., Dwivedi, S. L., Vetriventhan, M., & Reddy, K. N. (2015). Pearl millet. In M. Singh & H. D. Upadhyaya (Eds.), *Genetic and genomic resources for grain cereals improvement* (pp. 253–289). Oxford Academic Press.

Pucher, A., Sy, O., Sanogo, M. D., Angarawai, I. I., Zangre, R., Ouedraogo, M., Boureima, S., Hash, C. T., & Haussmann, B. I. (2016). Combining ability patterns among west African pearl millet landraces and prospects for pearl millet hybrid breeding. *Field Crops Research*, *195*, 9–20.

Rai, K. N., Gupta, S. K., Sharma, R., Govindaraj, M., Rao, A. S., Shivade, H., & Bonamigo, L. A. (2014). Pearl millet breeding lines developed at ICRISAT: A reservoir of variability and useful source of non-target traits. *SAT e-Journal*, *12*, 1–13.

Rai, K. N., Kumar, K. A., Andrews, D. J., Rao, A. S., Raj, A. G. B., & Witcome, J. R. (1990). Registration of 'ICTP 8203' pearl millet. *Crop Science*, *30*, 959–959. https://doi.org/10.2135/cropsci1990.0011183x003000040048x

Rai, K. N., Rao, A. S., & Hash, C. T. (1995). Registration of pearl millet parental lines ICMA 8804 and ICMB 8804. *Crop Science*, *35*, 1242. https://doi.org/10.2135/cropsci1995.0011183X003500040108x

Ramya, A. R., Ahamed, M. L., Satyavathi, C. T., Rathore, A., Katiyar, P., Raj, A. B., Kumar, S., Gupta, R., Mahendrakar, M. D., Yadav, R. S., & Srivastava, R. K. (2018). Towards defining heterotic gene pools in pearl millet [*Pennisetum glaucum* (L.) R. Br.]. *Frontiers in Plant Science*, *8*, 1934.

Reddy, N., Rao, K., & Ahmed, I. (2004). Geographical patterns of diversity in pearl millet germplasm from Yemen. *Genetic Resources and Crop Evolution*, *51*, 513–517.

Satyavathi, C. T., Ambawat, S., Khandelwal, V., & Srivastava, R. K. (2021). Pearl millet: A climate-resilient nutricereal for mitigating hidden hunger and provide nutritional security. *Frontiers in Plant Science*, *12*, 659938. https://doi.org/10.3389/fpls.2021.659938

Satyavathi, C. T., Tiwari, S., Bharadwaj, C., Rao, A. R., Bhat, J., & Singh, S. P. (2013). Genetic diversity analysis in a novel set of restorer lines of pearl millet [*Pennisetum glaucum* (L.) R. Br] using SSR markers. *Vegetos*, *26*, 72–82.

Sehgal, D., Skot, L., Singh, R., & Srivastava, R. K. (2015). Exploring potential of pearl millet germplasm association panel for association mapping of drought tolerance traits. *PloS One*, *10*, e0122165. https://doi.org/10.1371/journal.pone.0122165

Semalaiyappan, J., Selvanayagam, S., Rathore, A., Gupta, S. K., Chakraborty, A., Gujjula, K. R., Haktan, S., Viswanath, A., Malipatil, R., Shah, P., Govindaraj, M., Ignacio, J. C., Reddy, S., Singh, A. K., & Thirunavukkarasu, N. (2023). Development of a new AgriSeq 4K mid-density SNP genotyping panel and its utility in pearl millet breeding. *Frontiers in Plant Science*, *13*, 1068883. https://doi.org/10.3389/fpls.2022.1068883

Sharma, R., Sharma, S., & Gate, V. L. (2020). Tapping *Pennisetum violaceum*, a wild relative of pearl millet (*Pennisetum glaucum*), for resistance to blast (caused by *Magnaporthe grisea*) and rust (caused by *Puccinia substriata var. Indica*). *Plant Disease*, *104*, 1487–1491. https://doi.org/10.1094/PDIS-08-19-1602-RE

Sharma, R., Upadhyaya, H. D., Manjunatha, S. V., Rai, K. N., Gupta, S. K., & Thaku, R. P. (2013). Pathogenic variation in the pearl millet blast pathogen, *Magnaporthe grisea* and identification of resistance to diverse pathotypes. *Plant Disease*, *97*, 189–195. https://doi.org/10.1094/PDIS-05-12-0481-RE

Sharma, R., Upadhyaya, H. D., Sharma, S., Gate, V. L., & Raj, C. (2015). New sources of resistance to multiple pathotypes of *Sclerospora graminicola* in the pearl millet mini core germplasm collection. *Crop Science*, *55*, 1619–1628. https://doi.org/ 10.2135/cropsci2014.12.0822

Sharma, S., Sharma, R., Govindaraj, M., Mahala, R. S., Satyavathi, C. T., Srivastava, R. K., Gumma, M. K., & Kilian, B. (2021). Harnessing wild relatives of pearl millet for germplasm enhancement: Challenges and opportunities. *Crop Science*, *61*, 177–200. https://doi.org/10.1002/csc2.20343

Sharma, Y. K., Yadav, S. K., Khairwal, I. S., Bajri-chomu, D., & Bajri-osian, D. (2007). Evaluation of pearl millet germplasm lines against downy mildew incited by *Sclerospora graminicola* in western Rajasthan. *SAT e-Journal*, *3*(1), 1–2.

Singh, S., & Gupta, S. K. (2019). Formation of heterotic pools and understanding relationship between molecular divergence and heterosis in pearl millet [*Pennisetum glaucum* (L.) R. Br.]. *PloS One*, *14*, e0207463.

Singh, S., Gupta, S. K., Thudi, M., Das, R. R., Vemula, A., Garg, V., Varshney, R. K., Rathore, A., Pahuja, S. K., & Yadav, D. V. (2018). Genetic diversity patterns and heterosis prediction based on SSRs and SNPs in hybrid parents of pearl millet. *Crop Science*, *58*, 2379–2390. https://doi.org/10.2135/cropsci2018.03.0163

Singh, S. D., Andrews, D. J., & Rai, K. N. (1987). Registration of ICML 11 rust-resistant pearl millet germplasm. *Crop Science*, *27*, 367–368.

Singh, S. D., & Navi, S. S. (2000). Genetic resistance to pearl millet downy mildew ll. Resistance in wild relatives. *The Journal of Mycology and Plant Pathology*, *30*, 167–171.

Srivastava, R. K., Bollam, S., Pujarula, V., Pusuluri, M., Singh, R. B., Potupureddi, G., & Gupta, R. (2020a). Exploitation of heterosis in pearl millet: A review. *Plants*, *8*, 807.

Srivastava, R. K., Singh, R. B., Pujarula, V. L., Bollam, S., Pusuluri, M., Chellapilla, T. S., Yadav, R. S., & Gupta, R. (2020b). Genome-wide association studies and genomic selection in pearl millet: Advances and prospects. *Frontiers in Genetics*, *10*, 1389. https://doi.org/10.3389/fgene.2019.01389

Stich, B., Haussmann, B. I., Pasam, R., Bhosale, S., Hash, C. T., Melchinger, A. E., & Parzies, K. K. (2010). Patterns of molecular and phenotypic diversity in pearl millet [*Pennisetum glaucum* (L.) R. Br.] from west and Central Africa and their relation to geographical and environmental parameters. *BMC Plant Biology*, *10*, 216. https:// doi.org/10.1186/1471-2229-10-216

Sumanth, M., Sumathi, P., Vinodhana, N. K., & Sathya, M. (2013). Cytosterile diversification and downy mildew resistance in pearl millet [*Pennisetum glaucum* (L.) R. Br]. *Indian Journal of Genetics and Plant Breeding*, *73*, 325–327.

Teklewold, A., & Becker, H. C. (2005). Heterosis and combining ability in a diallel cross of Ethiopian mustard inbred lines. *Crop Science*, *45*, 2629–2635.

Thakur, R. P., King, S. B., Rai, K. N., & Rao, V. P. (1992). *Identification and utilization of smut resistance in pearl millet*. International Crops Research Center for the Semi-Arid Tropics.

UNESCO. (2023). *Biosphere reserves*. https://en.unesco.org/biosphere

Upadhyaya, H. D., Dwivedi, S. L., Sharma, S., Lalitha, N., Singh, S., Varshney, R. K., & Gowda, C. L. L. (2014a). Enhancement of the use and impact of germplasm in crop improvement. *Plant Genetic Resources, 12*, S155–S159. https://doi.org/10.1017/S1479262114000458

Upadhyaya, H. D., Gowda, C. L. L., Reddy, K. N., & Singh, S. (2009a). Augmenting the pearl millet core collection for enhancing germplasm utilization in crop improvement. *Crop Science, 49*, 573–580. https://doi.org/10.2135/cropsci2008.06.0378

Upadhyaya, H. D., & Ortiz, R. (2001). A mini core subset for capturing diversity and promoting utilization of chickpea genetic resources in crop improvement. *Theoretical and Applied Genetics, 102*, 1292–1298.

Upadhyaya, H. D., Reddy, K., Ramachandran, S., Kumar, V., & Ahmed, M. I. (2016). Adaptation pattern and genetic potential of Indian pearl millet named landraces conserved at the ICRISAT Genebank. *Indian Journal of Plant Genetic Resources, 29*, 97–113. https://doi.org/10.5958/0976-1926.2016.00015.2

Upadhyaya, H. D., Reddy, K., Singh, S., Ahmed, M. I., Kumar, V., & Ramachandran, S. (2014b). Geographical gaps and diversity in Deenanath grass (*Pennisetum pedicellatum* trin.) germplasm conserved at the ICRISAT genebank. *Indian Journal of Plant Genetic Resources, 27*, 93. https://doi.org/10.5958/0976-1926.2014.00001.1

Upadhyaya, H. D., Reddy, K. N., Ahmed, M. I., Dronavalli, N., & Gowda, C. L. L. (2012a). Latitudinal variation and distribution of photoperiod and temperature sensitivity for flowering in the world collection of pearl millet germplasm at ICRISAT genebank. *Plant Genetic Resources, 10*, 59–69. https://doi.org/10.1017/s1479262111000979

Upadhyaya, H. D., Reddy, K. N., Ahmed, M. I., Kumar, V., Gumma, M. K., & Ramachandran, S. (2017a). Geographical distribution of traits and diversity in the world collection of pearl millet [*Pennisetum glaucum* (L.) R. Br., synonym: *Cenchrus americanus* (L .) Morrone] landraces conserved at the ICRISAT genebank. *Genetic Resources and Crop Evolution, 64*, 1365–1381. https://doi.org/10.1007/s10722-016-0442-8

Upadhyaya, H. D., Reddy, K. N., Ahmed, M. I., Ramachandran, S., Kumar, V., & Singh, S. (2017b). Characterization and genetic potential of African pearl millet named landraces conserved at the ICRISAT genebank. *Plant Genetic Resources, 15*, 438–452. https://doi.org/10.1017/S1479262116000113

Upadhyaya, H. D., Reddy, K. N., Irshad Ahmed, M., & Gowda, C. L. L. (2010). Identification of gaps in pearl millet germplasm from Asia conserved at the ICRISAT genebank. *Plant Genetic Resources, 8*, 267–276. https://doi.org/10.1017/S1479262112000275

Upadhyaya, H. D., Reddy, K. N., Irshad Ahmed, M., & Gowda, C. L. L. (2012b). Identification of gaps in pearl millet germplasm from east and southern Africa conserved at the ICRISAT genebank. *Plant Genetic Resources, 10*, 202–213. https://doi.org/10.1017/S1479262112000275

Upadhyaya, H. D., Reddy, K. N., Irshad Ahmed, M., Gowda, C. L. L., & Haussmann, B. I. G. (2009b). Identification of geographical gaps in the pearl millet germplasm conserved at ICRISAT genebank from west and Central Africa. *Plant Genetic Resources, 8,* 45–51. https://doi.org/10.1017/S147926210999013X

Upadhyaya, H. D., Reddy, K. N., Singh, S., Gowda, C. L. L., Ahmed, M. I., & Kumar, V. (2014c). Diversity and gaps in *Pennisetum glaucum subsp. monodii* (Maire) Br. germplasm conserved at the ICRISAT genebank. *Plant Genetic Resources, 12,* 226–235. https://doi.org/10.1017/S1479262113000555

Upadhyaya, H. D., Reddy, K. N., Singh, S., Gowda, C. L. L., Irshad Ahmed, M., & Ramachandran, S. (2013). Latitudinal patterns of diversity in the world collection of pearl millet landraces at the ICRISAT genebank. *Plant Genetic Resources, 12,* 91–102. https://doi.org/10.1017/S1479262113000348

Upadhyaya, H. D., Yadav, D., Reddy, K. N., Gowda, C. L. L., & Singh, S. (2011). Development of pearl millet minicore collection for enhanced utilization of germplasm. *Crop Science, 51,* 217–223. https://doi.org/10.2135/cropsci2010.06.0336

Varshney, R. K., Shi, C., Thudi, M., Mariac, C., Wallace, J., Qi, P., Zhang, H., Zhao, Y., Wang, X., Rathore, A., & Srivastava, R. K. (2017). Pearl millet genome sequence provides a resource to improve agronomic traits in arid environments. *Nature Biotechnology, 35,* 969–976.

Vernooy, R., Shrestha, P., & Sthapit, B. (2015). *Community seed banks: Origins, evolution and prospects.* Routledge.

Wilson, J. P., Hess, D. E., & Hanna, W. W. (2000). Resistance to *Striga hermonthica* in wild accessions of the primary gene pool of *Pennisetum glaucum. Phytopathology, 90,* 1169–1172. https://doi.org/10.1094/PHYTO.2000.90.10.1169

Wilson, J. P., Hess, D. E., Hanna, W. W., Kumar, K. A., & Gupta, S. C. (2004). *Pennisetum glaucum* subsp. *monodii* accessions with Striga resistance in West Africa. *Crop Protection, 23,* 865–870. https://doi.org/10.1016/j.cropro.2004.01.006

Yadav, O. P. (2004). CZP 9802 – A new drought-tolerant cultivar of pearl millet. *Indian Farming, 54,* 15–17.

Yadav, O. P., Gupta, S. K., Govindaraj, M., Sharma, R., Varshney, R. K., Srivastava, R. K., Rathore, A., & Mahahla, R. S. (2021). Genetic gains in pearl millet in India: Insights into historic breeding strategies and future perspective. *Frontiers in Plant Science, 12,* 645038. https://doi.org/10.3389/fpls.2021.645038

Yadav, O. P., & Rai, K. N. (2013). Genetic improvement of pearl millet in India. *Agricultural Research, 2,* 275–292.

Yadav, O. P., Rai, K. N., & Gupta, S. K. (2012). Pearl millet: Genetic improvement and tolerance to abiotic stresses. In N. Tuteja, S. S. Gill, & R. Tuteja (Eds.), *Improving crop productivity in sustainable agriculture* (pp. 261–288). Wiley.

Yadav, O. P., Weltzien-Rattunde, E., & Bidinger, F. R. (2003). Genetic variation in drought response of landraces of pearl millet (*Pennisetum glaucum* (L.) R. Br.). *Indian Journal of Genetics and Plant Breeding, 63,* 37–40.

3

Pearl Millet Genetic Improvement for Food and Nutrition Security

Mahalingam Govindaraj and Mahesh Pujar

Chapter Overview

Global food production practices are challenged by climate change factors, such as increased temperature, increased CO_2 emissions, and drought. Pearl millet [*Pennisetum glaucum* (L.) R. Br.] is an excellent dryland crop that withstands varied edaphic and climatic conditions (42 °C temperature and 350 mm rainfall). Pearl millet is rich in protein (15%) and essential minerals (296 mg/100 g P; 307 mg/100 g K; 42 mg/100 g Ca; 137 mg/100 g Mg; 75 mg/kg Fe, and 40 mg/kg Zn). Therefore, pearl millet not only serves as a major staple food in the semi-arid tropics of India and Africa but also serves as the cheapest source of household nutrition. The availability of genetic variation for nutrition in pearl millet leads to additional expansion to fulfill maximum daily requirements. Biofortification improves grain mineral and vitamin concentration. HarvestPlus, in collaboration with International Crops Research Institute for the Semi-Arid Tropics and other public and private partners, has disseminated several biofortified hybrids and open-pollinated varieties that will aid in the reduction of malnutrition in India and sub-Saharan Africa. The application of modern genomic technology and speed breeding would substantially accelerate biofortification traits mainstreaming. The advanced pearl millet breeding program, therefore, should include essential nutrition traits in the products and their commercialization consortia. Nutri-dense varieties and hybrids must ensure value propositions in the smart-food products and its supply chain for human well-being.

Introduction

The green revolution efforts in selected crops contributed to overcoming food insecurity in the 1960s. The use of new intensified monocropping, higher inputs,

Pearl Millet: A Resilient Cereal Crop for Food, Nutrition, and Climate Security, First Edition. Edited by Ramasamy Perumal, P. V. Vara Prasad, C. Tara Satyavathi, Mahalingam Govindaraj, and Abdou Tenkouano.

and irrigation contributed to the loss of nutritionally diverse crops, such as millets. Unlike in previous decades, when the focus was solely on increasing productivity, the current challenge is to address food security and nutritional security together with increasing populations posing challenges in the face of a changing climate. These challenges can be met by focusing on food crop diversification, which has the potential to improve nutritional and food security, particularly in Africa and Asia, where malnutrition is prevalent. Food and nutrition insecurity is expected to increase due to climate vulnerability for food production.

Pearl millet is a C_4 summer grass ($2n = 14$). It has high photosynthetic efficiency and dry matter production capability and exhibits high tolerance to heat, drought, salinity, and acid soils. It is easily cultivable and produces a high yield in low rainfall regions, where other crops are challenging to grow (FAO, 2004). Because of these characteristics, pearl millet serves as an important staple food source for more than 90 million people living in India and Africa. Pearl millet, a domesticated Saharan Desert grass (NRC, 1996), is descended from a wild West African grass that spread to East Africa and then to India. Pearl millet is now grown on more than 30 million ha around the world, with the majority of the crop grown in Africa (>18 million ha) and Asia (>10 million ha) (Raheem et al., 2021). Ninety million people across the world rely on pearl millet for daily food and income, primarily in northwestern India and sub-Saharan Africa (Srivastava et al., 2020a).

India is the largest pearl millet growing country, with 7 million ha cultivated and has the highest average annual millet grain production of 8.61 million t (Directorate of Millets Development, 2020). Pearl millet is the fourth most widely cultivated cereal crop in India, after rice (*Oryza sativa* L.), wheat (*Triticum aestivum* L.), and maize (*Zea mays* L.). Rajasthan, Maharashtra, Uttar Pradesh, Gujarat, and Haryana are the major pearl millet growing states in India, accounting for approximately 90% of total production. In India, the primary season for pearl millet cultivation is the rainy season, also known as the *kharif* crop (June/July to September/October). It is also grown as a summer crop (February–May) in parts of Gujarat, Rajasthan, and Uttar Pradesh. From November to February, the states of Maharashtra and Gujarat grow pearl millet in very small areas during the postrainy (*rabi*) season (Satyavathi et al., 2021a). In Africa, the two main areas of pearl millet cultivation are the west/central Africa region (Nigeria, Niger, Chad, Mali, and Senegal) and the east/southern Africa region (Sudan, Ethiopia, and Tanzania). West/central Africa accounts for 90% of the total area in Africa's (and thus is the world's) largest pearl millet–producing region (Jukanti et al., 2016). Pearl millet was cultivated in 18.50 million ha of land in Africa, with a production of 11.36 million t. This crop represented 30% of the continent in various agro-ecologies. Overall, 49% of the world's pearl millet is grown in Africa (FAO, 2019).

Globally, one in three individuals, or more than two billion people, experience hidden hunger (FAO, 2019). Deficiencies in Fe and Zn, the two mineral elements most frequently missing in human diets, are among the top 10 risk factors for disease, particularly in developing nations, ranking fifth and sixth, respectively (WHO, 2002). India was placed 94th out of 107 nations that qualified for the global hunger index (Gramber et al., 2020), and, according to the most recent statistics available, almost 40% of children under 5 years of age are stunted and 21% of children under 5 years of age are seriously undernourished. Pearl millet, a highly nutritious crop, is a rich natural source of micronutrients like Fe and Zn. Developing high-Fe and high-Zn pearl millet hybrids and varieties through either genetic or agronomic biofortification would significantly contribute to mitigating malnutrition in these regions.

Multiple Uses of Pearl Millet

Pearl millet is known by several different names. There are many distinctive colors of pearl millet grains, including ivory, cream, yellow, gray, deep gray, grayish brown, purple, and purplish black (IBPGR/ICRISAT, 1993). Most often, it is consumed as a thick porridge, or "toh." Additionally, the grains are milled into flour to make fermented foods like "Kisra and gallettes," steam-cooked dishes known as "couscous," and unfermented breads and cakes known as "roti." It is also used in non-alcoholic beverages and snacks (Girgi & O'Kennedy, 2007). Additionally, it is used in the preparation of the fried cake known as "masa" in northern Nigeria, while roasted young ears are a kid-favorite snack.

In addition to being grown for grain, pearl millet is also cultivated as a forage crop for livestock grazing, silage preparation, hay production, and the cutting of green fodder (Newman et al., 2010). The advantage of using this crop as livestock feed is that it may be fed to animals at any stage of crop development without negative consequences (Arya et al., 2013). This has been practiced in the countries like the United States (Sheahan, 2014) and during the summer in Australia (Hanna, 1996), Canada (Brunette et al., 2014), and Mexico (Urrutia et al., 2015). It is also used as triple cropping method in southern Kyushu, Japan (Li et al., 2019); Iran (Aghaalikhani et al., 2008); Central Asia (Nurbekov et al., 2013); and Brazil (Dias-Martins et al., 2018). It has become a significant fodder crop in northwest India during the summer (Amarender Reddy et al., 2013) and recently has been used as cover crop over 5 million ha in Brazil (De Assis et al., 2018; Dias-Martins et al., 2018). Climate change and global warming have prompted a reconsideration of previously used fuel energy sources in order to reduce greenhouse gas emissions. This has resulted in an emphasis on the usage of various biofuels. Pearl millet lignocellulosic feed stock with high cellulose (41.6%) and hemicellulose (22.32%) concentration is a significant source of biofuel production that may potentially contribute to an increase in farmer income (Packiam et al., 2018).

Nutritional Importance of Pearl Millet

Pearl millet has recently been recognized as a nutri-cereal due to its tremendous nutritional properties, which have the potential to play an important role in meeting the daily nutritional requirements of people living in resource-poor areas such as Africa and India, where it is traditionally grown as a major staple food crop. Pearl millet is a significant source of energy, providing 361 Kcal/100 g, which is comparable to other important cereals like wheat (346 Kcal/100 g), rice (345 Kcal/100 g), maize (125 Kcal/100 g), and sorghum (349 Kcal/100 g), as per the Nutritive Value of Indian Foods (Gopalan & Deosthale, 2003). In pearl millet, the glycemic index (55) is the lowest, and its fiber concentration is the highest (1.2 g/100 g) compared with other cereals (Mani et al., 1993). Along with these, amino acids such as lysine and tryptophan are abundant in pearl millet grains. Furthermore, pearl millet grains are naturally rich in important minerals, such as Fe, Zn, Ca, Mn, K, and Mg, as well as important vitamins, such as thiamine and niacin (Table 3.1). The relevance of pearl millet from the nutritional standpoint is demonstrated by the fact that it is grown for its grain (a staple food crop) in parts of India and throughout most of sub-Saharan Africa, where it serves as a large source of Fe and Zn as well as greater source of energy. In places such as

Table 3.1 Nutrition Profile of Commercially Available Pearl Millet Foods in India.

Nutrition profile	Value
Carbohydrates (Kcal)	61.8
Protein (g)	10.9
Fat (g)	5.43
Energy (kcal)	347
Dietary fiber (g)	11.5
Ca (mg)	27.4
P (mg)	289
Fe (mg)	6.4
Mg (mg/100 g)	124
Zn (mg/100 g)	2.7
Thiamin (mg)	0.25
Niacin (mg)	0.9
Riboflavin	0.2
Folic acid (mg/100 g)	36.1

Note: From NIN, Hyderabad, 2018.

Maharashtra, Gujarat, and Rajasthan states of India, pearl millet accounts for 20%–60% of all cereal consumed. As a result, it accounts for 19%–63% of Fe and 16%–56% of Zn intake from all food sources (Parthasarathy Rao et al., 2006).

A collection of 87 genotypes, including varieties, hybrids, and germplasms, that constituted the national pearl millet collection in India were screened for important nutritional profiles (Goswami et al., 2022). Relatively good variation was observed for starch (50–63 g/100 g) and amylose (19–28 g/100 g) concentrations in grains. In addition, the glucose and sucrose altered considerably with little variance (<1 g/100 g). The protein content in pearl millet grains varied significantly from 8 to 18 g/100 g; with the exception of three genotypes (DR-1, CZP16-923, and KUDRI), the remaining 84 genotypes contained >10 g of protein per 100 g of grains. Some of the high-protein sources among the germplasm collections include IC-537953 (18.15 g/100 g), IC-420368 (17.33 g/100 g), IC-537983 (16.66 g/100 g), IC-420371 (16.53 g/100 g), IC-537985 (16.53 g/100 g), IC-420652 (16.29 g/100 g), and IC-420362 (16.01 g/100 g). Phytic acid, which is an antinutritional element, varied from 0.54 to 1.43 g/100 g, whereas the presence of phenol concentration was negligible and varied from 0.04 to 0.21 GAE g/100 g. The variability for lipid was quite high and ranged from 5% to 9% in grains. Relatively smaller variations were observed for palmitic acid (20%–32%), stearic acid (3%–85%), linoleic acid (32%–47%), and oleic acid (22%–33%).

Priority Nutrients Improvement

Enriching the staple food crops with vital nutritional features such as protein, vitamins, and some important micronutrients (e.g., Fe and Zn) would contribute considerably to overcoming malnutrition caused by a lack of one or more of these nutritional elements. This is especially crucial for those living in resource-poor areas such as Africa, as well as small-holder agricultural families that cannot afford various foods such as meat, fruits, and vegetables to meet their daily nutritional requirements. Processed foods made from pearl millet have a higher average Fe concentration of 6.4 mg/100 g than other main staple crops, such as wheat (3.9 mg/g), sorghum (3.9 mg/g), and rice (0.6 mg/g) (NIN, 2018). Similarly, the average Zn concentration of pearl millet grains is 2.7 mg/100 g, which is comparable to wheat (2.8 mg/100 g) but greater than sorghum (1.9 mg/100 g) and rice (1.2 mg/100 g). HarvestPlus, a biofortification program, mostly targeted for Fe and Zn. These micronutrient concentration targets for biofortified crops are based on the following factors: the target populations' nutritional requirements (particularly women, adolescent girls, and young children); per capita crop consumption; retention of the micronutrient in the crop following processing, cooking, and/or storing; and the proportion of a micronutrient absorbed by the body (or bioavailability) (www.http://HarvetsPlus.org).

The occurrence of a greater natural variation for Fe and Zn micronutrient concentration in pearl millet grains was discovered during the initial phase of screening for distinct nutritional characteristics in pearl millet. Emphasis was also placed on the yellow-seeded pearl millet initially, which is colloquially known as "golden pearl millet" due to the presence of β-carotene in the seeds. Beta-carotene is essential for human health; however, it was later discovered that β-carotene was not present in sufficient quantities in pearl millet grains (up to 1.73 g/g), making further genetic improvement through conventional breeding methods impossible. Golden pearl millet lines had other clinical issues, such as seedling emergence, precise β-carotene assays, and time-consuming and expensive phenotyping (Muthusamy et al., 2014). However, because efforts to find more sources within the germplasm were futile, β-carotene research was halted, and pearl millet breeding was refocused on improving grain Fe and Zn concentration.

Studies showed that the variability for the grain Fe concentration in pearl millet was much higher than for the Zn concentration (Govindaraj et al., 2020). However, the significant association observed between these two micronutrients in pearl millet ($r = \geq 0.70$) indicated that selecting for high Fe would not affect grain Zn concentration and that Zn concentration can be improved further as an associated trait (Govindaraj et al., 2019, 2020; Pujar et al., 2020a; Rai et al., 2012). Anemia caused by Fe deficiency is more prevalent and has more severe health consequences than anemia caused by Zn deficiency, which is a condition with a very clear cause-and-effect relationship. Considering these variables, the biofortification of pearl millet at the International Crops Research Institute for the Semi-Arid Tropics (ICRISAT) began to focus largely on screening germplasm and breeding for Fe concentration, with Zn serving as a secondary feature. This priority was also raised in the various pearl millet technical meetings held at the national and international levels. The recognition of improving Fe concentration as a primary trait in pearl millet was subsequently transmitted to the breeding organizations of the partners, and all pearl millet breeding was redirected toward improving grain Fe concentration as well as other agronomic qualities. HarvestPlus defined target levels of Fe and Zn in pearl millet by considering the average quantity of Fe and Zn in cereal grains, their retention after processing, and resolving the concerns associated to the bioavailability of these mineral nutrients. In pearl millet, the average grain Fe concentration is 42–47 mg/kg; the target level is fixed to 72–77 mg/kg while Zn an associated trait in millet breeding.

Breeding for Priority Micronutrients (Fe and Zn)

Biofortification research in pearl millet, initially at ICRISAT supported by the HarvestPlus conducted in partnership with public and private sector research organizations, has shown large variability for Fe and Zn concentration, with good prospects of developing cultivars with higher levels of these micronutrients (Figure 3.1). The biofortification priority index indicates that pearl millet is a major

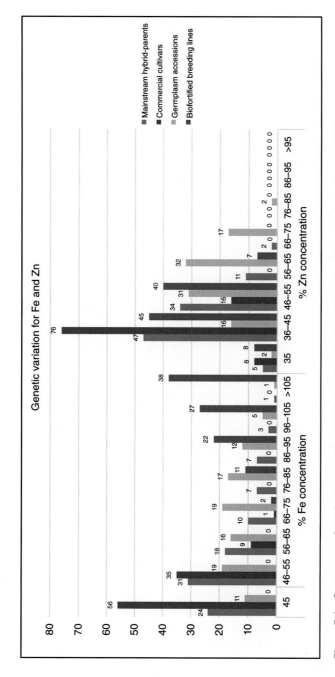

Figure 3.1 Percentage of entries grouped in grain Fe and Zn class in pearl millet. From Govindaraj et al. (2019).

target crop for Fe and Zn biofortification. Keeping this in view, the Indian Council of Agricultural Research (ICAR)-All India Coordinated Research Project on Pearl Millet has decided to alleviate malnutrition in the country through developing pearl millet hybrids and varieties with enhanced levels of Fe and Zn that would be within the reach of common citizens. Hence, a conscious and landmark decision was made during the 52nd Annual Group Meeting of All India Coordinated Research Project on Pearl millet during April 28–30, 2017, held at PAU, Ludhiana, to include minimum benchmark levels of 42 ppm of Fe and 32 ppm of Zn in the promotion criteria of the pearl millet test entries in the coordinated trials from 2017 onward in the presence of researchers from national, international, public, and private partners and stakeholders (Satyavathi et al., 2021b). Because pearl millet is highly cross-pollinated crop, open-pollinated varieties (OPVs) and hybrids are the two distinct cultivar options that can be developed and grown by farmers. Pearl millet as such is a high-Fe crop with a fairly high Zn concentration; however, commercially available cultivars have lower Fe and Zn concentrations than the germplasm and breeding lines. Because all efforts in the public and private sectors are geared toward hybrid development, the focus of the pearl millet biofortification research targeted for India is also on the development of high-Fe/Zn and high-yielding hybrids. More than 26,000 accessions of pearl millet have been conserved in the world's germplasm collections; among these, the so-called "core collections," representing 10% of the total collections, have revealed enormous variability for Fe (30–90 mg/kg) and Zn (30–74 mg/kg) in pearl millet, which provides an excellent means of selection for improvement (Govindaraj et al., 2020). Iniadi, a major reservoir of Fe and Zn, is indigenous to Togo and adjoining West African countries. The grain Fe and Zn concentration of iniadi germplasm ranged from 51 to 121 mg/kg and from 46 to 87 mg/kg, respectively (Rai et al., 2015). The early pre-breeding biofortification crosses at ICRISAT, which largely relied on the selected iniadi source for Fe and Zn biofortification, further provided evidence of this. A recent study that included germplasm collections, commercially available hybrids, OPVs, and landraces found Fe levels ranging from 36.52 ppm (NANDI-72) to 124.89 ppm (IC-537960) and Zn concentrations ranging from 15.62 ppm (Chanana Bajri-1) to 62.3 ppm (KBH-58) (Goswami et al., 2022).

The genetics investigation in pearl millet using the available variety has revealed that the Fe and Zn in pearl millet are mostly governed by additive gene activity (Govindaraj et al., 2013, 2016b; Kanatti et al., 2014a, 2016; Velu et al., 2011). As a result, intra-population improvement is likely to be successful, and producing high-Fe and high-Zn hybrids would entail the introduction of these micronutrients into both female (A/B-lines/seed parent) and male (R-line/pollinator) hybrid parental lines. Pearl millet grain Fe is improved, as is grain Zn concentration as an associated attribute, thanks to the presence of a very strong correlation between Fe and Zn ($r = 0.43$–0.90; $P < 0.01$) (Govindaraj et al., 2013; Kanatti et al., 2016;

Pujar et al., 2020a; Rai et al., 2012). Furthermore, while breeding for high Fe and Zn, strict consideration must be given to maintaining grain yield and other agronomically important traits (grain size, flowering, and maturity) because they are the preferred traits by farmers. A moderate to high correlation was observed between the two micronutrients, and 1000-grain weight (Kanatti et al., 2014a; Pujar et al., 2020a; Velu et al., 2007, 2008a, 2008b) suggests that the improvement of Fe and Zn concentrations in grains would not affect grain size. Studies connecting Fe and/or Zn with grain yield in pearl millet, however, have discovered both positive and weak to moderately negative relationships (Kanatti et al., 2014b; Pujar et al., 2020a; Rai et al., 2015). Because the negative correlation is always either very weak or nonsignificant, improvements for Fe and Zn in pearl millet will not lower the grain yield.

Breeding for the Improvement of Other Essential Nutrition

Breeding for Grain Protein

Exploration of pearl millet beyond its targeted nutritional feature (Fe and Zn) has revealed significant potential for supplementing the grains with other critical mineral elements as well as other nutrients such as protein and essential amino acids (EAAs). To ascertain the nutritional makeup of the pearl millet germplasm, a study was carried out by Goswami et al. (2022). The results of this study showed that the protein concentration ranged from 8.07 ± 0.22 g/100 g (Kudri) to 18.15 ± 0.68 g/100 g (IC-537953) within the germplasm collection. Similarly, when the several hybrid parents developed at ICRISAT were assessed over two contrasting seasons (2017 rainy and 2018 summer), a similarly considerable variation ranging from 6% to 18% for protein was also discovered (Pujar et al., 2020a). A positive correlation between grain Fe and Zn concentration and grain protein concentration further motivates breeders to focus on all three traits at the same time (Pujar et al., 2020a).

Breeding for protein should also take into consideration protein quality. Pearl millet protein quality depends on the amino acid concentration that especially includes the dietary indispensable amino acids, protein digestibility, and bioavailability of the amino acids. An investigation by Vinutha et al. (2022) on the amino acid profile among the pearl millet test genotypes (commercially cultivated and landraces) revealed the presence of higher glutamic acid concentration (17.58–22.614 g/100 g), followed by leucine (7.714–11.865 g/100 g) and aspartic acid (6.265–10.774 g/100 g). Furthermore, the percentage daily value [DV(%)] of EAAs was determined for adults (Table 3.2). The findings showed that, with the exception of lysine, the DV(%) of EAAs from pearl millet genotypes for an adult weighing 60 kg are sufficient (Vinutha et al., 2022).

Table 3.2 Percentage Daily Value (%DV) of Essential Amino Acids from Pearl Millet that an Adult Weighing 60 kg Can Obtain Based on per Capita Consumption of 100 g of Pearl Millet.

Genotype	Essential amino acids							
	His	Lys	Leu	LLE	Thr	Met+Cys	Phen+Tyr	Val
	–%DV–	–%DV–	–%DV–	–%DV–	–%DV–	–%DV–	–%DV–	–%DV–
HHB-67 (IMP)	70.24	31.11	169.6	93.68	88.29	50.89	86.15	78.89
PC-443	64.16	21.22	149.7	92.08	86.91	43.89	79.68	68.02
MPMH-17	54.94	12.15	144	156.3	105	51.827	87.37	92.46
86M86	55.94	22.22	132.3	104.4	80.71	51.84	81.14	72.02
Dhanshakti	67.47	13.22	118	106.5	82.14	48.44	86.25	80.44
HHB-299	54.94	12.15	144	156.3	105	51.87	87.37	92.46
AHB-1200	70.77	20.74	150.9	103.3	82.16	34.58	82.11	89.99
GHB-558	56.32	1.81	127.1	101.2	78.27	45.56	87.37	76.71
Chadi bajri	57.28	13.76	139.6	138.8	104.9	40.66	70.19	76.51
Dhodhasar local	58.66	10.55	132	109.9	82.19	42.18	75.47	81.81
PC-701	54.94	12.15	14.42	156.4	104.9	51.27	87.67	92.56
PC-612	63.49	11.93	158.5	103.2	82.46	61.85	84.34	87.26
Damodar bajra	45.34	22.56	139.7	85.23	88.33	33.56	78.48	82.73
Chanana bajra	61.88	13.87	166.2	141.9	81.26	44.56	82.01	89.89
Gadhwal ki Dhani	70.67	13.76	110.2	177	96.92	44.47	126.64	96.86

Note: From Vinutha et al. (2022).
Abbreviations: Cys, cystine; His, histidine; Leu, leucine; Lysine, Lysleucine; Met, methionine; Phe, phenylalanine; Thr, threonine; Tyr, tyrocine; Val, valine.

Breeding for Grain Minerals

In addition to being grown largely for grain output, pearl millet is also genetically biofortified to increase the Fe and Zn concentration of the grains, providing a major source of Fe and Zn to the people who are dependent on it. However, it has been discovered that pearl millet provides adequate opportunity for the enhancement of other necessary minerals. Given this, a detailed examination of the mineral concentration of pearl millet cultivars, particularly those in the public domain for usage, including OPVs, hybrids, landraces, and germplasm, was carried out. The mineral elements studied included P, Ca, Fe, K, Mg, Na, Zn, Se, Ni,

Mo, Mn, Cu, and Co. The analyzed mineral nutrients were expressed in parts per million (ppm) (Goswami et al., 2022). Phosphorous concentrations ranged from 151.14 ppm (Kaveri Super Boss) to 403.46 ppm (Kudri). Calcium concentration ranged from 192.5 ppm (MP-7792) to 405.59 ppm (MPMH-17). Potassium concentration ranged from 2495.57 ± 102.63 ppm (MPMH-17) to 5257.51 ± 192.35 ppm (Chanana Bajri-1). Magnesium concentration ranged from 1424.02 ± 43.44 ppm (HHB-146) to 2663.47 ± 92.71 ppm (IC-537983). Sodium concentration ranged from 23.25 ± 0.69 ppm (DR-3) to 49.97 ± 1.51 ppm (RHB-121). Manganese concentration ranged from 8.21 ± 0.24 ppm (Peeli Bajri) to 17.77 ± 0.65 ppm (Chanana Bajri-3). Copper concentration ranged from 1.29 ± 0.04 ppm (PUSA-1601) to 7.29 ± 0.23 ppm (KBH-3580). Cobalt concentration ranged from 0.19 ± 0.006 ppm (CZP16-923) to 0.51 ± 0.016 ppm (HHB-226). Nickel concentration ranged from 0.29 ± 0.012 ppm (MP-7792) to 1.91 ± 0.079 ppm (MPMH-17). Molybdenum concentration ranged from 0.59 ± 0.016 ppm (PUSA-605) to 0.95 ± 0.029 ppm (AHB-1200). Selenium concentration ranged from 0.001 ± 0 ppm in (NANDI-52) to 0.53 ± 0.015 ppm (PC-383).

The development of new hybrid or OPV cultivars with enhanced mineral nutrition depends on the accessibility of a wide range of breeding material, particularly among inbred lines. Therefore, an evaluation of several hybrid parental lines derived from population progenies, germplasm progenies, core collections representing various origins, and agronomic features was carried out at ICRISAT in 2020 (unpublished data). The results were encouraging. Calcium concentration in grains showed fourfold differences ranging from 108.047 to 394.167 mg/kg. Similarly, B concentrations showed fourfold differences, ranging from 1.503 to 5.728 mg/kg. Magnesium levels varied by two orders of magnitude, ranging from 908.33 to 1861.67 mg/kg. Sodium levels varied from 8.151 to 42.64 mg/kg, with a fivefold variation. Potassium levels varied from 3816.667 to 6966.66 mg/kg, whereas differences for P concentrations varied from 2966.56 to 4433.33 mg/kg. Between the hybrid parental lines, there was almost a threefold variance for Cu (2.59–8.608 mg/kg) and Se (0.047–0.143 mg/kg). Variability for minerals like S ranged from 955 to 1753.33 mg/kg; Mn ranged from 5.66 to 21.709 mg/kg, accounting for about two- and fourfold variations, respectively. About sixfold variation for Mo (0.263–1.464 mg/kg), eightfold variations each for Ni (0.324–2.590 mg/kg) and Cd (0.008–0.061 mg/kg), and fourfold variation for Pb (0.022–0.085 mg/kg) were observed. The highest of almost 23-fold variation (though magnitude was less) was observed for Co, ranging from 0.078 to 1.758 mg/kg. These findings encourage breeding for essential grain minerals along with targeted Fe and Zn in future pearl millet cultivars. Among the hybrid parents evaluated so far, ICMB 100075 and ICMB 100672 had higher levels of multiple grain mineral elements like Ca, Mg, P, K, B, and S, along with Fe and Zn

Table 3.3 Source of High Mineral Nutrients in Pearl Millet Inbred Lines Available at ICRISAT.

Genotype/inbred	Mineral element	Concentration
		mg/kg
ICMB 100648	Fe	120
ICMR 100978	Zn	87
ICMB 100075	Ca	394.167
ICMB 01444	Mg	1861.667
ICMP 100412	P	4433.333
ICMB 100075	K	6966.667
IPC 1329	S	1753.333
ICMB 08555	Na	42.642
ICMR 100978	Cu	8.608
ICMP 100239	Se	0.143
ICMB 100659	Mn	21.709
ICMR 13777	B	5.728
ICMR 100445	Mo	1.464
ICMR 101013	Co	1.758
ICMB 100630	Ni	2.590
IPC 795	Cd	0.061
ICMB 100666	Pb	0.085

Note: Unpublished data, 2020.

concentration, and can readily be used in the breeding program. The sources of high mineral concentration among ICRISAT-bred pearl millet hybrid parental lines are provided in the Table 3.3; those among commercially cultivated germplasm and landraces are provided in the Table 3.4.

Although the variability for these minerals is considerably high, the association of these traits with Fe and Zn along with economic traits would ultimately determine the way forward to breed these either concurrently or independently. The preliminary analysis of the correlation between these minerals and important economic traits (e.g., 1000 grain weight, days to 50% flowering, and grain yield) have shown mixed results (Figure 3.2). Grain yield is an important economic trait, and minerals like Ca, Mg, P, Cu, S, and Mo showed significant but low to moderate association with grain yield. The other minerals showed nonsignificant associations with grain yield.

Table 3.4 Source of High Mineral Nutrients in Pearl Millet Germplasm, Open-Pollinated Varieties, Hybrids, and Landraces.

Genotype	Mineral element	Concentration
		ppm
IC-537960	Fe	124.899
KBH-58	Zn	62.3
MPMH-17	Ca	405.59
IC-537983	Mg	2663.47
Kudri	P	403.460
Chanana Bajra-1	K	5257.51
RHB-121	Na	49.97
KBH-3580	Cu	7.29
PC-383	Se	0.53
Chanana Bajri-3	Mn	17.77
AHB-1200	Mo	0.95
HHB-226	Co	0.51
MPMH-17	Ni	1.91

Note: From Goswami et al. (2022).

High-throughput Phenotyping

Among the various analytical techniques available, inductively coupled plasma–optical emission spectrometry is regarded as the most accurate method for estimating Fe and Zn in pearl millet grain samples because it takes into account any potential soil contamination. High-throughput breeding for a trait, however, is not feasible. Given the high cost per sample and the destructive nature of the analytical protocol, it would take a long time. Additionally, a sophisticated laboratory and technical labor are needed. As a result, energy dispersive X-ray fluorescence (ED-XRF) spectrometry has become the most precise and high-throughput method for studying pearl millet (Govindaraj et al., 2016a). The risk of losing irreplaceable germplasm seeds can be decreased by screening early-stage germplasms with few seeds and using the same seeds for breeding programs after using a nondestructive Fe and Zn analytical approach. Studies on the correlation between XRF spectrometry and inductively coupled plasma have shown a significant positive association (Paltridge et al., 2012), suggesting the viability of using XRF spectrometry in early-stage screening of breeding material, most likely rejecting lines with low Zn concentration. In pearl millet, the use of ED-XRF spectrometry to develop biofortified crops for Fe and Zn has proved successful.

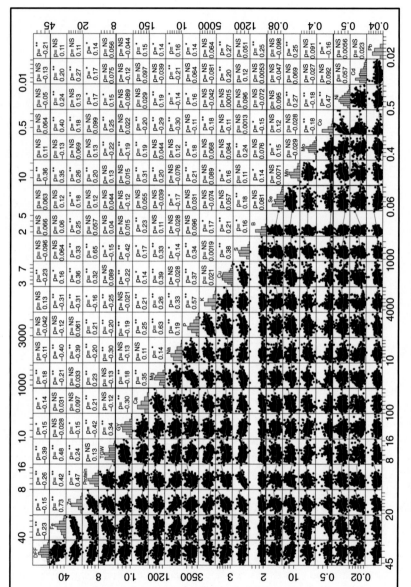

Figure 3.2 Correlation among different nutrient mineral elements present in pearl millet grains.

Performance of Biofortified Hybrids and OPVs

In accordance with genetic studies, the grain Fe and Zn concentration in pearl millet is mostly influenced by additive gene action. As a result, enhancement of these micronutrients would benefit greatly from population improvement (inter- and intrapopulation) strategies such as mass selection, recurrent selection for general combining ability, or composite breeding. OPVs developed from such methods have greater buffering effect against newly emerged plant epidemic or other extreme stresses. Earlier, a population's grain Fe and Zn concentrations were improved after three generations of progeny-based selection in an OPV (ICTP 8203), and one of the improved variants was deemed superior and named ICTP 8203 Fe 10-2. This was re-released under the name "Dhanashakti" following testing at the national level in more than 40 trials, where it was discovered to have superior Fe (71 mg/kg Fe) and yield (2.2 t/ha) values. In comparison to the original population, this variety produced grains with a 11% higher yield and an almost 9% higher Fe concentration (Rai et al., 2014).

Enabling the exploitation of heterosis, being highly cross-pollinated (>85%) (Burton, 1983), and the presence of the well-established male sterile system in pearl millet have greatly benefited in the generation of new hybrids. Hybrids have become popular over the period by virtue of their high yield. In India, about 5 million ha of pearl millet–cultivated land is occupied by hybrids. Unlike the grain yield trait, Fe and Zn are governed by additive gene action; hence, to develop high Fe and Zn hybrids, both the parents (A- and R- lines) must have high Fe and Zn concentrations. Several biofortified hybrids have been developed under the HarvestPlus-funded pearl millet biofortification program (Table 3.5). The biofortified high-Fe and high-Zn hybrids that are currently available in the market are developed from the *iniadi* source (Govindaraj et al., 2013; Rai et al., 2012; Velu et al., 2011). ICRISAT developed the first high-Fe biofortified hybrid ICMH 1201, popularly known as Shakti 1201, marketed by ShaktiVardhak Seed Company. This hybrid has 75 mg/kg of Fe and 40 mg/kg of Zn concentration in grains. This hybrid has 30% higher grain yield compared with the Dhanashakti (the first biofortified pearl millet cultivar) in addition to the similar levels of Fe and Zn concentration in grains. By virtue of this, ICMH 1201 quickly spread to over 35,000 farmers in Maharashtra and Rajasthan states of India (Govindaraj & Rai, 2016). Through several feeding trials conducted, it is promising to see that biofortified cultivars like Dhanashakti and Shakti 1201 are capable of meeting more than the daily requirement of Fe of 0.84 mg for men and can easily meet the 70% of the daily requirement of 1.65 mg for nonpregnant, nonlactating women and 42% of the daily requirement of 2.8 mg for pregnant women. These numbers were based on 240 g of grain consumed per day at a bioavailability of 7.5%. The grain consumption rate described above will also provide 80% of the daily necessary amount of 12 mg of Zn.

Table 3.5 Biofortified Hybrids and Open-pollinated Varieties (OPVs) of Pearl Millet Released in Africa and India Under the Biofortification Breeding Program.

SN	Cultivar name	Commercial name	Hybrid/ OPV	Year of release	Fe	Zn	Grain yield	Country name
					mg/kg		t/ha	
1	ICTP 8203 10–2	Dhanashakti	OPV	2013	71	40	2.2	India
2	ICMH 1201	Shakti-1201	hybrid (TLS)	NA	75	40	3.2	India
3	ICMH 1203	HHB 299	hybrid	2017	73	41	3.3	India
4	DHBH 1211/ MH 2078	Phule Mahashakti	hybrid	2018	87	41	2.9	India
5	Chakti	Chakti	OPV	2018	60	45	1.5	Niger (Africa)
6	MH 2072	AHB 1200 Fe	hybrid	2018	77	39	3.2	India
7	ABV 04	ABV 04	OPV	2018	70	63	2.9	India
8	MH 2185	AHB 1269 Fe	hybrid	2019	91	43	3.3	India
9	MH 2174	RHB 234	hybrid	2019	84	41	3.2	India
10	MH 2173	RHB 233	hybrid	2019	83	46	3.2	India
11	MH 2179	HHB 311	hybrid	2019	83	39	3.2	India
12	Moti Shakti	GHB1225	hybrid	2020	72	43	3.0	India

Abbreviations: TLS, Truthful labelled seed.

Bioavailability of Fe and Zn

Crop biofortification activities prioritize the investigation of Fe and Zn bioavailability among newly produced hybrids or varieties because only a small fraction of these collected minerals in the edible component are bioavailable. Bioavailability is the proportion of the nutrient in meal that can be absorbed and utilized by the body for metabolic activities (Welch & Graham, 2004). An Fe absorption test including a composite test meal that includes millet paste produced from regular, Fe-biofortified, and Fe-fortified pearl millet flour was conducted on 20 Beninese women who had low Fe levels (Cercamondi et al., 2013). The Fe absorption rate in regular millet-based meals was 7.5%, which was shown to be in concordance with that of Fe-biofortified millet consumption, implying that Fe-biofortified millet consumption can boost Fe absorption in women by twofold. The fact that the average Zn absorption in children was higher (0.95 ± 0.47 mg/d) than the regular absorption rate (0.67 ± 0.24 mg/d) was equally intriguing and encouraging. Another study conducted among children within the age group of 12–16 years in Maharashtra showed that the consumption of Fe-biofortified meal resulted in almost 1.6 times higher chances of recovering the required Fe concentration in the body (Finkelstein et al., 2015). A similar study into Fe absorption in 40 Indian (Karnataka) toddlers (aged 2 years) revealed that Fe-fortified (0.67 ± 0.48) pearl millet provided three times the mean Fe absorption of Fe compared with regular-based millet (0.23 ± 0.15 mg/d) (Kodkany et al., 2013). According to these findings, consumption of high-Fe and high-Zn biofortified pearl millet grains will readily provide more than 80% of the daily Fe requirements and 100% of the Zn requirements.

Nutrition Trait Mainstreaming

By using conventional breeding techniques, targeted breeding has so far proved crucial in raising the concentration of the vital micronutrients Fe and Zn in pearl millet. As a result, numerous biofortified hybrids of pearl millet and a few OPVs have been made available for sale in India and Africa. However, the current targeted breeding would be hardly successful in achieving HarvestPlus' ambitious goal of reaching 1 billion consumers by the year 2030. As a result, strategies have been developed to introduce these Fe and Zn traits into mainstream breeding. Under mainstream breeding, all the breeding lines and the elite parental lines are gradually introgressed with the Fe and Zn micronutrient traits at ICRISAT. Over 50% of the breeding lines for pearl millet now have the necessary Fe concentration because of these encouraging early results. It is anticipated that cultivars from the ICRISAT and National Agricultural Research System (NARS) pipelines will be biofortified without compromising agronomic performance. Once Fe and Zn traits have been stabilized in the breeding pipelines, only minimal "maintenance breeding" will be required. To promote these traits further, the ICAR has endorsed the inclusion of the minimum levels of Fe (42 mg/kg) and Zn (32 mg/kg) that must be bred into future pearl millet cultivars across the country set up by All-India Coordinated Research Project on Pearl Millet (AICRP Pearl Millet, 2018). As a result, in the future, any cultivar or hybrid that is released will contain minimum levels of Fe and Zn. High-throughput phenotyping platforms such as ED-XRF spectrometry would greatly assist in mainstream breeding of Fe and Zn by virtue of contributing to rapid and large-scale screening of the breeding lines for Fe and Zn. Furthermore, using genomic selection, marker-aided selection, and speed breeding would most likely improve and accelerate the mainstreaming process.

Enabling Genomics Technologies for Nutrition Traits (Fe and Zn)

To improve pearl millet breeding, a variety of genomic and molecular methods have been developed and used for quantitative trait loci (QTL)/gene identification, genetic diversity, and marker-assisted backcrossing (Ambawat et al., 2020;

Anuradha et al., 2017; Bollam et al., 2018; Kumar et al., 2018; Serba & Yadav, 2016; Singhal et al., 2018; Srivastava et al., 2020a, 2020b). It has been demonstrated that marker-aided selection and gene introgression into acceptable genetic backgrounds are particularly effective for improving crops because they lessen the laborious process of phenotypic evaluation and selection. 'Improved HHB 67', a Downy mildew–resistant cultivar (Hash et al., 2006) developed through marker-assisted backcrossing, is one such example of successful application of genomics technologies in pearl millet. Apart from this, several major QTL were identified for abiotic stresses like heat and drought. Similarly, exclusive studies were conducted to identify the major QTL for Fe and Zn micronutrients (Anuradha et al., 2017; Kumar et al., 2016; Mohan et al., 1997). The introduction of high-throughput genotyping platforms will significantly lower the cost of genotyping, allowing for the screening of many breeding pathways. A recent study on genome-wide association mapping identified 18 SNPs for Fe and 43 for Zn (Pujar et al., 2020b). Further confirmation of these described markers would be extremely beneficial for their incorporation into national and international biofortification breeding programs. Only a few SNP-based diagnostic markers for Fe/Zn characteristics in pearl millet have been found and recommended (Table 3.6). In addition, Fe and Zn candidate genes have been discovered in the gene families PglZIP, PglNRAMP, and PglFER (Mahendrakar et al., 2020). The results of this study can be used to functionally analyze different Fe and Zn metabolism gene homologs and possibly in pearl millet molecular breeding.

Table 3.6 Diagnostic Markers Developed by ICRISAT for Screening Grain Fe and Zn Concentrations in Pearl Millet.

Trait designation	Chr	Allele 1	Alternate allele	SNP/ INDEL	Donor sources	Susceptible controls
Grain Fe/Zn concentration	n.a.	C	T	C/T	863B	n.a.
Grain Fe/Zn concentration	n.a.	A	T	A/T	863B	n.a.
Grain Fe/Zn concentration	n.a.	A	G	A/G	863B	n.a.
Grain Fe/Zn concentration	n.a.	A	T	A/T	863B	n.a.

Note: From CGIAR (2024).
Abbreviations: Chr, chromosome; LG, linkage group; n.a., not available.

Product Development and Distribution Pathway

The ICRISAT played the lead role in the past and continues to do so in developing improved pearl millet lines through the population improvement scheme. ICRISAT focuses on development of OPVs in Africa and hybrids in Indian regions as per the demand that prevails. The Hybrid Parent Research Consortium, consisting of private and public partners headed by ICRISAT in India, has greatly contributed toward developing new hybrid products in India. Under this, ICRISAT primarily focuses on the continuous improvement of restorer (R-) and seed parent (B/A-line) heterotic pools by introducing new alleles through trait/diversity introgression onto elite parents. The introduction of novel germplasm must be based on yield, Fe and Zn levels, and a rigorous selection index. The elite hybrid parental lines that have been developed are then shared among the partners to develop new hybrids. When a hybrid is developed, it will be disseminated with the help of state agricultural universities and NARS.

Adoption and Commercialization

Traditional pearl millet hybrids are notably developed in terms of high grain production and short duration, which are economically valuable in terms of existing market demand. Therefore, the goal is to give farmers better crop and/or utilization options. Biofortified hybrids with better targeted micronutrients like Fe and Zn are evaluated in terms of health and nutritional effects. Consumption of biofortified pearl millet improves one's diet and overall health over time. Nutrition traits like Fe and Zn are invisible and have no effect on sensory properties. As a result, crop profiles must ensure value propositions for all value chain partners and take processor requirements into account when producing goods. The establishment of a sustainable market would be substantially aided by raising awareness of the health advantages of consuming biofortified pearl millet, stimulating demand, and promoting the product. Establishing a solid supply chain, especially for remote and difficult-to-connect areas, and ensuring identity storage, adherence to quality standards, and traceability would help. Farmers would be encouraged to adopt pearl millet on a large scale if the government were to get involved in policy formation and regulation. Millets are being promoted through the diffusion of technology, the provision of high-quality seeds through millet seed hubs, the creation of awareness, the establishment of a minimum support price and inclusion in the public distribution system. At the most basic level, government and government-aided schools in Karnataka and Telangana have implemented millet-based midday meal programs to promote children's intake of millet. The Indian government made careful, strategic efforts to increase the popularity of millets among producers and consumers, including designating millets, including pearl millet, as Nutri-Cereals in 2018, which was also recognized as the "National Year of Millets." The UN General Assembly also approved a resolution sponsored by India declaring 2023 the "International Year of Millets" to encourage millet consumption around the globe.

The consumer's ultimate preference and the demand it generates play a significant role in the commercialization of pearl millet products. Taste, accessibility, and perishability of the food products contribute to the success of any processed good in gaining consumer acceptance. One of the major problems associated with pearl millet lies in the off-flavor characteristics of flour. Furthermore, lack of good dough-making quality limits its use in bakery products and chapatti (flat Indian bread), making it unpopular among consumers. Considering this, the Division of Biochemistry, ICAR-Indian Agricultural Research Institute, New Delhi, has developed the soft bajra atta by fortifying the pearl millet flour with wheat gluten, which enhances the viscoelastic properties for improved dough and chapatti making quality, making pearl millet flour as good as wheat flour, making pearl millet a sustainable alternative to wheat in many baked and cooked preparations. Numerous pearl millet flour–based cooked products have been created in recent years and are easily accessible in the market (Figure 3.3).

Figure 3.3 Value-added pearl millet products.

Marketing Challenges

Although pearl millet is a nutrient-dense cereal, its flour is underutilized due to off flavor developed during storage. This is also the probable reason why there are not many processed products or value-added products available. Focus should be on developing a pearl millet cultivar with low rancidity. Lipids and the lipid-hydrolyzing enzymes such as lipase and lipoxygenase are the key determinants of off flavor development in pearl millet. A study on landraces and some selected popular commercially available pearl millet cultivars showed that landraces such as Jafrabadi, Chanana Bajra-2, Chadi Bajri, and Damodara Bajri showed longer shelf life, wherein the key substrate linoleic acid for lipoxygenase enzyme was found to be lower than that compared with Dhanashakti. This shows that landraces can play key roles reducing the rancidity problem in pearl millet (Vinutha et al., 2022). Some of the biofortified pearl millet hybrids and OPVs have been identified with low to moderate rancid groups for further investment and commercial scaling, which will increase market potential. Industries are already marketing pearl millet flour using different vacuum and heat treatment packaging techniques in India (Figure 3.3).

Conclusion

Pearl millet, because of its hardy nature, has emerged as a major staple food crop, serving millions of people living in dry regions. Its high nutritional status, with Fe and Zn, has great potential for improving human nutrition where it is grown as a staple food. In countries where Fe and Zn malnutrition is prevalent, particularly among children and women, moving to a millet-based diet (particularly biofortified millet) can help to significantly reduce malnutrition. A diet that include biofortified high-Fe and high-Zn pearl millet has been found to easily meet ~80% of the daily necessary Fe and Zn concentrations. Modern genomics and speed breeding techniques, along with mainstream Fe and Zn breeding approaches, would help to expedite the pace of genetic gain for high grain production and enhanced Fe and Zn concentrations. NARS partners will eventually have access to additional pearl millet lines with high yield, high Fe and Zn concentrations, and other agronomic traits. Preliminary diagnostic markers to screen for Fe and Zn lines can be used for their active application to select high-Fe and high-Zn lines. Also, some markers among these can be used for negative selection (e.g., to discard the lines with low Fe and Zn concentrations, similar to XRF screening). NARS can then choose and scale biofortified pearl millet cultivars to farmers and consumers. Because Fe and Zn are quantitatively inherited with invincible traits, a value proposition for all supply chain partners must be guaranteed by the crop profile, and processor needs must be considered while developing a product. Greater focus needs to be given to millet flour shelf life for developing diversified processed food industries.

References

Aghaalikhani, M., Ahmadi, M. E., & Modarres Sanavy, A. M. (2008). Forage yield and quality of pearl millet (*Pennisetum americanum*) as influenced by plant density and nitrogen rate. *Pajouhesh Sazandegi*, *77*, 19–27.

AICRP Pearl Millet. (2018). *Proceedings of the 53rd Annual Group Meeting of ICAR—All India Coordinated Research Project on Pearl Millet*. http://www.aicpmip.res.in/pw2018.pdf

Amarender Reddy A., Yadav, O. P., Malik, D. P., Singh, I. P., Ardeshna, N. J., Kundu, K. K., Gupta, S. K., Sharma, R., Sawargaonkar, G., Moses Shyam, D., & Sammi Reddy, K. (2013). *Utilization pattern, demand and supply of pearl millet grain and fodder in western India*. International Crops Research Institute for the Semi-Arid Tropics

Ambawat, S., Singh, S., Shobhit, M. R. C., & Satyavathi, C. T. (2020). Biotechnological applications for improvement of the pearl millet crop. In S. K. Gahlawat, S. Punia, A. K. Siroha, K. S. Sandhu, & M. Kaur (Eds.), *Pearl millet: Properties, functionality and its applications* (pp. 115–138). Taylor and Francis.

Anuradha, N., Satyavathi, C. T., Bharadwaj, C., Nepolean, T., Sankar, S. M., Singh, S. P., Meena, M. C., Singhal, T., & Srivastava, R. K. (2017). Deciphering genomic regions for high grain iron and zinc concentration using association mapping in pearl millet. *Frontiers in Plant Science*, *8*, 412. https://doi.org/10.3389/fpls.2017.00412

Arya, R. K., Kumar, S., Yadav, A. K., & Kumar, A. (2013). Grain quality improvement in pearl millet: A review. *Forage Research*, *38*, 189–201.

Bollam, S., Pujarula, V., Srivastava, R. K., & Gupta, R. (2018). Genomic approaches to enhance stress tolerance for productivity improvements in pearl millet: Genomic approaches. In S. S. Gosal & S. H. Wani (Eds.), *Biotechnologies of crop improvement* (Vol. 3, pp. 239–264). Springer.

Brunette, T., Baurhoo, B., & Mustafa, A. F. (2014). Replacing corn silage with different forage millet cultivars: Effects on milk yield, nutrient digestion, and ruminal fermentation of lactating dairy cows. *Journal of Dairy Science*, *97*, 6440–6449. https://doi.org/10.3168/jds.2014-7998

Burton, G. W. (1983). Breeding pearl millet. *Plant Breed Reviews*, *1*, 162–182.

Cercamondi, C. I., Egli, I. M., Mitchikpe, E., Tossou, F., Zeder, C., Hounhouigan, J. D., & Hurrell, R. F. (2013). Total iron absorption by young women from iron-biofortified pearl millet composite meals is double that from regular millet meals but less than that from post-harvest iron-fortified millet meals. *The Journal of Nutrition*, *143*, 1376–1382. https://doi.org/10.3945/jn.113.176826

CGIAR. (2024). *Excellence in breeding platform*. https://excellenceinbreeding.org

De Assis, R. L., de Freitas, R. S., & Mason, S. C. (2018). Pearl millet production practices in Brazil: A review. *Experimental Agriculture*, *54*, 699–718.

Dias-Martins, A. M., Pessanha, K. L., Pacheco, S., Rodrigues, J. A. S., & Carvalho, C. W. P. (2018). Potential use of pearl millet (*Pennisetum glaucum* [L.] R. Br.) in Brazil: Food security, processing, health benefits and nutritional products. *Food Research International, 109*, 175–186. https://doi.org/10.1016/j.foodres.2018.04.023

Directorate of Millets Development. (2020). http://www.aicpmip.res.in/aboutus.html

FAO. (2004). Pennesitum americanum *(L.) Leeke: species description*. FAO. http://www.fao.org/ag/AGP/AGPC/doc/Gbase

FAO. (2019). Food and agriculture Organization of the United Nations.

Finkelstein, J. L., Mehta, S., Udipi, S. A., Ghugre, P. S., Luna, S. V., Wenger, M. J., Murray-Kolb, L. E., Przybyszewski, E. M., & Haas, J. D. A. (2015). Randomized trial of iron-biofortified pearl millet in school children in India. *The Journal of Nutrition, 145*, 1576–1581. https://doi.org/10.3945/jn.114.208009

Girgi, M., & O'Kennedy, M. M. (2007). Pearl millet. In E. C. Pua & M. R. Davey (Eds.), *Transgenic crops IV: Volume 59. Biotechnology in agriculture and forestry* (pp. 119–127). Springer.

Gopalan, C., & Deosthale, Y. G. (2003). *Nutritive value of Indian foods*. National Institute of Nutrition.

Goswami, S., Vinutha, T., Tomar, M., Sachdev, A., Kumar, R. R., Veda, K., Bansal, N., Sangeetha, V., Rama Prashat, G., Singh, S. P., Khandelwal V., Manoj Kumar, C., Satyavathi, T., Praveen, S. & Tyagi, A. (2022). *Nutritional profiling of pearl millet genotypes*. Indian Council of Agricultural Research.

Govindaraj, M., & Rai, K. N. (2016). Breeding biofortified pearl millet cultivars with high iron density. *Indian Farming, 65*, 53–55.

Govindaraj, M., Rai, K. N., Cherian, B., Pfeiffer, W. H., Kanatti, A., & Shivade, H. (2019). Breeding biofortified pearl millet varieties and hybrids to enhance millet markets for human nutrition. *Agriculture, 9*, 106. https://doi.org/10.3390/agriculture9050106

Govindaraj, M., Rai, K. N., Kanatti, A., Upadhyaya, H. D., Shivade, H., & Rao, A. S. (2020). Exploring the genetic variability and diversity of pearl millet core collection germplasm for grain nutritional traits improvement. *Scientific Reports, 10*, 21177. https://doi.org/10.1038/s41598-020-77818-0

Govindaraj, M., Rai, K. N., Pfeiffer, W. H., Kanatti, A., & Shivade, H. (2016a). Energy-dispersive X-ray fluorescence spectrometry for cost-effective and rapid screening of pearl millet germplasm and breeding lines for grain iron and zinc density. *Communications in Soil Science and Plant Analysis, 47*, 2126–2134.

Govindaraj, M., Rai, K. N., & Shanmugasundaram, P. (2016b). Intra-population genetic variance for grain iron and zinc concentrations and agronomic traits in pearl millet. *The Crop Journal, 4*, 48–54.

Govindaraj, M., Rai, K. N., Shanmugasundaram, P., Dwivedi, S. L., Sahrawat, K. L., Muthaiah, A. R., & Rao, A. S. (2013). Combining ability and heterosis for grain iron and zinc densities in pearl millet. *Crop Science, 53*, 507–517.

Grebmer, V., Bernstein, K. J., Alders, R., Dar, O., Kock, R., Rampa, F., Wiemers, M., Acheampong, M., Hanano, A., Higgins, B., Ní Chéilleachair, R., Foley, C., Gitter, S., Ekstrom, K., & Fritschel, H. (2020). 2020 Global Hunger Index: One Decade to Zero Hunger: Linking Health and Sustainable Food Systems. *Bonn: Welthungerhilfe; and Dublin: Concern Worldwide.* https://www.globalhungerindex .org/pdf/en/2020.pdf

Gupta, S. K., Velu, G., Rai, K. N., & Sumalini, K. (2009). Association of grain iron and zinc content with grain yield and other traits in pearl millet (*Pennisetum glaucum* (L.) R. Br.). *Crop Improvement, 36,* 4–7.

Hanna, W. (1996). Improvement of millets: emerging trends. In V. L. Chopra, R. B. Singh, & A. Varma (Eds.), *Proceedings of the 2nd international crop science congress, New Delhi, India* (pp. 139–146). Oxford & IBH Publishing Co.

Hash, C. T., Sharma, A., Kolesnikova-Allen, M. A., Singh, S. D., Thakur, R. P., Bhasker Raj, A. G., Ratnaji Rao, M. N. V., Nijhawan, D. C., Beniwal, C. R., Sagar, P., Yadav, H. P., Yadav, Y. P., Srikant, S., Bhatnagar, S. K., Khairwal, I. S., Howarth, C. J., Cavan, G. P., Gale, M. D., Liu, C., . . . Witcombe, J. R. (2006). Teamwork delivers biotechnology products to Indian small-holder crop-livestock producers: Pearl millet hybrid "HHB 67 improved" enters seed delivery pipeline. *Journal of SAT Agricultural Research, 2.*

IBPGR/ICRISAT. (1993). *Descriptors for pearl millet (Pennisetum glaucum (L.) R. (Br.)).* IBPGR/ICRISAT

Jukanti, A. K., Gowda, C. L. L., Rai, K. N., Manga, V. K., & Bhatt, R. K. (2016). Crops that feed the world 11. Pearl millet (*Pennisetum glaucum* L.): An important source of food security, nutrition and health in the arid and semi-arid tropics. *Food Security, 8,* 307–329. https://doi.org/10.1007/s12571-016-0557-y

Kanatti, A., Rai, K. N., Radhika, K., & Govindaraj, M. (2016). Genetic architecture of open-pollinated varieties of pearl millet for grain iron and zinc densities. *The Indian Journal of Genetics and Plant Breeding, 76,* 299–303. https://doi .org/10.5958/0975-6906.2016.00045.6

Kanatti, A., Rai, K. N., Radhika, K., Govindaraj, M., Sahrawat, K. L., & Rao, A. S. (2014a). Grain iron and zinc density in pearl millet: Combining ability, heterosis and association with grain yield and grain size. *Springerplus, 3,* 763.

Kanatti, A., Rai, K. N., Radhika, K., Govindaraj, M., Sahrawat, K. L., Srinivasu, K., & Shivade, H. (2014b). Relationship of grain iron and zinc concentration with grain yield in pearl millet hybrids. *Crop Improvement, 41,* 91–96.

Kodkany, B. S., Bellad, R. M., Mahantshetti, N. S., Westcott, J. E., Krebs, N. F., Kemp, J. F., & Hambidge, K. M. (2013). Biofortification of pearl millet with iron and zinc in a randomized controlled trial increases absorption of these minerals above physiologic requirements in young children. *The Journal of Nutrition, 143,* 1489–1493. https://doi.org/10.3945/jn.113.176677

Kumar, S., Hash, C., Nepolean, T., Mahendrakar, M., Satyavathi, C., Singh, G., Rathore, A., Yadav, R. S., Gupta, R., & Srivastava, R. K. (2018). Mapping grain iron and zinc concentration quantitative trait loci in an iniadi-derived immortal population of pearl millet. *Genes, 9*, 248. https://doi.org/10.3390/genes9050248

Kumar, S., Hash, C. T., Thirunavukkarasu, N., Singh, G., Rajaram, V., Rathore, A., Senapathy, S., Mahendrakar, M. D., Yadav, R. S., & Srivastava, R. K. (2016). Mapping quantitative trait loci controlling high iron and zinc concentration in self and open pollinated grains of pearl millet [*Pennisetum glaucum* (L.) R. Br.]. *Frontiers in Plant Science, 7*, 1636. https://doi.org/10.3389/fpls.2016.01636

Li, B., Ishii, Y., Idota, S., Tobisa, M., Niimi, M., Yang, Y., & Nisihimura, K. (2019). Yield and quality of forages in a triple cropping system in southern Kyushu, Japan. *Agronomy, 9*, 277. https://doi.org/10.3390/agronomy9060277

Mahendrakar, M. D., Parveda, M., Kishor, P. K., & Srivastava, R. K. (2020). Discovery and validation of candidate genes for grain iron and zinc metabolism in pearl millet [*Pennisetum glaucum* (L.) R. Br.]. *Scientific Reports, 10*, 16562.

Mani, U. V., Prabhu, B. M., Damle, S. S., & Mani, I. (1993). Glycemic index of some commonly consumed foods in Western India. *Asia Pacific Journal of Clinical Nutrition, 2*, 111–114.

Mohan, M., Nair, S., Bhagwat, A., Krishna, T. G., Yano, M., & Bhatia, C. R. (1997). Genome mapping, molecular markers and marker-assisted selection in crop plants. *Molecular Breeding, 3*, 87–103.

Muthusamy, V., Hossain, F., Thirunavukkarasu, N., Choudhary, M., Saha, S., Bhat, J. S., Prasanna, B. M., & Gupta, H. S. (2014). Development of β-carotene rich maize hybrids through marker-assisted introgression of β-carotene hydroxylase allele. *PLoS One, 9*, e113583. https://doi.org/10.1371/journal.pone.0113583

Newman, Y., Jennings, E., Vendramini, J., & Blount, A. (2010). *Pearl millet (Pennisetum glaucum): Overview and management.* In *Institute of Food and Agricultural Sciences, University of Florida.*

NIN. (2018). National Institute of Nutrition, *Hyderabad.* http://www.aicpmip.res.in/Micronutrient%20Pearl%20Millet.pdf

NRC. (1996). *Lost crops of Africa, vol. 1: Grains.* National Research Council/National Academies Press.

Nurbekov, A., Jamoliddinov, A., Joldoshev, K., Rischkowskv, B., Nishanov, N., Rai, K. N., Gupta, S. K., & Rao, A. S. (2013). Potential of pearl millet as a forage crop in wheat-based double cropping system in Central Asia. *Journal of SAT Agricultural Research, 11*, 1–5.

Packiam, M., Subburamu, K., Desikan, R., Uthandi, S., Subramanian, M., & Soundarapandian, K. (2018). Suitability of pearl millet as an alternate lignocellulosic feedstock for biofuel production in India. *Journal of Applied and Environmental Microbiology, 6*, 51–58. https://doi.org/10.12691/jaem-6-2-4

Paltridge, N. G., Palmer, L. J., Milham, P. J., Guild, G. E., & Stangoulis, J. C. R. (2012). Energy-dispersive X-ray fluorescence analysis of zinc and iron concentration in rice and pearl millet grain. *Plant and Soil, 361*, 251–260.

Parthasarathy Rao, P., Birthal, P. S., Reddy, B. V. S., Rai, K. N., & Ramesh, S. (2006). Diagnostics of sorghum and pearl millet grains-based nutrition in India. *Sat e-Journal, 46*, 93–96.

Pujar, M., Gangaprasad, S., Govindaraj, M., Gangurde, S. S., Kanatti, A., & Kudapa, H. (2020a). Genetic variation and diversity for grain iron, zinc, protein and agronomic traits in advanced breeding lines of pearl millet [*Pennisetum glaucum* (L.) R. Br.] for biofortification breeding. *Genetic Resources and Crop Evolution, 67*, 2009–2022. https://doi.org/10.1007/s10722-020-00956-x

Pujar, M., Gangaprasad, S., Govindaraj, M., Gangurde, S. S., Kanatti, A., & Kudapa, H. (2020b). Genome-wide association study uncovers genomic regions associated with grain iron, zinc and protein concentration in pearl millet. *Scientific Reports, 10*, 19473. https://doi.org/10.1038/s41598-020-76230-y

Raheem, D., Dayoub, M., Birech, R., & Nakiyemba, A. (2021). The contribution of cereal grains to food security and sustainability in Africa: Potential application of UAV in Ghana, Nigeria, Uganda, and Namibia. *Urban Science, 5*, 8. https://doi .org/10.3390/urbansci5010008

Rai, K. N., Govindaraj, M., & Rao, A. S. (2012). Genetic enhancement of grain iron and zinc concentration in pearl millet. *Quality Assurance & Safety of Crops and Food, 4*, 119–125.

Rai, K. N., Patil, H. T., Yadav, O. P., Govindaraj, M., Khairwal, I. S., Cherian, B., Rajpurohit, B. S., Rao, A. S., & Kulkarni, M. P. (2014). Dhanashakti: A high-iron pearl millet variety. *Indian Farming, 64*, 32–34.

Rai, K. N., Velu, G., Govindaraj, M., Upadhyaya, H. D., Rao, A. S., Shivade, H., & Reddy, K. N. (2015). Iniadi pearl millet germplasm as a valuable genetic resource for high grain iron and zinc densities. *Plant Genetic Resources, 13*, 75–82.

Satyavathi, C. T., Ambawat, S., Khandelwal, V., Govindaraj, M., & Neeraja, C. N. (2021b). *Micronutrient rich pearl millet for nutritionally secure India*. ICAR-All India Coordinated Research Project on Pearl Millet.

Satyavathi, C. T., Ambawat, S., Khandelwal, V., & Srivastava, R. K. (2021a). Pearl millet: A climate-resilient Nutricereal for mitigating hidden hunger and provide nutritional security. *Frontiers in Plant Science, 12*, 659938. https://doi.org/10.3389/ fpls.2021.659938

Serba, D. D., & Yadav, R. S. (2016). Genomic tools in pearl millet breeding for drought tolerance: Status and prospects. *Frontiers in Plant Science, 7*, 1724. https://doi .org/10.3389/fpls.2016.01724

Sheahan, C. M. (2014). *Plant guide for pearl millet (Pennisetum glaucum)*. USDA-NRCS.

Singhal, T., Satyavathi, C. T., Kumar, A., Sankar, S. M., Singh, S. P., Bharadwaj, C., Arvind, J., Anuradha, N., Meena, M. C., & Singh, N. (2018). Genotype × environment interaction and genetic association of grain iron and zinc concentration with other agronomic traits in RIL population of pearl millet. *Crop and Pasture Science, 69*, 1092–1102. https://doi.org/10.1071/CP18306

Srivastava, R. K., Bollam, S., Pujarula, V., Pusuluri, M., Singh, R. B., Potupureddi, G., & Gupta, R. (2020a). Exploitation of heterosis in pearl millet: A review. *Plants, 9*, 807. https://doi.org/10.3390/plants9070807

Srivastava, R. K., Singh, R. B., Pujarula, V., Srikanth, B., Pusuluri, M., Satyavathi, C. T., Yadav, R. S., & Gupta, R. (2020b). Genome wide association studies and genomic selection in pearl millet: Advances and prospects. *Frontiers in Genetics, 10*, 1389. https://doi.org/10.3389/fgene.2019.01389

Urrutia, J. M., Hernández, A. A. J., Rosales, C. A. N., & Cervantes, B. J. F. (2015). Forage production and nutritional concentration of silage from three varieties of pearl millet (*Pennisetum glaucum*) harvested at two maturity stages. *Journal of Animal and Plant Sciences, 27*, 4161–4169.

Velu, G., Rai, K. N., Muralidharan, V., Kulkarni, V. N., Longvah, T., & Raveendran, T. S. (2007). Prospects of breeding biofortified pearl millet with high grain iron and zinc content. *Plant Breeding, 126*, 182–185.

Velu, G., Rai, K. N., Muralidharan, V., Longvah, T., & Crossa, J. (2011). Gene effects and heterosis for grain iron and zinc density in pearl millet (*Pennisetum glaucum* (L.) R. Br). *Euphytica, 180*, 251–259.

Velu, G., Rai, K. N., Sahrawat, K. L., & Sumalini, K. (2008a). Variability for grain iron and zinc contents in pearl millet hybrids. *Journal of SAT Agricultural Research, 6*, 1–4.

Velu, G., Rai, K. N., & Sharawat, K. L. (2008b). Variability for grain iron and zinc content in a diverse range of pearl millet populations. *Crop Improvement, 35*, 186–191.

Vinutha, T., Goswami, S., Kumar, R. R., Tomar, M., Veda, K., Sachdev, A., Navita, B., Sangeetha, V., Rama, P., Singh, S. P., Vikas, K., Manoj, K., Satyavathi, C. T., Praveen, S. & Tyagi, A. (2022). *Processing of pearl millet grains to develop nutri-smart food*. Indian Council of Agricultural Research.

Welch, R. M., & Graham, R. D. (2004). Breeding for micronutrients in staple food crops from a human nutrition perspective. *Journal of Experimental Botany, 55*, 353–364.

WHO. (2002). *The world health report 2002*. http://www.who.int/whr/2002

4

Genetic and Genomic Approaches for Accelerated Pearl Millet Breeding

Srikanth Bollam, Mahesh Mahendrakar, Sandeep Marla, Vinutha Kanuganahalli, Krishnam Raju Addanki, and Rakesh K. Srivastava

Chapter Overview

Genetic and genomic strategies present a transformative potential to catalyze a remarkable shift in pearl millet breeding, offering a swift trajectory toward the creation of superior cultivars that hold the key to alleviating food and nutrition insecurities in regions reliant on this climate-smart nutricereal. The advent of high throughput genotyping and genome sequencing has significantly propelled these advancements. Leveraging such technologies, alongside the availability of robust genomic resources for pearl millet such as whole genome resequencing, recently developed pan-genome, and refined reference genome assemblies, promises an in-depth dissection of crucial agronomic, quality, adaptation, and biotic and abiotic stress tolerance traits. These innovative genomic resources capitalize on cutting-edge molecular tools, such as advanced genotyping technologies, marker-assisted selection (MAS), and genomic selection (GS), effectively revolutionizing the breeding programs. Decoding the millet's genetic blueprint, these methods allow for precise identification and exploitation of favorable traits within its vast genetic diversity.

MAS enables breeders to identify genetic markers linked to traits, expediting the selection and precise transfer of beneficial genes across breeding lines. Complementing this, GS uses whole genome information to predict the breeding value of plants, enabling the identification of superior genotypes without exhaustive phenotypic evaluations. The synergy between genetic and genomic approaches, together with accelerated breeding strategies, has the potential to enhance the resilience, productivity, and nutritional quality of pearl millet. Collaborative efforts among multidisciplinary teams of geneticists, breeders, biotechnologists, and agronomists are amplifying the impact of these advancements. By sharing resources, expertise, and data, these collaborations

Pearl Millet: A Resilient Cereal Crop for Food, Nutrition, and Climate Security, First Edition.
Edited by Ramasamy Perumal, P. V. Vara Prasad, C. Tara Satyavathi, Mahalingam Govindaraj, and Abdou Tenkouano.
© 2025 American Society of Agronomy, Inc., Crop Science Society of America, Inc., Soil Science Society of America, Inc. Published 2025 by John Wiley & Sons, Inc.

foster innovation and accelerate the translation of genomic insights into practical breeding applications. This chapter reviews the developed resources and insights into these areas and discusses efficient strategies for future research.

Introduction

Pearl millet [*Pennisetum glaucum* (L.) R. Br.] is the sixth most important cereal crop and the earliest cultivated cereal in the world. It is a principal staple diet to address hidden hunger and micronutrient deficiencies that led to malnutrition for more than 90 million poor people living in the arid and semi-arid regions of Africa, Central Asia, and Latin America. Globally, pearl millet is cultivated in ~31 million hectares (Mha) and accounts for about half the global millet production (Srivastava et al., 2020a). It is a warm season crop with high tillers and short duration, and it belongs to the family Poaceae and sub-family Panicoideae. Pearl millet a small-seeded, highly cross-pollinated (due to its protogynous nature) crop with a genome size of ~2,400 Mb (Vadez et al., 2012). Pearl millet's utility is diverse; it is used as grain for human consumption, for processed products, for starch and alcohol, and as feed for poultry and dairy. Stover is used as forage and fodder, biofuel, fuel for cooking, and building material (Bollam et al., 2018). Pearl millet is a C4 crop with Kranz anatomy in leaves, is able to fix inorganic CO_2 and water utilization very efficiently and helps in global terrestrial carbon fixation together with other C4 crops (Choudhary et al., 2020). It can survive and thrive under adverse environmental conditions, such as drought, high air temperatures, high saline, and poor soils with low water holding capacities. Climate resilience and adaptive features of this crop make it mitigate the adverse effects of global climate change in the semi-arid tropics (Bollam et al., 2018). Apart from its hardy nature, the minimal water and fertilizer requirements to grow and produce make it a crop of choice for the resource-poor and marginal farmers of the world. Among all cereals, pearl millet grain is an inexpensive source of protein and mineral nutrients and can be considered a gifted crop for poor people living in dryland areas (Kumar et al., 2016).

Pearl millet grain possesses exceptional nutritional values. It is rich in protein, energy, carbohydrates, minerals, and micronutrients like Fe and Zn. Pearl millet grain is a great source of different vitamins (vitamin A, thiamine, niacin, and riboflavin) and has higher levels of α-amylase and dietary fiber. It is rich in antioxidants such as coumaric and ferulic acids. It is gluten free and retains alkaline properties even after cooking and hence is suitable for people with gluten allergies. It is not only a good source of fats with better digestibility but also has unsaturated fatty acids (Satyavathi et al., 2021a). A higher amount of slowly digestible starch and resistant starch and niacin in pearl millet grains makes it a low glycemic index crop, which also reduces the blood cholesterol levels of people suffering from diabetes (Kumar et al., 2016).

Climate change and the growing demand for nutritious food by the increasing global population challenge the agriculture sector and global food security efforts. Being a climate change–ready crop, pearl millet has the potential to address these challenges. Several decades of efforts on pearl millet genetic improvements and breeding technologies resulted in the development of hybrids with yield advantage, the identification of tolerant lines for different abiotic and biotic stresses, and nutritional traits (Yadav & Rai, 2013). Recent advances in genomic resources and modern technologies available in pearl millet can offer immense opportunities for its further improvement by accelerating breeding schemes. This chapter reviews the recent advances and their interventions in pearl millet genetic improvement efforts and the development of genomic tools, technologies, and accelerated breeding programs to address the global challenges.

Genetic Improvement of Biotic, Abiotic, and Nutritional Traits in Pearl Millet

Pearl millet is a resilient crop with the ability to withstand diverse abiotic and biotic stresses. However, despite its inherent tolerance, pearl millet encounters various challenges in effectively managing these stressors. The advances in the genetic improvement of pearl millet for prevalent biotic and abiotic stresses and for nutritional and forage quality-related traits are discussed below.

Biotic Stresses (Downy Mildew and Blast)

Pearl millet production is severely hampered by few biotic stresses as compared to other crops.

Major biotic stresses include fungal infections such as downy mildew (DM) and blast, which cause economic losses in major pearl millet–growing regions of the world.

Downy Mildew
Downy mildew, also known as green ear disease, is a systemic infection by an obligate biotrophic pseudo-fungus *Sclerospora graminicola*. DM is rated as the number one biotic constraint potentially resulting in heavy economic losses of pearl millet production in India and Western Africa. DM disease affects pearl millet panicles and is clearly identified by leaf chlorosis, leaf inflorescence, and seed set failure. The occurrence of DM has been moderately adaptable on diverse

hybrids, and more than 90% incidence has been noted on some crosses in "farmers" fields (Thakur et al., 2007). The spread of DM disease is supported by high relative humidity (85%–90%) with moderate temperatures (20–30 °C) (Thakur et al., 2008). This disease was first reported on pearl millet in India by Butler (1907). Yield losses in pearl millet due to DM usually do not exceed 20%, but this loss can reach alarmingly higher levels when a single genetically uniform pearl millet cultivar is repeatedly and extensively grown. During 1970–1971, the annual grain yield of the popular Indian pearl millet hybrid (HB3) was estimated at ~8.2 million t (Singh, 1995). In 1970–1971, a severe yield loss of HB3 (~4.6 million t) was observed (Raj & Sharma, 2022). The corresponding changes in the population structure of the pathogen over a period played a role in the destruction of the crop. Different phenotypic methods (Thakur et al., 2008), screening techniques, and resources were developed to identify the virulent traits for DM at different ICRISAT centers. The draft genome sequence of *S. graminicola* pathotype 1 from India (Siddaiah et al., 2017) will be a useful resource for breeding programs to develop DM-resistant varieties. Using morpho-physiological, biochemical, and molecular markers together to screen germplasm lines against disease at the laboratory with field experiments may authenticate the results (Verma et al., 2021a).

Foliar Blast

Foliar blast disease of pearl millet is caused by the fungal phytopathogen *Magnaporthe grisea*, often referred to as the gray spot of leaf and stem, which has resulted in economic losses worldwide. Blast disease was first reported in 1942 in Kanpur, Uttar Pradesh, in India (Mehta et al., 1953). The favorable conditions for blast disease are high relative humidity, long period of leaf wetness, and a temperature range of 17–28 °C. The symptoms of the disease are grayish, water-soaked foliar lesions, resulting in chlorosis and premature drying of young leaves, causing losses of both grain yield (Timper et al., 2002; Wilson et al., 1989) and forage yield (Wilson & Gates, 1993). Blast disease affects mainly the aerial parts of the plant at all growth stages from seedling stage to panicle, which results in neck and grain blast; it becomes more severe during high humid conditions with high plant density (Thakur et al., 2011). Initially it was considered a minor disease, but in the recent past it has increased at an alarming rate, particularly in commercial hybrids on both forage and grain yield. Blast disease pathogen is highly variable, and it varies from plant to plant and from field to field during a particular season. The continuous monitoring of pathogens from one host to the other helps in designing the best management methods to control the disease. Several efforts have been made to understand the inheritance of resistance to *M. grisea*. In pearl millet, the resistance to blast was derived from the *P. glaucum* ssp. *Monodii* accession where the rust gene (*Rr 1*) was also identified (Hanna et al., 1987).

Abiotic Stresses

Drought Stress

Drought stands as a major abiotic constraint to pearl millet production in Southern Asia and sub-Saharan Africa, arising from inadequate rainfall and its unpredictable distribution. Consequently, extensive initiatives have been undertaken to map and identify regions prone to drought, enabling the delineation of target populations of environments or mega-environments (Gupta et al., 2013). These regions exhibit substantial variability in terms of the timing, intensity, and duration of drought episodes (van Oosterom et al., 1995). Under abiotic stress, the topic of drought tolerance in pearl millet has remained a pivotal point of strategic research, leading to a comprehensive understanding of its response to drought. Notably, drought occurring during the seedling stage has been observed to result in diminished plant stands, significantly reducing overall yield (Soman et al., 1987). Conversely, drought experienced during the vegetative stage of growth has minimal adverse effects on productivity (Bidinger et al., 1987). In fact, there is often a notable increase in panicle number, reflecting a compensatory mechanism to counter the impact of drought on the main shoot (van Oosterom et al., 2003). However, it is during the crucial grain-filling stage that pearl millet exhibits the highest sensitivity to drought stress. At this stage, both grain number and grain size experience significant reductions when the crop is subjected to drought stress (Fussell et al., 1991).

Investigations into the physiology, phenology, and morphology of pearl millet have facilitated the dissection of drought tolerance mechanisms, shedding light on the intricate yield formation process under drought conditions (van Oosterom et al., 2003; Yadav et al., 2011). This understanding has enabled plant researchers to efficiently work on specific target traits suitable for diverse drought environments. However, using physiological traits as selection indices for drought tolerance poses significant challenges, particularly when dealing with larger genotype populations in nursery breeding. To overcome this, the strategy of exploiting drought escape mechanisms by prioritizing early maturity has been successfully implemented, resulting in substantial genetic gains in drought-prone regions of northwestern India (Yadav et al., 2011). Furthermore, the manipulation of easily measurable morphological traits, such as high tillering, small grain size, and shorter grain filling periods, has been successful in breeding programs. These traits exhibit abundant natural variation, facilitating their integration into breeding efforts (Yadav et al., 2017, 2021).

The significance of using adapted germplasm in drought-tolerant breeding programs has been underscored because the observed performance under drought stress largely stems from the adaptation to stress conditions (Yadav et al., 2009, 2021). By hybridizing adapted landraces with elite genetic material, novel gene combinations are generated, integrating both stress adaptation and high productivity (Patil et al., 2020; Presterl & Weltzien, 2003; Yadav & Rai, 2011). Notably, genomic regions associated with drought tolerance and other responsive traits have been successfully detected and mapped (Serba & Yadav, 2016; Yadav et al., 2002, 2004). These associated regions are being actively used in gene editing technology to enhance drought tolerance in breeding programs (Bidinger et al., 2007; Sharma et al., 2014).

Research on the impact of drought on pearl millet encompasses various genetic analysis aspects, such as agronomic (Bidinger et al., 1987, Mahalakshmi et al., 1988, Yadav et al., 2002, 2004), physiological (Ghatak et al., 2016; Kusaka et al., 2005), biochemical (Choudhary et al., 2021), and expression studies (Choudhary et al., 2015; Dudhate et al., 2018; Jaiswal et al., 2018; Shivhare et al., 2020a, 2020b). The approaches used to induce drought stress in these studies vary significantly, ranging from the use of osmotically stressing chemicals to field trials. The extensive availability of diverse pearl millet accessions and the potential for incorporating desirable traits from wild species offer ample opportunities for enhancing drought tolerance/resistance in pearl millet through genetic engineering. Furthermore, thorough analysis of the insights gained from quantitative trait analysis and transcriptomic studies provide recommendations on areas that warrant further attention and research.

Heat Stress

The optimum temperature range for the normal growth of pearl millet is around 33–34 °C. Pearl millet is negatively affected by temperatures above this range, affecting both the seedling and reproductive stages. According to a study by Knox et al. (2011), climate change models predict a decline of 6%–17% in pearl millet yield in sub-Saharan Africa and Southern Asia by the year 2050. In regions such as India as well as western and southern Africa, the surface temperatures of the soil often exceed 45 °C, occasionally reaching 60 °C. These high temperatures pose a significant challenge because they contribute to poor plant stands, particularly affecting the vulnerability of pearl millet seedlings during their initial 10-day period (Peacock et al., 1993). Thus, it is crucial to prioritize the improvement of heat tolerance during the seedling stage. Studies have reported genetic variations in seedling survival under elevated soil surface temperatures (Peacock et al., 1993), and effective results have been achieved through the selection of seedlings with higher emergence rates using artificial screening techniques (Soman & Peacock, 1985).

During the last decade, pearl millet has emerged as a highly productive and remunerative crop in the hot and dry summer season in the northern and western parts of India (Yadav & Rai, 2013). With higher air temperatures (often >42 °C) coinciding with flowering, the crop suffers from reproductive sterility, leading to drastic reductions in seed set and lower grain yield (Djanaguiraman et al., 2018; Gupta et al., 2015). Heat tolerance at the reproductive stage has emerged as an important target trait to enhance genetic gains. Heat-tolerant breeding materials that have been used to enhance heat tolerance under the field conditions (Gupta et al., 2016) have facilitated the pyramiding of heat tolerance in high-yielding hybrids. When screening a pearl millet heat stress cDNA library using the plaque hybridization method, a clone from the pearl millet heat stress–responsive Expressed Sequence Tags (EST) database was used as a DNA probe because it had the highest degree of similarity to the *PgHsp70* (*Pennisetum glaucum Heat shock proteins*) gene (Reddy et al., 2010). In another report, the *PgHsp90* gene was discovered, characterized, and cloned, and the sequence was investigated for heat stress in pearl millet with three exons and three introns (Reddy et al., 2011). Using qRT-PCR analysis for gene expression in response to abiotic stress, Nitnavare et al. (2016) identified another gene from pearl millet, *PgHsp10*, and characterized it for heat stress with three exons and two introns. Sun et al. (2020) reported that, from a total of 2,792 transcription factors, 1,223 transcriptional regulators (including 318 transcription factors and 149 transcriptional regulators) were expressed under heat stress. A total of 6,920 genes were identified exclusively for heat stress using RNA and Pacbio sequence data from the pearl millet reference genome. In a recent study, physiological and transcriptional changes brought on by heat stress were observed in the roots of pearl millet. These studies also observed that trehalose accumulates in roots after 3–7 hours under heat stress conditions as this peroxidase activity gradually increases (Sun et al., 2021). This study also identified and validated key transcription factors (7 heat shock, 16 basic region/leucine zipper, and 18 basic helix-loop-helix) under heat stress for differential expression.

Salinity Stress

In arid and semi-arid environments, like all abiotic stress, salt stress shows a broad range of detrimental effects on a variety of crops (Varshney et al., 2017). The regular use of irrigated water containing salts in the form of NaCl or NaHCO$_3$ causes salinization, a significant problem in these areas. Salinity stress initially increases the uptake of Na$^+$ and Cl ions in various plant portions, which significantly inhibits the growth of plants at the reproductive stage and eventually decreases productivity (Dwivedi et al., 2012). Salinity also causes plants to produce more malondialdehyde and reactive oxygen species, which can negatively affect growth. The activity of antioxidant enzymes such as peroxidase, ascorbate peroxidase, superoxide dismutase, and catalase has been reported to be higher in salt-tolerant species and to endure and survive under salinity stress (Khan et al., 2021; Kumar et al., 2021). Reduced shoot N and increased Na and K content also play key roles in salt stress tolerance in pearl millet (Dwivedi et al., 2016). Several genes were identified related to salinity in pearl millet: dehydrin (*PgDHN*), *PgNHX1* (Na+/H+ antiporter), *PgVDAC* (voltage-dependent anion channel), and late embryogenesis abundant (*PgLEA*) (Agarwal et al., 2010; Reddy et al., 2012;

Verma et al., 2007). Shinde et al. (2018) studied the transcriptomic profiling using tolerant (ICMB 01222) and susceptible (ICMB 081) lines and identified 11,627 DEGs. These DEGs represent different ion transporters, transcription factors, and controlled metabolic pathways deployed in the development of pearl millet productivity under salinity stress conditions (Shinde et al., 2018). Using small RNA sequencing target prediction, 14 novel miRNA were identified in peal millet (Shinde et al., 2020).

Nutritional Traits

Pearl millet is a highly nutritious crop and offers great opportunity to achieve food and nutritional security. It is rich in dietary fiber, protein, and essential micronutrients like Fe, Zn, Ca, and Mg. As a gluten-free grain, it offers a valuable dietary option for individuals with gluten intolerance (Saleh et al., 2013). It is rich in omega-3 fatty acids compared with other cereals (Srivastava et al., 2021). Pearl millet's resilience to harsh climates and ability to grow in semi-arid regions make it an important crop for food security in areas prone to drought and with limited agricultural resources. Its nutritional value and adaptability make pearl millet a significant contributor to ensuring a diverse and nutritious food supply, particularly in regions where it is cultivated and consumed. The recommended daily requirement (RDA) for Fe and Zn would be met by breeding micronutrient-rich pearl millet varieties, and their consumption would also support the development of biofortified pearl millet lines. Recent reports found that around 250 g of biofortified pearl millet can provide 100% and 85% of the daily RDA for Zn and Fe, respectively (Satyavathi et al., 2021a). To increase the micronutrient content of farm harvest through "pulling" nutrients from the earth and "pushing" them into edible plant parts in their bioavailable forms, this technique involves focusing on and changing the pathways through which minerals flow through the soil. The strategy of pulling and pushing can help address the problem of malnutrition, which hampers human development. Biofortification, as opposed to existing methods such as by supplementation, fortification, or dietary diversification, offers a practical solution to provide essential nutrients to underprivileged communities in remote and rural areas with limited access to well-rounded and nourishing food (Bouis et al., 2011).

Forage Quality

Forage quality studies on pearl millet have been conducted to assess the nutritive value of the plant for livestock feed. Studies on forage quality have focused on several aspects, including dry matter yield, crude protein content, fiber content, mineral composition, and digestibility. These factors are important for determining the nutritional value of the forage and its potential to meet the requirements of different livestock species. Evaluation and identification of promising pearl millet germplasm for grain and fodder traits were performed by Khairwal et al. (2007). A set of 2,375 germplasm lines, 180 landraces, and 504 accessions of core collection of pearl millet for grain and fodder yield and their related traits evaluation resulted in identification of IP 3416, IP 17880, and IP 104 for higher green fodder yield per plant. Other potential lines identified were IP 15304, IP 3616, IP 6125, IP 13885, IP 6146, IP 6897.

Efforts toward improving the production and utilization of sorghum and pearl millet as livestock feed/dual-purpose crops were undertaken by Blümmel et al. (2003). They observed that in pearl millet, genotypic variation in grain and stover yield and quality was expressed better under high fertilizer application compared with low fertilizer application. Except for cell wall digestibility, no significant genotypic differences were found for fodder quality measurements in pearl millet. Digestible organic matter intake of stover was not significantly related to stover yield (Blümmel et al., 2003). Dry matter yield is an important factor in determining the productivity of pearl millet as a forage crop. Several studies have shown that pearl millet can produce high yields of dry matter, especially in areas with low rainfall or in drought-prone regions. However, dry matter yield can vary depending on factors such as soil fertility, rainfall, and management practices (Blümmel et al., 2003). Crude protein content is another important factor in forage quality. Pearl millet has a relatively high crude protein content compared with other warm-season grasses, with values ranging from 6% to 12% (Harinarayana et al., 2005). However, protein content can be influenced by factors such as stage of growth, fertilization, and environmental conditions (Harinarayana et al., 2005).

Fiber content, including neutral detergent fiber (NDF) and acid detergent fiber (ADF), is also an important component of forage quality. High fiber content can reduce digestibility and limit the intake of forage by livestock (Blümmel et al., 2003). Studies have shown that pearl millet has relatively low fiber content, with NDF and ADF values ranging from 55% to 68% and from 30% to 40%, respectively (Vinutha et al., 2021). Finally, digestibility is a key factor in determining the nutritional value of forage. Several studies have shown that pearl millet has high digestibility, with values ranging from 60% to 70%. However, digestibility can be influenced by factors such as stage of growth, fiber content, and environmental conditions, including voluntary feed intake (Blümmel et al., 2003). In summary, forage quality studies on pearl millet have shown that it is a productive and nutritious crop for livestock feed. The crop has high dry matter yield, relatively high crude protein content, low fiber content, and high mineral content. These factors make pearl millet an important crop for livestock production in arid and semi-arid regions of the world.

Genetics of Brown Midrib in Pearl Millet

Brown midrib (*bmr*) pearl millet has a brown-colored midrib or stem. Pearl millet *bmr* mutants were identified independently in two different programs (Gupta, 1995). Purdue brown midrib (Pbmr) was generated at Purdue University using the chemical mutagen diethyl sulfate (Cherney et al., 1988), while the second brown midrib line SDML 89107 was developed from a pearl millet selection from ICRISAT, Zimbabwe. Classical genetics revealed that *bmr* is most likely controlled by a single recessive gene, and allelism tests revealed the gene controlling *bmr* in Pbmr and SDML 89107 lines is allelic (Gupta, 1995). The *bmr* trait in pearl millet is valued for its improved digestibility and forage quality, particularly for livestock feed (Gupta & Govintharaj, 2023; Harinarayana et al., 2005). The *bmr* trait is associated with reduced lignin content in the plant, which makes it easier for animals to digest and extract nutrients from the forage. The increased digestibility of *bmr* pearl millet has led to its popularity in livestock production, where it is often used as a forage crop for grazing, hay, or silage (Degenhart et al., 1995). It is known for its high yield potential, drought tolerance, and ability to grow in marginal soil conditions, making it suitable for various regions. A study by Degenhart et al. (1995) showed that brown mid-rib lines yielded 77% as much forage as normal lines.

Farmers and researchers have also developed hybrids and cultivars of *bmr* pearl millet with improved agronomic traits and higher forage quality (Blümmel et al., 2003; Gupta & Govintharaj, 2023). These advancements have contributed to its widespread cultivation and use as a valuable feed resource in many parts of the world. There have been numerous studies conducted on *bmr* pearl millet, particularly focusing on forage quality, agronomic characteristics, and its impact on livestock production. Several studies have investigated the nutritional composition of *bmr* pearl millet, comparing it with other forage crops (Akin et al., 1991; Oskey et al., 2023). These studies have shown that *bmr* pearl millet generally has higher digestibility, lower lignin content, and improved energy value compared with non-*bmr* pearl millet and other forage crops, like sorghum and corn (Cherney et al., 1988; Sattler et al., 2010).

Research has explored the impact of *bmr* pearl millet on livestock performance, including weight gain, milk production, and feed conversion efficiency (Bernard & Tao, 2020; Cherney et al., 1990). Studies have generally found positive results, with animals fed *bmr* pearl millet–based diets exhibiting improved growth rates and feed efficiency compared with animals fed non-*bmr* pearl millet or other forage crops (Degenhart et al., 1995). Studies have focused on evaluating the agronomic traits of *bmr* pearl millet, such as yield potential, drought tolerance, disease resistance, and adaptation to different environmental conditions. These studies aim to develop high-yielding and resilient *bmr* pearl millet cultivars that can meet the demands of farmers and provide reliable forage resources. Pearl millet has been shown to have higher water use efficiency than any crop; however, a trade-off on biomass yield was observed (Machicek et al., 2019). The suitability of pearl millet for silage and hay production has been studied, examining factors such as fermentation quality, dry matter yield, and nutrient preservation during storage. These studies help optimize the use of pearl millet as preserved forage for live-stock feeding (Blümmel, 2017). Although pearl millet displayed lower perfor-mance compared with sorghum and maize, there is good potential to use the crop for fodder yield where water is a limiting factor (Vinutha et al., 2021).

Genomic Approaches

Development of Heterotic Gene Pools in Pearl Millet

The development of high-yielding hybrids with desired traits is the major goal of global pearl millet improvement programs. The introduction of hybrid technology has proven to be a successful strategy because it has led to a quantum jump in pearl millet productivity (Yadav & Rai, 2013). Its outcrossing nature and narrow genetic base are the major limiting factors for pearl millet hybrid breeding schemes. The availability of predefined germplasm for a trait of interest is the foundation for exploiting the hybrid vigor in crop plants. To achieve these, pheno-typic and/or genotypic evaluation of the germplasm for genetic variability and clustering them into different heterotic pools is a prerequisite (Ramya et al., 2018). Heterotic pool-backed hybrid parental material generation increases the effi-ciency of breeding programs by saving crossing time and field evaluation costs for selecting heterotic crosses (Teklewold & Becker, 2005). These heterotic pools ena-ble breeders to generate a higher number of cross-combinations within a short period and minimize the time and cost on redundant hybrids advancement.

For a successful hybrid breeding program in cross-pollinated crops, heterotic pools with decent genetic variability and strategies for improved combining abil-ity are needed for accelerated genetic gain. In pearl millet, very few studies used genomic tools to define the genetic pools. ICRISAT, together with several National Agricultural Research Services and private-sector partners, initiated efforts to develop the genetic pools by using African- and Asian-based hybrid parental lines to maximize the genetic variability of pearl millet (Satyavathi et al., 2013). Heterosis breeding in pearl millet relies on genetically distinguishable pools of A- (Cytoplasmic Male Sterile), B- (Maintainer), and R- (Restorer) lines to keep up the momentum of genetic gain. The first report on the identification of heterotic gene pools by using genomic markers in pearl millet was published by ICRISAT researchers (Ramya et al., 2018). A set of 342 hybrid parental lines of pearl millet, comprising 160 B-lines and 182 R-lines, together with a control (world reference germplasm line-Tift 23D2B1-P1-P5) were part of the study, where EST simple sequence repeat (SSR) and genomic SSR markers were used to define the heterotic gene pools. This study identified 9 and 11 clusters in B-lines and R-lines, respectively. In another study, two clusters were defined by using 40 SSR markers in a set of 190 (95 each of B-line and R-line) pearl millet lines (Anuradha et al., 2018). Singh and Chhabra (2018) reported different genomic groups among 150 (75 each of B-line and R-line) advanced hybrid parental lines of pearl millet by using 56 SSR and 75,000 single nucleotide polymorphism (SNP) markers. In a study by Singh and Gupta (2019), a total of eight heterotic gene pools were defined by using 150 hybrid parental lines covering 75 each of the B-lines and R-lines, with 56 SSR markers distributed across the pearl millet chromosomes. In another study, 580 hybrid parents of pearl millet comprising 320 R-lines and 260 B-lines, developed from six breeding schemes in India, together with genotyping information (RAD-genotyping-by-sequencing [GBS] generated ~0.9 million SNPs), resulted in 12 R-line and 7 B-line groups, with some outliers (Gupta et al., 2020). Genetic variability in pearl millet for essential micronutrients was evaluated in a set of 212 core pearl millet germplasm accessions, which resulted in eight distinct clusters (Govindaraj et al., 2020). In a more recent study, 45 pearl millet populations that originated from and were bred either in Africa or Asia were examined to derive information on heterotic pools. These populations were clustered into seven groups (G1–G7) once genotyped by 29 highly polymorphic SSRs (Patil et al., 2021).

Earlier, a few other diversity analysis-related studies using molecular markers clustered pearl millet germplasm into different pools. Pucher et al. (2016) studied patterns of combining ability in West African population hybrids but could not define strong heterotic groups. In another study, a small set of 18 pearl millet accessions were grouped into three clusters using 28 SSR markers (Kapadia, 2016). Genetic relatedness of 511 pearl millet accessions was evaluated using 83,875 SNPs, and the researchers reported a high genetic diversity in germplasm from Senegal compared with other regions of Africa and Asia (Hu et al., 2015). In another study, 45 novel restorer lines bred in India and Africa having decent combining ability, together with 50 SSR markers, were assessed to study genetic diversity and formed eight distinct groups (Satyavathi et al., 2013). Singh et al. (2013) also performed genetic diversity analysis of 20 pearl millet cultivars with 21 SSR markers and reported three different groups. Sumanth et al. (2013) evaluated 42 pearl millet inbred lines (22 B-lines and 20 R-lines) with 17 SSR markers to study the genetic diversity for defining the clusters. A set of 145 inbred lines derived from 122 landraces were evaluated, with 20 SSR markers to study the genetic diversity, which defined the lines that originated from West and Central African countries into a group (Stich et al., 2010). Kapila et al. (2008) reported genetic diversity among 72 pearl millet lines (70 maintainers and 2 pollinators) originating from sub-Saharan Africa and India evaluated using 34 SSR primers. The 72 lines formed five clusters and corroborated with their pedigree and individual traits. In another study, genetic diversity analysis of 46 wild and 421 cultivated pearl millet accessions from Niger were analyzed using 25 SSR markers and identified clear differentiation between wild and cultivated lines of Niger (Mariac et al., 2006).

Trait Mapping and Validation

Downy Mildew

Molecular diversity analysis is becoming a method of choice for the improvement of agriculturally important plant genetic resources (Mandloi et al., 2022; Rathore et al., 2023; Singhal et al., 2022). Parmar et al. (2022) investigated a set of pearl millet genotypes and revealed the existence of ISSR marker loci associated with the presence of disease resistance or susceptible genes, which may be further useful for marker-assisted breeding (MAB) for the improvement of crop against DM disease. Raj & Sharma (2022) conducted a study to determine the genetic structure of *S. graminicola* populations and reported that the mating system plays an important role in the evolution of plant pathogens during selection pressure from the resistant host or harsh environmental conditions. Hadimani et al. (2023) reported for the first time on the identification and characterization of effector protein–encoding functional genes in the DM pathogen, which can be used in screening oomycetes of DM diseases in other crops across the globe. DM resistance is a quantitative character that shows the dominance over susceptibility and recessive traits, and the resistance of a host plant is governed by several genes along with modifiers (Dwivedi et al., 2012; Hash & Witcombe, 2001). DNA markers for more than 60 different quantitative trait loci (QTLs) for DM resistance in different DM pathotypes were identified (Breese et al., 2002). Using MAS, some DM-resistant QTLs have been transferred into commercial R-lines (H 77/833–2, ICMP 451) and B-lines (843B, 81B), which are effective against Indian pathotypes of *S. graminicola* (Hash et al., 2006). Resistance QTLs for DM have been mapped using DM pathogen and F_4 mapping populations from India, Nigeria, Niger, and Senegal, namely, LGD-1-B-10 x Blast ICMP 85410UK and 7042(S)-1 x and P 7–3 by field and greenhouse screening in India and the United Kingdom. Two DM-resistant QTLs were detected on linkage group (LG)1 and LG2, which are consistent in both field and glass house screens of India and the United Kingdom.

Restriction fragment length polymorphism (RFLP) markers were used to construct the map in F_2 populations and QTL mapping for DM resistance on LGs (Breese et al., 2002; Gulia et al., 2007; Liu et al., 1994). Jogaiah et al. (2014) developed sequence characterized amplified region markers from inter-SSRs associated with pearl millet DM resistance LG. Parmar et al. (2022) used gene-linked ISSR marker(s) for screening of 48 pearl millet genotype(s) against DM based on disease indexing under field conditions, and seven genotypes (i.e., DMRBL 2, ICHPR 18-4, ICMB-98222, ICMB-92888, DHLB-33, ICMB-02444, and ICMR-06222) were found to be highly resistant. Pearl millet hybrid ICMH 88088 was developed by ICRISAT, and ICMH451 and Pusa 23 (Singh, 1995) are popular and highly DM-resistant hybrids with high yields. ICRISAT developed ICMV1 and ICMV2 for Senegal and IKMP2, IKMP3, and IKMV 8201 for Burkina Faso for DM resistance (Singh et al., 1993). Molecular markers–based selection has demonstrated its value in accelerating the process of precision breeding to create agriculturally significant varieties for disease resistance and several other features (Tripathi et al., 2021). HHB67 is the first success story of MAB in field crops in the public domain with a DM-resistant version in Indian popular hybrids, widely grown in North India (Hash et al., 2006). HHB67 Improved and the recently released HHB67 Improved 2–7 are the further improved versions of HHB 67 using MAS and MAB strategies, where the male line was introgressed with additional DM resistance QTLs on LG3, LG4, and LG6 from two DM-resistant donor parents (P1449 and 863B). Anther success story of MAS and MAB is the recently released GHB538 Improved (Maru Sona), a MAB product of the popular pearl millet hybrid GHB538, which was first released for cultivation in A1 zones of India (dry regions of Rajasthan, Gujarat, and Haryana states) during 2004–2005. It is also Gujarat's first hybrid cultivar to be developed through genomics-assisted breeding. The ICRISAT research team introgressed the male parent of GHB538 with three DM resistance QTLs from LG1, LG3, and LG4 to create Maru Sona. Compared with the original version, Maru Sona exhibited remarkable resistance to DM (82%).

Blast

The information related to inheritance of disease resistance facilitates the introgression of resistance genes in elite cultivars. The study in the monodii accession for blast resistance showed three independent and dominant genes (Hanna & Wells, 1989). The blast resistance gene in Tift 85DB, which is derived from monodii, is effective against three *Pyricularia grisea* isolates. Two major QTLs for blast resistance have been identified on LG1 and LG6 using SSR markers (Sanghani et al., 2018). Sharma et al. (2020) conducted a study to determine the level of blast disease resistance on 305 pearl millet accessions collected from 13 countries using five *M. grisea* isolates. It has been observed that 182 accessions showed resistance against five isolates. Maganlal (2018) reported the markers to tolerance against blast and improved several other agronomic traits (S3_0019-S3_4763) on LG3 and (BRP8, BRP11, and BRP7) LG1. Prakash et al. (2019) reported the draft genome sequence information of blast fungus *Magnaporthe* isolate PMg_Dl in pearl millet with the genome size of 47.90 (Mb) consisting of 9,649 coding DNA sequences. Among these, a total of 849 coding DNA sequences were annotated to be involved in pathogenicity, virulence, and effector genes. Sharma et al. (2020) conducted a study for screening under greenhouse conditions using 305 accessions of *Pennisetum violaceum*, a wild relative of pearl millet, against five pathotype isolates, and 17 accessions were found resistant to all five pathotypes. Molecular markers–based selection has demonstrated its value in accelerating the process of precision breeding to create agriculturally significant varieties for disease resistance and several other features (Tripathi et al., 2021). The elite B-lines and R-lines developed at ICRISAT (863B, ICMB 01333, ICMB 01777, ICMB 02111, ICMB 03999, ICMB 93222, ICMB 97222, ICMR 06222, ICMR 06444, and ICMR 07555) have been identified with high levels of resistance to blast (Thakur et al., 2009). The pearl millet reference genome sequence of Tift 23D2B1-P1-P5 and resequencing of 994 pearl millet genotypes (Varshney et al., 2017) helps breeding programs to map resistance genes for biotic and abiotic stresses and to develop resistance lines. Two hybrid parent lines (ICMB 06444 and ICMB 97222) were shown to have high levels of blast resistance (Sharma et al., 2013). These lines can be used for the development of high-yielding, blast-resistant pearl millet hybrids in India. Recently, HHB 146 hybrid, which was developed by the introgression of LG4 blast resistance QTL from 863B-P2, has shown a high level of blast resistance (Bollam et al., 2018).

Drought Stress

Over the past decade, there has been a growing trend in the use of omics genomic tools to enhance drought tolerance in pearl millet. This shift can be attributed to the availability of affordable high-throughput genotyping facilities and the increased interest among researchers to imitate the successful results accomplished through these approaches in extensively studied rice and wheat crops. Notably, the research efforts in pearl millet received significant enhancement with the publicly available draft genome sequence (https://cegresources.icrisat .org/data_public/PearlMillet_Genome/v1.1/), which provided genomic studies for improving the plant's ability to withstand abiotic stresses (Varshney et al., 2017).

Quantitative Studies

The first genetic map was constructed for drought tolerance with RFLP markers in pearl millet (Liu et al., 1994). Subsequently, marker-based investigations were conducted using the RFLP map as a foundational tool for molecular marker–based information. Under the DFID-JIC-ICRISAT project, the makers were developed and implemented. Several studies were conducted later by different researchers (Busso et al., 1995; Devos et al., 2000; Devos et al., 2000; Liu et al., 1994), which used the RFLP map for their research endeavors. Significant success was made through identification QTL on LG2, which was governed with grain yield per se and drought tolerance in grain yield, with phenotypic variance accounting for 23%. With limited water conditions, drought-tolerant lines demonstrated improved maintenance of harvest index (HI) and biomass productivity (Yadav et al., 2002). Yadav et al. (2004), using a testcross population, reported 79 F_3s for grain yield, stover yield, HI, and panicle HI in three, four, five, and three genomic regions identified, respectively, on different LGs (Table 4.1). Bidinger et al. (2007) also worked on drought and identified three QTLs on LG2, LG3, and LG4, which contributed to increased grain yield under varying post-flowering moisture conditions. LG2 and LG3 explained 13%–25% phenotypic variation explained for grain yield across the different environmental conditions. Furthermore, these identified QTLs were found to be co-located with HI, grain number, and other yield-related traits under terminal drought conditions. In addition, the drought tolerant LG2 is also associated with increasing plant growth under salt stress conditions (Sharma et al., 2011, 2014). The transpiration rate is lower in drought-tolerant genotypes compared with sensitive lines in pearl millet, reported by identified QTLs on LG2 (Aparna et al., 2015; Kholova et al., 2012). The utilization of genomic tools has facilitated the development of conserved-intron scanning primer and SNP markers, which have been instrumental in identifying candidate genes associated with drought-tolerant QTLs in diverse accessions of pearl millet (Sehgal et al., 2012). For water use efficiency, four genomic regions were identified on a major LG2 QTL interval contributing to terminal drought stress in pearl millet (Tharanya et al., 2018). In a study by Jangra et al. (2019), a drought-tolerant QTL from the donor 863 B was introgressed into the male parent HBL 11 of the popular pearl millet hybrid HHB-226, imparting tolerance to terminal drought stress. Several QTLs have been reported to be associated with drought tolerance in pearl millet (Table 4.1).

The deployment of NGS technologies for transcriptomic analysis has emerged as a highly efficient and cost-effective approach in studying gene expression in crop plants at different developmental stages with a wide range of expression levels (Srivastava et al., 2022). A relatively recent study by Jaiswal et al. (2018) reported by transcriptome analysis under drought conditions using the de novo assembly of pearl millet revealed various key genes, transcription factors, and gene regulatory networks governed for water stress conditions in pearl millet. The publicly available pearl millet genome sequence and whole genome resequencing (WGRS) data (Varshney et al., 2017) unravel the genetic basis of drought stress tolerance in pearl millet advanced genomic levels (Dudhate et al., 2018). Recently, Shivhare et al. (2020b) identified key candidate genes that regulated drought-responsive conditions with PRLT2/89-33 (a drought-tolerant pearl millet line) and validated in vegetative and reproductive stages by using a comparative transcriptomics approach. Another significant approach to tackle the improvement of genetic gain in terms of yield under harsh climatic conditions is genomic selection. This approach completely relies on precision phenotypic and genotypic data points and is a marker-aided selection (Meuwissen et al., 2001).

Table 4.1 Summary of QTLs/Candidate Genes Identified Associated with Drought Stress

References	Markers	Abiotic stress	Linkage group and candidate gene	Population
Yadav et al. (2002)	RFLP	Terminal drought stress (40 DAS)	LG2 (32%), LG5 (14%)	H77/833-2 × PRLT2/89-33
			LG2 and LG6	
			LG2, LG3, LG4, and LG6	
			LG2	
			LG2 and LG6	
Bidinger et al. (2007)	RFLP		LG2, LG3, and LG4	ICMB 841×863B
			LG1 (34%)	
			LG1 and LG3	
			LG1	
Kholova et al., (2012)	SSR and DaRT markers	Vegetative seedling stage	LG1, LG2, LG4, LG6, and LG7	PLRT 2/89-33 and H 77/833-2
			LG2, LG3, and LG7	
Aparna et al. (2015)	RFLP	~25 DAS	LG3 (12.9%)	ICMB 841 × 863B
			LG1, LG2, LG4, and LG6 (major)	
			LG6 (24.5%) and LG7 (24.7%)	
			LG1, LG2, LG5, LG6, and LG7	
Debieu et al. (2018)	GBS	Early drought stress (21 DAS)	Chr 3 (19%)	ICML IS 11001-11215, SL 1-5, Souna 3, and ICMB88004
			Chr 2 (14%)	
			Chr 6 (14%)	
Jangra et al. (2019)	SSR	Drought-tolerant QTL introgression	LG2 and LG5	Introgression into hybrid HHB 226 from 863B, male parent
Shivhare et al. (2020a)	Transcriptome study	Vegetative and reproductive stages	LG2 Salt stress protein	PRLT2/89-33
			LG2 Glutamine synthetase root isozymes,	
			LG2 Lateral root formation,	
			LG3 Involved in primary metabolism and cellular processes	

Abbreviations: DaRT, diversity arrays technology; DAS, days after sowing; GBS, genotyping-by-sequencing; LG, linkage group; QTL, quantitative trait loci; RFLP, restriction fragment length polymorphism; SSR, simple sequence repeat

Nutritional Traits

Several studies have reported that a large variability was observed for grain Fe and Zn content in germplasm, breeding lines, and cultivars. Yadav et al. (2017) reported in germplasm lines that 41 genotypes had >80 ppm Fe and 42 had >60 ppm Zn, whereas 33 lines were observed to be greater sources of both nutrients. An earlier study on 297 Iniadi germplasm from Western Africa (Togo, Eastern Ghana, Southern Burkina Faso, and Western Benin) has shown huge variability for Fe and Zn content (Rai et al., 2008a, 2008b, 2008c). Several accessions studied at ICRISAT showed a broad range of variability in grain nutrient content in diverse breeding materials such as Iniadi germplasm accessions (51–121 ppm Fe, 46–87 ppm Zn), population progenies (18.0–135.0 ppm Fe, 22.0–92.0 Zn), inbred parents (30.3–102.0 ppm Fe, 27.4–84.0 ppm Zn), hybrids derived from diverse inbreds (25.8–80.0 ppm Fe, 22.0–70 ppm Zn), and commercial hybrids (31.0–61.0 ppm Fe, 32.0–54.0 ppm Zn) (Govindaraj et al., 2019). The other study on open-pollinated grains of the recombinant inbred line population observed that grain Fe ranged from 22 to 154 ppm and Zn ranged from 19 to 121 ppm (Mahendrakar et al., 2019). Analysis of 281 advanced breeding lines exhibited substantial variability for Fe (35–116 ppm) and Zn (21–80 ppm) (Pujar et al., 2020). A pearl millet cultivar, HHB 67 Improved 2, developed through genomics-assisted breeding has been recently released in India and is highly cultivated in all over India. This hybrid also contributes immensely toward food and nutritional security.

Several studies of genetic resources have contributed to developing and mapping the QTLs and tagging markers for grain Fe and Zn content in pearl millet. This comprised reverse genetic stocks like the TILLING (Targeting Induced Local Lesions in Genomes) populations as well as forward genetic stocks including bi-parental mapping populations and association mapping panels. The TILLING populations are being developed at ICRISAT in the three genetic backgrounds, including Tift 23DB1-P1-P5 (world reference germplasm) for functional genomic related to grain Fe, Zn, other nutritional and agronomic traits (Kumar et al., 2016, 2018; Singhal et al., 2021). The association studies revealed several marker–trait associations (Anuradha et al., 2017; Pujar et al., 2020). Recently, Mahendrakar et al. (2020) reported unique genes on seven LGs (details in Table 4.2).

Transcriptomics studies may open the door to gene discovery and to understanding the metabolic pathway of beneficial nutritional products by revealing the underlying mineral balance (Hirai et al., 2004). For pearl millet, information on the gene-regulating steps of the Fe and Zn metabolic pathways is limited. The potential genes for pearl millet's grain Fe and Zn metabolism were recently found using transcriptome investigations (Mahendrakar et al., 2020). The spatiotemporal identification and characterization of grain Fe concentration (GFeC) and grain Zn concentration (GZnC) has been carried out using expression analysis for a total of 29 unique genes at different stages of development. In high and

Table 4.2 Summary Table for Grain Fe and Zn Content Studies in Pearl Millet

Study no.	Study	QTLs/MTAs/candidate genes	Phenotypic variance	References
1	Identification and validation of candidate genes underlying GFeC and GZnC	5 candidate genes include *PglFER1*, *PglZIP2*, *PglZIP4*, *PglNramp5*, and *PglZIP9* controlling with GFeC and GZnC	–	Mahendrakar et al. (2020)
2	Genome-wide association studies for Fe and Zn	78 MTAs were identified, of which 18 were associated with Fe, 43 with Zn, and 17 with protein	–	Pujar et al. (2020)
3	Identification of multilocation QTLs for grain Fe and Zn	14 QTLs for Fe and 8 for Zn, Fe, and Zn co-mapped on LG2 and LG3	Fe 19.66% and Zn 25.95%	Singhal et al. (2021)
4	To map QTLs linked with GFeC and GZnC in an iniadi-derived immortal pearl millet mapping population	Total of 11 QTLs detected for GFeC: 8 for GZnC and 3 co-mapped for both GFeC-GZnC observed, 1 on LG1 and 2 on LG7	GFeC: 9.0%–31.9% (cumulative 74%); GZnC: 9.4%–30.4% (cumulative 65%)	Kumar et al. (2018)
5	Genome-wide association mapping of QTLs for Fe and Zn	SSRs associated with both grain GZnC and GZnC on LG3, LG5, and LG7	11.4%, 13.3%, and 11.3% R^2, respectively	Anuradha et al. (2017)
6	Mapping of QTLs controlling high GFeC and GZnC in self- and open-pollinated grains of pearl millet	In selfed seed, 2 co-localized QTL for GFeC and GZnC on LG3. In open pollinated seeds 2 QTLs for GFeC on LG3 and LG5, and 2 QTLs for GZnC on LG3 and LG7	Fe: 19.0%, Zn: 36.0%, Fe: 16.0%, Zn: 42.0%	Kumar et al. (2016)

Abbreviations: GFeC, grain Fe concentration; GZnC, grain Zn concentration; LG, linkage group; MTA, marker-trait association; QTL, quantitative trait loci; SSR, simple sequence repeat.

low genotypes for GFeC and GZnC, tissue- and stage-specific expressions of the Fe and Zn genes were observed (Mahendrakar et al., 2020). The *PglZIP*, *PglNRAMP*, *PglYSL*, and *PglFER* family genes were identified as candidates for GFeC and GZnC by the results. The most promising candidate gene for GFeC was discovered to be the ferritin-like gene (*PglFER-1*). By annotating QTL regions for grain Fe and Zn concentrations, genetic regions underpinning GFeC and GZnC were uncovered. For GFeC and GZnC, expressed genes were linked with significant QTLs that were co-localized on LG7 (Mahendrakar et al., 2020). The study created a framework for functional dissection and offered insight into several Fe and Zn metabolism gene homologs.

The cereal crop species can be used to assess the effectiveness of genomic selection in increasing the level of genetic gain in crops and the application of genomic models (Srivastava et al., 2020a). To predict grain yield for test crosses and the performance of single cross hybrids derived from a CMS system in each trial (Liang et al., 2018), an initial GS study on pearl millet was conducted by Varshney et al. (2017) at ICRISAT, Patancheru. Genomic sequencing and high-throughput genotyping approaches have made it possible to create fast and affordable genotypic assays, enabling the use of genomic prediction and GS in plant breeding. A complex trait's genomics-estimated breeding values (GEBVs) are calculated using DNA marker data by GS, and the end choice is made only on these values without the use of additional field-based phenotypic data points (Crossa et al., 2017). A high degree of prediction accuracy has been noted in terms of the Pearson correlation coefficient between the predicted and observed values, and efficiency in various contexts was improved using two rates of heredity (Varshney et al., 2017).

Accelerated Breeding Approaches

Speed Breeding

Speed breeding is a novel plant breeding method that shortens generation time and accelerates crop improvement programs by manipulating growth conditions such as light, photo period, and temperature and promotes rapid generation cycling of the plants (Watson et al., 2018). Speed breeding has been successfully applied in breeding schemas of different cereals, like wheat, rice, maize, barley, and sorghum, offering various advantages. Pearl millet is a short-day species and flowers early with short days. Although for optimum plant growth in fields a day-length of 12 hours and 28–30 °C temperature is suitable (Bidinger et al., 1987), an experiment conducted by Ong and Everard (1979) showed that induction of short days had no adverse effect on head development. When the number of short days was reduced from a minimum of 4–14 days, the treatment showed considerable reduction in the time to anthesis, plant height, and final number of leaves produced. A minimum of four short days was essential for floral initiation and another four for crop development.

Other ways to achieve speed breeding include reducing germination time by altering temperature, which increased linearly from 10 to 34 °C (Ong & Monteith, 1985). Altering the light intensity and duration will also reduce the time required to complete the life cycle of a crop. However, a major limitation in pearl millet is that speed breeding is not suitable for short-day plants (Swami et al., 2023). Nevertheless, obligatory short-day plants, in another clade, have observed flowering in response to low-intensity light under long-day conditions. This kind of treatment is considered as stress-induced flowering, and 4 weeks of such treatment has shown flowering in these plants. Parallel adaptation of protocol from other studies performed in rice and soybean was successfully demonstrated by Jähne et al. (2020). A key message from this experiment is that, although rice (monocot) and soybean (dicot) are short-duration crops, the light quality requirements were contradictory. A well-defined protocol for pearl millet must be developed based on its physiological response toward light and other factors that can trigger early flowering.

Genomic Selection

Genomic selection is an emerging approach that helps select promising lines among experimental material in crop improvement programs. GS has the capacity to improve the productivity of the lines by accelerating genetic gains and considerably reducing the breeding cycles. This methodology requires genome-wide markers to predict the GEBV of the test population to select prospective lines, thus accelerating the breeding cycles (Spindel et al., 2015). It is proven to be a cost-effective strategy with huge potential to improve complex quantitative traits with narrow heritability and reduces phenotyping, line, and hybrid development costs (Crossa et al., 2017). GS requires an efficient training population having both whole genome marker data and precise phenotyping data. By using statistical and machine learning algorithms, GS predicts the GEBVs (marker–phenotype relationship) of test populations having only genotypic data. Selected lines with high-quality GEBVs will be deployed into crossing programs (Meuwissen et al., 2001). GS has proven to be a reliable alternative to traditional MAS because it facilitates predicting the promising lines by tracking several minor effect genes along with key genes using the whole genome marker data (Srivastava et al., 2020a and 2020b). With the availability of whole genome sequencing data, it is established to be a prospective tool to improve the line selection procedures in breeding programs in the majority of the crops (Asoro et al., 2011; Crossa et al., 2010; Muleta et al., 2019; Ornella et al., 2012; Poland et al., 2012; Spindel et al., 2015).

Pearl millet offers plenty of prospects to apply GS in crop improvement programs because it has a huge number of phenotypic datasets on different traits in both hybrid and open-pollinated production systems. The recent developments in pearl millet genomics research, such as the pearl millet genome assembly (Varshney et al., 2017), pangenome analysis (Yan et al., 2023), and improved assembly of pearl millet reference (Salson et al., 2023), have made it possible to use GS schemes to boost the line selection process in pearl millet breeding programs. Varshney et al. (2017) used the GS tool in pearl millet for the first time, using WGRS data of Pearl Millet Inbred Germplasm Association Panel lines together with phenotyping data of yield traits to predict the promising lines for hybrid breeding programs. The GS model Ridge Regression Best Linear Unbiased Prediction (RR-BLUP) developed by Endelman (2011) was trained on grain yield data (five seasons) of 64 pearl millet hybrids and predicted a total of 170 hybrid combinations; some of these combinations were already deployed into breeding programs for pearl millet productivity improvement (Varshney et al., 2017). In another study, Liang et al. (2018) assessed a set of 276 inbred pearl millet lines developed by ICRISAT using two genotyping datasets: one generated by conventional RAD-seq (15,306 SNPs) and one generated by tunable GBS (32,463 SNPs); four GS schemes were implemented using RR-BLUP method. GS generated a median range of prediction accuracies for this study (Liang et al., 2018). Jarquin et al. (2020), using the same dataset with the four environments, tested grain yield data of 276 pearl millet hybrids and 33 inbred parents. The results highlighted the advantages of incorporating the parental inbred phenotypic data in the calibration sets. Accounting for genotype × environment interaction improved the prediction ability. It also suggested that the use of genotypic data derived from a tunable GBS platform improved prediction accuracy (Jarquin et al., 2020). The applicability of GS in pearl millet breeding programs could be expanded by including hybrids with diverse genetic backgrounds. However, the application of GS in breeding programs has some challenges, such as the cost of genotyping and the lack of standard procedures where GS can be applied (Crossa et al., 2017).

Molecular Trait Discovery for Pearl Millet Improvement

Knowledge of the genetic and molecular basis of traits of interest is important to develop sustainable crop improvement strategies, mainly for fine-tuning gene expression to improve a trait of interest or develop durable insect/disease management strategies. Despite the availability of a pearl millet reference genome and sequence data of ~900 pearl millet diverse accessions (Varshney et al., 2017), knowledge of the genetic and molecular basis of pearl millet agronomic, nutritional, forage, and insect/disease resistance traits is limited. Pearl millet plant height gene *d2* coding for ABCB1 auxin transporter (Parvathaneni et al., 2019), the ortholog of maize *br2*/sorghum *dw3*, cloned using fine mapping, is an important agronomic gene characterized. TILLING, generating nucleotide substitutions or indels in the genome, is a useful genetic resource for gene identification and characterization (Mo et al., 2018; Till et al., 2004; Xin et al., 2008). However, generating TILLING populations is time consuming.

Modern biotechnology tools, such as CRISPR-Cas (clustered regularly interspaced short palindromic repeats and CRISPR-associated protein) gene-editing, provide a platform to characterize the genetic basis of crop improvement traits (Jinek et al., 2012; Pixley et al., 2022; Wang & Doudna, 2023; Zhu et al., 2020). CRISPR-Cas9 DNA editing eliminates gene activity by generating single nucleotide changes or indels precisely in the targeted gene of interest, enabling rapid identification of the genes controlling the trait of interest. For pearl millet molecular trait discovery, CRISPR-Cas9 DNA editing can be used as a reverse genetics/gene-first approach, for example, generating gene-knockouts of pearl millet orthologs of maize *bm1*/sorghum *bmr6* (cinnamyl alcohol dehydrogenase 2) (Halpin et al., 1998; Sattler et al., 2009) and maize *bm3*/sorghum *bmr12* (caffeic O-methyltransferase) (Saballos et al., 2012; Vignols et al., 1995) to determine if mutations in these genes confer the *bmr* trait. Knowledge of the genetic basis of traits facilitates the development of allelic-specific markers for MAB.

CRISPR-Cas9 gene editing requires transforming pearl millet with CRIPSR-Cas9 DNA and a 20-bp target single-guide RNA sequence complementary to the reference genome (Shan et al., 2013). A major limiting step for CRISPR-Cas9 gene editing is the transformation of pearl millet with CRISPR-Cas9 machinery. Recent studies using growth-regulating factors *BABY BOOM* (*BBM*), *WUSCHEL* (*WUS*), and GROWTH REGULATING FACTOR (*GRF4*) and GRF-INTERACTING FACTOR (*GIF1*), which improved transformation efficiency and reduced the regeneration time of transformed plants in wheat, rice, and sorghum (Aregawi et al., 2022; Debernardi et al., 2020; Lowe et al., 2016), can be leveraged to improve pearl millet transformation and regeneration. After DNA editing, genetic segregation of Cas9 DNA can be done by selfing or crossing and selecting plants without the Cas9 DNA to remove Cas9 DNA from the desired mutants (Chen et al., 2019). Eliminating Cas9 DNA from the gene-edited plants makes the plants transgene free; these gene-edited plants may not require the stringent regulatory approvals needed for transgenic crops. In addition to gene knockouts, variants of CRISPR-Cas9, such as base editors and prime editors, generate SNPs and insert small fragments (Lin et al., 2020; Zong et al., 2017), and RNA-targeting CRISPR-Cas13 that targets mRNA for RNA silencing (Sharma et al., 2022) can be used for gene identification. In the case of food crops with restrictions on transgenic technology, such as wheat, pearl millet, and sorghum, CRISPR gene-editing technology can be used as a research tool for gene identification and characterization. Overall, gene editing is a valuable biotechnology tool for pearl millet molecular trait discovery, generating novel genetic diversity, and creating desired alleles in elite genetic backgrounds to bypass marker-assisted backcrossing for trait introgression.

Future Thrusts and Conclusions

Pearl millet is one of the hardiest cereals on the planet. It can grow favorably in environments where no other crop may give economic returns. It is regarded as a smart crop, with lesser environmental footprints in terms of water, energy, and carbon. Recent advancements in the pearl millet genomic resources, such as the availability of reference genome assemblies, WGRS, pangenomes, and improved reference genome assemblies, together with good quality genetic resources in the form of heterotic gene pools, enable pearl millet researchers to quickly identify and deploy the high-quality trait specific loci into the desired background. It may also help foster precise genomic selection schemas by combining linked markers for improved grain and fodder productivity and nutritional and stress tolerance-related genomic regions, thus accelerating the precision and efficiency of breeding programs. Accelerated breeding technologies, such as speed breeding, and whole genome prediction tools aided by machine learning algorithms will pave the way for a quantum jump in the trait improvement programs for a more resilient and prosperous agricultural sector, ensuring sustainable agri-food systems for future generations.

References

Agarwal, P., Agarwal, P. K., Joshi, A. J., Sopory, S. K., & Reddy, M. K. (2010). Overexpression of PgDREB2A transcription factor enhances abiotic stress tolerance and activates downstream stress-responsive genes. *Molecular Biology Reports*, *37*, 1125–1135.

Akin, D. E., Rigsby, L. L., Hanna, W. W., & Gates, R. N. (1991). Structure and digestibility of tissues in normal and brown midrib pearl millet (*Pennisetum glaucum*). *Journal of the Science of Food and Agriculture*, *56*(4), 523–538.

Anuradha, N., Satyavathi, C. T., Bharadwaj, C., Nepolean, T., Sankar, S. M., Singh, S. P., Meena, M. C., Singhal, T., & Srivastava, R. K. (2017). Deciphering genomic regions for high grain iron and zinc content using association mapping in pearl millet. *Frontiers in Plant Science*, *8*, 412.

Anuradha, N., Satyavathi, C. T., Bharadwaj, C., Sankar, M., Singh, S. P., & Pathy, T. L. (2018). Pearl millet genetic variability for grain yield and micronutrients in the arid zone of India. *Journal of Pharmacognosy and Phytochemistry*, *7*(1), 875–878.

Aparna, K., Nepolean, T., Srivastsava, R. K., Kholová, J., Rajaram, V., Kumar, S., Rekha, B., Senthilvel, S., Hash, C. T., & Vadez, V. (2015). Quantitative trait loci associated with constitutive traits control water use in pearl millet [*Pennisetum glaucum* (L.) R. Br.]. *Plant Biology, 17*(5), 1073–1084.

Aregawi, K., Shen, J., Pierroz, G., Sharma, M. K., Dahlberg, J., Owiti, J., & Lemaux, P. G. (2022). Morphogene-assisted transformation of *Sorghum bicolor* allows more efficient genome editing. *Plant Biotechnology Journal, 20*, 748–760.

Asoro, F. G., Newell, M. A., Beavis, W. D., Scott, M. P., & Jannink, J. L. (2011). Accuracy and training population design for genomic selection on quantitative traits in elite North American oats. *Plant Genome Journal, 4*, 132.

Bernard, J. K., & Tao, S. (2020). Lactating dairy cows fed diets based on corn silage plus either brown midrib forage sorghum or brown midrib pearl millet silage have similar performance. *Applied Animal Science, 36*(1), 2–7.

Bidinger, F. R., Mahalakshmi, V., & Rao, G. D. P. (1987). Assessment of drought resistance in pearl millet (*Pennisetum americanum* (L.) Leeke). II. Estimation of genotype response to stress. *Australian Journal of Agricultural Research, 38*(1), 49–59.

Bidinger, F. R., Nepolean, T., Hash, C. T., Yadav, R. S., & Howarth, C. J. (2007). Quantitative trait loci for grain yield in pearl millet under variable post flowering moisture conditions. *Crop Science, 47*(3), 969–980.

Blümmel, M. (2017). *Improved sorghum and pearl millet forage cultivars for intensifying dairy systems.* International Livestock Research Institute.

Blümmel, M., Zerbini, E., Reddy, B. V. S., Hash, C. T., Bidinger, F., & Khan, A. A. (2003). Improving the production and utilization of sorghum and pearl millet as livestock feed: Progress towards dual-purpose genotypes. *Field Crops Research, 84*(1-2), 143–158.

Bollam, S., Pujarula, V., Srivastava, R. K., & Gupta, R. (2018). Genomic approaches to enhance stress tolerance for productivity improvements in pearl millet. *Biotechnologies of Crop Improvement, 3*, 239–264.

Bouis, H. E., Hotz, C., McClafferty, B., Meenakshi, J. V., & Pfeiffer, W. H. (2011). Biofortification: A new tool to reduce micronutrient malnutrition. *Food and Nutrition Bulletin, 32*(1, Suppl 1), S31–S40.

Breese, W. A., Hash, C. T., Devos, K. M., Howarth, C. J., & Leslie, J. F. (2002). Pearl millet genomics and breeding for resistance to downy mildew. In J. F. Leslie (Ed.), *Sorghum and millets diseases 2000* (pp. 243–246). Iowa State Press.

Busso, C. S., Liu, C. J., Hash, C. T., Witcombe, J. R., Devos, K. M., De Wet, J. M. J., & Gale, M. D. (1995). Analysis of recombination rate in female and male gametogenesis in pearl millet (*Pennisetum glaucum*) using RFLP markers. *Theoretical and Applied Genetics, 90*(2), 242–246.

Butler, E. J. (1907). Some diseases of cereals caused by Sclerospora graminicola. Agricultural Research Institute.

Chen, K., Wang, Y., Zhang, R., Zhang, H., & Gao, C. (2019). CRISPR/Cas genome editing and precision plant breeding in agriculture. *Annual Review of Plant Biology*, *70*, 667–697.

Cherney, D. J. R., Patterson, J. A., & Johnson, K. D. (1990). Digestibility and feeding value of pearl millet as influenced by the brown-midrib, low-lignin trait. *Journal of Animal Science*, *68*(2), 4345–4351.

Cherney, J. H., Axtell, J. D., Hassen, M. M., & Anliker, K. S. (1988). Forage quality characterization of a chemically induced brown-midrib mutant in pearl millet. *Crop Science*, *28*(5), 783–787.

Choudhary, M., Jayanand, & Padaria, J. C. (2015). Transcriptional profiling in pearl millet (*Pennisetum glaucum* LR Br.) for identification of differentially expressed drought responsive genes. *Physiology and Molecular Biology of Plants*, *21*, 187–196.

Choudhary, M. L., Tripathi, M. K., Gupta, N., Tiwari, S., Tripathi, N., Parihar, P., & Pandya, R. K. (2021). Screening of pearl millet [*Pennisetum glaucum* [L] R Br] germplasm lines against drought tolerance based on biochemical traits. *Current Journal of Applied Science and Technology*, *40*, 1–12.

Choudhary, S., Guha, A., Kholova, J., Pandravada, A., Messina, C. D., Cooper, M., & Vadez, V. (2020). Maize, sorghum, and pearl millet have highly contrasting species strategies to adapt to water stress and climate change-like conditions. *Plant Science*, *295*, 110297.

Crossa, J., Campos, G. D. L., Pérez, P., Gianola, D., Burgueno, J., Araus, J. L., Makumbi, D., Singh, R. P., Dreisigacker, S., Yan, J., & Arief, V. (2010). Prediction of genetic values of quantitative traits in plant breeding using pedigree and molecular markers. *Genetics*, *186*(2), 713–724.

Crossa, J., Pérez-Rodríguez, P., Cuevas, J., Montesinos-López, O., Jarquín, D., De Los Campos, G., Burgueño, J., González-Camacho, J. M., Pérez-Elizalde, S., Beyene, Y., Dreisigacker, S., Singh, R., Zhang, X., Gowda, M., Roorkiwal, M., Rutkoski, J., & Varshney, R. K. (2017). Genomic selection in plant breeding: Methods, models, and perspectives. *Trends in Plant Science*, *22*(11), 961–975.

Debernardi, J. M., Tricoli, D. M., Ercoli, M. F., Hayta, S., Ronald, P., Palatnik, J. F., & Dubcovsky, J. (2020). A GRF–GIF chimeric protein improves the regeneration efficiency of transgenic plants. *Nature Biotechnology*, *38*, 1274–1279.

Debieu, M., Sine, B., Passot, S., Grondin, A., Akata, E., Gangashetty, P., Vadez, V., Gantet, P., Foncéka, D., Cournac, L., Hash, C. T., Kane, N. A., Vigouroux, T., & Laplaze, L. (2018). Response to early drought stress and identification of QTLs controlling biomass production under drought in pearl millet. *PLoS One*, *13*(10), e0201635.

Degenhart, N. R., Werner, B. K., & Burton, G. W. (1995). Forage yield and quality of a brown mid-rib mutant in pearl millet. *Crop Science*, *35*(4), 986.

Devos, K. M., Pittaway, T. S., Reynolds, A., & Gale, M. D. (2000). Comparative mapping reveals a complex relationship between the pearl millet genome and those of foxtail millet and rice. *Theoretical and Applied Genetics*, *100*, 190–198.

Djanaguiraman, M., Perumal, R., Ciampitti, I. A., Gupta, S. K., & Prasad, P. V. V. (2018). Quantifying pearl millet response to high temperature stress: Thresholds, sensitive stages, genetic variability and relative sensitivity of pollen and pistil. *Plant, Cell & Environment, 41*(5), 993–1007.

Dudhate, A., Shinde, H., Tsugama, D., Liu, S., & Takano, T. (2018). Transcriptomic analysis reveals the differentially expressed genes and pathways involved in drought tolerance in pearl millet [Pennisetum glaucum (L.) R. Br]. *PLoS One, 13*(4), e0195908.

Dwivedi, S. L., Ceccarelli, S., Blair, M. W., Upadhyaya, H. D., Are, A. K., & Ortiz, R. (2016). Landrace germplasm for improving yield and abiotic stress adaptation. *Trends in Plant Science, 21*(1), 31–42.

Dwivedi, S. L., Upadhyaya, H. D., Senthilvel, S., Tom Hash, C., Fukunaga, K., Diao, X., Santra, D., Baltensperge, D., & Prasad, M. (2012). Millets genet genomic res. In J. Janick (Ed.), *Plant Breeding Reviews* (Vol. 35, pp. 247–375). John Wiley & Sons.

Endelman, J. B. (2011). Ridge regression and other kernels for genomic selection with R package rrBLUP. *The plant genome, 4*(3).

Fussell, L. K., Bidinger, F. R., & Bieler, P. (1991). Crop physiology and breeding for drought tolerance: Research and development. *Field Crops Research, 27*(3), 183–199.

Ghatak, A., Chaturvedi, P., Nagler, M., Roustan, V., Lyon, D., Bachmann, G., Postl, W., Schröfl, A., Desai, N., Varshney, R. K., & Weckwerth, W. (2016). Comprehensive tissue-specific proteome analysis of drought stress responses in *Pennisetum glaucum* (L.) R. Br. (*Pearl millet*). *Journal of Proteomics, 143*, 122–135.

Govindaraj, M., Rai, K. N., Cherian, B., Pfeiffer, W. H., Kanatti, A., & Shivade, H. (2019). Breeding biofortified pearl millet varieties and hybrids to enhance millet markets for human nutrition. *Agriculture, 9*(5), 106.

Govindaraj, M., Rai, K. N., Kanatti, A., Upadhyaya, H. D., Shivade, H., & Rao, A. S. (2020). Exploring the genetic variability and diversity of pearl millet core collection germplasm for grain nutritional traits improvement. *Scientific Reports, 10*(1), 1–13.

Gulia, S. K., Wilson, J. P., Carter, J., & Singh, B. P. (2007). Progress in grain pearl millet research and market development. In *Issues in new crops and new uses* (pp. 196–203). ASHS Press.

Gupta, S. C. (1995). Inheritance and allelic study of brown midrib trait in pearl millet. *Journal of Heredity, 86*(4), 301–303.

Gupta, S. K., Ameta, V. L., Pareek, S., Mahala, R. S., Jayalekha, A. K., Deora, V. S., Verma, Y. S., Boratkar, M., Atkari, D., & Rai, K. N. (2016). *Genetic enhancement for flowering period heat tolerance in peart millet (Pennisetum glaucum L.(R.) Br.).* ICRISAT.

Gupta, S. K., & Govintharaj, P. (2023). Inheritance and allelism of brown midrib trait introgressed in agronomically promising backgrounds in pearl millet (*Pennisetum glaucum* (L.) R. Br.). *Czech Journal of Genetics and Plant Breeding, 59*(3), 176–187.

Gupta, S. K., Patil, K. S., Rathore, A., Yadav, D. V., Sharma, L. D., Mungra, K. D., Patil, H. T., Gupta, S. K., Kumar, R., Chaudhary, V., & Das, R. R. (2020). Identification of heterotic groups in South-Asian-bred hybrid parents of pearl millet. *Theoretical and Applied Genetics, 133*(3), 873–888.

Gupta, S. K., Rai, K. N., Singh, P., Ameta, V. L., Gupta, S. K., Jayalekha, A. K., Mahala, R. S., Pareek, S., Swami, M. L., & Verma, Y. S. (2015). Seed set variability under high temperatures during flowering period in pearl millet (*Pennisetum glaucum* L.(R.) Br.). *Field Crops Research, 171*, 41–53.

Gupta, S. K., Rathore, A., Yadav, O. P., Rai, K. N., Khairwal, I. S., Rajpurohit, B. S., & Das, R. R. (2013). Identifying mega-environments and essential test locations for pearl millet cultivar selection in India. *Crop Science, 53*(6), 2444–2453.

Hadimani, S., De Britto, S., Udayashankar, A. C., Geetha, N., Nayaka, C. S., Ali, D., Alarifi, S., Ito, S.-I., & Jogaiah, S. (2023). Genome wide characterization of effector protein-encoding genes in *Sclerospora graminicola* and its validation in response to pearl millet downy mildew disease stress. *Journal of Fungi, 9*, 431. https://doi.org/10.3390/jof9040431

Halpin, C., Holt, K., Chojecki, J., Oliver, D., Chabbert, B., Monties, B., Edwards, K., Barakate, A., & Foxon, G. A. (1998). Brown-midrib maize (bm1)—A mutation affecting the cinnamyl alcohol dehydrogenase gene. *Plant Journal for Cell and Molecular Biology, 14*, 545–553.

Hanna, W. W., & Wells, H. D. (1989). Inheritance of Pyricularia leaf spot resistance in pearl millet. *The Journal of Heredity, 80*(2), 145–147.

Hanna, W. W., Wells, H. D., & Burton, G. W. (1987). Registration of pearl millet inbred parental lines, Tift 85D2A1 and Tift 85D2B 1. *Crop Science, 27*(6), 1324–1325.

Harinarayana, G., Melkania, N. P., Reddy, B. V. S., Gupta, S. K., Rai, K. N., & Sateesh Kumar, P. (2005). Forage potential of sorghum and pearl millet. In *Sustainable development and management of drylands in the twenty-first century: Proceedings of the Seventh International Conference on the Development of Dryland, Syria* (pp. 292–321). ICRISAT.

Hash, C. T., Thakur, R. P., Rao, V. P., & Raj, A. B. (2006). Evidence for enhanced resistance to diverse isolates of pearl millet downy mildew through gene pyramiding. *International Sorghum and Millets Newsletter, l47*, 134–138.

Hash, C. T., & Witcombe, J. R. (2001). Pearl millet molecular marker research. *International Sorghum and Millets Newsletter, l42*, 8–15.

Heffner, E. L., Jannink, J. L., & Sorrells, M. E. (2011). Genomic selection accuracy using multifamily prediction models in a wheat breeding program. *The Plant Genome, 4*(1).

Hirai, M. Y., Yano, M., Goodenowe, D. B., Kanaya, S., Kimura, T., Awazuhara, M., et al. (2004). From the Cover: Integration of Transcriptomics and Metabolomics for Understanding of Global Responses to Nutritional Stresses in *Arabidopsis thaliana. Proc. Natl. Acad. Sci.*, *101*(27), 10205–10210. https://doi.org/10.1073/pnas.0403218101

Hu, Z., Mbacké, B., Perumal, R., Guèye, M. C., Sy, O., Bouchet, S., Prasad, P. V., & Morris, G. P. (2015). Population genomics of pearl millet (*Pennisetum glaucum* (L.) R. Br.): Comparative analysis of global accessions and Senegalese landraces. *BMC Genomics*, *16*(1), 1048.

Jähne, F., Hahn, V., Würschum, T., & Leiser, W. L. (2020). Speed breeding short-day crops by LED-controlled light schemes. *Theoretical and Applied Genetics*, *133*(8), 2335–2342.

Jaiswal, S., Antala, T. J., Mandavia, M. K., Chopra, M., Jasrotia, R. S., Tomar, R. S., & Kumar, D. (2018). Transcriptomic signature of drought response in pearl millet (*Pennisetum glaucum* (L.)) and development of web-genomic resources. *Scientific Reports*, *8*(1), 3382.

Jangra, S., Rani, A., Yadav, R. C., Yadav, N. R., & Yadav, D. (2019). Introgression of terminal drought stress tolerance in advance lines of popular pearl millet hybrid through molecular breeding. *Plant Physiology Reports*, *24*, 359–369.

Jarquin, D., Howard, R., Liang, Z., Gupta, S. K., Schnable, J. C., & Crossa, J. (2020). Enhancing hybrid prediction in pearl millet using genomic and/or multi-environment phenotypic information of inbreds. *Frontiers in Genetics*, *10*, p.496995.

Jinek, M., Chylinski, K., Fonfara, I., Hauer, M., Doudna, J. A., & Charpentier, E. (2012). A programmable dual-RNA–guided DNA endonuclease in adaptive bacterial immunity. *Science*, *337*, 816–821.

Jogaiah, S., Sharathchandra, R. G., Raj, N., Vedamurthy, A. B., & Shetty, H. S. (2014). Development of SCAR marker associated with downy mildew disease resistance in pearl millet (*Pennisetum glaucum* L.). *Molecular Biology Reports*, *41*(12), 7815–7824.

Kapadia, V. N., 2016. Estimation of Heterosis for Yield and Its Relevant Traits in Forage Pearl Millet [Pennisetum glaucum LR Br.]. *International Journal of Agriculture Sciences*, ISSN, pp. 0975–3710.

Kapila, R. K., Yadav, R. S., Plaha, P., Rai, K. N., Yadav, O. P., Hash, C. T., & Howarth, C. J. (2008). Genetic diversity among pearl millet maintainers using microsatellite markers. *Plant Breeding*, *127*(1), 33–37.

Khairwal, I. S., Yadav, S. K., Rai, K. N., Upadhyaya, H. D., Kachhawa, D., Nirwan, B., Bhattacharjee, R., Rajpurohit, B. S., & Dangaria, C. J. (2007). Evaluation and identification of promising pearl millet germplasm for grain and fodder traits. *Journal of SAT Agricultural Research*, *5*(1), 1–6.

Khan, M. A., Hamayun, M., Asaf, S., Khan, M., Yun, B. W., Kang, S. M., & Lee, I. J. (2021). Rhizospheric *Bacillus* spp. rescues plant growth under salinity stress via regulating gene expression, endogenous hormones, and antioxidant system of *Oryza sativa* L. *Frontiers in Plant Science*, *12*, 665590.

Kholová, J., Nepolean, T., Hash, C. T., Supriya, A., Rajaram, V., Senthilvel, S., Kakkera, A., Yadav, R., & Vadez, V. (2012). Water saving traits co-map with a major terminal drought tolerance quantitative trait locus in pearl millet [*Pennisetum glaucum* (L.) R. Br.]. *Molecular Breeding, 30*, 1337–1353.

Knox, J. W., Hess, T. M., Daccache, A., & Perez Ortola, M. (2011). *What are the projected impacts of climate change on food crop productivity in Africa and South Asia?* Cranfield University.

Kumar, P., Subbarao, P. M. V., Kala, L. D., & Vijay, V. K. (2021). Thermogravimetry and associated characteristics of pearl millet cob and eucalyptus biomass using differential thermal gravimetric analysis for thermochemical gasification. *Thermal Science and Engineering Progress, 26*, 101104.

Kumar, S., Hash, C. T., Nepolean, T., Mahendrakar, M. D., Satyavathi, C. T., Singh, G., Rathore, A., Yadav, R. S., Gupta, R., & Srivastava, R. K. (2018). Mapping grain iron and zinc content quantitative trait loci in an iniadi-derived immortal population of pearl millet. *Genes, 9*(5), 248.

Kumar, S., Hash, C. T., Thirunavukkarasu, N., Singh, G., Rajaram, V., Rathore, A., Senapathy, S., Mahendrakar, M. D., Yadav, R. S., & Srivastava, R. K. (2016). Mapping quantitative trait loci controlling high iron and zinc content in self and open pollinated grains of pearl millet [*Pennisetum glaucum* (L.) R. Br.]. *Frontiers in Plant Science, 7*, 1636.

Kusaka, M., Ohta, M., & Fujimura, T. (2005). Contribution of inorganic components to osmotic adjustment and leaf folding for drought tolerance in pearl millet. *Physiologia Plantarum, 125*(4), 474–489.

Liang, Z., Gupta, S. K., Yeh, C. T., Zhang, Y., Ngu, D. W., Kumar, R., Patil, H. T., Mungra, K. D., Yadav, D. V., Rathore, A., Srivastava, R. K., Gupta, R., Yang, J., Varshney, R. V., Schnable, P. S., & Schnable, J. C. (2018). Phenotypic data from inbred parents can improve genomic prediction in pearl millet hybrids. *G3: Genes, Genomes, Genetics, 8*(7), 2513–2522.

Lin, Q., Zong, Y., Xue, C., Wang, S., Jin, S., Zhu, Z., Wang, Y., Anzalone, A. V., Raguram, A., Doman, J. L., Liu, D. R., & Gao, C. (2020). Prime genome editing in rice and wheat. *Nature Biotechnology, 38*(5), 582–585.

Liu, C. J., Witcombe, J. R., Pittaway, T. S., Nash, M., Hash, C. T., Busso, C. S., & Gale, M. D. (1994). An RFLP-based genetic map of pearl millet (*Pennisetum glaucum*). *Theoretical and Applied Genetics, 89*(4), 481–487.

Lowe, K., Wu, E., Wang, N., Hoerster, G., Hastings, C., Cho, M.-J., Scelonge, C., Lenderts, B., Chamberlin, M., Cushatt, J., Wang, L., Ryan, L., Khan, T., Chow-Yiu, J., Hua, W., Yu, M., Banh, J., Bao, Z., Brink, K., . . . Gordon-Kamm, W. (2016). Morphogenic regulators *Baby boom* and *Wuschel* improve monocot transformation. *The Plant Cell, 28*, 1998–2015.

Machicek, J. A., Blaser, B. C., Darapuneni, M., & Rhoades, M. B. (2019). Harvesting regimes affect brown midrib sorghum-sudangrass and brown midrib pearl millet forage production and quality. *Agronomy, 9*(8), 416.

Maganlal, S. J., Sanghani, A. O., Kothari, V. V., Raval, S. S., Kahodariya, J. H., Ramani, H. R., Vadher, K. J., Gajera, H. P., Golakiya, B. A., & Mandavia, M. K. (2018). The SSR based linkage map construction and identification of QTLs for blast (Pyricularia grisea) resistance in pearl millet (Pennisetum glaucum (L.) r. br.). *Journal of Pharmacognosy and Phytochemistry, 7*(2), 3057–3064.

Mahalakshmi, V., Bidinger, F. R., & Rao, G. D. P. (1988). Timing and intensity of water deficits during flowering and grain-filling in pearl millet. *Agronomy Journal, 80*(1), 130–135.

Mahendrakar, M. D., Kumar, S., Singh, R. B., Rathore, A., Potupureddi, G., Kishor, P. K., Gupta, R., & Srivastava, R. K. (2019). Genetic variability, genotype × environment interaction and correlation analysis for grain iron and zinc contents in recombinant inbred line population of pearl millet [*Pennisetum glaucum* (L). R. Br.]. *Indian Journal of Genetics and Plant Breeding, 79*(03), 545–551.

Mahendrakar, M. D., Parveda, M., Kishor, P. K., & Srivastava, R. K. (2020). Discovery and validation of candidate genes for grain iron and zinc metabolism in pearl millet [*Pennisetum glaucum* (L.) R. Br.]. *Scientific Reports, 10*(1), 16562.

Mandloi, S., Tripathi, M. K., Tiwari, S., & Tripathi, N. (2022). Genetic diversity analysis among late leaf spot and rust resistant and susceptible germplasm in groundnut (*Arachis hypogea* L.). *Israel Journal of Plant Sciences, 69*(1), 1–9. https://doi.org/10.1163/22238980-bja10058

Mariac, C., Luong, V., Kapran, I., Mamadou, A., Sagnard, F., Deu, M., Chantereau, J., Gerard, B., Ndjeunga, J., Bezançon, G., & Pham, J. L. (2006). Diversity of wild and cultivated pearl millet accessions (*Pennisetum glaucum* [L.] R. Br.) in Niger assessed by microsatellite markers. *Theoretical and Applied Genetics, 114*(1), 49–58.

Mehta, P. R., Singh, B., & Mathur, S. C. (1953). A new leaf spot disease of Bajra (*Pennisetum typhoides* Stapf and Hubbard) caused by a species of *Pyricularia*. *Indian Phytopathology, 5*(2), 142–143.

Meuwissen, T. H., Hayes, B. J., & Goddard, M. E. (2001). Prediction of total genetic value using genome-wide dense marker maps. *Genetics, 157*, 1819–1829.

Mo, Y., Howell, T., Vasquez-Gross, H., de Haro, L. A., Dubcovsky, J., & Pearce, S. (2018). Mapping causal mutations by exome sequencing in a wheat TILLING population: A tall mutant case study. *Molecular Genetics and Genomics, 293*, 463–477.

Muleta, K. T., Pressoir, G., & Morris, G. P. (2019). Optimizing genomic selection for a sorghum breeding program in Haiti: A simulation study. *G3: Genes, Genomes, Genetics, 9*(2), 391–401.

Nitnavare, R. B., Yeshvekar, R. K., Sharma, K. K., Vadez, V., Reddy, M. K., & Reddy, P. S. (2016). Molecular cloning, characterization and expression analysis of a heat shock protein 10 (Hsp10) from *Pennisetum glaucum* (L.), a C 4 cereal plant from the semi-arid tropics. *Molecular Biology Reports, 43*, 861–870.

Ong, C. K., & Everard, A. (1979). Short day induction of flowering in pearl millet (*Pennisetum typhoides*) and its effect on plant morphology. *Experimental Agriculture*, *15*(04), 401. https://doi.org/10.1017/s0014479700013053

Ong, C. K., & Monteith, J. L. (1985). Response of pearl millet to light and temperature. *Field Crops Research*, *11*, 141–160. https://doi.org/10.1016/0378-4290(85)90098-x

Ornella, L., Singh, S., Perez, P., Burgueño, J., Singh, R., Tapia, E., Bhavani, S., Dreisigacker, S., Braun, H. J., Mathews, K., & Crossa, J. (2012). Genomic prediction of genetic values for resistance to wheat rusts. *The Plant Genome*, *5*(3).

Oskey, M., Velasquez, C., Peña, O. M., Andrae, J., Bridges, W., Ferreira, G., & Aguerre, M. J. (2023). Yield, nutritional composition, and digestibility of conventional and brown midrib (BMR) pearl millet as affected by planting and harvesting dates and interseeded cowpea. *Animals*, *13*(2), 260.

Parmar, A., Tripathi, M. K., Tiwari, S., Tripathi, N., Parihar, P., & Pandya, R. K. (2022). Characterization of pearl millet [*Pennisetum glaucum* (L.) R Br.] genotypes against downy mildew disease employing disease indexing and ISSR markers. *Octa Journal of Biosciences*, *10*(2), 134–142.

Parvathaneni, R. K., Spiekerman, J. J., Zhou, H., Wu, X., & Devos, K. M. (2019). Structural characterization of ABCB1, the gene underlying the d2 dwarf phenotype in pearl millet, *Cenchrus Americanus* (L.) Morrone. *G3: Genes, Genomes, Genetics*, *9*, 2497–2509.

Patil, K. S., Gupta, S. K., Marathi, B., Danam, S., Thatikunta, R., Rathore, A., Das, R. R., Dangi, K. S., & Yadav, O. P. (2020). African and Asian origin pearl millet populations: Genetic diversity pattern and its association with yield heterosis. *Crop Science*, *60*(6), 3035–3048.

Patil, K. S., Mungra, K. D., Danam, S., Vemula, A. K., Das, R. R., Rathore, A., & Gupta, S. K. (2021). Heterotic pools in African and Asian origin populations of pearl millet [*Pennisetum glaucum* (L.) R. Br.]. *Scientific Reports*, *11*(1), 12197.

Peacock, J. M., Soman, P., Jayachandran, R., Rani, A. U., Howarth, C. J., & Thomas, A. (1993). Effects of high soil surface temperature on seedling survival in pearl millet. *Experimental Agriculture*, *29*(2), 215–225.

Pixley, K. V., Falck-Zepeda, J. B., Paarlberg, R. L., Phillips, P. W. B., Slamet-Loedin, I. H., Dhugga, K. S., Campos, H., & Gutterson, N. (2022). Genome-edited crops for improved food security of smallholder farmers. *Nature Genetics*, *54*, 364–367.

Poland, J., Endelman, J., Dawson, J., Rutkoski, J., Wu, S., Manes, Y., Dreisigacker, S., Crossa, J., Sánchez-Villeda, H., Sorrells, M., & Jannink, J. L. (2012). Genomic selection in wheat breeding using genotyping-by-sequencing. *The Plant Genome*, *5*(3).

Prakash, G., Kumar, A., Sheoran, N., Aggarwal, R., Satyavathi, C. T., Chikara, S. K., Ghosh, A., & Jain, R. K. (2019). First draft genome sequence of a pearl millet blast pathogen, *Magnaporthe grisea* strain PMg_Dl, obtained using PacBio single-molecule real-time and illumina NextSeq 500 sequencing. *Microbiology Resource Announcements*, *8*(20), e01499-18.

Presterl, T., & Weltzien, E. (2003). Exploiting heterosis in pearl millet for population breeding in arid environments. *Crop Science, 43*(3), 767–776.

Pucher, A., Sy, O., Sanogo, M. D., Angarawai, I. I., Zangre, R., Ouedraogo, M., Boureima, S., Hash, C. T., & Haussmann, B. I. (2016). Combining ability patterns among West African pearl millet landraces and prospects for pearl millet hybrid breeding. *Field Crops Research, 195*, 9–20.

Pujar, M., Govindaraj, M., Gangaprasad, S., Kanatti, A., & Shivade, H. (2020). Genetic variation and diversity for grain iron, zinc, protein and agronomic traits in advanced breeding lines of pearl millet [*Pennisetum glaucum* (L.) R. Br.] for biofortification breeding. *Genetic Resources and Crop Evolution, 67*, 2009–2022.

Rai, K. N., Gowda, C. L. L., Reddy, B. V. S., & Sehgal, S. (2008a). Adaptation and potential uses of sorghum and pearl millet in alternative and health foods. *Comprehensive Reviews in Food Science and Food Safety, 7*(4), 320–396.

Rai, K. N., Hash, C. T., Singh, A. K., & Velu, G. (2008b). Adaptation and quality traits of a germplasm-derived commercial seed parent of pearl millet. *Plant Genetic Resources Newsletter, 154*, 20–24.

Rai, P. K., Jaiswal, D., Singh, R. K., Gupta, R. K., & Watal, G. (2008c). Glycemic properties of *Trichosanthes dioica* leaves. *Pharmaceutical Biology, 46*(12), 894–899.

Raj, C. & Sharma, R., 2022. Sexual Compatibility Types in F1 Progenies of Sclerospora graminicola, the Causal Agent of Pearl Millet Downy Mildew. *J. Fungi, 8*, 629.

Ramya, A. R., Ahamed, M. L., Satyavathi, C. T., Rathore, A., Katiyar, P., Raj, A. B., Kumar, S., Gupta, R., Mahendrakar, M. D., Yadav, R. S., & Srivastava, R. K. (2018). Towards defining heterotic gene pools in pearl millet [*Pennisetum glaucum* (L.) R. Br.]. *Frontiers in Plant Science, 8*, 1934.

Rathore, M. S., Tiwari, S., Tripathi, M. K., Gupta, N., Yadav, S., Singh, S., & Tomar, R. S. (2023). Genetic diversity analysis of groundnut germplasm lines in respect to early and late leaf spot diseases and biochemical traits. *Legume Research, 46*(11), 1439–1444. https://doi.org/10.18805/LR-4833

Reddy, P. S., Mallikarjuna, G., Kaul, T., Chakradhar, T., Mishra, R. N., Sopory, S. K., & Reddy, M. K. (2010). Molecular cloning and characterization of gene encoding for cytoplasmic Hsc70 from *Pennisetum glaucum* may play a protective role against abiotic stresses. *Molecular Genetics and Genomics, 283*, 243–254.

Reddy, P. S., Reddy, G. M., Pandey, P., Chandrasekhar, K., & Reddy, M. K. (2012). Cloning and molecular characterization of a gene encoding late embryogenesis abundant protein from *Pennisetum glaucum*: Protection against abiotic stresses. *Molecular Biology Reports, 39*, 7163–7174.

Reddy, P. S., Thirulogachandar, V., Vaishnavi, C. S., Aakrati, A., Sopory, S. K., & Reddy, M. K. (2011). Molecular characterization and expression of a gene encoding cytosolic Hsp90 from *Pennisetum glaucum* and its role in abiotic stress adaptation. *Gene, 474*(1–2), 29–38.

Saballos, A., Sattler, S. E., Sanchez, E., Foster, T. P., Xin, Z., Kang, C. H., Pedersen, J. F., & Vermerris, W. (2012). Brown midrib2 (Bmr2) encodes the major 4-coumarate:coenzyme A ligase involved in lignin biosynthesis in sorghum (*Sorghum bicolor* (L.) Moench). *Plant Journal, 70,* 818–830.

Saleh, A. S., Zhang, Q., Chen, J., & Shen, Q. (2013). Millet grains: Nutritional quality, processing, and potential health benefits. *Comprehensive Reviews in Food Science and Food Safety, 12*(3), 281–295.

Salson, M., Orjuela, J., Mariac, C., Zekraouï, L., Couderc, M., Arribat, S., Rodde, N., Faye, A., Kane, N. A., Tranchant-Dubreuil, C., & Vigouroux, Y. (2023). An improved assembly of the pearl millet reference genome using Oxford Nanopore long reads and optical mapping. *G3: Genes, Genomes, Genetics, 13*(5), p.jkad051.

Sanghani, J. M., Sanghani, A. O., Kothari, V. V., Raval, S. S., & Kahodariya, J. H. (2018). The SSR based linkage map construction and identification of QTLs for blast (*Pyricularia grisea*) resistance in pearl millet (*Pennisetum glaucum* (l.) r. br). *Journal of Pharmacognosy and Phytochemistry, 7*(2), 3057–3064.

Sattler, S. E., Funnell-Harris, D. L., & Pedersen, J. F. (2010). Brown midrib mutations and their importance to the utilization of maize, sorghum, and pearl millet lignocellulosic tissues. *Plant Science, 178*(3), 229–238.

Sattler, S. E., Saathoff, A. J., Haas, E. J., Palmer, N. A., Funnell-Harris, D. L., Sarath, G., & Pedersen, J. F. (2009). A nonsense mutation in a cinnamyl alcohol dehydrogenase gene is responsible for the Sorghum brown midrib6 phenotype. *Plant Physiology, 150,* 584–595.

Satyavathi, C. T., Ambawat, S., Khandelwal, V., Govindaraj, M., & Neeraja, C. N. (2021a). *Micronutrient rich pearl millet for nutritionally secure India.* ICAR-All India Coordinated Research Project on Pearl Millet.

Satyavathi, C. T., Tiwari, S., Bharadwaj, C., Rao, A. R., Bhat, J., & Singh, S. P. (2013). Genetic diversity analysis in a novel set of restorer lines of pearl millet [*Pennisetum glaucum* (L.) R. Br] using SSR markers. *Vegetos, 26*(1), 72–82.

Sehgal, D., Rajaram, V., Armstead, I. P., Vadez, V., Yadav, Y. P., Hash, C. T., & Yadav, R. S. (2012). Integration of gene-based markers in a pearl millet genetic map for identification of candidate genes underlying drought tolerance quantitative trait loci. *BMC Plant Biology, 12,* 1–13.

Serba, D. D., & Yadav, R. S. (2016). Genomic tools in pearl millet breeding for drought tolerance: Status and prospects. *Frontiers in Plant Science, 7,* 1724.

Shan, Q., Wang, Y., Li, J., Zhang, Y., Chen, K., Liang, Z., Zhang, K., Liu, J., Xi, J. J., Qiu, J.-L., & Gao, C. (2013). Targeted genome modification of crop plants using a CRISPR-Cas system. *Nature Biotechnology, 31,* 686–688.

Sharma, P. C., Sehgal, D., Singh, D., Singh, G., & Yadav, R. S. (2011). A major terminal drought tolerance QTL of pearl millet is also associated with reduced salt uptake and enhanced growth under salt stress. *Molecular Breeding, 27,* 207–222.

Sharma, P. C., Singh, D., Sehgal, D., Singh, G., Hash, C. T., & Yadav, R. S. (2014). Further evidence that a terminal drought tolerance QTL of pearl millet is associated with reduced salt uptake. *Environmental and Experimental Botany*, *102*, 48–57.

Sharma, R., Upadhyaya, H. D., Manjunatha, S. V., Rai, K. N., Gupta, S. K., & Thakur, R. P. (2013). Pathogenic variation in the pearl millet blast pathogen *Magnaporthe grisea* and identification of resistance to diverse pathotypes. *Plant Disease*, *97*(2), 189–195.

Sharma, S., Sharma, R., Pujar, M., Yadav, D., Yadav, Y., Rathore, A., Mahala, R. S., Singh, I., Verma, Y., Deora, V. S., Vaid, B., Jayalekha, A. K., & Gupta, S. K. (2020). Use of wild Pennisetum species for improving biotic and abiotic stress tolerance in pearl millet. *Crop Science*, *61*(1), 289–304.

Sharma, V. K., Marla, S., Zheng, W., Mishra, D., Huang, J., Zhang, W., Morris, G. P., & Cook, D. E. (2022). CRISPR guides induce gene silencing in plants in the absence of Cas. *Genome Biology*, *23*, 6.

Shinde, H., Dudhate, A., Anand, L., Tsugama, D., Gupta, S. K., Liu, S., & Takano, T. (2020). Small RNA sequencing reveals the role of pearl millet miRNAs and their targets in salinity stress responses. *South African Journal of Botany*, *132*, 395–402.

Shinde, H., Tanaka, K., Dudhate, A., Tsugama, D., Mine, Y., Kamiya, T., Gupta, S. K., Liu, S., & Takano, T. (2018). Comparative de novo transcriptomic profiling of the salinity stress responsiveness in contrasting pearl millet lines. *Environmental and Experimental Botany*, *155*, 619–627.

Shivhare, R., Asif, M. H., & Lata, C. (2020a). Comparative transcriptome analysis reveals the genes and pathways involved in terminal drought tolerance in pearl millet. *Plant Molecular Biology*, *103*, 639–652.

Shivhare, R., Lakhwani, D., Asif, M. H., Chauhan, P. S., & Lata, C. (2020b). De novo assembly and comparative transcriptome analysis of contrasting pearl millet (*Pennisetum glaucum* L.) genotypes under terminal drought stress using illumina sequencing. *The Nucleus*, *63*, 341–352.

Siddaiah, C. N., Satyanarayana, N. R., Mudili, V., Gupta, V. K., Gurunathan, S., Rangappa, S., Huntrike, S. S., & Srivastava, R. K. (2017). Elicitation of resistance and associated defense responses in *Trichoderma hamatum* induced protection against pearl millet downy mildew pathogen. *Scientific Reports*, *7*, 43991.

Singh, A. K., Rana, M. K., Singh, S., Kumar, S., Kumar, D. & Arya, L., 2013. Assessment of genetic diversity among pearl millet [Pennisetum glaucum (L.) R Br.] cultivars using SSR markers. *Range Management and Agroforestry*, 34(1), pp. 77–81.

Singh, J. & Chhabra, A.K., 2018. Genetic variability and character association in advance inbred lines of pearl millet under optimal and drought condition. *Ekin Journal of Crop Breeding and Genetics*, 4(2), pp. 45–51.

Singh, S., & Gupta, S. K. (2019). Formation of heterotic pools and understanding relationship between molecular divergence and heterosis in pearl millet [*Pennisetum glaucum* (L.) R. Br.]. *PLoS One*, *14*(5), e0207463.

Singh, S. D. (1995). Downy mildew of pearl millet. *Plant Disease, 79*, 545–550. https://doi.org/10.1094/PD-79-0545

Singh, S. D., King, S. B., & Werder, J. (1993). *Downy mildew disease of pearl millet.* International Crops Research Institute for the Semi-Arid Tropics.

Singhal, T., Satyavathi, C., Tara Singh, S. P., Mallik, M., Mukesh, S. S., & Bharadwaj, C. (2022). Mapping and identification of quantitative trait loci controlling test weight and seed yield of pearl millet in multi agro-climatic zones of India. *Field Crops Research, 288*, 108701. https://doi.org/10.1016/j.fcr.2022.108701

Singhal, T., Satyavathi, C. T., Singh, S. P., Kumar, A., Sankar, S. M., Bhardwaj, C., Mallik, M., Bhat, J., Anuradha, N., & Singh, N. (2021). Multi-environment quantitative trait loci mapping for grain iron and zinc content using bi-parental recombinant inbred line mapping population in pearl millet. *Frontiers in Plant Science, 12*, 659789.

Soman, P., Jayachandran, R., & Bidinger, F. R. (1987). Uneven Variation in Plant-to-Plant Spacing in Pearl Millet 1. *Agronomy Journal, 79*(5), 891–895.

Soman, P., & Peacock, J. M. (1985). A laboratory technique to screen seedling emergence of sorghum and pearl millet at high soil temperature. *Experimental Agriculture, 21*(4), 335–341.

Spindel, J., Begum, H., Akdemir, D., Virk, P., Collard, B., Redona, E., . . . & McCouch, S. R. (2015). Genomic selection and association mapping in rice (Oryza sativa): effect of trait genetic architecture, training population composition, marker number and statistical model on accuracy of rice genomic selection in elite, tropical rice breeding lines. *PLoS genetics, 11*(2), e1004982.

Srivastava, R. K., Bollam, S., Pujarula, V., Pusuluri, M., Singh, R. B., Potupureddi, G., & Gupta, R. (2020a). Exploitation of heterosis in pearl millet: A review. *Plants, 9*(7), 807.

Srivastava, R. K., Satyavathi, C. T., Mahendrakar, M. D., Singh, R. B., Kumar, S., Govindaraj, M., & Ghazi, I. A. (2021). Addressing iron and zinc micronutrient malnutrition through nutrigenomics in pearl millet: Advances and prospects. *Frontiers in Genetics, 12*, 723472.

Srivastava, R. K., Singh, R. B., Pujarula, V. L., Bollam, S., Pusuluri, M., Chellapilla, T. S., & Gupta, R. (2020b). Genome-wide association studies and genomic selection in pearl millet: Advances and prospects. *Frontiers in Genetics, 10*, 1389.

Srivastava, R. K., Yadav, O. P., Kaliamoorthy, S., Gupta, S. K., Serba, D. D., Choudhary, S., & Varshney, R. K. (2022). Breeding drought-tolerant pearl millet using conventional and genomic approaches: Achievements and prospects. *Frontiers in Plant Science, 13*, 781524.

Stich, B., Haussmann, B. I., Pasam, R., Bhosale, S., Hash, C. T., Melchinger, A. E., & Parzies, H. K. (2010). Patterns of molecular and phenotypic diversity in pearl millet [*Pennisetum glaucum* (L.) R. Br.] from West and Central Africa and their relation to geographical and environmental parameters. *BMC Plant Biology*, *10*(1), 1–10.

Sumanth, M., Sumathi, P., Vinodhana, N. K., & Sathya, M. (2013). Cytosterile diversification and downy mildew resistance in pearl millet [*Pennisetum glaucum* (L.) R. Br]. *Indian Journal of Genetics and Plant Breeding*, *73*(03), 325–327.

Sun, M., Huang, D., Zhang, A., Khan, I., Yan, H., Wang, X., Zhang, X., Zhang, J., & Huang, L. (2020). Transcriptome analysis of heat stress and drought stress in pearl millet based on Pacbio full-length transcriptome sequencing. *BMC Plant Biology*, *20*(1), 1–15.

Sun, M., Lin, C., Zhang, A., Wang, X., Yan, H., Khan, I., Wu, B., Feng, G., Nie, G., Zhang, X., & Huang, L. (2021). Transcriptome sequencing revealed the molecular mechanism of response of pearl millet root to heat stress. *Journal of Agronomy and Crop Science*, *207*(4), 768–773.

Swami, P., Deswal, K., Rana, V., Yadav, D., & Munjal, R. (2023). Speed breeding— A powerful tool to breed more crops in less time accelerating crop research. In M. K. Khan, M. Hamurcu, S. Gezgin, A. Pandey, & O. P. Gupta (Eds.), *Abiotic stresses in wheat* (pp. 33–49). Academic Press.

Teklewold, A., & Becker, H. C. (2005). Heterosis and combining ability in a diallel cross of Ethiopian mustard inbred lines. *Crop Science*, *45*(6), 2629–2635.

Thakur, R. P., Rai, K. N., Khairwal, I. S., & Mahala, R. S. (2008). Strategy for downy mildew resistance breeding in pearl millet in India. *Journal of SAT Agricultural Research*, *6*.

Thakur, R. P., Rao, V. P., & Sharma, R. (2007). Evidence for temporal virulence change in pearl millet downy mildew pathogen populations in India. *Journal of SAT Agricultural Research*, *3*(1), 1–3.

Thakur, R. P., Rao, V. P., & Sharma, R. (2011). Influence of dosage, storage time and temperature on efficacy of metalaxyl-treated seed for the control of pearl millet downy mildew. *European Journal of Plant Pathology*, *129*(2), 353–359.

Thakur, R. P., Sharma, R., Rai, K. N., Gupta, S. K., & Rao, V. P. (2009). Screening techniques and resistance sources for foliar blast in pearl millet. *Journal of SAT Agricultural Research*, *7*, 1–5.

Tharanya, M., Kholova, J., Sivasakthi, K., Seghal, D., Hash, C. T., Raj, B., . . . & Vadez, V. (2018). Quantitative trait loci (QTLs) for water use and crop production traits co-locate with major QTL for tolerance to water deficit in a fine-mapping population of pearl millet (Pennisetum glaucum LR Br.). *Theoretical and applied genetics*, *131*, 1509–1529.

Till, B. J., Reynolds, S. H., Weil, C., Springer, N., Burtner, C., Young, K., Bowers, E., Codomo, C. A., Enns, L. C., Odden, A. R., Greene, E. A., Comai, L., & Henikoff, S. (2004). Discovery of induced point mutations in maize genes by tilling. *BMC Plant Biology*, *4*, 12.

Timper, P., Wilson, J. P., Johnson, A. W. & Hanna, W. W., 2002. Evaluation of pearl millet grain hybrids for resistance to Meloidogyne spp. and leaf blight caused by Pyricularia grisea. *Plant disease*, 86(8), pp. 909–914.

Tripathi, M. K., Yadav, D., Bhardwaj, R., Talwar, A. M., Tiwari, V. K., Kachole, U. G., Sravanti, K., Shanthi Priya, M., Athoni, B. K., Anuradha, N., Govindaraj, M., Nepolean, T., & Tonapi, V. A. (2021). Performance and stability of pearl millet varieties for grain yield and micronutrients in arid and semi-arid regions of India. *Frontiers in Plant Science*, *12*, 670201. https://doi.org/10.3389/fpls.2021.670201

Vadez, V., Hash, T., Bidinger, F. R., & Kholova, J. (2012). Phenotyping pearl millet for adaptation to drought. *Frontiers in Physiology*, *3*, 386.

Van Oosterom, E. J., Bidinger, F. R., & Weltzien, E. R. (2003). A yield architecture framework to explain adaptation of pearl millet to environmental stress. *Field Crops Research*, *80*(1), 33–56.

Van Oosterom, E. J., Mahalakshmi, V., Arya, G. K., Dave, H. R., Gothwal, B. D., Joshi, A. K., Joshi, P., Kapoor, R. L., Sagar, P., Saxena, M. B. L., Singhania, D. L., & Vyas, K. L. (1995). Effect of yield potential, drought escape and drought tolerance on yield of pearl millet (*Pennisetum glaucum*) in different stress environments. *Indian Journal of Agricultural Sciences*, *65*(9), 629–635.

Varshney, R. K., Shi, C., Thudi, M., Mariac, C., Wallace, J., Qi, P., Zhang, H., Zhao, Y., Wang, X., Rathore, A., & Srivastava, R. K. (2017). Pearl millet genome sequence provides a resource to improve agronomic traits in arid environments. *Nature Biotechnology*, *35*(10), 969.

Verma, D., Singla-Pareek, S. L., Rajagopal, D., Reddy, M. K., & Sopory, S. K. (2007). Functional validation of a novel isoform of Na+/H+ antiporter from *Pennisetum glaucum* for enhancing salinity tolerance in rice. *Journal of Biosciences*, *32*(3), 621–628.

Verma, R., Tripathi, M. K., Tiwari, S., Pandya, R. K., Tripathi, N., & Parihar, P. (2021a). Screening of pearl millet [*Pennisetum glaucum* (L.) R. Br.] genotypes against blast disease on the basis of disease indexing and gene-specific SSR markers. *International Journal of Current Microbiology and Applied Sciences*, *10*(02), 1108–1117.

Vignols, F., Rigau, J., Torres, M. A., Capellades, M., & Puigdomènech, P. (1995). The brown midrib3 (bm3) mutation in maize occurs in the gene encoding caffeic acid O-methyltransferase. *Plant Cell*, *7*, 407–416.

Vinutha, K. S., Khan, A. A., Ravi, D., Prasad, K. V. S. V., Reddy, Y. R., Jones, C. S., & Blümmel, M. (2021). Comparative evaluation of sorghum and pearl millet forage silages with maize. *Animal Nutrition and Feed Technology*, *21*, 1–14.

Wang, J. Y., & Doudna, J. A. (2023). CRISPR technology: A decade of genome editing is only the beginning. *Science*, *379*, eadd8643.

Watson, A., Ghosh, S., Williams, M. J., Cuddy, W. S., Simmonds, J., Rey, M. D., . . . & Hickey, L. T. (2018). Speed breeding is a powerful tool to accelerate crop research and breeding. *Nature plants, 4*(1), 23–29.

Wilson, J. P., & Gates, R. N. (1993). Forage yield losses in hybrid pearl millet due to leaf blight caused primarily by *Pyricularia grisea. Phytopathology, 83*(7), 739–744.

Wilson, J. P., Wells, H. D., & Burton, G. W. (1989). Inheritance of resistance to *Pyricularia grisea* in pearl millet accessions from Burkina Faso and inbred Tift 85DB. *The Journal of Heredity, 80*(6), 499–501.

Xin, Z., Li Wang, M., Barkley, N. A., Burow, G., Franks, C., Pederson, G., & Burke, J. (2008). Applying genotyping (TILLING) and phenotyping analyses to elucidate gene function in a chemically induced sorghum mutant population. *BMC Plant Biology, 8*, 103.

Yadav, O. P., Gupta, S. K., Govindaraj, M., Sharma, R., Varshney, R. K., Srivastava, R. K., Rathore, A., & Mahala, R. S. (2021). Genetic gains in pearl millet in India: Insights into historic breeding strategies and future perspective. *Frontiers in Plant Science, 12*, 396.

Yadav, O. P., & Rai, K. N. (2011). Hybridization of Indian landraces and African elite composites of pearl millet results in biomass and stover yield improvement under arid zone conditions. *Crop Science, 51*(5), 1980–1987.

Yadav, O. P. & Rai, K. N., 2013. Genetic improvement of pearl millet in India. *Agricultural Research, 2*, pp. 275–292.

Yadav, O. P., Singh, D. V., Vadez, V., Gupta, S. K., Rajpurohit, B. S., & Shekhawat, P. S. (2017). Improving pearl millet for drought tolerance–Retrospect and prospects. *Indian Journal of Genetics and Plant Breeding, 77*(4), 464–474.

Yadav, R. S., Allen, D. K., Mathews, R., & Macduff, J. (2011). QTL analysis of components of nitrogen use efficiency in a perennial ryegrass mapping family under high and low nitrate supply in flowing solution culture. In *29th Fodder Crops and Amenity Grasses Section Meeting*. EUCARPIA.

Yadav, R. S., Hash, C. T., Bidinger, F. R., Cavan, G. P., & Howarth, C. J. (2002). Quantitative trait loci associated with traits determining grain and stover yield in pearl millet under terminal drought-stress conditions. *Theoretical and Applied Genetics, 104*, 67–83.

Yadav, R. S., Hash, C. T., Bidinger, F. R., Devos, K. M., & Howarth, C. J. (2004). Genomic regions associated with grain yield and aspects of post-flowering drought tolerance in pearl millet across stress environments and tester background. *Euphytica, 136*, 265–277.

Yadav, R. S., Sehgal, D., Nepolean, T., Vadez, V., Hash, C. T., Sharma, P. C., & Khairwal, I. S. (2009). Translational genomics in breeding increased drought tolerance in pearl millet. In *Biospectrum 2009, International Symposium 'Second Green Revolution: Priorities, Programmes, Social and Ethical Issues'*. Rajiv Gandhi Centre for Biotechnology.

Yan, H., Sun, M., Zhang, Z., Jin, Y., Zhang, A., Lin, C., Wu, B., He, M., Xu, B., Wang, J., & Qin, P. (2023). Pangenomic analysis identifies structural variation associated with heat tolerance in pearl millet. *Nature Genetics*, *55*(3), 507–518.

Zhu, H., Li, C., & Gao, C. (2020). Applications of CRISPR–Cas in agriculture and plant biotechnology. *Nature Reviews Molecular Cell Biology*, *21*(11), 661–677.

Zong, Y., Wang, Y., Li, C., Zhang, R., Chen, K., Ran, Y., Qiu, J.-L., Wang, D., & Gao, C. (2017). Precise base editing in rice, wheat and maize with a Cas9-cytidine deaminase fusion. *Nature Biotechnology*, *35*, 438–440.

Kapadia, V.N., 2016. Estimation of Heterosis for Yield and Its Relevant Traits in Forage Pearl Millet [Pennisetum glaucum LR Br.]. *International Journal of Agriculture Sciences, ISSN*, pp.0975-3710.

5

Germplasm Utilization and Pre-breeding in Pearl Millet

P. Sanjana Reddy

Chapter Overview

Pearl millet [*Pennisetum glaucum* (L.) R. Br.] is the world's hardiest cereal crop catering to the food and nutritional security for people in the arid and semi-arid tropics. The crop is resilient to biotic and abiotic stresses and poor soil conditions. Efforts to increase the yield potential of pearl millet have also increased its vulnerability to stresses. Pre-breeding is a necessary step in linking genetic diversity to utilization. Pre-breeding activities range from the evaluation of plant genetic resources to identify donors for desirable agronomic traits to the transfer of these target traits into well-adapted genetic backgrounds by hybridizations to generate populations that can be used for actual breeding programs. A total of 56,580 accessions (including possible duplicates) of pearl millet are conserved in 70 gene banks of 46 countries. The revised core comprising 2,094 accessions, a minicore collection of pearl millet comprising 238 accessions, and a pearl millet germplasm association panel comprised of 250 accessions are available and widely used for research studies. About 8,460 accessions of pearl millet have been evaluated by the National Agricultural Research Systems program of India, and 20,844 accessions from 51 countries were evaluated for 23 traits at ICRISAT. Development of ICTP 8203, a widely cultivated variety in India, from an *iniadi* landrace from northern Togo is a milestone in demonstrating the utility of the landrace. The wild germplasm, *Pennisetum glaucum* ssp. Monodii, has been exploited as a new source of cytoplasmic-nuclear male sterility, and *Pennisetum purpureum* Schum., also called Napier grass, has been used to develop plants with high biomass–yielding perennial forage. The chapter explains several such advances made in trait exploitation making use of germplasm, landraces and wild species to develop nutrient-dense, biotic and abiotic stress-resistant grain, and forage pearl millet cultivars that are suitable for harsh agro-ecologies and low-input situations in Asia and Africa.

Pearl Millet: A Resilient Cereal Crop for Food, Nutrition, and Climate Security, First Edition.
Edited by Ramasamy Perumal, P. V. Vara Prasad, C. Tara Satyavathi, Mahalingam Govindaraj, and Abdou Tenkouano.

Importance of Genetic Resources in Pearl Millet

Crop improvement helps sustain human life, in which genetic resources are the key components utilized by plant breeders. Biodiversity helps in sustaining production under changing climates and global warming and is not limited to any geographic area or a country to meet its requirements. Over time, conventional breeding has led to a reduction in genetic diversity that has resulted in yield plateaus and an increase in vulnerability to biotic and abiotic stresses. Cereal landraces conserved in marginal lands have enhanced the scope for incorporating adaptation traits and addressing yield constraints. The ability to adapt to ever-changing pest and disease pressures and climatic conditions has been greatly aided by crop wild relatives (Dempewolf et al., 2017). Pre-breeding with crop wild relatives is essential to preserve genetic diversity in the natural and current forms and to ensure global food security. Apart from this, genetic diversity could be obtained from domesticated or semi-domesticated cultivars that are currently in use or that have become obsolete, landraces, or historic varieties (Rao, 2004).

Many countries store their germplasm in gene banks. However, only roughly 30 countries have safe long-term storage of their germplasm due to the shortage of long-term maintenance provisions in many countries (Kell et al., 2017). The international collections hosted by the 11 Consultative Group on International Agricultural Research (CGIAR) Centers include over 760,000 accessions of crops, forages, and trees that were originally obtained from 207 countries as well as pre-bred materials. The Genebanks CRP was conceived as a partnership between these 11 centers (represented by the CGIAR Consortium) and the Crop Trust. The objective of the program is to conserve the diversity of plant genetic resources in CGIAR-held collections and to make this diversity available to breeders and researchers in a manner that meets high internationally agreed-upon genebank standards; that maximizes cost efficiency, security, reliability, and sustainability over the long term; and that is supportive of and consistent with the Treaty.

Pearl millet [*Pennisetum glaucum* (L.) R. Br.] is a C4 cross-pollinated millet crop usually grown in marginal environments, including the hottest and driest regions in India and Africa. It is the sixth important cereal crop after wheat (*Triticum aestivum* L.), maize (*Zea mays* L.), rice (*Oryza sativa* L.), barley (*Hordeum vulgare* L.), and sorghum [*Sorghum bicolor* (L.) Moench] in the world. Pearl millet is grown in an area of 30.93 m ha (FAOSTAT, 2021), providing food to almost 90 million poor people (Shivhare & Lata, 2017). In terms of global statistics on millets, representing mostly pearl millet, India, with 9.76 m ha under cultivation, occupies 31.6% of the total global millet area, followed by Niger, Sudan, Mali, Nigeria, and Chad. India is also the largest producer, with 13.21 m t, which is 43.9% of global millet production (30.09 m t). The other major pearl millet–producing countries are China (2.7 m t), Niger (2.15 m t), Nigeria (1.92 m t), Sudan (1.5 m t), and Mali (1.49 m t).

The inherent characteristics of the crop to survive under extreme heat and being tolerant to drought and poor soil conditions makes it the only option in the arid and semi-arid tracts where other cereal crops fail to grow. Pearl millet is less susceptible to biotic stresses. Pearl millet grain is gluten free and has an impressive nutritional profile, thereby addressing nutritional security in marginal areas of the world.

Despite impressive progress made in pearl millet with respect to developing and deploying high-yielding cultivars by public and private sector partnerships in addition to improved agronomic management practices, the productivity of pearl millet continues to be as low as ~1.2 t/ha (Ramu et al., 2023). Pearl millet possesses a great deal of genetic diversity in terms of agronomic and yield traits, adaptability, and quality attributes. Plant genetic resources must be collected, conserved, characterized, evaluated, documented, and distributed in order to be available for use in plant breeding programs. This chapter discusses the genetic diversification that has taken place through pre-breeding in pearl millet.

Domestication, Progenitors, Gene Pool, and Genetic Diversity

Origin and Domestication

Due to the existence of large genetic diversity within the species and the presence of cross-fertile wild species, it is widely believed that pearl millet originated in Africa, which is considered as the primary center of diversity for pearl millet. According to Harlan (1975), the Sahelian Zone, which stretches from western Sudan to Senegal, appears to be the center of origin. Pearl millet has reportedly been cultivated for thousands of years (Brunken, 1977; Burton & Powell, 1968; Rachie & Majmudar, 1980). In his study on the origin of crops, Vavilov (1949/1950) proposed Ethiopia as the center of domestication for pearl millet. Two theories contradict the Ethiopian origin: (a) a wild progenitor (*Pennisetum americanum* ssp. *monodii*) found in east of Sudan has never been collected in the Ethiopian highlands, and (b) the pearl millet of Ethiopia lacks sufficient morphological diversity. Hence, Ethiopia is seen as the result of post-domestication introduction. Between 5000 and 4000 BCE, the Mande people lived close to the headwaters of the Niger River, where they domesticated a number of Western African crops, including pearl millet (Murdock, 1959). Western Africa, south of the Saharan desert and north of the woodland zone, is currently home to the most morphological diversity (Anand Kumar & Appa Rao, 1987). From Western Africa where it was first domesticated, it went to Eastern Africa, and about 2,000 years ago, it reached Southern Africa and India. India is regarded as the secondary center for pearl millet diversity.

According to the shape of the seeds, cultivated pearl millet can be divided into four major races: *typhoides, nigritarum, globosum,* and *leonis* (Brunken, 1977). However, there are also intermediate forms for any of these basic types. The race *typhoides* is the most diverse and extensively spread of the four races. It is the only fundamental race discovered outside of Africa and is the dominant race grown in southern Africa (Appa Rao et al., 1986) and India (Kumari et al., 2016). The race *nigritarum* can be found in northern Nigeria and western Sudan. According to Bono (1973), the race *globosum* is the most prevalent in central Nigeria, Niger, Ghana, Togo, and Benin. Sierra Leone is the home for the *leonis* race. The grain's substantially elongated characteristic may be an adaptation to the areas of Sierra Leone with comparatively high rainfall.

Taxonomic Classification

The scientific binomials of pearl millet have been shuffled by various taxonomists. During the early 1900s, pearl millet was referred as *Pennisetum typhoideum, Penicillaria spicata, Panicum spicatum,* and *Pennisetum alopecuroides* (Chase, 1921). The binomials *P. typhoideum* L. C. Rich. and *P. americanum* were used by several workers by the mid-nineteenth century. In the 1960s, pearl millet was referred as *Pennisetum typhoides* (Burton & Powell, 1968), and the name *P. glaucum* (L.) R. Br. Was adopted by Hitchcock and Chase (1951) in *Manual of the Grasses of the United States*. Scientists currently engaged in research on pearl millet use this name.

Based on its clustered spikelets that are encircled by an involucre, taxonomists place the genus *Pennisetum* in the tribe Paniceae, close to the genus Cenchrus. Gymnothrix, Brevivalvula, Penicillaria, Heterostachya, and Eu-Pennisetum are the five sections that make up the genus. The section Penicillaria contains all Pennisetum species with penicillate anther tips and pinnate involucral bristles (Leeke, 1907). The genus *Pennisetum* contains two species that are reproductively isolated, according to Brunken (1977). These include *P. americanum* (L.) Leeke, a diploid (2n = 14) that is indigenous to the semi-arid tropics of Africa and India, and *Pennisetum purpureum* Schumach, a perennial tetraploid (2n = 28) species that is found throughout the wet tropics of the world. Three subspecies make up the species *americanum*. The cultivated pearl millet falls within the subspecies *americanum*. The name *P. americanum* has been changed to *P. glaucum* (L.) R. Br. According to the current taxonomical classification, cultivated pearl millet is placed under family *Poaceae*, subfamily *Panicoideae*, tribe *Paniceae*, subtribe *Panicinae*, section *Penicillaria*, genus *Pennisetum*, and species *glaucum*.

Another system of classification proposed by Harlan and de Wet (1971), based on genepools, groups the total variation into primary, secondary, and tertiary genepools. All cultivars of *P. glaucum*, its subspecies, wild progenitor *violaceum* (monodii), and weedy form *stenostachyum* are included in the primary genepool because these taxa naturally cross with the cultivated species. The allotetraploid, rhizomatous perennial species *P. purpureum* Schum., also called as Napier grass, and the apomictic and octaploid species *Pennisetum squamulatum* Fresen. are included in the secondary genepool. They can be easily crossed with cultivated pearl millet, but their hybrids are sterile. The species that are cross-incompatible with pearl millet are all included in the tertiary genepool (Sharma et al., 2020a).

Pearl Millet Genetic Resources and Diversity

The collection, conservation and maintenance of germplasm helps in maintaining genetic resources and catering to the needs of crop improvement programs globally. The Rockefeller Foundation and the Indian Agricultural Research Institute made the first attempts to pool world pearl millet germplasm in 1959–1962. Later, in an intensive campaign led by ICRISAT, 10,764 accessions were gathered from 65 organizations that included different national and international institutes, universities, national agricultural research systems, etc. This includes 2,178 accessions from Institute Francais de Recherche Scientifique pour le development en Cooperation 2,022 accessions from the Rockefeller Foundation, New Delhi, India; and 974 accessions from the International Bureau for Plant Genetic Resources (IBPGR), Rome, Italy. The gaps in the collection and the priority areas for systematic collection of germplasm in different countries were identified and focused, which led to the collection of 10,830 pearl millet samples during 76 collection missions in 28 countries. By the end of 2006, the genebank at ICRISAT had 21,594 accessions from 51 countries, including 750 accessions belonging to 24 species of genus *Pennisetum*. This can be classified as samples from institutions (10,201 accessions), farmers' fields (6,537 accessions), commercial markets (1,681 accessions), farmer's stores (1,357 accessions), threshing floors (479 accessions), and wild species (750 accessions). Biological status of accessions indicates the presence of landraces (18,447), breeding materials (2,268), advanced cultivars (129), and wild relatives (750) (Upadhyaya et al., 2007). The pearl millet germplasm landraces were characterized in batches of 500–1,000 every year at ICRISAT farm, Patancheru (17.53 N lat, 78.27 E long, 545 m asl), in alfisols, during the rainy and summer seasons from 1974 through 2013, for 21 morphoagronomic traits following descriptors for pearl millet (IBPGR and ICRISAT, 1993). Landraces with early flowering (33–40 days) were predominant in Pakistan, Ghana, Togo, and India; with very late flowering (121–159 days) in Sierra Leone and the Central African Republic; with short plant height (80–100 cm) in India, Zambia, and Sudan; with tallness (401–490 cm) in Chad, Burkina Faso, Nigeria, and the Central African Republic; with high tillering (11–35) in India and Yemen; with high panicle exertion (11–29 cm) in Ghana, Chad, India, and Yemen; with long panicles (75–135 cm) in Nigeria and Niger; with thick panicles (41–58 mm) in

Namibia, Togo, and Zimbabwe; and with large seeds (16–19 g per 1,000) in Togo, Benin, Ghana, and Burkina Faso. Collections from Ghana for flowering (36–150 days); Burkina Faso for plant height (80–490); India and Yemen for total (1–35) and productive (1–19) tillers per plant; Niger for panicle exertion (−45 to 21.0 cm), panicle length (9–135 cm), and thickness (12–55 mm); and Zimbabwe for 1,000-seed weight (3.5–19.3 g) were important sources for trait diversity (Upadhyaya et al., 2016). The variability among these accessions studied over years and pooled by ICRISAT is given in Table 5.1.

In the USDA Germplasm Resources Information Network, 1,283 active collections are being maintained, and around 75% of them belong to Zimbabwe, India, Nigeria, and Burkina Faso.

The National Bureau of Plant Genetic Resources (NBPGR), India, conserves around 9,000 accessions of pearl millet under long-term conservation in its genebank. Most of these accessions are indigenous (8,827), with only 168 accessions from other countries (Kumari et al., 2016)

Thus, a total of 56,580 accessions (including possible duplicates) of pearl millet are available worldwide that are conserved in 70 genebanks of 46 countries. Most of these germplasm accessions include landraces, followed by breeding/research material and wild relatives (Yadav et al., 2017).

Table 5.1 Range of Variation for Important Agronomic Characters of Pearl Millet Germplasm Assembled at ICRISAT Genebank in The Rainy Season (R) and Summer Season (S).

Character	Minimum	Maximum	Mean	Variance
Time to flower, days (R)	33.0	159.0	72.8 ± 0.17	569.75
Time to flower, days (S)	32.0	138.0	71.4 ± 0.08	123.47
Plant height, cm (R)	30.0	490.0	246.2 ± 0.46	4,427.63
Plant height, cm (S)	25.0	425.0	160.1 ± 0.25	1,311.57
Total tillers, n (R)	1.0	35.0	2.7 ± 0.01	3.13
Productive tillers, n (R)	1.0	19.0	2.1 ± 0.01	1.28
Panicle exertion, cm (R)	−45.0	29.0	3.7 ± 0.05	43.34
Panicle length, cm (R)	5.0	135.0	28.2 ± 0.07	112.90
Panicle length, cm (S)	4.0	125.0	25.4 ± 0.07	109.62
Panicle width, mm (R)	8.0	58.0	24.0 ± 0.03	22.75
Panicle width, mm (S)	9.0	61.0	22.9 ± 0.04	25.73
1,000-seed weight, g (S)	1.50	21.30	8.5 ± 0.02	4.98

Note: Adapted from Upadhyaya et al. (2007)

Critical assessment of collection for geographical and trait-diversity gaps using various GIS tools has revealed several gaps in germplasm collection from Asian and African continents. Almost all cultivated accessions have been characterized for 23 morpho-agronomic characters following prescribed pearl millet descriptors. A large variation exists for phenotypic and phenological traits among available germplasm. In general, Indian pearl millet landraces have mainly contributed to earliness, high tillering, high harvest index, and local adaptation, whereas African landraces have been a good source of bigger panicles, large seed size, and disease resistance. Systematic evaluation and screening of germplasm has led to the identification of specific sources of better grain quality, resistance to diseases, and tolerance to abiotic stresses (e.g., drought and heat). These germplasm sources continue to play a critical role in crop improvement programs across the world. Formation of trait-specific gene pools as well as core and minicore collections is likely to enhance the use of genetic resources to a greater degree (Yadav et al., 2017).

Core and Mini-core Collections

At ICRISAT, Hyderabad, the 16,000 accessions were initially stratified according to geographical distribution, followed by hierarchical clustering on 11 quantitative traits using Ward's method, which resulted in 25 distinct groups. Approximately 10% accessions were then randomly selected from each of these 25 distinct groups to form a core collection of 1,600 accessions by 1998. This collection was augmented by adding 501 accessions representing 4,717 accessions assembled and characterized over the next 9 years. The revised core consisted of 2,094 accessions. A minicore collection of pearl millet comprising 238 accessions was constituted by using data on 10 qualitative and 8 quantitative traits of 2,094 core collection accessions from 46 countries (Upadhyaya et al., 2011).

The Pearl Millet Germplasm Association Panel, which includes 250 landraces, elite cultivars, and mapping population parents collected from a wide geographical range in Africa and Asia, was recently established from the global pearl millet germplasm collection of ICRISAT (Yadav et al., 2011).

Trait-specific Germplasm

Trait-specific evaluation of pearl millet genetic resources is important for enhancing utilization by plant breeders and refining genetic resource management strategy. A total of 221 Indian pearl millet collections from the National Genebank were characterized and evaluated for 27 agro-morphological descriptors. This study led to identification of unique germplasm for specific traits, such as disease tolerance, early maturity, spike length, yield, dual purpose type, popping characteristics, and antioxidant properties. Frequency distribution analysis showed the predominance of cylindrical and compact spike, grey seeds, and earliness (<40 days to spike emergence). Four accessions of pearl millet germplasm—IC309064, IC393365, IC306465, and IC283866—were observed to be multiple disease resistant (Kumari et al., 2016) (Table 5.2).

Table 5.2   Trait-specific germplasm identified from several studies.

Trait	Germplasm
Early flowering and maturity	IC343664, IC343689, IC343661, IC309064 (<40), IP 13520, IP 9532, IP 18132, IP 17720
Productive tillers, *n*	IC309870, IC283763, IC326053 (>9); IP 20349, IP 21155 (>12)
Seed size, g	IC283725, IC283734, IC283681, IC283692, IC283693, IC283842 (>1.4)
Spike length, cm	IC393365, IC283740, IC332702 (>26)
Spike girth, cm	IC283693, IC283842, IC367638 (>3), IP 16403, IP 12845 IP 13363, IP 19386, IP 19448 (>5 cm)
Yield per plant, g	IC283737, IC306463, IC283842, IC332702 (>160)
Green fodder potential	IP 257995, IP13528, IP13566, IC332706, and IC275069
Dual purpose	IC283705, IC283745, IC283885, and IC335901
Resistant to blast and downy mildew disease	IC309064, IC393365, IC306465, and IC283866
Popping yield, %	IC283896, IC306465, IC283848, IC283882, IC283734 (>80) (Kumari et al., 2016), IC 283734, IC 283745, IC 283763, IC 283842, IC 283908, 841-B, and PPMI 301
Popped size, mm	IC283705, IC283734, IC283740, IC326053, IC343664, IC283844 (>8)
Popping expansion	IC283705, IC335902, IC283740, IC283734, IC283841, IC283848 (>3)
Puffing index	IC283705, IC343703, IC283842, IC283734, IC335902 (>9)
Flavonoids, mg/100 g	IC283693, IC283702, IC332707 (>216)
Total phenolics, mg/100 g	IC283693, IC332706, IC332700 (>307)
Ferric ion reducing antioxidant power, m mol TE/g	IC283847, IC332706, IC332715 (>10.34)
Salinity tolerance	IP 6101, IP 3732, IP 3757, IP 6102
Thermo tolerance	IP 21517, IP 3175
Drought tolerance	IP 8955, IP 9406
Sweet stalk	IP 3471, IP 14439, IP 13817, IP 12128
Purplish-black seed color	IP 10759, IP 13324
Yellow endosperm	IP 15536, IP 15533

Note: Adapted from Kumari et al. (2016), Chauhan et al. (2015), and Sehgal et al. (2015).

Utilization of Germplasm

Utilization of Wild Species

Introgression of genome of wild germplasm is used for creating diversity in cultivated germplasm. After several generations of backcrossing and selection, larger introgressions carrying favorable traits, as well as cryptic introgressions, are present throughout the genome. The success depends both genotype and ploidy level. With clearly defined objectives and availability of good screening or selection techniques, a team approach is used for testing and evaluating large populations. With highly heritable characteristics and multiple cycles per year, there is a greater chance of success when using wild species to improve cultivated species. Difficulty levels of transfer increase from using wild subspecies in the primary gene pool to using wild species in the tertiary gene pool.

The ICRISAT genebank maintains 739 accessions of 23 wild species, while the USDA-ARS Southern Regional Plant Introduction Station, Griffin, Georgia, USA, holds over 90 accessions of 21 wild species (USDA-ARS, 2002), of which the majority of lines belong to *P. glaucum* ssp. *monodii* (syn. *Pennisetum violaceum*), followed by *Pennisetum pedicellatum* and *Pennisetum polystachion* (Rao et al., 2003). There has been a general lack of interest in using wild species because of the large amount of genetic variability available in pearl millet landraces.

Pennisetum glaucum ssp. *monodii* has been exploited as a new source of cytoplasmic-nuclear male sterility (CMS). CMS sources have been identified in populations derived from crosses between GP1 wild relative [*P. glaucum* ssp. *monodii* (=*P. violaceum*)] as the female parent and cultivated pearl millet as the male parent. The discovery of A1 CMS at Tifton, Georgia, USA (Burton, 1958), initiated the era of hybrid cultivar development in pearl millet [*P. glaucum* (L). R. Br.], which led to the release of the first grain hybrid in India in 1965 (Athwal, 1965). The germplasm accessions from Senegal were identified as source of A_v (Marchais & Pernes, 1985). Hanna (1989) identified an A4 CMS system at Tifton, Georgia, USA, in a wild grassy *P. glaucum* (L.) R. Br. ssp. *monodii* (Maire) Brunken. The cytoplasm from *monodii* influences dry matter yields, providing a unique opportunity to increase productivity of commercial forage hybrids (Hanna, 1997). A few accessions of *P. glaucum* ssp. *monodii* (PS 202, 637, 639, and 727) are good sources of resistance to a cereal parasitic weed, *Striga hermonthica*, in sub-Saharan West Africa. PS 202 is also a source of resistance to one of the most devastating fungal diseases of pearl millet, downy mildew (DM) (Wilson et al., 2004). By pollination *Pennisetum schweinfurthii* × pearl millet hybrids with pearl millet pollen, the cytoplasm of *P. schweinfurthii* has also been transmitted to pearl millet (Hanna & Dujardin, 1985).

Pennisetum purpureum (napiergrass) is a rhizomatous perennial with vigorous growth, pest resistance, and forage yield. It crosses with cultivated pearl millet to produce sterile triploid hybrids, which are further propagated vegetatively. Genes for controlling earliness, long inflorescence, leaf size, and stiff stalk and restorer genes of the A_1 CMS system have also been transferred from *P. purpureum* (Dujardin & Hanna, 1989). The species *Pennisetum mezianum* is used for drought tolerance, *Pennisetum orientale* for drought tolerance and forage, *Pennisetum schweinfurthii* for large seeds, and *Pennisetum pedicellatum* and *Pennisetum polystachion* for DM resistance (Rai et al., 1997). Recently, new sources of blast resistance and flowering-stage heat tolerance have been developed in cultivated pearl millet backgrounds using wild *P. glaucum* ssp. *violaceum* (Sharma et al., 2020b).

Apomictic but highly sterile hybrids were reported between pearl millet and *Pennisetum setaceum* (triploid) and *P. orientale* (tetraploid). The high sterility and low expression of apomixis in these species prevented their use as a source of genes for apomixis, despite the fact that interspecific hybrids are robust. The hexaploid obligate apomictic species *P. squamulatum* was crossed successfully to a tetraploid pearl millet, and several partially male sterile, obligate apomictic interspecific derivatives were produced (Hanna, 1992). *Pennisetum squamulatum*, a hexaploid species ($2n = 6x = 54$), was the third species to be tried in the germplasm transfer program. The hybrid between pearl millet × *P. squamulatum* was male and female sterile. When tetraploid pearl millet was used in a crossing and backcrossing program, hundreds of partially male-fertile, obligate apomictic, interspecific hybrids and backcrossed derivatives could be produced (Dujardin & Hanna, 1983, 1985a). Research on *P. squamulatum* revealed that this species was an obligate apomict (Dujardin & Hanna, 1984). As "bridges" to transfer genetic material, double-cross hybrids between *P. squamulatum*, napiergrass, and pearl millet have also been created (Dujardin & Hanna, 1984, 1985b). Altering complete genome sets has been the main goal of creating an apomictic pearl millet. However, it is important to consider transferring genes, chromosomal fragments, or even a single chromosome to pearl millet using gamma radiation, cell culture, and cytological and genetic methods.

Utilization of Landraces

Indian landraces provide excellent sources of early maturity, better tillering, and shorter plants. In contrast, African sources, particularly those from the West African region, are excellent sources of large head volume and seed size, higher degrees of resistance to diseases, and better seed quality. The increasing use of the African germplasm at ICRISAT and in almost all Indian National Programs has substantially contributed to the diversification of the genetic base of breeding programs (Harinarayana & Rai, 1989; Harinarayana et al., 1988). Donor parents like 863B (IP 22303), P 1449-2 (IP 21168), ICMB 90111 (IP 22319), ICMP 451 (IP 22442), and IP 18293 were identified as sources for resistance against different pathotypes of DM disease in India (Upadhyaya et al., 2007).

Before the establishment of the All India Coordinated Millets Improvement Project (AICMIP) in 1965 in India, mass selection led to the development of several varieties (Co 2, Co 3, AKP 1, AKP 2, RSJ, RSK, N-28-15-1) from Indian landraces and Co 1 and S 530 from African landraces. Pearl millet germplasm assembled at the Indian Agricultural Research Institute, New Delhi, during the 1960s led to the development of varieties Improved Ghana and Pusa Moti by mass selection. The pearl millet hybrid era in India started in 1962 with the introduction of the male-sterile line Tift 23A from Tifton, Georgia, USA. Five hybrids (HB 1, HB 2, HB 3, HB 4, and HB 5), based on this line, were released during 1965–1969, and three hybrids (BJ 104, BK 560, and CJ 104) were released in 1977. During 1981–1984, another 10 hybrids were released, of which only three were of significance (MBH 118, GHB 27, and GHB 32). Following the breakdown of 5141 A and J 104 to DM, four hybrids (ICMH 451, ICMH 501, MH 182, HHB 30) were released in 1986 and three others (ICMH 423, MH 169, HHB 50) in 1987. Of these, only ICMH 451 has been cultivated widely.

The *iniadi* landrace is widely distributed in northern Togo and adjoining western Benin, southern Burkina Faso, and eastern Ghana regions (Andrews & Anand Kumar, 1996). Besides early maturity (75–85 days), this landrace is among the germplasm least sensitive to longer days for flowering; has fewer tillers (1–2 per plant); has compact, conical panicles with good exertion; and is typically lustrous, mostly dark grey with globular grains of large size (1,000-seed weight, 13–20 g). The widespread use of this landrace in pedigree and population breeding in India, southern and eastern Africa, western and central Africa, and the United States, has led to the development and release of several open-pollinated varieties (OPVs) and parental lines of hybrids (Andrews & Anand Kumar, 1996). Development of ICTP 8203 by random mating five S2 progenies selected at ICRISAT from an *iniadi* landrace from northern Togo is a milestone in demonstrating the utility of this landrace. This variety was released in India in 1988 (Rai et al., 1990). It was cultivated on 0.6–0.7 million ha at the peak of its adoption in Maharashtra State, India (Bantilan et al., 1998). It is still cultivated on 300,000 ha and retains its initial high level of DM resistance (<2% disease incidence in farmers' fields over two decades).

Among the progenies that contributed to the development of ICTP 8203, one S1 progeny (designated as Togo 13-4) during the inbreeding and selection process was found to be an excellent maintainer of an A_1-system male-sterile line, 81A. It also had a high level of DM resistance in the disease nursery (<5% disease incidence) at ICRISAT, Patancheru. An A/B pair of the male-sterile F_1 (81A×Togo 13-4) and the maintainer (B) line (Togo 13-4) was established in 1981 summer (dry) season. Individual plants selected from the S_1 progeny were further selfed and backcrossed onto individual plants of the sterile hybrid. Selfing and selection in the maintainer progeny with concurrent backcrossing onto individual plants of the sterile backcross progeny was continued for six generations. It led to the identification of a family comprising six male-sterile BC_6 progenies and six maintainer S_8 lines, of which one pair was designated as 863A and 863B. This A/B pair was made available to research programs in the National Agricultural Research System and the private sector in India in 1987, who made extensive use of it in breeding hybrids. Line 863A or its sub-selections are the seed parents of three commercial hybrids developed and marketed by three private seed companies in India.

Despite the development of elite breeding material, the study and exploitation of landraces in pearl millet breeding programs continued. Stich et al. (2010) reported patterns of diversity in flowering, photoperiod response, panicle length, and population structure differentiation in 145 inbreds derived from 122 Western and Central African landraces. These landraces also exhibited exceptional buffering capacity against variable environmental conditions, while the landraces from Cameroon, Togo, and Ghana were found to be good sources of earliness and bigger seeds. On the other hand, landraces from Yemen were found to be potential sources of variation for early maturity, short stature, large seeds, and cold tolerance. Similarly, Yadav and Bidinger (2008) evaluated 169 landraces from India for their grain and stover yield and discerned significant differences among them for the examined parameters, such as early flowering, profuse tillering, more panicles, and larger seeds. In fact, several landraces outperformed controls in terms of grain and stover yield and were thus considered as potential resources for developing dual-purpose hybrids adapted for arid climatic conditions.

Climate change has resulted in extremes in the rainfall pattern. Drought, caused by the low rainfall and its erratic distribution, is the primary abiotic production constraint in South Asia and sub-Saharan Africa. Efforts have therefore been made for mapping and delineation of drought-prone regions to define the target population of environments or mega-environments (Gupta et al., 2013), which are highly variable in terms of their timing, intensity, and duration of drought (van Oosterom et al., 1995). The role of adapted germplasm has also been emphasized for drought breeding because the measured performance under drought stress is largely a result of adaptation to stress conditions (Yadav et al., 2009, 2012). Hybridization of adapted landraces with elite genetic material creates new gene combinations that lead to amalgamating of adaptation to stress environments and high productivity (Patil et al., 2020; Presterl & Weltzien, 2003; Yadav & Rai, 2011).

Utilization of Germplasm Lines

A small fraction of germplasm has been utilized so far primarily due to the huge number of germplasm accessions and the presence of undesired traits in the unadapted genetic background. These twin problems have been largely circumvented. Development of core and minicore collections (Upadhyaya et al., 2011) is prompting breeders to use the desired germplasm in broadening the genetic base of commercial cultivars, which is essential to reduce the chances of disease epidemics and to mitigate the effects of climate change.

Germplasm with sweet stalk, yellow endosperm, high Fe and Zn, new dwarfing genes, CMS lines, early-maturing lines, and sources of resistance to abiotic and biotic stresses are widely used in crop improvement programs in different countries. Several new and useful traits, such as narrow leaf, glossy leaf, brown midrib leaf, and leaf color variants, have been used extensively in academic studies (Upadhyaya et al., 2007). Development of effective field screening techniques at ICRISAT have led to the identification of excellent sources of resistance in the germplasm and breeding lines, originating mostly from West Africa. In recent years, systematic efforts were made by ICRISAT and NBPGR to evolve and accelerate the germplasm collection, with multi-locational evaluation at some selected locations representing different agro-climatic conditions and cataloguing of genetic resources. About 8,460 accessions of pearl millet have been evaluated from 1986 to 1989 at NBPGR Headquarters, New Delhi; NBPGR Regional Station, Jodhpur; and AICMIP, Pune, for a set of 14 descriptors, and these include promising fodder types (267 accessions), grain types (156 accessions), dual-purpose types (53 accessions), early-maturing types (54 accessions), and bold seed types (20 accessions). Upadhyaya et al. (2007) evaluated 20,844 accessions from 51 countries for 23 traits, including number of flowering days, plant height, tiller number, 100-seed weight, panicle length and shape, seed shape and color, and reported significant variations among the accessions studied. Therefore, germplasm collections (both cultivated and wild) and their characterization using both conventional and advanced techniques such as TILLING need to be done for expansion of its cultivated gene pool.

Pre-breeding

In order to introduce novel features and genetic diversity into elite germplasm, pre-breeding is a crucial step in the plant breeding process. Pre-breeding may not always be distinguished clearly from breeding in many breeding programs. The objective of pre-breeding is to successfully introduce new genetic variation into the elite cultivar without causing linkage drag while maintaining the favorable genes. Hence, it is a lengthy process, particularly if it entails discovery and translational research, wide crossing with wild relatives, strategic crossing, and progeny selection (Sukumaran et al., 2022).

For defining product profiles, lists of the essential and value-added traits have been determined for each market segment in Asia and Africa because pearl millet is grown for various purposes in various agroecologies. High yield in desired maturity background, improved terminal drought and heat tolerance, resistance to DM and blast, Fe and Zn fortification, low rancidity, lodging tolerance, and forage yield and quality are some of the important traits in pearl millet breeding programs. Pre-breeding can help to improve these traits by supplying new and varied sources of genetic variability to quickly address the current production constraints as well as by introgressing the genes and alleles for particular traits from wild species, which are absent in the cultivated gene pool. Pre-breeding initiatives are carried out in association with breeders, physiologists, and pathologists and will be crucial to improving the genetic gain in pearl millet.

Pre-breeding for Tolerance to Abiotic Stresses

Although there has been significant improvement in overall production and productivity, largely as a result of a strong focus on yield and yield components, biotic and abiotic stresses continue to be problematic and result in significant yield losses due to the frequent breakdown of resistance to changing pathogen virulences and the lack of use of a diverse gene pool. Abiotic stresses have negative impacts on the growth and yield potential of pearl millet, despite being well adapted to drought, salinity, high temperature, poor soil conditions (Shivhare & Lata, 2017). In order to incrementally and sustainably improve breeding programs, a constant influx of novel genes and alleles conferring resistance or tolerance to significant biotic and abiotic challenges is necessary. The desirable gene frequency for defense against stresses frequently decreases as a result of the extensive use of elite×elite germplasm in breeding programs to create high-yielding lines (Cobb et al., 2019).

The precision and speed of introgression will increase with the adoption of diagnostic markers, particularly for characteristics with significant genotype-by-environment interactions. There are more than 29 million genome-wide single nucleotide polymorphisms (SNPs) in the pearl millet plant as a result of the Pearl Millet 1,000 Genome Resequencing Project (Varshney et al., 2017). Additionally, 31 wild *Pennisetum* accessions that were collected from samples in the Sahel region, from Senegal to Sudan, were sequenced as part of this effort. Important details on the nature of some wild *Pennisetum* species' genetic diversity as well as the relative contraction and expansion of genes relevant to adaptation and fitness during the course of pearl millet's domestication history were revealed by the resequencing. The diversity of genes and alleles present in pearl millet crop wild relatives as well as the relative diversity of wild germplasm can be determined with the available genomic resources. Microsatellite markers have been shown to be effective for research on genetic diversity by Mariac et al. (2006). In comparison to 421 cultivated lines from Niger, more allelic diversity and a higher number of alleles were found in 46 wild accessions. Before the introgression begins, it is possible to choose advantageous genes and alleles for trait introgression and to functionally characterize the provided genes at the molecular level. In order to map traits and place them in the right genetic contexts, marker systems based on next-generation sequencing, such as whole-genome resequencing SNPs and genotyping-by-sequencing SNPs, may be useful. By applying genome-wide background selection, these markers can also aid in removing the negative linkage drag that is typically seen in the wide cross derivatives. The genetic diversity of pearl millet, which was lost during the domestication process, can be increased by using the repository of markers that is currently available. Genebanks can use molecular markers to identify intra- and inter-accession diversity among several regeneration cycles and characterize them. The crop wild relatives of pearl millet can also be used to increase the allelic richness of the farmed germplasm by introducing variation at the gene, allele, or haplotype levels using next-generation genotyping tools and high-end genomic resources. To create pearl millet cultivars resistant to the biotic and abiotic challenges, these genomic technologies must be applied in a multidisciplinary manner and may also be used in precise trait mapping, introgressions, and speed breeding.

When it comes to finding sources of tolerance to drought, salinity, and high temperatures, pearl millet has, by comparison, garnered the most attention among the many millets (Dwivedi et al., 2012). For the purpose of finding germplasm that is tolerant to diverse abiotic stresses, ICRISAT has even constructed phenotypic screens (ICRISAT, 2008/2009; Krishnamurthy et al., 2007). The next sections describe the identification and use of pearl millet germplasm for abiotic stress resistance.

Drought

A lot of research has been done to comprehend the reaction and adaptation of pearl millet to water deficiency at various developmental growth stages. The crop's growth phases determine the additional effects of low water stress circumstances. According to Farooq et al. (2009) and Lata et al. (2015), water stress during the germination stage or seedling emergence stage results in seedling death, which leads to poor crop establishment. The main reason for pearl millet's low yield in arid and semi-arid environments is severe moisture stress during the seedling stage (Carberry et al., 1985; Soman & Peacock, 1985; Soman et al., 1987). After the establishment of the seedlings, it has been discovered that the influence of the drought stress has little bearing on the grain yield of pearl millet (Lahiri & Kumar, 1966).

It has been demonstrated that the impact of low water stress is directly related to leaf formation and secondary root growth and that the effect of drought stress on the germination of seedlings depends on the availability of water (Gregory, 1983). Using different polyethylene glycol (PEG)6000 concentrations, the effects of moisture stress were also examined. It was discovered that these effects had a substantial impact on a number of seedling metrics, including germination rate, root and shoot length, and root/shoot ratio (Govindaraj et al., 2010). It was also proven to be a quick, easy, and affordable way to screen pearl millet germplasm at the germination and early seedling growth stages using PEG6000. Due to its asynchronous tillering and quick growth rate, which enable it to recover quickly, drought stress at the vegetative phase exhibits little to nearly zero reduction in crop growth and production (Bidinger et al., 1987; Mahalakshmi et al., 1987). The main shoot's flowering time is delayed by further drought stress during the vegetative phase. Because of its phenological plasticity, it can delay entering the most delicate flowering stage until the stress has subsided (Henson & Mahalakshmi, 1985). However, pearl millet grain and stover yield as well as yield stability are most severely impacted by post-flowering drought stress or terminal drought stress (Bidinger & Hash, 2004; Kholová & Vadez, 2013; Mahalakshmi et al., 1987; Winkel et al., 1997).

Successful growth of pearl millet under drought stress is more of a reflection of its adaptation to drought stress than genetic yield potential (Bidinger et al., 1994). Because adaptation is much less understood physiologically and genetically, it is more difficult to improve for adaptation than for yield potential. Pearl millet landraces that evolved in dry areas as a result of natural and man-made selection over thousands of years demonstrate better adaptability to drought stress (Yadav, 2010, 2014; Yadav et al., 2000). Efforts were made to utilize these landraces in pearl millet conventional breeding approaches. Cycles of mass selection in genetically heterogeneous landraces have been found to increase yield considerably (Bidinger et al., 1995; Yadav & Bidinger, 2007; Yadav & Manga, 1999) and have been revealed as a valuable germplasm source to breed drought-tolerant lines (Yadav, 2004) and to develop inbred pollinator lines for hybrid breeding (Yadav et al., 2009, 2012).

Numerous landraces were discovered in various parts of West Africa, Ghana, and northwestern India that may be a rich source of variety for pearl millet's ability to withstand abiotic stress. Due to their higher grain and stover yields compared with conventionally bred lines, these landraces are preferentially grown in the desert regions of northwestern India (Gujarat, Rajasthan, and Haryana), where drought is a frequent occurrence (Bidinger et al., 2009). For instance, CZP9802, the first open-pollinated pearl millet variety derived from Rajasthani landraces, has a high yield (14%–33% higher grain yield and 18%–36% higher stover yield) when compared to the controls Pusa 266 and ICTP 8203 (Dwivedi et al., 2012; Yadav, 2004). It is also highly adapted to drought. Due to its extraordinary characteristics of flowering in just 48 days and reaching maturity in about 75 days, CZP9802 was identified as a suitable cultivar for arid zones of India and has the ability to escape terminal drought stress (Yadav et al., 2004). Okashana1, another early-maturing type, is popular in Namibia (Daisuke, 2005). The development of several pearl millet cultivars around the world, including ICMV88904 (released as ICMV221), was greatly aided by a relatively photoperiod-insensitive, early-maturing West African landrace known as *Iniadi* with compact, conical panicles and bold grains (Andrews & Anand Kumar, 1996; Witcombe et al., 1997). The plant variety ICMV221 is terminal drought tolerant and DM resistant and has higher grain yield potential. It is grown in India and numerous African nations.

Earliness and short and rapid grain filling period are the two most important traits that have been manipulated for enhancing tolerance to terminal drought. Early flowering essentially determines grain productivity in water stress (Bidinger et al., 1987; Fussell et al., 1991; van Oosterom et al., 1995). Genetic variation for earliness is widely available in the germplasm (Rai et al., 1997; Yadav et al., 2017), and phenotype-based selection was accomplished (Rattunde et al., 1989). The frequently exploited basis of earliness is the Iniadi-type landraces collected from western Africa (Andrews & Anand Kumar, 1996). Promising lines with the early flowering trait have been developed from Iniadi landraces and adopted in Indian and African agro-ecologies.

A huge amount of genetic variation in the global germplasm of pearl millet is available for traits that are constitutive and responsive to drought. Murty et al. (1967) characterized 1,532 accessions and reported variation in days to flowering among Indian (52–77 days) and exotic (53–85 days) collections. In another study, days to flowering showed much wider variability (33–140 days) among a world collection of 16,968 pearl millet accessions (Harinarayana et al., 1988). The germplasm line IP 4021 (Bhilodi), collected from Gujarat, was the earliest to flower, with 33 days to 50% flowering in the rainy season and 34 days in the summer season. Furthermore, screening efforts identified two promising drought-tolerant lines among 509 accessions that were screened (Harinarayana et al., 1988). Landraces identified from Rajasthan, India, exhibited wide variability for different characters. The SR 15, SR 17, and SR 54 landraces were identified for earliness. Three other unique landraces that showed earliness were Chadi from Rajasthan, Bhilodi from Gujarat, and Pittaganti from Eastern Ghats of India (Anand Kumar & Appa Rao, 1987). Furthermore, germplasm accessions IP 4066, IP 9496, and IP 9426 were identified as promising early lines. In a detailed study conducted on a wide range of environmental conditions, 105 landraces were evaluated, a wide range in drought response was observed, and 15 landraces (IP 3243, IP 3228, IP 3424, IP 3296, IP 3362, IP 3180, IP 3272, IP 3303, IP 3252, IP 3258, IP 11141, IP 3318, IP 3123, IP 3363, and IP 3244) having a high degree of drought tolerance were identified for use in developing drought-tolerant cultivars (Yadav et al., 2003).

Differential response of landraces and elite genetic materials across contrasting drought and non-drought environments provided the need of amalgamating the drought tolerance of landraces and the high-yielding potential of elite genetic material (Yadav, 2006, 2010; Yadav & Rai, 2011). Crosses between adapted landraces and elite genetic materials showed enhanced adaptation to drought and high productivity (Patil et al., 2020; Presterl & Weltzien, 2003; Yadav, 2008; Yadav & Rai, 2011). Selection of specific landraces and elite materials for hybridization is governed by combinations of phenotypic traits (e.g., tillering, panicle size, seed size, height, and earliness), drought adaptation, and yield potential in two groups of materials (Patil et al., 2020).

The development of hybrids with strategic use of a diverse range of germplasm has been a top priority in pearl millet breeding, especially in India. However, it is generally argued that hybrids are more suitable for optimum environments than heterogeneous OPVs because the latter has a population-buffering effect to provide stable performance in unpredictable drought environments. A comprehensive study reported a yield advantage of hybrids ranging from 19% to 35% over OPVs (Yadav et al., 2012). The current seed delivery system for pearl millet in India also favors hybrids, but this is not the case in Africa. Nevertheless, hybrids are more likely to play a key role than composites in enhancing pearl millet productivity in water stress environments.

DNA markers help determine the degree and quantum of genomic diversity in pearl millet germplasm to identify promising lines for hybridization and analyze population structure and QTLs linked with environmental stress tolerance (Vadez et al., 2012). An attempt was made to enhance the drought tolerance of an elite inbred pollinator (H 77/833-2) line using a donor PRLT 2/89-33 line and the elite inbred seed parent maintainer genotype ICMB 841 using the 863B line as a donor. Marker-assisted back-crossing methods have played a significant role in the considerable boost of grain yields, production, and quality in pearl millet, which has been achieved in India over the past many years. Development of a marker flanking QTLs governing yield and related traits under limited soil water contents has been a major focus in the last several decades (Yadav, 2010, 2014).

Salinity

Nearly 20% of the world's cultivated land and half of the world's irrigated areas are affected by salinity (Shrivasata & Kumar, 2015). Salt stress impacts are particularly severe in arid and semiarid areas (Jha et al., 2021). Pearl millet is considered as fairly tolerant to salinity and thereby as an alternate option for salinity-affected areas. It has been indicated that pearl millet is more tolerant to salt stress than maize, wheat, and rice, with high genotypic variation in salt tolerance (Dong & Zheng, 2006). The presence of high levels of tolerance in other *Pennisetum* species (Muscolo et al., 2003) and within *P. glaucum* (Dua, 1989) opens the door to understanding traits associated with tolerance and incorporating these tolerant genotypes into suitable management programs to increase the productivity of soils affected by salt water and saline water. This indicates that pearl millet has abundant genetic resources for enhancing productivity in salinity-prone environments.

Studies describing salt-tolerant genotypes and the precise processes underlying salt tolerance, however, are few (Sabir et al., 2011; Wang et al., 2007).

Through various physiological traits (e.g., compartmentation, the synthesis and accumulation of compatible solutes in the cytoplasm, vigor to provide dilution of salt concentration by growth, and effective exclusion of Na^+), plants can overcome the toxic effect of excessive ions. In terms of the overall plant response to salinity, pearl millet has been shown to exhibit significant genotypic variation (Dua, 1989). In general, decreased shoot N content, increased K^+ and Na^+ concentrations, and whole plant salinity tolerance were related. There is a lot of room for improvement in pearl millet's ability to tolerate salt, and shoot Na^+ concentration could be thought of as a potential nondestructive selection criterion for screening during the vegetative stage (Krishnamurthy et al., 2007).

Early vegetative growth stage has been reported to be more sensitive to salt stress, as compared to the adult stage (Cardamone et al., 2018). Osmotic adjustment and efficient scavenging of free radicals can be considered as key mechanisms controlling salt tolerance among pearl millet genotypes (Jha et al., 2021).

Large genotypic variation has been observed in pearl millet toward salinity stress tolerance. A wide range of pearl millet breeding lines has been evaluated extensively for salt tolerance (Krishnamurthy et al., 2007; Mukhopadhyay et al., 2005; Ribadiya et al., 2018; Toderich et al., 2018; Yakubu et al., 2010). Typically, the landraces and wild relatives of a crop species exhibit genetic diversity and are known to harbor novel genes for environmental adaptation and other agronomically important traits. Therefore, these genotypes can be used as valuable genetic resources for developing abiotic stress tolerance (Hoang et al., 2016; Manga, 2015; Quan et al., 2018). In a recent study, 12.90%–22.43% reduction in pearl millet grain yield was observed under salinity levels of 8–12 dS/m (Yadav et al., 2020). An average reduction of approximately three- to fourfold in shoot biomass productivity and grain yield was reported in 15 pearl millet accessions (Krishnamurthy et al., 2011). In another study, 11 pearl millet lines showed significant reductions of 19.1% and 41.3% in biomass and grain yield, respectively, under salinity (Toderich et al., 2018), whereas a mean reduction of 47%–86% in grain yield and 51% in fodder yield was observed under high-saline conditions in pearl millet (Choudhary et al., 2019; Kulkarni et al., 2006; Ribadiya et al., 2018). Growing cultivars tolerant to soil salinity/sodicity can boost millet output and bring more land under this crop. The germplasm line IP 22269 was identified as tolerant to salt stress in field experiments performed in northern Tunisia (Hajlaoui et al., 2021).

Heat Tolerance

Crop yields are expected to reduce by 17% for every degree centigrade increase in average growing season temperature (Lobell & Asner, 2003). By the end of the twenty-first century, an increase in temperature up to a maximum of 4 °C is predicted due to global climatic change (Driedonks et al., 2016), exposing the plants to heat stress. Heat stress tolerance is the ability of a plant to evade the negative impact of heat stress and attain economic yields comparable to yields under normal conditions (Wahid et al., 2007). Although pearl millet is a heat-tolerant crop, the germination rate, establishment, initial growth, and photosynthesis rate increases to a temperature of about 35 °C; above that, normal growth is affected (Arya et al., 2014). Heat stress tolerance at seedling stage in pearl millet was studied extensively by Yadav et al. (2011). They have listed the morphological traits governing seedling heat tolerance, such as seedling heat tolerance index, seed to seedling heat tolerance index, emergence rate, leaves/seedling, seedling height, and seedling fresh and dry weights. Hybrids were found to perform better than the parents with respect to the heat tolerance and yield traits (Joshi et al., 1997). The genotypes CVJ 2-5-3-1-3 and 77/371//BSECP CP-1 were identified as the best general combiners for heat tolerance indices. The heat tolerance at seedling stage had no significant impact on its later growth and development (Yadav 2006). Significant genetic diversity was found among the 105 landraces of pearl millet (*P. glaucum*) collected from northwestern India by Yadav et al. (2004). The landraces are grown by farmers over a period of time and have acquired the necessary adaptation genes and thus could be potential source for heat tolerance. The manifestation of heterosis in the landrace-based topcross hybrids varied for different traits, and significant heterosis for biomass, grain yield, and stover yield was observed in specific male-sterile seed parent × landrace-based pollinator combinations (Yadav & Bidinger, 2008). Large genetic variation for tolerance to heat at reproductive stage has been observed, and heat-tolerant lines have been identified, such as ICMB 92777, ICMB 05666, ICMB 00333, ICMB 01888, ICMB 02333, and ICMB 03555 among the maintainer lines and ICMV 82132, MC 94, ICTP 8202, and MC-bulk among the populations (Yadav et al., 2012). Based on 3- to 4-year field screening (2009–2012), five hybrid seed parents (ICMB 92777, ICMB 05666, ICMB 00333, ICMB 02333, and ICMB 03555) and a germplasm accession IP 19877 with 61%–69% seed set as compared to 71% seed set in a heat-tolerant commercial hybrid 9444 (used as a control) was identified (Gupta et al., 2015). Hybrids outperformed the others in terms of seed set and grain yield, for which CZH 233, CZP 9603, CZI 2011/5, and CZMS 21A were the best performing genotypes (Aravind et al., 2017).

Pre-breeding for Biotic Stresses

The sustainability of pearl millet cultivation is seriously threatened by the vulnerability of high-yielding lines to newly developing diseases and pests.

Downy Mildew

Downy mildew (DM) [caused by *Sclerospora graminicola* (Sacc.) J. Schröet] is the most widespread and most important disease causing heavy economic losses in

India, Burkina Faso, Chad, Mali, Nigeria, Niger, Mali, Senegal, Sudan, Tanzania, and Zambia (Singh et al., 1993). In India, the disease is more prevalent on uniform F1s than open pollinated varieties (Thakur et al., 1999). Variation in virulence has previously been documented for *S. graminicola* isolates from several countries (Thakur et al., 2004a). Host-driven virulence selection (Thakur et al., 1992) and evolution of host-specific virulence have also been observed in the pathogen populations (Thakur et al., 1999). Furthermore, variation in the virulence of single-zoospore isolates (Thakur et al., 1998a, 1998b) and single-oospore isolates (Thakur & Shetty, 1993) and DNA polymorphism among isolates (Sastry et al., 1995) have also been clearly documented. Greenhouse and field screening of a large number of germplasm accessions and breeding lines has led to the identification of several resistance sources (Singh et al., 1997), which have been extensively used to develop DM-resistant hybrids. The most resistant pearl millet lines IP 18292 and IP 18293 that were resistant (<10% incidence) at 11 of the 17 locations and moderately resistant (>10%) at Bagauda, Cinzana, Durgapura, Kamboinse, Mandor, and Sadore showed the highest relative variations across locations and years, indicating their resistance is limited to certain DM pathotypes. In contrast, although the average resistance levels of lines 700651 and P 310-17 are slightly lower, these two lines showed much lower variation across different locations and years. Therefore, stable sources of resistance are needed in breeding programs (Thakur et al., 2004b). Resistance in lines IP 18292, IP 18293, 700651, and P 310-17 was more stable than in other lines regardless of the location or seasons (Thakur et al., 2006).

Large numbers of germplasm accessions and breeding lines have been screened at ICRISAT and at the AICPMIP centers, and a number of resistant germplasm/lines have been identified (Singh, 1995; Singh et al., 1997). Resistance stability of several of these lines, including P7 (ICML 12), SDN 503 (ICML 13), 700251 (ICML 14), 700516 (ICML 15), 700651 (ICML 16), and 7042R (ICML 22), has been confirmed through multilocational testing (Singh et al., 1994). Several other lines and germplasm accessions, including P 310-17, P1449-3, IP 18292, IP 18293, IP 18294, IP 18295, and IP 18298, with high levels of resistance have been identified (Singh et al., 1997; Thakur et al., 2004b). Some of these sources have been strategically used to some extent in resistance breeding at ICRISAT and AICPMIP centers. In general, there is enough geographical, morphological, and genetic diversity in germplasm and breeding lines for DM resistance (Singh, 1995; Singh et al., 1997).

Eight pathotype isolates (Sg 384, Sg 409, Sg 445, Sg 457, Sg 510, Sg 519, Sg 526, and Sg 542) of *S. graminicola* were collected from different geographical locations in India. The most virulent pathotype isolate of *S. graminicola* is selected from several isolates collected from a particular pearl millet–growing area and characterized for virulence diversity. For the identification of new and diverse sources of DM resistance, a pearl millet mini-core collection comprising 238 accessions was screened against eight pathotypes (Sg 384, Sg 409, Sg 445, Sg 457, Sg 510, Sg 519, Sg 526, and Sg 542) of *S. graminicola* collected from different geographical locations in India. Significant differences for DM reaction were observed among pathotypes, mini-core accessions, and their interactions. Of the 238 accessions, 68 accessions were resistant ($\leq$10% DM incidence) to pathotype Sg 510, followed by 40 accessions resistant to Sg 457. Resistance to pathotypes Sg 519, Sg 526, Sg 384, Sg 445, and Sg 542 was observed in 15, 27, 29, 30, and 34 mini-core accessions, respectively. Resistance to two or more pathotypes was observed in 62 accessions. Several of these accessions also exhibited desirable agronomic traits. The multiple pathotype resistant germplasm accessions having desirable agronomic characteristics and collected from different agro-ecologies would be useful in breeding programs to develop pearl millet hybrids resistant to difficult-to-manage, highly virulent pathotypes of *S. graminicola*. The multiple-pathotype ($n = 6$–7) resistant accessions (IP 9645, IP 11943, IP 14542, IP 14599, IP 21438, and IP 14537) identified in this study could be used as donor parents for pyramiding resistance alleles in the parental lines of commercially successful pearl millet hybrids. Some of the DM-resistant mini-core accessions have also been found resistant to another important disease, pearl millet blast caused by *Pyricularia grisea* (teleomorph: *Magnaporthe grisea*). Mini-core accessions IP 11247 and IP 12650, having resistance to two pathotypes of *S. graminicola*, also carry resistance to pathotype Pg 118 of *M. grisea* (Sharma et al., 2013). The DM-resistant accessions IP 3646, IP 9692, and IP 17396 have been reported to be resistant to *M. grisea* pathotype Pg 45. Resistance to multiple pathotypes of *M. grisea* has also been observed in some of the selected DM-resistant accessions. Mini-core accession IP 11010 was found resistant to two pathotypes (Pg 118 and Pg 119) of *M. grisea*, and IP 21187 was resistant to four pathotypes (Pg 53, Pg 56, Pg 118, and Pg 119) (Sharma et al., 2013). Once considered a minor disease of pearl millet, blast has now emerged as a serious threat to pearl millet production in India. Because DM and blast are the two major diseases of pearl millet, mini-core accessions such as IP 11010 and IP 21187, having resistance to multiple pathotypes of both DM and blast, would be very useful in the pearl millet breeding program. The results of this study clearly indicate that the mini-core collections can be used as a starting point to screen for desirable traits in a crop species. Mini-core collections have been successfully used to identify sources of disease resistance (Sharma et al., 2010, 2012), salinity (Serraj et al., 2004), drought tolerance (Kashiwagi et al., 2005; Upadhyaya, 2005), and multiple traits of economic importance (Upadhyaya et al., 2014). The agronomically desirable multiple-pathotype–resistant germplasm accessions identified in this study have originated from different agro-ecologies and are therefore expected to be highly divergent and useful in breeding programs to develop pearl millet hybrids with resistance against difficult-to-manage, highly virulent pathotypes of *S. graminicola*.

Blast

Blast or leaf spot disease (caused by *P. grisea* Sacc. [syn. *M. grisea*]) has emerged as a very destructive disease of pearl millet (Rai et al., 2012). Hot humid weather especially during rainy days provides the most conducive conditions for the pathogen to grow and spread. Monitoring of virulence of pathogen populations and screening of genetic resources led to the identification of stable resistant lines to develop blast-resistant hybrid parent lines (Sharma et al., 2013; Yella Goud et al., 2016).

Five *M. grisea* pathotypes (Pg118, Pg119, Pg56, Pg53, and Pg45) have been identified based on reaction type. In an attempt to identify sources of resistance in pearl millet mini-core collection to different pathotypes, accessions were evaluated in greenhouse conditions against five *M. grisea* pathotypes. Among 238 accessions, 32 were reported to be resistant to at least one pathotype. Three accessions (IP 7846, IP 11036, and IP 21187) exhibited resistance to four of five pathotypes. Twenty-one of these accessions originated in India; therefore, germplasm accessions from India appear to be promising sources of blast resistance and could be evaluated against diverse pathotypes of *M. grisea* to discover supplementary sources of blast resistance (Sharma et al., 2013). In United States, sources of blast resistance in pearl millet have been recognized, and efforts have been made to integrate resistance into improved cultivars and elite breeding lines (Hanna et al., 1987; Wilson & Hanna, 1992). Wild species of *Pennisetum* may be useful sources of genes for disease resistance. Pearl millet land races also provide an abundant source of genetic diversity for resistance. Resistance to leaf blast in pearl millet was derived from *P. glaucum* ssp. *monodii* accession from Senegal in which the Rr1 rust gene was found (Hanna et al., 1987). Blast resistance in *P. glaucum* ssp. *monodii* was found to be controlled by three independent dominant genes (Hanna & Wells, 1989), although Tift 85DB, with resistance derived from *P. glaucum* ssp. *monodii*, was shown to have a single resistance gene (Wilson et al., 1989). This resistance was efficient against diverse isolates tested in the United States. However, Tift 85DB has been reported to be susceptible to Indian isolate (Gupta et al., 2012), indicating that the pearl millet–infecting populations of *M. grisea* are different in India from those reported from the United States. Resistance in an elite parent line (ICMB 06222) from pearl millet fields at ICRISAT, Patancheru, India, to isolate Pg45 is reported to be governed by a single dominant gene (Gupta et al., 2012). *Magnaporthe grisea* is extremely variable, with several strains specialized in their host selection, and hence strains infecting rice are reported not to infect pearl millet and vice versa. Nevertheless, unlike rust and DM pathogens, *M. grisea* does not pass through a sexual stage to survive from season to season, implying there is fewer probability of developing novel genetic recombinants. Consequently, breeding for robust resistance to blast in pearl millet might be easier than that to rust or DM (Thakur et al., 2009). Several other sources of *Magnaporthe* leaf spot resistance have been identified from Burkina Faso landraces. Each has been characterized as having dominant, single-gene resistance that is independent of the *monodii* resistance gene. Thakur et al. (2009) developed field and greenhouse screening techniques to identify sources of hybrid parental lines with resistance to blast disease. Among 211 elite hybrid parental lines, which include 126 designated B-lines, 65 potential

R-lines, and 20 designated R-lines were evaluated for blast resistance in a disease nursery, 45 lines recognized as blast resistant (score ≤3.0 on a scale of 1–9) were further screened using a greenhouse screening technique. Twenty-five (8 designated B-lines, 14 potential R-lines, and 3 designated R-lines) of the 45 lines were found resistant to blast disease under greenhouse screening. Fifteen pearl millet genotypes were screened against foliar blast under artificial conditions. Three lines (PPMI 1087, PPMI 1089, and PPMI 660) showed high resistance (score of 0.0–0.4 on a scale of 0–9), and two entries (PPMI 1084 and J 108) were resistant (score of 1.0–1.3) (Prakash et al., 2016).

A set of 93 genotypes that includes 45 maintainer lines and 48 restorer lines were screened against three isolates by Uniform Blast Nursery under artificial screening conditions and field conditions in Jaipur. The line ICMB 95444 had a mean disease score of 7.94 in the artificial screening across three pathotypes and 8.6 in the field screening at Jaipur, indicating that ICMB 95444 is susceptible to blast. Among all the genotypes, IC414K14B5, IC594K16B5, IC6912K18B, RBB037, ICMR06444, and TS19K461 had low mean severity scores. The genotype IC36844K18R had low blast score (2.35) in hotspot, compared with artificial screening (4.9). Ten genotypes (IC16576K16R, IC15354K18R, IC16714K18R, IC16752K18R, IC16975K18R, IC17002K18R, IC17057K18R, IC17137K18R, IC25251R18R, and IC25254R18R) were found to be resistant to Jaipur isolates in the Uniform Blast Nursery, but the resistant levels were slightly low in hotspots (Rao et al., 2021).

Rust

Rust (caused by *Puccinia substriata* var. *indica* Ramachar & Cumm) is generally considered as a disease of less importance in grain crop; however, it is of great importance in fodder crop, where it reduces both quantity and quality. Rust infection of pearl millet forage has been reported to cause up to 51% reduction in digestible dry matter yield (Monson et al., 1986).

The field screening in the late rainy season under high disease pressure and greenhouse screening led to the identification of stable resistance sources (Sharma et al., 2020b; Singh et al., 1990). Resistance to rust has been reported in some pearl millet germplasm accessions and breeding lines (Singh et al., 1997; Wilson, 1993). However, lines that were resistant in India became susceptible in the United States, indicating the existence of different physiological races in India and the United States (Tapsoba & Wilson, 1996; Wilson, 1991). About 214 advanced breeding lines, including 126 designated B-lines, 23 designated R-lines, and 65 potential R-lines, were evaluated for rust resistance in the disease nursery during the summer season 2008–2009 under natural epiphytotic conditions. Eight lines (one B-line, seven R-lines) that showed resistance (≤10% rust severity) in the field screen were evaluated in the greenhouse by artificial inoculation of potted seedlings to confirm their resistance. One B-line (ICMB 96222) and three R-lines

(ICMR 0699, ICMP451-P8, and ICMP 451-P6) were resistant, whereas the other four R-lines were susceptible (Sharma et al., 2009). Two exotic gene pools of pearl millet, Tiff #2 (a bulk population of 740 accessions of cultivated pearl millet mainly from Africa; Hanna, 1990) and Tiff #5 [a bulk population of 114 accessions of wild *P. glaucum* ssp., *monodii* (Maire) Brunken and weedy *P. glaucum* ssp., *stenostachyum* (Klotzsch) Stapf and Hubb. from Niger, Senegal, and Mali; (Hanna et al., 1993) were improved for rust resistance through recurrent selection (Tapsoba et al., 1997).

Ergot and Smut

Ergot (caused by *Claviceps fusiformis* Lov.) and smut (caused by *Moesziomyces penicillariae* Bref. Vanky) are important floral diseases, and grain yield losses are proportional to their severity. Both pathogens are soil borne and infect the host at flowering through stigma. Hybrid cultivars based on CMS lines are generally more susceptible to this disease than open-pollinated varieties. Short protogyny and rapid pollination were found to be associated with ergot resistance (Thakur & Williams, 1980). During the 1977–1984 period, 2,752 germplasm accessions from 19 countries and some unknown sources, obtained from the world collection of ICRISAT's Genetic Resources Program, were screened. No accession was found to have an acceptable level of resistance to ergot. However, 27 accessions originating from India (10), Nigeria (11), Togo (3), and Uganda (3) had varying frequencies of plants, with 0%–10% ergot severity and >75% selfed seed set. More than 8,350 lines from breeding programs of ICRISAT and AICPMIP were screened, wherein it was found that hybrids in particular were susceptible. The screening of breeding materials for several consecutive years did not provide positive results. A total of 283 ergot-resistant lines and populations were identified and designated as ICRISAT Millet Ergot Resistant numbers. A number of lines (ICMP) and populations (ICMIPES) were identified as having stable resistance over years and across locations (Thakur et al., 1985, 1993). Very high levels of ergot resistance in the germplasm accessions were not observed; hence, ergot-resistant lines were developed by intermating less susceptible plants and selecting and rescreening resistant progenies for several generations under high disease pressure (Thakur et al., 1993). Smut resistance is a dominant trait and is easily transferable. However, quantitative resistance involving additive and non-additive gene effects has also been reported. A large number of lines have been found as resistant to smut and DM (Thakur et al., 1992, 2011).

Insect Pests

Gahukar and Reddy (2019) reported that pearl millet is infested by over 150 insect pests during its growth and development. Out of these, shoot fly and stem borer are comparatively more serious pests, attacking at vegetative as well as at ear head stages of the crop. The stem borer (*Coniesta ignefusalis* Hampson) has been reported to be a major pest of pearl millet in the Sahelian and sub-Saharan regions. NS-JIL-01 accession was found to be highly resistant, with a significant least leaf damage score of 0.25 (Abubakar et al., 2020). However, insect pests do not pose a serious problem to pearl millet production as compared to diseases.

Germplasm Utilization for Grain Micronutrients

Studies estimate that, of the world population, 60%–80% are Fe deficient, >30% are Zn deficient, and about 15% are Se deficient (Combs, 2001; Kennedy et al., 2003). Improving essential nutrient content in staple food crops through biofortification breeding can overcome the micronutrient malnutrition problem. Genetic improvement depends on the availability of genetic variability in the primary gene pool. In a set of 297 *iniadi* germplasm accessions originating from northern Togo (146) and adjoining eastern Ghana (83), southern Burkina Faso (63), and western Benin (5) regions of Western Africa, 27 accessions (20 from Togo and 7 from Ghana) having a Fe density of 95–121 mg/kg (1 SE of difference above that for ICTP 8203) and a Zn density of 59–87 mg/kg were selected as a valuable germplasm resource for genetic improvement of these two micronutrients in pearl millet (Rai et al., 2014). In a study involving core germplasm collection of diverse origin evaluated in field trials at ICRISAT, India, the accessions differed significantly for all micronutrients, with over twofold variation for Fe (34–90 mg/kg), Zn (30–74 mg/kg), and Ca (85–249 mg/kg). High estimates of heritability (>0.81) were observed for Fe, Zn, Ca, P, Mo, and Mg. The lower magnitude of genotype-by-environment interaction observed for most of the traits implies strong genetic control for grain nutrients. The top 10 accessions for each nutrient and 15 accessions from five countries for multiple nutrients were identified. For Fe and Zn, 39 accessions, including 15 with multiple nutrients, exceeded the Indian cultivars, and 17 of them exceeded the biofortification breeding target for Fe (72 mg/kg). Most of these nutrients were positively and significantly associated among themselves and with days to 50% flowering and 1,000-grain weight, indicating the possibility of their simultaneous improvement in superior agronomic background. The identified core collection accessions rich in specific and multiple-nutrients would be useful as the key genetic resources for developing biofortified and agronomically superior cultivars (Govindaraj et al., 2020).

Earlier research reported up to 24.3% protein content in germplasm (Jambunathan & Subramanian, 1988) and up to 19.8% in elite breeding lines (Singh et al., 1987). However, no serious efforts were made to improve protein content because of its negative correlations with grain yield (Singh & Nainawatee, 1999). Although selecting micronutrient-dense lines among existing breeding populations and varieties within breeding programs is the first approach, the potential for micronutrient enhancement through deliberate selection from germplasm collections is much greater than by selection within available breeding lines or varieties (Gerloff & Gabelman, 1983; Graham & Welch, 1996).

Application of Genomic Tools for Plant Genetic Resources Utilization and Pre-breeding

Genomics has assisted pre-breeding through various technologies including sequencing and re-sequencing platforms, reference genome sequences, high-throughput genotyping platforms, SNP arrays, genome editing (GE) tools, etc., and are discussed in detail by Singh et al. (2020).

Genome Sequencing

There has been an increase in the number of assembled plant genomes of different crops, including wild relatives facilitated by inexpensive sequencing and resequencing technologies. Because a single reference genome does not represent the total diversity within a species, resequencing of cultivars, landraces, and wild accessions is required to harness the total genetic variation and to identify the superior alleles for the target traits. Genome information availability has generated many next-generation sequencing-based platforms for allele mining and candidate gene identification. Sequencing-based approaches provide an opportunity to identify novel variations for a large number of genes through genotype–phenotype associations. Genetic and genomic sequence information is now readily available for pearl millet. The genome size of pearl millet ~1.79 Gb, representing 38,579 genes, 88,256 SSRs, and 4,50,000 SNPs, is a valuable resource for constructing precision genetic maps (Varshney et al., 2017). The Pearl Millet inbred Germplasm Association Panel (PMiGAP) developed at ICRISAT in partnership with Aberystwyth University was re-sequenced and is a repository of approximately 29 million genome-wide SNPs. PMiGAP has been used to map traits related to drought tolerance, grain Fe and Zn content, N use efficiency, components of endosperm starch, grain yield, etc. (Srivastava et al., 2020).

Molecular Markers and Genetic Maps

The availability of molecular markers linked to specific traits enhances pre-breeding efficiency and effectiveness through marker-assisted selection. Molecular markers that are linked to the genes of a desired trait, known as diagnostic markers, can be indirectly used for selection of target traits (Singh et al., 2020). Developing marker trait associations between 250 SSR and 17 genetic markers with grain Fe and Zn content for 130 diversified lines across different environments revealed that the Xicmp3092 marker had a strong association with grain Fe content on LG 7, and markers Xpsmp2086, Xpsmp2213 and Xipes0224 showed association with grain Zn content on LG 4 and LG 6, respectively (Srivastava et al., 2020).

Genome-wide Association Studies

Genome-wide association studies (GWASs) could overcome several constraints of conventional linkage mapping and provide a powerful complementary strategy for dissecting complex traits. GWASs make use of past recombinations in diverse association panels to identify genes linked to phenotypic traits at higher resolution than QTL analysis. GWASs have become a powerful tool for QTL mapping in plants because a broad range of genetic resources may be accessed for marker trait association without any limitation on marker availability. Different approaches used for GWASs include SNP marker arrays and the SNP chips approach. The availability of high-density SNP marker arrays has opened a way for cost-effective GWASs using natural populations. Because the cost of sequencing is continuously declining, genotyping-by-sequencing (also known as the next-generation genotyping method) is becoming more common for discovering novel plant SNPs that can be used for GWASs (Arruda et al., 2016). A set of 250 full-sib progenies and 34 SSR markers were used for GWASs, and results revealed the strong association of the *Xpsmp*2248_162 marker allele at linkage group (LG)6, with earlier flowering time and reduced plant height. Marker allele *Xpsmp*2224_157 on LG7 was strongly associated with plant height. For panicle length, *Xpsmp*2077_136, *Xpsmp*2233_260, and *Xpsmp*2224_157 were strongly associated with LG2, LG5, and LG7, respectively, whereas the *Xpsmp*2237_230 marker allele showed strong positive association on LG7 with grain yield. For stover dry matter yield, the *Xicmp*3058 193 marker allele showed strong positive correlation on LG6 (Srivastava et al., 2020).

Genomics Selection

The genomics-assisted breeding approach known as genomic selection (GS) is a better approach that simultaneously uses large genotypic data (genome wide) (exceeding phenotypic data), phenotypic data, and modeling using statistical tools to predict the genomic estimated breeding values (GEBVs) for each individual (Crossa et al., 2017; Meuwissen et al., 2001). In genomic selection, a statistical model is generated using a representative sample of the breeding population known as the "training population." This model is subsequently used to calculate the allelic effects of all marker loci (i.e., genomic assisted breeding values without having phenotypic data), and these values can be used for preselection of trait-specific genotypes (Heffner et al., 2011). Xu et al. (2012) and Spindel et al. (2016) highlighted that coupling of genome-wide data with genomic selection offered great specificity and predictability, which can be used to accelerate pre-breeding. Using GS, complex traits can be improved rapidly through generation of reliable phenotypes by shortening the selection cycle. GS can facilitate selection of complex traits, such as grain yield (Saint Pierre et al., 2016) and tolerance to abiotic and biotic stress. In genomic selection, genetic diversity specific to the population or family (species) of interest is captured through markers developed through genotyping by sequencing, which minimized the ascertainment bias. GS is superior with respect to fixing all the genetic variation and when selecting individuals with higher GEBV without any phenotyping. In pearl millet, the high outcrossing rates, heterozygous nature, presence of inbreeding depression, and residual heterozygosity pose bottlenecks in inbred line development programs for the development of association mapping panels and for parental line/cultivar development as used in the training sets for GS.

Genome Editing

Recent advancements in genomics have also made feasible the editing of genomes and their use in crop improvement programs. Pre-breeding involves genetic transformation through recombination, and GE tools provide an alternative. To replace conventional genetic engineering, a number of GE technologies have been developed during last two decades, including antisense, RNA interference, virus-induced gene silencing, oligonucleotide directed mutagenesis, zinc finger nuclease, transcription activator-like effects nucleases, and clustered regularly interspaced short palindromic repeats/Cas9 (CRISPR/Cas9) (Sauer et al., 2016). These GE technologies can accelerate pre-breeding programs through beneficial knockout mutations (e.g., identification of genes for disease resistance or suppressing of unwanted traits linked with desired traits in wild species) because products of these technologies are not considered a genetically modified organism (Huang et al., 2016).

Conclusions

Pre-breeding has been emphasized in recent years due to the reduction in variability and susceptibility to biotic and abiotic stresses in the elite lines and the requirement for nutrition enhancement. In pearl millet it played a major role in the supply of useful variability from promising landraces and wild relatives to the breeding pipeline. Also in recent years, various genetic and genomic tools were developed to be used by pearl millet workers worldwide. Pre-breeding has led to the identification and development of traits, genes, sources, and intermediate products that have led to the development of products. However, with the availability of genetic and genomic resources, there is a lot of opportunity for genetic enhancement catering to the product profiles in pearl millet.

References

Abubakar, A., Falus, O. A., Olayemi, I. K., Adebola, M. O., Daudu, Y. O. A., & Dangana, M. C. (2020). Evaluation of pearl millet (*Pennisetum glaucum* L. (R. Br.)) landraces for resistance to stem borer (*Coniesta ignefusalis* Hampson.) infestation. *Notulae Scientia Biologicae*, *12*, 807–817. https://doi.org/10.15835/12410818

Anand Kumar, K., & Appa Rao, S. (1987). Diversity and utilization of pearl millet germplasm. In J. R. Witcombe & S. R. Beckerman (Eds.), *Proceedings of the International Pearl millet workshop* (pp. 69–82). ICRISAT.

Andrews, D. J., & Anand Kumar, K. (1996). Use of the West African pearl millet landrace *Iniadi* in cultivar development. *Plant Genetic Resources Newsletter*, *105*, 15–22.

Appa Rao, S., Mengesha, M. H., & Rajagopal Reddy, C. (1986). New sources of dwarfing genes in pearl millet (*Pennisetum americanum*). *Theoretical and Applied Genetics*, *73*, 170–174.

Aravind, J., Manga, V. K., Bhatt, R. K., & Pathak, R. (2017). Differential response of pearl millet genotypes to high temperature stress at flowering. *Journal of Environmental Biology*, *38*, 791–797.

Arruda, M. P., Brown, P., Krill, A., Brown-Guedira, G., Thurber, C., Foresman, B., & Kolb, F. (2016). Genome-wide association mapping of fusarium head blight resistance in wheat using genotyping-by-sequencing. *Plant Genome*, *9*, 1–14.

Arya, R. K., Singh, M. K., Yadav, A. K., Kumar, A., & Kumar, S. (2014). Advances in pearl millet to mitigate adverse environment conditions emerged due to global warming. *Forage Research*, *40*, 57–70.

Athwal, D. S. (1965). Hybrid bajara-1 marks a new era. *Indian Farming*, *15*, 6–7.

Bantilan, M. C. S, Subba Rao, K. V., Rai, K. N., & Singh, S. D. (1998). Research on high yielding pearl millet – background for an impact study in India. In *Assessing joint research impacts: Proceedings of an International Workshop in Joint Impact Assessment of NARS/ICRISAT Technologies for the Semi-Arid Tropics* (pp. 52–61). ICRISAT.

Bidinger, F. R., and Hash, C. T. (2004). "Pearl millet," in Physiology and Biotechnology Integration for Plant Breeding, eds H. T. Nguyen and A. Blum (New York, NY: Marcel Dekker), 225–270.

Bidinger, F. R., Mahalakshmi, V., & Rao, G. D. P. (1987). Assessment of drought resistance in pearl millet [*Pennisetum americanum* (L.) Leeke]. II. Estimation of genotype response to stress. *Australian Journal of Agricultural Research, 38*, 49–59.

Bidinger, F. R., Mahalakshmi, V., Talukdar, B. S., & Sharma, R. K. (1995). Improvement of landrace cultivars of pearl millet for arid and semi-arid environments. *Annals of Arid Zone, 34*, 105–110.

Bidinger, F. R., Weltzien, E., Mahalakshmi, V., Singh, S. D., & Rao, K. P. (1994). Evaluation of landrace top cross hybrids of pearl millet for arid zone environments. *Euphytica, 76*, 215–226.

Bidinger, F. R., Yadav, O. P., and Weltzien, R. E. (2009). Genetic improvement of pearl millet for the arid zone of northern India: lessons from two decades of collaborative ICRISAT-ICAR research. Expt. Agric. 45, 107–115. doi: 10.1017/S0014479708007059

Bono, M. (1973). Contribution a la morpho-systematique des *Pennisetum* annuels cultives pour leur grain en afrique occidentale francophone. *Agronomia Tropical, 2*, 229–355.

Brunken, J. N. (1977). A systematic study of *Pennisetum* sect. *Pennisetum* (gramineae). *American Journal of Botany, 64*, 161–176. https://doi.org/10.2307/2442104

Burton, G. W. (1958). Cytoplasmic male-sterility in pearl millet (*Pennisetum glaucum*) (L.) R. Br. *Agronomy Journal, 50*, 230. https://doi.org/10.2134/agronj1958.00021962005000040018x

Burton, G. W., & Powell, J. B. (1968). Pearl millet breeding and cytogenetics. In *Advances in Agronomy* (Vol. *20*, pp. 49–89).

Carberry, P. S., Cambell, L. E., and Bidinger, F. R. (1985). The growth and development of pearl millet as affected by plant population. Field Crops Res. 11, 193–220. doi: 10.1016/0378-4290(85)90102-9

Cardamone, L., Cuatrín, A., Grunberg, K., & Tomás, M. A. (2018). Variability for salt tolerance in a collection of *Panicum coloratum* var. makarikariense during early growth stages. *Tropical Grasslands-Forrajes Tropicales, 6*, 134–147.

Chase, A. (1921). The linnaean millet using concept of pearl millet. *American Journal of Botany, 8*, 41–49.

Chauhan, S.S., Jha, S.K., Jha, G.K., Sharma, D.K., Satyavathi, C.T. and Kumari, J. 2015. Germplasm screening of pearl millet (Pennisetum glaucum) forpopping characteristics. Indian Journal of Agricultural Sciences 85 (3): 344–348.

Choudhary, S., Vadez, V., Hash, C. T., & Kavi Kishor, P. B. (2019). Pearl millet mapping population parents: Performance and selection under salt stress across environments varying in evaporative demand. *Proceedings of the National Academy of Sciences, India Section B: Biological Sciences, 89*, 201–211. https://doi.org/10.1007/s40011-017-0933-1

Cobb, J. N., Juma, R. U., Biswas, P. S., Arbelaez, J. D., Rutkoski, J., Atlin, G., Hagen, T., Quinn, M., & Ng, E. H. (2019). Enhancing the rate of genetic gain in public-sector plant breeding programs: Lessons from the breeder's equation. *Theoretical and Applied Genetics, 132,* 627–645. https://doi.org/10.1007/s00122-019-03317-0

Combs, G. F. (2001). Selenium in global food systems. *The British Journal of Nutrition, 85,* 517–547.

Crossa, J., Pérez-Rodríguez, P., Cuevas, J., Montesinos-López, O., Jarquín, D., de los Campos, G, Burgueño, J., González-Camacho, J. M., Pérez-Elizalde, S., Beyene, Y., Dreisigacker, S., Singh, R., Zhang, X., Gowda, M., Roorkiwal, M., Rutkoski, J., & Varshney, R. K. (2017). Genomic selection in plant breeding: Methods, models, and perspectives. *Trends in Plant Science, 22,* 961–975.

Daisuke, U. N. O. (2005). Farmer's selection of local and improved pearl millet varieties in Ovamboland, Northern Namibia. Afr. Study Monogr. 30, 107–117.

Dempewolf, H., Baute, G., Anderson, J., Kilian, B., Smith, C., & Guarino, L. (2017). Past and future use of wild relatives in crop breeding. *Crop Science, 57,* 1070–1082. https://doi.org/10.2135/cropsci2016.10.0885

Dong, Y. C., & Zheng, D. S. (2006). *Crops and their wild relatives in China.* China Agriculture Press.

Driedonks, N., Rieu, I., & Vriezen, W. H. (2016). Breeding for plant heat tolerance at vegetative and reproductive stages. *Plant Reproduction, 29,* 67–79. https://doi.org/10.1007/s00497-016-0275-9

Dua, R. P. (1989). Salinity tolerance in pearl millet. *Indian Journal of Agriculture Research, 23,* 9–14.

Dujardin, M., & Hanna, W. W. (1983). Apomictic and sexual pearl millet x *Pennisetum squamulatum* hybrids. *Journal of Heredity, 74,* 277–279.

Dujardin, M., & Hanna, W. W. (1984). Microsporogenesis, reproductive behavior, and fertility of five *Pennisetum* species. *Theoretical and Applied Genetics, 67,* 197–201.

Dujardin, M., & Hanna, W. W. (1985a). Cytology and reproduction of reciprocal backcrosses between pearl millet and sexual and apomictic hybrids of pearl millet x *Pennisetum squamulatum. Crop Science, 25,* 59–62.

Dujardin, M., & Hanna, W. W. (1985b). Cytology and reproductive behavior of pearl millet napier grass hexaploids x *Pennisetum squamulatum* trispecific hybrids. *Journal of Heredity, 76,* 382–384.

Dwivedi, S., Upadhyaya, H., Senthilvel, S., and Hash, C. (2012). "Millets: Genetic, and Genomic Resources," in Plant Breeding Reviews, ed. J. Janick (Hoboken, NJ: John Wiley, and Sons, Inc.), 247–375.

Farooq, M., Wahid, A., Kobayashi, N., Fujita, D., and Basra, S. M. A. (2009). Plant drought stress; effects, mechanisms and management. Agron. Sustain. Dev. 29, 185–212. doi: 10.1051/agro:2008021

FAOSTAT (2021): Food and Agriculture Organization of the United Nations. FAOSTAT, http://faostat,fao,org/site/339/default,aspx

Fussell, L. K., Bidinger, F. R., & Bieler, P. (1991). Crop physiology and breeding for drought tolerance. Research and development. *Field Crops Research, 27,* 183–199.

Gabelman, W. H., & Gerloff, G. C. (1983). The search for and interpretation of genetic controls that enhance plant growth under deficiency levels of a macronutrient. *Plant and Soil, 72,* 335–350.

Gahukar, R. T., & Reddy, G. V. P. (2019). Management of economically important insect pests of millet. *Journal of Integrated Pest Management, 10*(1), 28. https://doi .org/10.1093/jipm/pmz026

Govindaraj, M., Rai, K. N., Kanatti, A., Upadhyaya, H. D., Shivade, H., & Rao, A. S. (2020). Exploring the genetic variability and diversity of pearl millet core collection germplasm for grain nutritional traits improvement. *Scientific Reports, 10,* 21177 [Erratum: 11(1), 9757]. https://doi.org/10.1038/s41598-020-77818-0.

Govindaraj, M., Shanmugasundaram, P., Sumathi, P., and Muthiah, A. R. (2010). Simple, rapid and cost effective screening method for drought resistant breeding in pearl millet. Electron. J. Plant Breed. 1, 590–599.

Graham, R. D., & Welch, R. M. (1996). *Breeding for staple-food crops with high micronutrient density.* In *Agricultural Strategies for Micronutrients Working Paper* (Vol. 3). International Food Policy Research Institute.

Gregory, P. J. (1983). Response to temperature in a stand of pearl millet (Pennisetum typhoides S. & H.): III. Root development.]. J. Exp. Bot. 34, 744–756. doi: 10.1093/ jxb/34.6.744

Gupta, S. K., Rai, K. N., Singh, P., Ameta, V. L., Gupta, S. K., Jayalekha, A. K., Mahala, R. S., Pareek, S., Swami, M. L., & Verma, Y. S. (2015). Seed set variability under high temperatures during flowering period in pearl millet (*Pennisetum glaucum* L. (R.) Br.). *Field Crops Research, 171,* 41–53.

Gupta S. K., Rathore A., Yadav O. P., Rai K. N., Khairwal I. S., Rajpurohit B. S., et al. (2013). Identifying mega-environments and essential test locations for pearl millet cultivar selection in India. *Crop Sci.* 53, 2444–2453. 10.2135/cropsci2013.01.0053

Gupta, S. K., Sharma, R., Rai, K. N., & Thakur, R. P. (2012). Inheritance of foliar blast resistance in pearl millet (*Pennisetum glaucum* L. (R.) Br.). *Plant Breeding, 131,* 217–219.

Hajlaoui, H. R., Akrimi, R., & Hajlaoui, F. (2021). Screening field grown pearl millet (*Pennisetum glaucum* L.) genotypes for salinity tolerance in the north of Tunisia. *IOP Conference Series: Earth and Environmental Science, 904,* 012046. https://doi .org/10.1088/1755-1315/904/1/012046

Hanna, W. W. (1989). Characteristics and stability of a new cytoplasmic-nuclear male sterile source in pearl millet. *Crop Science, 29,* 1457–1459. https://doi.org/10.2135/ cropsci1989.0011183X002900060026x

Hanna, W. W. (1990). Registration of Tift #2 S-1 pearl millet germplasm. *Crop Science, 30,* 1376.

Hanna, W. W. (1992). Utilization of germplasm from wild species. In W. W. Hanna (Ed.), *Desertified grass lands. Their biology and management* (pp. 251–257). The Linnean Society of London.

Hanna, W. W. (1997). Registration of Tift 8593 pearl millet genetic stock. *Crop Science, 37,* 1412. https://doi.org/10.2135/cropsci1997.0011183X003700040100x

Hanna, W. W., & Dujardin, M. (1985). *Interspecific transfer of apomixis in* Pennisetum. Presented at the XV International Grassland Congress, Kyoto, Japan.

Hanna, W. W., & Wells, H. D. (1989). Inheritance of *Pyricularia* leaf spot resistance in pearl millet. *Journal of Heredity, 80,* 145–147.

Hanna, W. W., Wells, H. D., & Burton, G. W. (1987). Registration of pearl millet inbred parental lines, Tift 85D2A1 and Tift 85D2B1. *Crop Science, 27,* 1324–1325.

Hanna, W. W., Wilson, J. P., Wells, H. D., & Gupta, S. C. (1993). Registration of Tift #5 S-1 pearl millet germplasm. *Crop Science, 30,* 1417–1418.

Harinarayana, G., Appa Rao, S., & Mengesha, M. H. (1988). Prospects of utilising genetic diversity in pearl millet. In R. S. Paroda, R. K. Arora, & K. P. S. Chandel (Eds.), *Plant genetic resources: Indian perspective* (pp. 170–182). National Bureau of Plant Genetic Resources.

Harinarayana, G. & Rai, K. N. (1989). Use of pearl millet germplasm and its impact on crop improvement in India. In *Collaboration on genetic resources. Proceedings of a Workshop on Germplasm Exploration and Evaluation in India* (pp. 89–92). ICRISAT.

Harlan, J. R. (1975). Geographic patterns of variation in some cultivated plants. *The Journal of Heredity, 66,* 184–191.

Harlan, J.R.; De Wet, J.M.J. Toward a Rational Classification of Cultivated Plants. *Taxon* 1971, *20,* 509–517.

Heffner, E. L., Jannink, J. L., Iwata, H., Souza, E., & Sorrellls, M. E. (2011). Genomic selection accuracy for grain quality traits in biparental wheat populations, Geographic patterns of variation in some cultivated plants. *Crop Science, 51,* 2597–2606.

Henson, I. E., and Mahalakshmi, V. (1985). Evidence for panicle control of stomatal behaviour in water stressed plants of pearl millet. Field Crops Res. 11, 281–290. doi: 10.1016/0378-4290(85)90109-1

Hitchcock, A. S., & Chase, A. (1951). *Manual of the grasses of the United States* (2nd ed. (USDA Misc. Publ. 200)). US Government Printing Office.

Hoang, T. M. L., Tran, T. N., Nguyen, T. K. T., Williams, B., Wurm, P., Bellairs, S., & Mundree, S. (2016). Improvement of salinity stress tolerance in rice: Challenges and opportunities. *Agronomy, 6,* 54.

Huang, S., Weigel, D., Beachy, R. N., & Li, J. (2016). A proposed regulatory framework for genome-edited crops. *Nature Genetics, 48,* 109–111.

IBPGR and ICRISAT (1993). Descriptors for Pearl Millet [Pennisetum glaucum (L.) R. Br.]. Rome: International Board for Plant Genetic Resources.

ICRISAT (2008/2009). MTP Project 5: Producing More and Better Food at Lower Cost of Staple Cereal and Legume Hybrids in the Asian SAT (Sorghum, Pearl Millet, and Pigeonpea) through Genetic Improvement. Patancheru: ICRISAT, 158–159

Jambunathan, R., & Subramanian, V. (1988). Grain quality and utilization of sorghum and pearl millet. In J. M. J. de Wet & T. A. Preston (Eds.), *Biotechnology in tropical crop improvement* (pp. 133–139). ICRISAT.

Jha, S., Singh, J., Chouhan, C., Singh, O., & Srivastava, R. K. (2021). Evaluation of multiple salinity tolerance indices for screening and comparative biochemical and molecular analysis of pearl millet [*Pennisetum glaucum* (L.) R. Br.] genotypes. *Journal of Plant Growth Regulation, 41*, 1820–1834. https://doi.org/10.1007/s00344-021-10424-0

Joshi, A. K., Pandya, J. N., Mathukia, R. K., Pethani, K. V., & Dave, H. R. (1997). Seed germination in pearl millet hybrids and parents under extreme temperature conditions. *GAU Research Journal, 23*, 77–83.

Kashiwagi, J., Krishnamurthy, L., Upadhyaya, H. D., Krishna, H., Chandra, S., Vadez, V., & Serraj, R. (2005). Genetic variability of drought-avoidance root traits in the mini-core germplasm col-lection of chickpea (*Cicer arietinum* L.). *Euphytica, 146*, 213–222. https://doi.org/10.1007/s10681-005-9007-1

Kell, S., Marino, M., & Maxted, N. (2017). Bottlenecks in the PGRFA use system: Stakeholders' perspectives. *Euphytica, 213*, 170.

Kennedy, G., Nantel, G., & Shetty, P. (2003). The scourge of "hidden hunger": Global dimensions of micronutrient deficiencies. *Food Nutrition and Agriculture, 32*, 8–16.

Kholová, J., and Vadez, V. (2013). Water extraction under terminal drought explains the genotypic differences in yield, not the anti-oxidant changes in leaves of pearl millet (Pennisetum glaucum). Funct. Plant Biol. 40, 44–53.

Krishnamurthy, L., Serraj, R., Rai, K. N., Hash, C. T., & Dakheel, A. J. (2007). Identification of pearl millet [*Pennisetum glaucum* (L.) R. Br.] lines tolerant to soil salinity. *Euphytica, 158*, 179–188. https://doi.org/10.1007/s10681-007-9441-3

Krishnamurthy, L., Zaman-Allah, M., Purushothaman, R., Irshad Ahmed, M., & Vadez, V. (2011). Plant biomass productivity under abiotic stresses in SAT agriculture. In M. D. Matovic (Ed.), *Biomass - Detection, production and usage* (pp. 247–264)). IntechOpen. https://doi.org/10/5772/17279

Kulkarni, V. N., Rai, K. N., Dakheel, A. J., Ibrahim, M., Hebbara, M., & Vadez, V. (2006). Pearl millet germplasm adapted to saline conditions. *SAT eJournal, 2*, 1–4.

Kumari, J., Bag, M. K., Pandey, S., Jha, S. K., Chauhan, S. S., Jha, G. K., Gautam, N. K., & Dutta, M. (2016). Assessment of phenotypic diversity in pearl millet [*Pennisetum glaucum* (L.) R. Br.] germplasm of Indian origin and identification of trait-specific germplasm. *Crop & Pasture Science, 67,* 1223–1234. https://doi.org/10.1071/CP16300

Lahiri, A. N., and Kumar, V. (1966). Studies on plant-water relationship III: further studies on the drought mediated alterations in the performance of bulrush millet. Proc. Natl. Inst. Sci. India B. 32, 116–129.

Lata, C., Muthamilarasan, M., and Prasad, M. (2015). "Drought stress responses and signal transduction in plants," in The Elucidation of Abiotic Stress Signaling in Plants, ed. G. K. Pandey (New York, NY: Springer), 195–225

Leeke, P. (1907). *Untersuchungen uber Abstammung und Heimat der Negerhirse (Pennisetum americanum (L.) K. S. Schum.).* Schweizerbart.

Lobell, D. B., & Asner, G. P. (2003). Climate and management contribution to recent trends in US agricultural yields. *Science, 299,* 1032.

Mahalakshmi, V., Bidinger, F. R., and Raju, D. S. (1987). Effect of timing of water deficit on pearl millet (Pennisetum americanum). Field Crops Res. 15, 327–339. doi: 10.1016/0378-4290(87)90020-7

Manga, V. K. (2015). Diversity in pearl millet [*Pennisetum glaucum* (L) R BR] and its management. *Indian Journal of Plant Sciences, 4,* 38–51.

Marchais, L., & Pernes, J. (1985). Genetic divergence between wild and cultivated pearl millets (*Pennisetum typhoides*). I. Male sterility. *Zeitschrift für Pflanzenzüchtung, 95,* 103–112.

Mariac, C., Luong, V., Kapran, I., Mamadou, A., Sagnard, F., Deu, M., Chantereau, J., Gerard, B., Ndjeunga, J., Bezançon, G., Pham, J. L., & Vigouroux, Y. (2006). Diversity of wild and cultivated pearl millet accessions [*Pennisetum glaucum* (L.) R. Br.] in Niger assessed by microsatellite markers. *Theoretical and Applied Genetics, 114,* 49–58. https://doi.org/10.1007/s00122-006-0409-9

Meuwissen, T. H. E., Hayes, B. J., & Goddard, M. E. (2001). Prediction of total genetic value using genome-wide dense marker maps. *Genetics, 157,* 1819–1829.

Monson, W. G., Hanna, W. W., & Gaines, T. P. (1986). Effects of rust on yield and quality of pearl millet forage. *Crop Science, 26,* 637–639.

Mukhopadhyay, R., Hash, C. T., Bhasker Raj, A. G., & Kavi Kishor, P. B. (2005). Assessment of opportunities to map pearl millet tolerance to salinity during germination and early seedling growth. *SAT eJournal, 1,* 1–4.

Murdock, G. P. (1959). *Africa: Its people and their cultural history.* McGraw Hill.

Murty, B. R., Upadhyaya, M. K., & Manchanda, P. L. (1967). Classification and cataloguing of a world collection of genetic stocks of *Pennisetum. Indian Journal of Genetics, 27,* 313–394.

Muscolo, A., Panuccio, M. R., & Sidari, M. (2003). Effects of salinity on growth, carbohydrate metabolism and nutritive properties of kikuyu grass (*Pennisetum clandestinum* Hochst). *Plant Science, 104,* 1103–1110.

Patil K. S., Gupta S. K., Marathi B., Danam S., Thatikunta R., Rathore A., et al. (2020). African and Asian origin pearl millet populations: genetic diversity pattern and its association with yield heterosis. *Crop Sci.* 60 3035–3048. 10.1002/csc2.20245

Prakash, G., Srinivasa, N., Mukesh Sankar, S., Singh, S. P., & Tara Satyavathi, C. (2016). Standardization of pearl millet blast (*Magnaporthe grisea*) phenotyping under artificial conditions. *Annals of Agricultural Research, 37*, 200–205.

Presterl T., Weltzien E. (2003). Exploiting heterosis in pearl millet population breeding in arid environments. *Crop Sci.* 43 767–776. 10.2135/cropsci2003.0767

Quan, R., Wang, J., Hui, J., Bai, H., Lyu, X., Zhu, Y., Zhang, H., Zhang, Z., Li, S., & Huang, R. (2018). Improvement of salt tolerance using wild rice genes. *Frontiers in Plant Science, 8*, 2269.

Rachie, K. O., & Majmudar, J. V. (1980). *Pearl millet*. The Pennsylvania State University Press.

Rai, K. N., Appa Rao, S., & Reddy, K. N. (1997). Pearl millet. In D. Fuccillo, L. Sears, & P. Stapleton (Eds.), *Biodiversity in trust, conservation and use of plant genetic resources in CGIAR centers* (pp. 243–258). Cambridge University Press.

Rai, K. N., Kumar, A. K., Andrews, D. J., Rao, A. S., Raj, A. G. B., & Witcombe, J. R. (1990). Registration of 'ICTP 8203' pearl millet. *Crop Science, 30*, 959.

Rai, K. N., Velu, G., Govindaraj, M., Upadhyaya, H. D., Rao, A. S., Shivade, H., & Reddy, K. N. (2014). Iniadi pearl millet germplasm as a valuable genetic resource for high grain iron and zinc densities. *Plant Genetic Resources: Characterization and Utilization, 13*, 75–82. https://doi.org/10.1017/S1479262114000665

Rai, K. N., Yadav, O. P., Gupta, S. K., Mahala, R. S., & Gupta, S. K. (2012). Emerging research priorities in pearl millet. *Journal of SAT Agricultural Research, 10*, 1–4. http://ejournal.icrisat.org/index.htm

Ramu, P., Srivastava, R. K., Sanyal, A., Sanyal, A., Fengler, K., Cao, J., Zhang, Y., Nimkar, M., Gerke, J., Shreedharan, S., Llaca, V., May, G., Peterson-Burch, B., Lin, H., King, M., Das, S., Bhupesh, V., Mandaokar, A., Maruthachalam, K., . . . Babu, R. (2023). Improved pearl millet genomes representing the global heterotic pool offer a framework for molecular breeding applications. *Communications Biology, 6*, 902. https://doi.org/10.1038/s42003-023-05258-3

Rao, K. B., Motukuri, S. R. K., Arun Kumar, K., Praveen Babu, C. H. V. N., & Pathak, V. (2021). Pearl millet blast pathogen virulence study and identification of resistance donors on virulent isolate. *Journal of Pure and Applied Microbiology, 15*, 752–758. https://doi.org/10.22207/JPAM.15.2.27

Rao, N. K. (2004). Plant genetic resources: Advancing conservation and use through biotechnology. *African Journal of Biotechnology, 3*, 136–145.

Rao, N. K., Reddy, L. J., & Bramel, P. J. (2003). Potential of wild species for genetic enhancement of some semi-arid food crops. *Genetic Resources and Crop Evolution, 50*, 707–721.

Rattunde, H. F., Singh, P., & Witcombe, J. R. (1989). Feasibility of mass selection in pearl millet. *Crop Science, 29*, 1423–1427.

Ribadiya, T. R., Savalia, S. G., Vadaliya, B. M., & Davara, M. A. (2018). Effect of salinity on yield, yield attributes and quality of pearl millet (*Pennisetum glaucum* L) varieties. *International Journal of Chemical Studies, 6*, 878–882.

Sabir, P. M., Ashraf, N. A., & Akram, A. (2011). Accession variation for salt tolerance in proso millet (*Panicum miliaceum* L.) using leaf proline content and activities of some key antioxidant enzymes. *Journal of Agronomy and Crop Science, 197,* 340–347.

Saint Pierre, C., Burgueño, J., Crossa, J., Fuentes Dávila, G., Figueroa López, P., Solís Moya, E., Ireta Moreno, J., Hernández Muela, V. M., Zamora Villa, V. M., Vikram, P., Mathews, K., Sansaloni, C., Sehgal, D., Jarquin, D., Wenzl, P., & Singh, S. (2016). Genomic prediction models for grain yield of spring bread wheat in diverse agro-ecological zones. *Scientific Reports, 6,* 27312.

Sastry, J. G., Ramakrishna, W., Sivaramkrishnan, S., Thakur, R. P., Gupta, V. S., & Ranjekar, P. K. (1995). DNA fingerprinting detects genetic variability in the pearl millet downy pathogen (*Sclerospora graminicola*). *Theoretical and Applied Genetics, 91,* 856–861.

Sauer, N. J., Mozoruk, J., Miller, R. B., Warburg, Z. J., Walker, K. A., Beetham, P. R., *et al.* (2016), "Oligonucleotide-Directed Mutagenesis for Precision Gene Editing," *Plant Biotechnology Journal,* vol. 14, no. 2, February, pp. 496–502.

Sehgal, D., Skot, L., Singh, R., Srivastava, R. K., Das, S. P., Taunk, J., et al. (2015). Exploring potential of pearl millet germplasm association panel for association mapping of drought tolerance traits. PLoS ONE 10:e0122165. doi: 10.1371/ journal. pone.0122165

Serraj, R., Krishnamurthy, L., & Upadhyaya, H. D. (2004). Screening chickpea mini-core germplasm for tolerance to salinity. *International Chickpea and Pigeonpea Newsletter, 11,* 29–32.

Sharma, R., Rao, V. P., Upadhyaya, H. D., Reddy, V. G., & Thakur, R. P. (2010). Resistance to grain mold and downy mildew in a mini-core collection of sorghum germplasm. *Plant Disease, 94,* 439–444. https://doi.org/10.1094/PDIS-94-4-0439

Sharma, R., Thakur, R. P., Rai, K. N., Gupta, S. K., Rao, V. P., Rao, A. S., & Kumar, S (2009). Identification of rust resistance in hybrid parents and advanced breeding lines of pearl millet. *Journal of SAT Agricultural Research, 7,* 1–4.

Sharma, R., Upadhyaya, H. D., Manjunatha, S. V., Rai, K. N., Gupta, S. K., & Thakur, R. P. (2013). Pathogenic variation in the pearl millet blast pathogen, *Magnaporthe grisea* and identification of resistance to diverse pathotypes. *Plant Disease, 97,* 189–195. https://doi.org/10.1094/PDIS-05-12-0481-RE

Sharma, R., Upadhyaya, H. D., Manjunatha, S. V., Rao, V. P., & Thakur, R. P. (2012). Resistance to foliar diseases in a mini-core collection of sorghum germplasm. *Plant Disease, 96,* 1629–1633. https://doi.org/10.1094/PDIS-10-11-0875-RE

Sharma, S., Sharma, R., Govindaraj, M., Mahala, R. S., Satyavathi, C. T., Srivastava, R. K., Gumma, M. K., & Kilian, B. (2020a). Harnessing wild relatives of pearl millet for germplasm enhancement: challenges and opportunities. *Crop Science, 61,* 177–200.

Sharma, S., Sharma, R., Pujar, M., Yadav, D., Yadav, Y. P., Rathore, A., Mahala, R. S., Singh, I., Verma, Y., Deora, V. S., Vaid, B., Jayalekha, A. K., & Gupta, S. K. (2020b).

Utilization of wild Pennisetum species for improving biotic and abiotic stress tolerance in pearl millet (*Pennisetum glaucum* L.). *Crop Science, 61*, 289–304. https://doi.org/10.1002/csc2.20408

Shivhare R and Lata C (2017) Exploration of Genetic and Genomic Resources for Abiotic and Biotic Stress Tolerance in Pearl Millet. Front. Plant Sci. 7:2069. doi: 10.3389/fpls.2016.02069

Shrivasata, P., & Kumar, R. (2015). Soil salinity: A serious environmental issue and plant growth promoting bacteria as one of the tools for its alleviation. *Saudi Journal of Biological Sciences, 22*, 123–131. https://doi.org/10.1016/j.sjbs.2014.12.001

Singh, F., & Nainawatee, H. S. (1999). Grain quality traits. In I. S. Khairwal, K. N. Rai, D. J. Andrews, & G. Harinarayana (Eds.), *Pearl millet breeding* (pp. 157–183). Oxford and IBH.

Singh, K., Gupta, K., Tyagi, V., & Rajkumar, S. (2020). Plant genetic resources in India: Management and utilization. *Vavilovskii Zhurnal Genet Selektsii, 24*, 306–314. https://doi.org/10.18699/VJ20.622

Singh, P., Singh, U., Eggum, B. O., Anand Kumar, K., & Andrews, D. J. (1987). Nutritional evaluation of high protein genotypes of pearl millet (*Pennisetum americanum* (L.) Leeke). *Journal of the Science of Food and Agriculture, 38*, 41–48. https://doi.org/10.1002/jsfa.2740380108

Singh, S. D. (1995). Downy mildew of pearl millet. *Plant Disease, 79*, 545–550. https://doi.org/10.1094/PD-79-0545

Singh, S. D., Alagarswamy, G., Talukdar, B. S., & Hash, C. T. (1994). Registration of ICML 22 photoperiod insensitive, downy mildew resistant pearl millet germplasm. *Crop Science, 34*, 1421.

Singh, S. D., King, S. B., & Malla, R.,. P. (1990). Registration of five pearl millet germplasm sources with stable resistance to rust. *Crop Science, 30*, 1165. https://doi.org/10.2135/cropsci1990.0011183X003000050061x

Singh, S. D., King, S. B., & Werder, J. (1993). *Downy mildew disease of pearl millet* (Information Bulletin No. 37). International Crops Research Institute for the Semi-Arid Tropics.

Singh, S. D., Wilson, J. P., Navi, S. S., Talukdar, B. S., Hess, D. E., and Reddy, K. N. (1997). Screening Techniques and Sources of Resistance to Downy Mildew and Rust in Pearl Millet. Patancheru: ICRISAT

Soman, P., and Peacock, J. M. (1985). A laboratory technique to screen seedling emergence of sorghum and pearl millet at high soil temperature. Exp. Agric. 21, 335–341. doi: 10.1017/S0014479700013168

Soman, P., Jayachandran, R., and Bidinger, F. R. (1987). Uneven variation in plant to plant spacing in pearl millet. Agron. J. 79, 891–895. doi: 10.2134/agronj1987.00021962007900050027

Spindel, J. E., Begum, H., Akdemir, D., Collard, B., Redoña, E., Jannink, J. L., & McCouch, S. (2016). Genome-wide prediction models that incorporate de novo GWAS are a powerful new tool for tropical rice improvement. *Heredity*, *116*, 395–408.

Srivastava, R. K., Singh, R. B., Pujarula, V. L., Bollam, S., Pusuluri, M., Chellapilla, T. S., Yadav, R. S., & Gupta, R. (2020). Genome-wide association studies and genomic selection in pearl millet: Advances and prospects. *Frontiers in Genetics*, *10*, 1389. https://doi.org/10.3389/fgene.2019.01389

Stich, B., Haussmann, B. I. G., Pasam, R., Bhosale, S., Hash, C. T., Melchinger, A. E., et al. (2010). Patterns of molecular and phenotypic diversity in pearl millet [Pennisetum glaucum (L.) R. Br.] from West and Central Africa and their geographical and environmental parameters. BMC Plant Biol. 10: 216. doi: 10.1186/1471-2229-10-216

Sukumaran, S., Rebetzke, G., Mackay, I., Bentley, A. R., & Reynolds, M. P. (2022). Pre-breeding strategies. In M. P. Reynolds & H. J. Braun (Eds.), *Wheat improvement*. Springer. https://doi.org/10.1007/978-3-030-90673-3_25

Tapsoba, H., & Wilson, J. P. (1996). Pathogenic variation in *Puccinia substriata* var. *indica* in the south-eastern United States and screening for resistance in pearl millet germplasm. *Plant Disease*, *80*, 395–397.

Tapsoba, H., Wilson, J. P., & Hanna, W. W. (1997). Improvement of resistance to rust through recurrent selection in pearl millet. *Crop Science*, *37*, 365–369.

Thakur, R. P., Pushpavathi, B., & Rao, V. P. (1998a). Virulence characterization of single-zoospore isolates of *Sclerospora graminicola* from pearl millet. *Plant Disease*, *82*, 747–751.

Thakur, R. P., Rai, K. N., King, S. B., & Rao, V. P. (1993). Identification and utilization of ergot resistance in pearl millet (Research Bulletin No. 17). ICRISAT.

Thakur, R. P., Rao, V. P., & Hash, C. T. (1998b). A highly virulent pathotypes of *Sclerospora graminicola* from Jodhpur, Rajasthan, India. *International Sorghum and Millets Newsletter*, *39*, 140–142.

Thakur, R. P., Rao, V. P., Sastry, J. G., Sivaramakrishnan, S., Amruthesh, K. N., & Barbind, L. D. (1999). Evidence for a new virulent pathotypes of *Sclerospora graminicola* on pearl millet. *Journal of Mycology and Plant Pathology*, *29*, 61–69.

Thakur, R. P., Rao, V. P., Williams, R. J., Chahal, S. S., Mathur, S. B., Pawar, N. B., Nafade, S. D., Shetty, H. S., Singh, G., & Bangar, S. G. (1985). Identification of stable resistance to ergot in pearl millet. *Plant Disease*, *69*, 982–985.

Thakur, R. P., Rao, V. P., Wu, B. M., Subbarao, K. V., Shetty, H. S., Singh, G., Lukose, C., Panwar, M. S., Sereme, P., Hess, D. E., Gupta, S. C., Datta, V. V., Panicker, S., Pawar, N. B., Bhangale, G. T., & Panchbhai, S. D. (2004a). Host resistance stability to downy mildew in pearl millet and pathogenic variability in *Sclerospora graminicola*. *Crop Protection*, *23*, 901–908.

Thakur, R. P., Sharma, R., Rai, K. N., Gupta, S. K., & Rao, V. P. (2009). Screening techniques and resistance sources for foliar blast in pearl millet. *Journal of SAT Agricultural Research*, *7*, 1–5.

Thakur, R. P., Sharma, R., & Rao, V. P. (2011). *Screening techniques for pearl millet diseases*. Information Bulletin No. 89. ICRISAT.

Thakur, R. P., Shetty, H. S., & Khairwal, I. S. (2006). Pearl millet downy mildew research in India: Progress and perspectives. *International Sorghum and Millets Newsletter*, *47*, 125–130.

Thakur, R. P., & Shetty, K. G. (1993). Variation in pathogenicity among single-oospore isolates of *Sclerospora graminicola*, the causal organism of downy mildew in pearl millet. *Plant Pathology*, *42*, 715–721.

Thakur, R. P., Shetty, K. G., & King, S. B. (1992). Selection for host specific virulence in asexual populations of *Sclerospora graminicola*. *Plant Pathology*, *41*, 626–632.

Thakur, R. P., Sivaramakrishnan, S., Kannan, S., Rao, V. P., Hess, D. E., & Magill, C. W. (2004b). Genetic and pathogenic variability among isolates of *Sclerospora graminicola*, the downy mildew pathogen of pearl millet. In P. Spencer-Phillips & M. Jeger (Eds.), *Advances in downy mildew* (Vol. 2, pp. 179–192). Kluwer Academic Publishers.

Thakur, R. P., & Williams, R. J. (1980). Pollination effects on pearl millet ergot. *Phytopathology*, *70*, 80–84.

Toderich, K., Shuyskaya, E., Rakhmankulova, Z., Bukarev, R., Khujanazarov, T., Zhapaev, R., Ismail, S., Gupta, S. K., Yamanaka, N., & Boboev, F. (2018). Threshold tolerance of new genotypes of *Pennisetum glaucum* (L) R Br to salinity and drought. *Agronomy*, *8*, e230.

Upadhyaya, H. D. (2005). Variability for drought resistance related traits in the mini-core collection of peanut. *Crop Science*, *45*, 1432–1440. https://doi.org/10.2135/cropsci2004.0389

Upadhyaya, H. D., Dwivedi, S. L., Vadez, V., Hamidou, F., Singh, S., Varshney, R. K., & Liao, B. (2014). Multiple resistant and nutritionally dense germplasm identified from mini core collection in peanut. *Crop Science*, *54*, 2120–2130. https://doi.org/10.2135/crop-sci2014.01.0040

Upadhyaya, H. D., Reddy, K. N., Ahmed, M. I., Kumar, V., Gumma, M. K., & Ramachandran, S. (2016). Geographical distribution of traits and diversity in the world collection of pearl millet [*Pennisetum glaucum* (L.) R. Br., synonym: *Cenchrus americanus* (L.) Morrone] landraces conserved at the ICRISAT genebank. *Genetic Resources and Crop Evolution*, *64*, 1365–1381. https://doi.org/10.1007/s10722-016-0442-8

Upadhyaya, H. D., Reddy, K. N., & Gowda, C. L. L. (2007). Pearl millet germplasm at ICRISAT genebank – Status and impact. *SAT eJournal*, *3*.

Upadhyaya, H. D., Yadav, D., Reddy, K. N., Gowda, C. L. L., & Singh, S. (2011). Development of pearl millet minicore collection for enhanced utilization of germplasm. *Crop Science*, *51*, 217–223. https://doi.org/10.2135/cropsci2010.06.0336

USDA-ARS. (2002). *Germplasm Resources Information Network - (GRIN)*. USDA-ARS National Germplasm Resources Laboratory. http://www.ars-grin.gov/var/apache/cgibin/npgs/html/site-holding.pl?S9

Vadez, V., Hash, T., & Kholova, J. (2012). Phenotyping pearl millet for adaptation to drought. *Frontiers in Physiology*, *3*, 386. https://doi.org/10.3389/fphys.2012.00386

Van Oosterom E. J., Mahalakshmi V., Arya G. K., Dave H. R., Gothwal B. D., Joshi A. K., et al. (1995). Effect of yield potential, drought escape and drought tolerance on yield of pearl millet (*Pennisetum glaucum*) in different stress environments. *Indian J. Agric. Sci.* 65 629–635.

Varshney, R. K., Shi, C., Thudi, M., Mariac, C., Wallace, J., Qi, P., Zhang, H., Zhao, Y., Wang, X., Rathore, A., Srivastava, R. K., Chitikineni, A., Fan, G., Bajaj, P., Punnuri, S., Gupta, S. K., Wang, H., Jiang, Y., Couderc, M., . . . Xu, X. (2017). Pearl millet genome sequence provides a resource to improve agronomic traits in arid environments. *Nature Biotechnology*, *35*, 969. https://doi.org/10.1007/s00122-011-1580-1

Vavilov, N. I. (1949/1950). The origin, variation, immunity and breeding of cultivated plants. *Chronica Botanica*, *13*, 1–366.

Wahid, A., Gelani, S., Ashraf, M., & Foolad, M. R. (2007). Heat tolerance in plants: An overview. *Environmental and Experimental Botany*, *61*, 199–223.

Wang, L., Wang, X. Y., Weng, Q. F., Wu, B. E., & Cao, L. P. (2007). Identification of salt tolerance in Chinese prosomillet germplasm. *Journal of Plant Genetic Resources*, *8*, 426–429.

Wilson, J. P. (1991). Detection of pathogenic races of *Puccinia substriata* var. *indica* in the U.S. *Phytopathology*, *81*, 815.

Wilson, J. P. (1993). Two new dominant genes for resistance to pearl millet rust. *Phytopathology*, *83*, 1395.

Wilson, J. P., Burton, G. W., Wells, H. D., Zongo, J. D., & Dicko, I. O. (1989). Leaf spot, rust, and smut resistance in pearl millet landraces from Central Burkina Faso. *Plant Disease*, *73*, 345–349.

Wilson, J. P., & Hanna, W. W. (1992). Disease resistance in wild *Pennisetum* species. *Plant Disease*, *76*, 1171–1175.

Wilson, J. P., Hess, D. E., Hanna, W. W., Kumar, K. A., & Gupta, S. C. (2004). *Pennisetum glaucum* subsp. *monodii* accessions with *Striga* resistance in West Africa. *Crop Protection*, *23*, 865–870. https://doi.org/10.1016/j.cropro.2004.01.006

Winkel, T., Renno, J. F., and Payne, W. A. (1997). Effect of the timing of water deficit on growth, phenology and yield of pearl millet (Pennisetum glaucum (L.) R. Br.) grown in Sahelian conditions. J. Exp. Bot. 48, 1001–1009. doi: 10.1093/jxb/48.5.1001

Witcombe, J. R., Rao, M. N. V. R., Raj, A. G. B., and Hash, C. T. (1997). Registration of 'ICMV 88904' pearl millet. Crop Sci. 37:10221023. doi: 10.2135/cropsci1997. 0011183X0037

Xu, Y., Lu, Y., Xie, C., Gao, S., Wan, J., & Prassana, B. M. (2012). Whole genome strategies for marker assisted plant breeding. *Molecular Breeding, 29*, 833–854.

Yadav, A. K., Narwal, M. S., & Arya, R. K. (2011). Genetic dissection of temperature tolerance in pearl millet (*Pennisetum glaucum*). *Indian Journal of Agricultural Sciences, 81*, 203–213.

Yadav, A. K., Narwal, M. S., & Arya, R. K. (2012). Study of genetic architecture for maturity traits in relation to supra-optimal temperature tolerance in pearl millet (*Pennisetum glaucum* (L.) R.Br.). *International Journal of Plant Breeding and Genetics, 6*, 115–128.

Yadav, O. P. (2004). CZP 9802—a new drought-tolerant cultivar of pearl millet. Indian Farming 54, 15–17.

Yadav, O. P. (2006). Heterosis in crosses between landraces and elite exotic populations of pearl millet (*Pennisetum glaucum* (L.) R. Br.) in arid zone environments. *Indian Journal of Genetics, 66*, 308–311.

Yadav, O. P. (2008). Performance of landraces exotic elite populations and their crosses in pearl millet (*Pennisetum glaucum*) in drought and non-drought conditions. *Plant Breeding, 127*, 208–210.

Yadav, O. P. (2010). Drought response of pearl millet landrace-based populations and their crosses with elite composites. *Field Crops Research, 118*, 51–57.

Yadav, O. P. (2014). Developing drought-resilient crops for improving productivity of drought-prone ecologies. *Indian Journal of Genetics, 74*, 548–552.

Yadav, O. P., & Bidinger, F. R. (2007). Utilization, diversification and improvement of landraces for enhancing pearl millet productivity in arid environments. *Annals of Arid Zone, 46*, 49–57.

Yadav, O. P., & Bidinger, F. R. (2008). Dual-purpose landraces of pearl millet (*Pennisetum glaucum*) as sources of high stover and grain yield for arid zone environments. *Plant Genetic Resources, 6*, 73–78.

Yadav O. P., Bidinger F. R., Singh D. V. (2009). Utility of pearl millet landraces in breeding dual-purpose hybrids for arid zone environments of India. *Euphytica* 166 239–247. 10.1007/s10681-008-9834-y

Yadav, O. P., & Manga, V. K. (1999). Diversity in pearl millet landraces of arid regions of Rajasthan. In A. S. Faroda, N. L. Joshi, S. Kathju, & A. Kar (Eds.), *Management of arid ecosystem* (pp. 57–60). Arid Zone Research Association of India & Scientific Publishers.

Yadav O. P., Rai K. N. (2011). Hybridization of Indian landraces and African elite composites of pearl millet results in biomass and stover yield improvement under arid zone conditions. *Crop Sci.* 51 1980–1987. 10.2135/cropsci2010.12.0731

Yadav, O. P., Upadhyaya, H., Reddy, K. N., Jukanti, A. K., Pandey, S., & Tyagi, R. K. (2017). Genetic resources of pearl millet: Status and utilization. *Indian Journal of Plant Genetic Resources, 30*, 31–47. https://doi.org/10.5958/0976-1926.2017.00004.3

Yadav O. P., Weltzien-Rattunde E., Bidinger F. R., Mahalakshmi V. (2000). Heterosis in landrace-based topcross hybrids of pearl millet across arid environments. *Euphytica* 112 285–295. 10.1023/A:1003965025727

Yadav, O. P., Weltzein, E., & Bidinger, F. R. (2004). Diversity among pearl millet landraces collected in north western India. *Annals of Arid Zone*, *43*, 45–53.

Yadav, O. P., Weltzien-Rattunde, E., & Bidinger, F. R. (2003). Genetic variation in drought response of landraces of pearl millet (*Pennisetum glaucum* (L.) R. Br.). *Indian Journal for Genetics and Plant Breeding*, *63*, 37–40.

Yadav, T., Kumar, A., Yadav, R. K., Yadav, G., Kumar, R., & Kushwaha, M. (2020). Salicylic acid and thiourea mitigate the salinity and drought stress on physiological traits governing yield in pearl millet- wheat. *Saudi Journal of Biological Sciences*, *27*, 2010–2017.

Yakubu, H., Ngala, A. L., & Dugje, I. Y. (2010). Screening of millet (*Pennisetum glaucum* L.) varieties for salt tolerance in semi-arid soil of northern Nigeria. *World Journal of Agricultural Sciences*, *6*, 374–380.

Yella Goud, T., Sharma, R., Gupta, S. K., Uma Devi, G., Gate, V. L., & Boratkar, M. (2016). Evaluation of designated hybrid seed parents of pearl millet for blast resistance. *Indian Journal of Plant Protection*, *44*, 83–87.

6

Pearl Millet Hybrid Development and Seed Production

O. P. Yadav, S. K. Gupta, R. S. Mahala, Ajay P. Ramalingam, Sabreena A. Parray, and Ramasamy Perumal

Chapter Overview

Pearl millet [*Pennisetum glaucum* (L.) R. Br.] is the sixth most important crop grown for food, fodder and forage. It is a potential crop for regions having low soil fertility, high pH, low soil moisture, high temperature, high salinity, and limited rainfall. Being a highly cross-pollinated species, pearl millet was researched comprehensively in 1960s to exploit heterosis. The discovery of cytoplasmic-nuclear male-sterility (CMS) in US proved a milestone in exploiting heterosis commercially in India. With development of fertility restorers of hybrids, a new era of commercially viable hybrid development started. With 25-30% yield advantage in hybrids in comparison to open-pollinating varieties, hybrid development has been a top priority. Hybrids having high yield potential with maturity duration of 75–85 days, and tolerance to different biotic and abiotic stresses remain high priority. Genetic and cytoplasmic diversification is the most important strategy to control downy mildew and other diseases. Genetic diversity in parental lines is critical to enhance genetic potential of hybrids. A large number of genetically diverse hybrids are developed and deployed in different ecological regions. Consequently, productivity has increased from 303 kg/ha during 1950–1954 to 1239 kg/ha during 2020-24 in India due to the widespread use of high-yielding and disease-resistant cultivars with improved production technology. A robust seed production and delivery system is in place to provide high-quality seeds of improved hybrids. During the last three decades, yield levels in seed production plots have doubled, mainly due to the development of high-yielding parental lines and improved crop management skills of farmers. Government policies and protection of hybrids through the Protection of Plant Variety and Farmers' Rights Authority (PPVFRA) have played a pivotal role for increased investment in research and development by the private sector. Good opportunities are coming up to extend advantage of hybrid technology to eastern, central and west African countries.

Pearl Millet: A Resilient Cereal Crop for Food, Nutrition, and Climate Security, First Edition.
Edited by Ramasamy Perumal, P. V. Vara Prasad, C. Tara Satyavathi, Mahalingam Govindaraj, and Abdou Tenkouano.
© 2025 American Society of Agronomy, Inc., Crop Science Society of America, Inc., Soil Science Society of America, Inc. Published 2025 by John Wiley & Sons, Inc.

Introduction

Climate change with increased CO_2 and environmental temperature are major challenges to the present and future food supply of the world. A balanced environment warrants good adaptation strategies for food crops. Research focus on the development of smart cereal crops that are resilient to the effect of climate change is crucial to meet growing food demand. Predicted future climatic changes will still exacerbate the challenge of variable temporal and spatial distribution of precipitation across different agro-climatic regions. Recent studies clearly highlighted that drought and high temperature are the major challenges for crop yield losses, and it is expected that these losses will become more common and severe because of climate change.

Pearl millet [*Pennisetum glaucum* (L.) R. Br.] is the sixth most important nutritionally rich cereal crop after wheat, maize, rice, barley, and sorghum (FAOSTAT, 2023) and is known to be climate-change resilient. Its shorter life cycle allows it to grow efficiently with minimal moisture in high-temperature environments (Nedumaran et al., 2014). It is cultivated on >30 million ha worldwide, with a majority of the cultivated area in Africa and the Indian subcontinent (Gupta et al., 2015). The nutritional advantages of pearl millet grain come from higher levels of protein; low starch; high fiber; and higher concentrations of micronutrients (Fe and Zn), vitamins, essential amino acids, and antioxidants compared with rice, wheat, maize, and sorghum (Hassan et al., 2021; Muthamilarasan et al., 2016).

In India, pearl millet is the fourth most important food crop after rice, wheat, and maize. Its grain is valued as human food and its stover as important ration for livestock in crop-livestock based integrated systems. The major pearl millet growing states in India are Rajasthan, Maharashtra, Gujarat, Uttar Pradesh, and Haryana, which account for >90% of pearl millet acreage in the country. Most pearl millet in India is grown in the rainy (*kharif*) season (June–September). It is also being increasingly cultivated during the summer (February–May) in parts of Gujarat, Rajasthan, and Uttar Pradesh and during the post-rainy (*rabi*) season (November–February) at a small scale in Maharashtra and Gujarat.

In the United States, pearl millet is one of the unexplored alternative annual cereal crops (Bhattarai et al. 2019). With its short life cycle, pearl millet can be grown as grain, forage, and/or cover crop in dryland production systems. The high photosynthetic efficiency and dry-matter production capacity of pearl millet make it a highly desirable crop for growers in adverse agro-climatic regions where other cereals are likely to fail (Yadav & Rai, 2013). Pearl millet can also be a suitable alternative crop for replacing drought-sensitive food/feed crops in regions with receding groundwater levels, such as the Ogallala Aquifer Region of the United States. The vegetative, reproductive, and physiological characteristics of pearl millet make it a crop well suited for growth under difficult conditions, including low soil fertility, high pH, low soil moisture, high temperature, high salinity, and limited rainfall, where other cereals such as maize, rice, sorghum, and wheat would fail.

In the United States, the Fe-deficit anemia rate is reported to be 10% for women of childbearing age due to losses from menstruation, while 9% of children ages 12–36 months are Fe deficient, and one-third of these children develop anemia. Although the overall rate of Fe deficiency anemia is low in the United States, low-income families are particularly at risk. Including pearl millet in their diet would not only help to overcome the need of the required daily Fe nutrition but would also help in diversifying their food habits. Furthermore, pearl millet is gluten free, which makes it an alternative energy source to wheat and corn in the United States because 1–3 million people in the United States suffer from celiac disease associated with gluten-associated diets (Green et al., 2015) and has been reported to enhance the market for pearl millet grain in the United States (Hassan et al., 2021). Pearl millet is also a suitable feed ingredient in poultry diets (Davis et al., 2003). Its suitability for poultry diets ensures the bioavailability of micronutrients from its grain for monogastric animals and humans.

Yield and forage quality of pearl millet are similar to those of forage sorghum but with greater nutrient and water use efficiency (WUE). The WUE of forage pearl millet (280 kg dry matter ha/mm) is slightly lesser than forage sorghum (310 kg dry matter ha/mm) (Chapman & Carter, 1976; de Lima, 1998). Unlike sorghum, pearl millet possesses more tillers and a thin stem and has more feed value with high palatability. The absence of prussic acid in forage pearl millet makes it safe at any stage of the crop for animal grazing as well as harvesting. Dry matter digestibility of forage pearl millet (64.0%–69.0%) is slightly lower than corn (73.0%) but is similar to forage sorghum (63.0%) (Hassanat, 2007). Leaf-to-stem ratio serves as an indicator of forage quality. Low values indicate the dilution of plant mass with stem mass and thus a greater likelihood of lignin accumulation and lower forage quality (Machicek, 2018). Pearl millet forage produces an excellent leaf-to-stem ratio, resulting in lignin around 3.5%–6.3%, compared with 4.9% in corn and 8.3% in sorghum. Pearl millet also has higher crude protein (9%–18%) concentrations than corn (8.6%) and sorghum (13.3%) (Hassanat, 2007) and higher digestibility than sorghum and corn (Harinarayana et al., 2005; Kumar, 1989). Production of pearl millet silage (31 t/ha) is higher than sorghum (19 t/ha) and corn (27 t/ha) (Kichel et al., 1999). Pearl millet with brown midrib (*bmr*) is low in lignin content, which improves forage digestibility (Cherney et al., 1990; Sattler et al., 2014; Tang et al., 2014).

Pearl millet can be also grown as a forage and cover crop in dryland production in the Great Plains states (Colorado, Iowa, Kansas, Minnesota, Montana, Nebraska, New Mexico, North Dakota, Oklahoma, South Dakota, Texas, and Wyoming) of the Unites States. Its wide adaptability for double cropping and suitability for fall grazing at any stage due to the absence of prussic acid make it good cover crop for water-limited environments. Therefore, the development of new pearl millet forage and grain hybrids with better nutritive value will increase adoption of this drought-tolerant forage crop in the Great Plains region of the United States.

The choice of type of cultivar as a commercial crop is largely determined by its pollination behavior. Therefore, the floral biology of pearl millet has historically attracted the attention of researchers. Godbole (1925) studied pearl millet flowers in detail in the 1920s and established that each individual spikelet in its panicle bears two types of florets: bisexual and staminate. Later, Ayyangar et al. (1933) concluded that the stigma emerged after the maturity of spikelets, irrespective of whether the panicle has fully emerged from the boot or is still inside the boot. The stigma maturity is attained earlier than anther dehiscence, the phenomenon termed as "protogynous flowering" (Figure 6.1). The time difference between stigma emergence and the anther dehiscence varies from 24 to 48 hours and is responsible for cross-pollination (>85%) in pearl millet.

Cross-pollination is largely affected by wind (Leuck & Burton, 1966) and to some extent by insects. Some selfing cannot be ruled out because of pollination of the stigma from the pollen of panicles produced on the tillers of the same plant or from pollen produced by the same panicle on which the stigma is born. These flowering features also make controlled pollination in pearl millet very convenient. It is very

Figure 6.1   Pearl millet panicle at full stigma emergence.

easy to produce an adequate quantity of selfed, crossed, or sibbed seed in pearl millet by controlled pollination even from a small number of plants, making breeding operations easy and quick. The outcrossing nature of pearl millet ultimately provided the basis of exploitation of heterosis, which is expressed as the phenotypic and functional superiority exhibited in the first filial generation over the parents. In fact, heterosis has fascinated plant breeders ever since it was first described by Darwin (1876) in the vegetable kingdom and later elaborated in maize (East, 1908; Shull, 1908). Heterosis remains one singularity that is most exploited in crops despite being partially understood even after more than 100 years.

Heterosis is highly prevalent in pearl millet. An outcrossing rate >85%; ease of inbred development; the discovery of cytoplasmic male-sterility (CMS) and fertility restorer genes; the lack of any negative association of CMS with growth and development, diseases, and insect pests; expression of a positive and high magnitude of heterosis in productivity of hybrids; and economical seed production have provided a perfect platform for commercial exploitation of heterosis in pearl millet for the benefit of the farming community. This article reports some historical attempts to exploit heterosis in pearl millet, examines the strategic research undertaken to further exploit it commercially, documents the innovations in hybrid seed production, and suggests strategies to further enhance the degree of heterosis to deal with challenges in the production of pearl millet.

Initial Efforts in Heterosis Exploitation

A significant breakthrough in agriculture has been the use of heterosis, which has resulted in a huge increase in crop yield globally. Comparing the current grain yield of cereals with the pre-hybrid yield, there has been an almost fivefold increase (Srivastava et al., 2020a). In maize, heterosis (Shull, 1908) contributed to a quadrupling of maize kernel yield with superior vigor, which created a huge change in maize cultivation. Hybrid rice cultivars have higher yield potential than conventional inbred varieties by about 10%–20% (Tewodros & Mohammed, 2016). Rice production is increased by 20 million t in far-east Asia, with 55% total area of rice cultivation with hybrid rice. Pearl millet breeding has accomplished heterosis with enormous progress with improved agronomic and high-yielding hybrids (Jukanti et al., 2016).

The hermaphrodite nature of flowers of small size in pearl millet had put serious restrictions on exploiting heterosis at the commercial level. There were some innovative attempts in the late 1940s to exploit heterosis through developing "chance hybrids" (Burton, 1948). The method included growing four inbreds of similar maturity in the mixture and allowing them to cross-pollinate to produce seeds that contained ~40% hybrid seed. The chance hybrids outyielded local varieties marginally but could not become popular due to a lack of efficient seed production programs and their limited genetic superiority.

The discovery of cytoplasmic-nuclear male-sterility (CMS) proved a milestone in exploiting heterosis commercially, with increased yield potential of up to 25%–30% and resistance to several biotic and abiotic stresses (Burton, 1958). Several male-sterile lines were developed at Tifton, Georgia, and released. Two of them (MS Tift 23A and MS Tift 18A) were introduced in India in the early 1960s at the Punjab Agricultural University (PAU), Ludhiana. It was a landmark decision to use these male-sterile lines in the Indian pearl millet breeding programs, which were used extensively at the PAU and the ICAR-Indian Agricultural Research Institute (IARI), New Delhi. Tift 23A was found to be more promising because of its short stature, profuse tillering, uniform flowering, and good combining ability (Burton, 1965). This innovation and discovery of good restorers in Indian breeding material laid the foundation of pearl millet hybrid breeding in the world.

With the availability of male-sterile line Tift 23A, efforts were made to cross a large number of existing inbred lines available at the PAU, Ludhiana. One such cross of Tift 23 A and Bajra Inbred Line (BIL) 3B provided almost double the yield compared with the best check, and this was released as Hybrid Bajra-1 (HB-1) in 1965 (Athwal, 1966). The release of HB-1 dramatically improved the grain yield productivity of the crop in dry areas of India. This initiated a new era of improvement of pearl millet. HB-1 has the distinction of being the first CMS-based hybrid of pearl millet in the world and of any crop in India. The three-line hybrid system (i.e., A-line or male sterile line, B-line or maintainer line, and R-line or restorer line) established itself as a commercially viable method of hybrid development in pearl millet in 1960s. Since then, the current hybrid breeding program has been continuously building upon this strong foundation created almost six decades ago (Yadav et al., 2021).

Research Strategies to Exploit Heterosis

Breeding strategies in pearl millet have evolved very comprehensively over several decades, taking into account understanding of its pollination behavior, challenges in its production, access to germplasm, and accumulated knowledge in the fields of its genetics, physiology, and pathology.

Cultivar Type

Before the discovery of CMS, breeding efforts in genetic improvement of pearl millet, which started as early as the 1930s, attempted to capitalize on existing genetic variation within traditional landraces by subjecting them to simple mass selection and were moderately successful (Athwal, 1961). The greater urgency for population improvement programs started with the establishment of the International Crops Research Institute for the Semi-Arid Tropics (ICRISAT) when a diverse range of germplasm from across the world was acquisitioned in the 1970s (Gill, 1991; Khairwal, 1999). Eventually, a large number of populations and trait-based composites of the broad genetic base were established, and a diverse range of elite breeding materials was developed (Rai & Kumar, 1994; Singh et al., 1980).

With the discovery of CMS, new possibilities in hybrid breeding were initiated. With investment by and involvement of the private sector in pearl millet hybrid breeding, especially in India due to supportive government policies, the focus of genetic improvement continued to rise in hybrid breeding. The availability of global genetic resources and the extent of diversity present prompted their utilization (Yadav et al., 2012b). For example, a wide range of germplasm material from India and Africa with diverse phenotypic characteristics (e.g., tillering, panicle size, earliness, grain size, and grain color) was strategically exploited to diversify the genetic base of both seed and restorer parents (Andrews & Kumar, 1996; Patil et al., 2020; Rai et al., 2009; Yadav et al., 2012a). In the last four decades, hybrid breeding has received a very high priority in India using genetically diverse parental lines targeting various production ecologies that have helped to intensify the genetic gains (Rai et al., 2009; Yadav et al., 2012b).

Use of Alternative Cyto-sterile Sources

When utilization of cytoplasmic-nuclear male-sterile line Tift 23A in the breeding programs marked the beginning of new phase in pearl millet improvement in India, this line was widely used in hybrid development using pollinators bred in India. As a result, five hybrids (HB 1, HB 2, HB 3, HB 4, and HB 5) based on Tift 23A were released between 1965 and 1969. There also existed limited variation in pollinator parents of hybrids (Dave, 1986). Cultivation of few hybrids with narrow genetic base on a large scale led to downy mildew (DM; caused by *Sclerospora graminicola*) epidemics offsetting impressive achievements made in hybrid development in the mid-1970s (Dave, 1986). Such epidemics reappeared whenever a few hybrids occupied a large area year after year (Singh, 1995).

However, no association of A1 cytoplasm was established with the DM epidemic (Yadav, 1996; Yadav et al., 1993), and the recurring DM epidemics in pearl millet hybrids in India prompted intensified efforts on genetic diversification of hybrid parental lines, especially after the 1980s. This involved both cytoplasmic and nuclear diversification of parental lines. In addition to A1 CMS source (Burton, 1965), A2 and A3 (Athwal, 1961, 1966), A4 (Hanna, 1989), and A5 sources (Rai, 1995) were reported. Extensive characterization of these sources established the instability of A2 and A3 sources, whereas A4 and A5 were found to be more promising (Rai et al., 1996). This was followed by the development of several lines based on A4 and A5 sources by ICRISAT, but utilization of these two sources remained restricted because of a lack of suitable restorers. ICRISAT initiated breeding efforts for developing restorers, especially for the A5 CMS system (Rai et al., 1996, 2009). Research programs in India have now started breeding both A- and R-lines and developing hybrids based on these CMS systems. The understanding of the genetics of A4 (Gupta et al., 2012) and A5 CMS (Gupta et al., 2018) helped in the well-organized and efficient utilization of these CMS sources. As a result, a large number of male-sterile lines were made available in the background of different CMS sources (Table 6.1). This has significantly contributed to the genetic diversification of hybrids, minimizing any risk that might be associated with genetic uniformity.

Approaches for Development of Superior Parental Lines

Parental lines having better yield potential and tolerance to different biotic and abiotic stresses along with good combining ability is the key to the development of commercially feasible hybrids. Genetic diversity in parental lines is one of the important components to enhance genetic gain in hybrid breeding. Single nucleotide polymorphism (SNP) markers are commonly used to map the genetic distance of core germplasm and designated inbred lines to ensure desired diversity in the breeding program. New germplasm, like landraces and germplasm from gene bank or improved open pollinated varieties (OPVs), can be strategically involved in hybrid parent breeding programs based on the traits required in seed and restorer parents to enhance genetic diversity. Optimal levels of genetic diversity should be maintained in the breeding program to enhance the heterosis because heterosis does not increase or, it may decrease, after a certain level of genetic diversity.

Table 6.1 Number of Male-sterile Lines Based on Four Cytoplasmic Male Sterility Sources Bred at ICRISAT During 1986–2023

Period	Number of designated A-lines				
	A_1 cytoplasm	A_4 cytoplasm	A_5 cytoplasm	Others	Total
1986–1995	26	1	0	0	27
1996–2004	32	31	4	1	68
2005–2011	28	28	7	0	63
2012–2022	38	43	15	0	96
Total	124	103	26	1	254

Germplasm with traits of importance, like desirable maturity period and panicle and grain traits contributing toward high yield, should be continually identified and involved in the breeding pipeline to develop superior parental lines. Recycling of inbreds can be a very effective strategy. Table 6.2 indicates the involvement of diverse germplasm, composites, and elite breeding lines to strengthen the development of seed and restorer parents at ICRISAT. The hybrid parental lines should also have good combining ability along with other desirable traits; hence, early-generation testing should be an essential part of the hybrid breeding pipeline. To enhance heterosis, the programs should develop heterotic pools that should be as divergent as possible because heterosis depends upon the differences in the allele frequency between two populations. In pearl millet, studies have indicated that seed (B-lines) and pollinator parents (R-lines) exist as two separate heterotic pools, thus leading to continuous enhancement of productivity (Gupta et al., 2020). New germplasm should be continuously introgressed into established heterotic pools, which helps in the development of superior hybrid parental lines. Many commercial breeding programs use a well-defined breeding matrix approach to optimize the number of progeny rows, the extent of diversity, and selection percentage during generation advancement based on general combining ability (GCA) and SCA data. They rely on prediction models to design new breeding crosses based on genetically estimated breeding value and phenotypically estimated breeding value data.

Table 6.2   Type of Material Used in The Development of Designated Maintainer (B) Lines (1981–2018) and Restorer (R) Lines (2006–2019) of Pearl Millet Bred at ICRISAT, India

| Type of genetic material | Number of lines derived | | Remarks |
	B-lines	R-lines	
Germplasm	3	2	Inbreeding and selection directly from germplasm
Composites	17	66	Includes composites and open-pollinated varieties
Germplasm × elite line	22	6	Includes early-generation breeding lines derived from crosses between germplasm and elite lines
Composite × elite line	37	6	Includes early-generation breeding lines derived from crosses of composites and elite lines
Elite line × elite line	124	42	Includes crosses between elite lines from advanced generations

Speed breeding is also gaining importance to accelerate inbred development process. Superior parental lines are those that not only make superior hybrids but also produce high seed yield per se. Hence, parent yield data are also used as one of the components in designing new breeding crosses.

Marker-assisted Selection

Substantial efforts have been made to develop and use genomic tools for increasing grain yield, yield stability, and productivity of hybrid cultivars in pearl millet. The use of marker-assisted backcross methods in pearl millet genetic improvement has attempted to enhance drought tolerance of elite inbred pollinator line H 77/833-2using donor PRLT 2/89-33 line and elite inbred seed parent maintainer genotype ICMB 841 using 863B line as a donor. Several validated QTLs have been introgressed into elite hybrid parental lines (A-/B-, R-), resulting in the improved version of the hybrids or essentially derived varieties. QTLs associated with DM resistance in pearl millet H 77/833-2 line and two more QTLs introgressed using marker-assisted backcross have resulted in the development of a male parental line with improved DM resistance. The popular hybrid HHB 67 Improved was bred by a cross between improved DM-resistant seed parent 843-22A and improved restorer parent H 77/833-2-202 line (Hash et al., 2006). HHB 67 Improved is currently grown in >10% of the entire crop area every year in India and is one of the most impactful research outputs in crop plants globally. Recently, efforts were made to further improve HHB 67 Improved hybrid. Several DM-resistant double QTL introgression lines were generated in the genetic background of restorer parent H 77/833-2-202. These improved introgression lines were used to produce HHB 67 Improved–like test-cross hybrids with enhanced levels of DM resistance (Srivastava et al., 2020b).

Strategic Use of Germplasm

Characterization of the global germplasm collection comprising nearly 23,000 accessions from 51 countries suggested the existence of large variations for phenotypic and phenological traits (Yadav & Rai, 2013). Generally, Indian pearl millet landraces possessed earliness, high tillering, high harvest index, and local adaptation, whereas African material has been a good source of bigger panicles, large seed size, and disease resistance. Such information provided the basis of the strategic use of germplasm in breeding of hybrid parental lines. These two groups of germplasm sources continue to play a critical role in crop improvement programs across the world. As a result, a large number of genetically diverse hybrids have been developed with different combinations of phenotypic traits that are important for adaptation to different ecological regions.

Classifying Target Environment

Delineation of production area into mega-environments has been another strategic move in genetic improvement programs given that a single pearl millet cultivar cannot be expected to perform well under all environmental conditions and a cultivar planted outside its adaptation zone would suffer yield reduction due to significant genotype × environment interactions. Therefore, breeding and evaluation require a subdivision of the testing environments into relatively more homogeneous groups of locations, called mega-environments, where specific genotypes can be targeted for an individual mega-environment. Using multi-location and multi-year data, Gupta et al. (2013) found that all varietal test locations across all the pearl millet growing zones can be represented by two mega-environment,s which is analogous to the All India testing procedure (Yadav et al., 2012b) of testing of selected hybrids with specific adaptation in their respective adaptation zones. Different pearl millet breeding programs in India develop their product profiles depending on the need of the target mega-environment that is delineated by geographical locations and rainfall patterns.

Trait Priorities

The importance of traits is dictated by the roles of specific traits in determining productivity and adaptation. Similarly, the choice of farmers' community for various phenotypic traits that influence crop appearance is equally significant.

Phenotypic Traits in Hybrids

High grain yield, disease resistance, and maturity duration of 75–85 days, as per the agroecological requirements, have been accorded the highest priority (Yadav & Rai, 2013). Because of the growing importance of dry stover for fodder purposes, there has been a considerable emphasis, in recent times, on breeding for dual-purpose cultivars producing both high stover and high grain yields (Yadav et al., 2012a). Some of the traits in hybrids common to all the environments include lodging resistance, compact panicles, good exertion, and good seed set.

Traits that have regional preferences include various maturity types, tillering, panicle size (a combination of panicle length and girth), and seed size and have been strategically manipulated. This was evidenced from a recent investigation where 122 commercial hybrids showed the existence of significant variation for flowering time (42–58 days), tillering (1.1–4.4 panicles/plant), individual grain size (7.6–17.3 mg), plant height (185–268 cm), and panicle length (20–33 cm), which highlighted the successful efforts of the Indian program of pearl millet improvement toward genetic diversification of hybrids (Yadav et al., 2017).

Phenotypic Traits in The Hybrid Parental Line

The d_2 dwarfing gene-based shorter height is the most dominant plant type developed in seed parent breeding (Rai & Hanna, 1990; Rai & Rao, 1991) because it reduces the risk of lodging in high-management conditions and helps in easy detection of off-type and pollen shedders in the seed production plots. The A-lines have been bred for complete and stable male sterility and B-lines for profuse pollen production ability across seasons and sites. In the breeding of A-lines, high grain yield potential, both as lines per se and in hybrids (i.e., combining ability), is the most important consideration. High yield, however, is achieved in combination with other agronomic and farmers' preferred traits.

New plant types such as A-lines with long panicles (30–80 cm compared with standard normal of 10–20 cm), thick panicles (40–50 mm diameter compared with standard normal of 20–30 mm diameter), and large seed size (17–20 g of 1,000-seed mass compared with standard normal of 9–12 g) are being developed in India. The foremost requirement in the restorer lines is to produce highly fertile hybrids and to produce profuse pollen that remains viable at air temperatures as high as 42–44 °C. It is desirable to breed pollinators that are taller than A-lines, usually in the range of 150–180 cm height with built-in attributes of panicle, maturity, and tillering that will be preferred by farmers. Besides being able to produce high-yielding hybrids, pollinators must have acceptable level of lodging resistance and should also possess adequate levels of resistance to various diseases.

Considering all these traits and the diverse nature of agro-ecologies in which pearl millet is grown, product profiles of hybrids and parental lines have been developed to fulfill the future needs of hybrid breeding programs on long-term basis (Table 6.3). There has been a clear distinction between the public and private sector hybrid breeding programs with respect to trait-based breeding. For instance, private-sector breeding programs have largely focused on relatively better endowed environments, giving greater emphasis to breeding dual-purpose hybrids. As a result, private-sector hybrids are generally taller, later in maturity, with longer panicles and fewer effective tillers per plant (AICPMIP, unpublished data). In contrast, public-sector hybrids are generally shorter, early in maturity, with smaller panicles and a higher number of effective tillers per plant. The private sector has also placed greater emphasis on breeding large-seeded hybrids. ICRISAT follows a trait-based breeding approach to develop a diverse range of materials to meet the above-mentioned diverse needs for various agronomic traits, with high grain yield as a common trait, as is the case with all the hybrid breeding programs. These trait-specific breeding lines and hybrid parents are disseminated to both sectors through constitution of trait-specific trials, which are evaluated in several locations every year by public-sector breeding programs under ICAR-ICRISAT Research Partnership Project. Under this project, ICRISAT

Table 6.3 Product Profiles of Pearl Millet Breeding Program for India

Product profile	Estimated area	Area/ effort	Target agro-ecologies	Product development goals
	ha	%		
Parent lines of medium to late maturing, dual-purpose hybrids with disease resistance, for adaptation to better endowed environments	4.0 m ha (rainy season)	65	Target: India: East Rajasthan, Central & South Gujarat, Haryana, Uttar Pradesh, Maharashtra, Peninsular India (400–700 mm/ annum)	Must-have traits: 1) Parents with high productivity and good GCA for grain yield to develop hybrids at least 10% increase in grain yield (3–4 t/ha) with improved fodder yield 2) 76–90 days maturity 3) Downy mildew ($\leq$10%) and blast resistance (<3.0 score) 4) High grain Fe ($\geq$60 ppm) and Zn ($\geq$35 ppm) content Long-term priority: 1) Lodging tolerance 2) Low rancidity 3) N-use efficiency 4) Stover digestibility
Parent lines of early maturing, dual purpose hybrids with disease resistance for adaptation to drought prone environments	Total area of 3.5 m ha	20	Target: India: Western Rajasthan and drier parts of Gujarat and Haryana (200–400 mm/ annum)	Must have traits: 1) Parents with high productivity and good GCA for grain yield to develop hybrids at least 10% increase in grain yield (2.0–2.5 t/ha) over representative checks with improved fodder yield (~2.0 t/ha) 2) 65–75 days maturity 3) Downy mildew ($\leq$10%) and blast resistance (<3.0 score) 4) Improved terminal drought tolerance 5) High grain Fe ($\geq$50 ppm) and Zn ($\geq$35 ppm) content and better fodder quality. Long-term priority: 1) Stover digestibility 2) Seedling-stage heat tolerance

(continued overleaf)

Table 6.3 (Continued)

Product profile	Estimated area	Area/ effort	Target agro-ecologies	Product development goals
	ha	%		
Parent lines of medium to late maturing, dual-purpose hybrids with disease resistance, for adaptation to high heat stress ecology in summer	0.6–0.8 m ha (and increasing)	5	Target: India: North Gujarat, South Rajasthan, Western UP (summer season)	Must-have traits: 1) Parents with high productivity and good GCA for grain yield to develop hybrids at least 10% increase in grain yield (5–6 t/ha) over representative checks with improved fodder yield 2) 85–90 days maturity 3) Downy mildew (≤10%) and blast resistance (<3.0 score) 4) Flowering-stage heat tolerance (summer season; >42 °C) 5) High grain iron (≥60 ppm) and zinc (≥35 ppm) content
Cultivars and hybrid parents exclusively for forage and high biomass	1.0 m ha	10	Target: India: Gujarat, Punjab, Rajasthan, UP, MP, Peninsular India (both summer and rainy season)	Must-have traits: 1) 10% increase in biomass yield over best check with comparable or improved fodder quality traits 2) Green biomass of 40–55 t/ha, dry biomass of 15–20 t/ha 3) Non-hairy leaves with leaf/stem ratio of 3–5 4) IVDMD of ≥50% with crude protein of ≥10% 5) Single cut (50–80 days); Multi-cut (50–110 days) 6) Downy mildew (≤10%), rust (≤10%), and blast resistance (<3.0 score)

Abbreviations: GCA, general combining ability; IVDMD, in vitro dry matter digestibility.

also constitutes and disseminates traits-based trials with respect to other traits, such as lodging resistance, compact panicles, bristled panicles, dwarf height, and stay-green.

Disease Resistance

Pearl millet production is confronted with relatively fewer biotic stresses as compared to other crops. Among the diseases, DM is the most important constraint, especially on genetically uniform hybrids. Other diseases include smut (caused by *Moesziomyces penicillariae*), rust (caused by *Puccinia substriata* var. *indica*), blast (caused by *Pyricularia grisea*), and ergot (caused by *Claviceps fusiformis*). Being a crop grown by resource-poor farmers, diseases and pests are best managed through host plant resistance (HPR) because it does not incur additional cost (Yadav et al., 2021). Disease management through HPR involves a sound knowledge of biology and epidemiology of diseases, availability of pure culture of the pathogens, effective inoculation techniques, greenhouse and field screening facilities, appropriate disease rating scales, and availability of resistance sources. From among various diseases, DM resistance has been accorded highest priority (Thakur et al., 2006).

Insect-pest Resistance

Although many insect pests have been reported in pearl millet, only a few of them are of some, albeit little, economic importance in causing losses to the crop. The insect-pest incidence on commercial cultivars and experimental test genotypes is closely monitored during breeding of cultivars (Yadav et al., 2021).

Drought Tolerance

Pearl millet is severely challenged by drought, caused by the low rainfall and its erratic distribution, in all production environments of South Asia and sub-Saharan Africa. The subject of drought tolerance in pearl millet has remained a very important strategic issue. Dissection of drought tolerance in terms of physiology, phenology, and morphology of the crop has led to the understanding of the yield formation process under drought (Van Oosterom et al., 2003; Yadav et al., 2011), helping breeders to identify and target specific traits in different drought environments. Morphological traits (e.g., high tillering, small grain size, and shorter grain-filling periods) that can be measured easily have been manipulated successfully in breeding programs because there is abundant variation for these traits (Yadav et al., 2017).

Heat Tolerance

Whereas drought is a common occurrence, heat stress assumes importance in specific regions. During the last decade, pearl millet has emerged as a highly productive and remunerative crop in the hot and dry summer season in the northern and western parts of India (Yadav & Rai, 2013). With higher air temperatures (often >42 °C) coinciding with flowering, the crop suffers from reproductive sterility, leading to drastic reductions in seed set and finally to lower grain yield (Djanaguiraman et al., 2018; Gupta et al., 2015). Heat tolerance at the reproductive stage has emerged as an important target trait to enhance genetic gains.

Grain Nutrient Density

The goal of core breeding has been to increase yield potential because pearl millet has been considered as a highly nutritious cereal with higher levels of proteins and several minerals than other major cereals. Improving grain nutritional traits is a recent addition to breeding objectives, in view of global recognition of widespread deficiencies of Fe and Zn. The major areas addressed include the assessment of the extent of genetic variation for grain Fe and Zn contents, identification of diverse seed-mineral dense germplasm, nature of the genotype × environment interaction, relationships between grain minerals and agronomic traits, and genetic control of micronutrients (Govindaraj et al., 2019a, 2019b).

Development of biofortified cultivars started with exploration of existing genetic variation within open-pollinating composite varieties because such material is genetically heterogeneous, with significant variation for quantitative traits. Based on this premise, ICTP 8203 was subjected to three generations of progeny-based selection to improve Fe content in its grain.

The focus was to develop a higher-Fe version of ICTP 8203 without compromising grain yield and without causing changes in other traits. Therefore, 11 S3 progenies with 92–165 ppm Fe content and 40%–65% selfed seed set were selected and random-mated in the late-summer planting to constitute an improved higher-Fe version of ICTP 8203, which was designated and tested as ICTP 8303 Fe10-2 and later named and released as Dhanashakti, which was the first biofortified cultivar released in any crop in India. Dhanashakti produced a mean grain yield of 2.2 t/ha, which was 11% higher than that of ICTP 8203 (1.97 t/ha) (Rai et al., 2014). Also it yielded 5.3 t/ha of dry stover yield, which was 13% higher than that of ICTP 8203 (4.7 t/ha). Although both varieties were similar for time taken to flowering (45 days), Dhanashakti was 7 cm taller than ICTP 8203, which could have contributed to its higher stover yield. Limited testing showed that both varieties were equally and highly resistant to DM and that Dhanashakti had slightly better resistance to blast and rust. This success prompted more efforts to develop biofortified hybrids.

Evaluations to identify sources for high Fe and Zn grain contents led to identification of a diverse range of material for high Fe and Zn in elite agronomic backgrounds. Several hybrid parental lines were identified as moderate-to-high Fe lines and good general combiners for use in hybrid breeding (Govindaraj et al., 2019a, 2019b). Eight hybrids (HHB 299, AHB 1200 Fe, AHB 1269 Fe, Phule Mahasakthi, RHB 233, RHB 234, HHB 311, and HHB 67 Improved 2 for high protein) and one composite variety (ABV 04) have been released for cultivation in different agro-ecologies (Yadava et al., 2020). It is important to note that higher adoption of biofortified pearl millet hybrids and varieties in the long run largely depends on higher Fe/Zn contents coupled with high yield, DM resistance, and drought tolerance.

Unlike grain yield, the performance of lines is significantly and positively correlated with the GCA for Fe and Zn in pearl millet, implying the lines selected for high Fe and Zn will also be high general combiners for these micronutrients (Govindaraj et al., 2013; Kanatti et al., 2014a,; Rai et al., 2012; Velu et al., 2011). It has been observed that inbreeding had either no significant effect or had marginally increased both micronutrients (Rai et al., 2017). In contrast to the low heritability and inbreeding depression of grain yield, micronutrient contents are highly heritable, and hybrids can be readily improved through hybrid parents breeding.

So far, the best source of high Fe and Zn contents in pearl millet is found to be *iniadi* germplasm (Govindaraj et al., 2013; Rai et al., 2012; Velu et al., 2011), which refers to an early-maturing and large-seeded landrace found in the adjoining parts of Togo, Ghana, Benin, and Burkina Faso (Andrews & Kumar, 1996). Considering the additive gene action and that one source of germplasm genes introgressed in both parental lines, it is expected to reduce genetic diversity between male and female groups for other important traits. This may also lead to reduced heterosis for yield traits, which are predominantly under non-additive gene control. Thus, genomics approaches for selective introgression of genes for Fe and Zn contents in the parental lines without disrupting the diversity for other traits can play a major role in future biofortification breeding. New sources, other than *iniadi*, of Fe and Zn contents in the germplasm collections are also being explored at ICRISAT for genetic diversification for high Fe and Zn.

Given that heterosis is largely explained either due to dominance or overdominance effects, there has been no better-parent heterosis for Fe and Zn reported in pearl millet since the predominance of additive gene action in the genetic control of Fe and Zn contents. This indicates that there would be little opportunity to exploit better-parent heterosis for improving these micronutrients.

However, development of hybrids with high Fe and Zn contents requires incorporation of high Fe and Zn genes into both parental lines of hybrids where the midparent heterosis of a hybrid is gradually increased. Therefore, to breed high-Fe/Zn hybrids, all potential parental lines should be characterized for these micronutrients, and only selected lines should be hybridized.

Available Grain and Forage Hybrids

More than 100 hybrids have been developed and deployed in different production environments across India in the last 25 years by both public and private sectors. In addition, a large number of private-sector hybrids are commercialized as truthfully labeled seed. This has enabled farmers to choose from a wide range of available cultivars with appropriate trait combinations that they consider fit to meet their requirement in different crop production environments of various states in India. However, hybrids for African regions have just started (Sattler & Haussmann, 2020). Similarly, forage hybrids are being targeted in the United States, Brazil, and Australia.

Besides grain production, pearl millet is grown as feed and forage crop for livestock grazing, silage, hay, and green fodder chopping and is fed to animals at any crop growth stage without adverse effect in a range of countries, such as the United States; in the summer season in Australia, Canada, and Mexico; and in triple-cropping systems in southern Kyushu, Japan, Iran, Central Asia, and Brazil (Govintharaj et al., 2021). Recently, it has emerged as an important fodder crop during the summer months of northwestern India (Amarender Reddy et al., 2013). Pearl millet has high tillering potential and quick regenerative ability, assuring the possibility of multi-cutting, which allows the year-round supply of green/dry forage. Almost all the released forage cultivars of pearl millet available in the global or Indian market are either OPVs or single-cross hybrids with a single/multi-cut habit. There is huge demand for multi-cut (two or three cuts per season) forage cultivars with high crude protein and in vitro organic matter digestibility traits to meet the year-round need of forage/feed. Recent studies have indicated that forage yields of three-way cross hybrids are higher than OPVs and top-cross hybrids, whereas forage quality traits are comparable. Also, seed production is economical in the case of three-way hybrids because F_1 sterile hybrids (female parent of the three-way top-cross hybrid) produce 40%–70% higher seed yield than the female inbred parent of a single-cross forage hybrid (Gupta et al., 2022). Considering these advantages, a few three-way forage hybrids have recently been released in the Indian market.

Hybrid Yield Potential Compared with Traditional Cultivars and Landraces

Two types of commercial cultivars are available in pearl millet, including hybrids and open-pollinated varieties. In addition, landraces are also grown in large parts of the African continent. In a comprehensive study that compared 142 hybrids and 84 composites in 94 drought-prone environments for 12 years, hybrids had an overall superiority of 25% in their yield performance and had comparable dry stover yield (Yadav et al., 2017). Several studies have also assessed comparative performance of landraces and improved genetic material in contrasting environments and established that the more severe the drought stress intensity, the better the relative performance of landrace-based materials compared with that of conventionally bred materials (Weltzien & Witcombe, 1989; Yadav & Bidinger, 2007, 2008; Yadav & Weltzien Rattunde, 2000).

It has also been shown that landraces possess good levels of drought adaptation but fail to capitalize on yield-enhancing conditions. On the other hand, elite genetic material has a greater yield potential expressed under better endowed conditions but lacks adaptation to severe drought stress conditions. Hybridization between such materials showed that crosses between them had adaptation range beyond that of their parental populations because they were better able than their landrace parents to capitalize on the additional resources of good growing seasons and simultaneously have a better capacity than their elite parents to tolerate drought. Such materials have been identified as suitable to drive hybrid parental lines for a wide range of conditions (Crossa & Gardner, 1987; Yadav & Rai, 2011). Farmers in west Africa typically cultivate a certain pearl millet landrace in a given area, but there is little variation across individual landraces (Ndoye & Gahukar, 1987). Pearl millet production is low in west Africa because most of the farmers still depend on OPVs and landraces that have lower grain yield (Nigeria, 471 kg/ha; Senegal, 694 kg/ha). In contrast, Indian farmers began growing pearl millet hybrid cultivars several decades ago, which, together with improved agronomic procedures, resulted in a massive grain output rise from 305 kg/ha (1951–1955) to 1,157 kg/ha (2010–2014) (Sattler et al., 2019). In the Sudano-Sahelian target zone, established cultivars are generally outperformed by pearl millet population hybrids produced from west African OPVs. Pucher et al. (2016) studied 20 West Africa pearl millet OPVs and their 100 offspring produced from a 10 × 10 factorial at six locations in West Africa in a year and reported a panmictic mid-parent heterosis and panmictic better-parent heterosis for grain yield up to 73% and 69% with increased means of 17% and 4%, respectively. Five improved Sahelian landraces were tested for their potential to combine at two sites over the course of 2 years in Niger using a diallel mating strategy and resulted a panmictic better-parent heterosis up to 81%, with an average of 44% for grain yield (Ouendeba et al., 1993).

Public- and Private-Sector Efforts for Hybrid Promotion

Pearl millet OPVs differ from hybrids primarily in their ability to self-perpetuate and are genetically and phenotypically heterogeneous populations. Seed harvested from an isolated plot of an OPV should have essentially the same yield potential and uniformity level as the crop from which it was harvested, at least for two to three crop cycles. In the case of hybrids, it is necessary to reconstitute the cultivar by crossing its parents to retain uniformity and yield potential of the cultivar. Therefore, the private sector has been exclusively involved in hybrid breeding and seed delivery, whereas the public sector has been undertaking both hybrid and OPV breeding. ICRISAT focused on population development and development of OPVs in the 1970s. Since the 1980s, the ICRISAT research program for Asia has put a greater emphasis on hybrid breeding (Yadav et al., 2012b).

Hybrid cultivars have several advantages over OPVs. They include higher yield potential at comparable maturity; the ability to combine desirable characters from different parents into a single cultivar; and greater uniformity, facilitating harvest operations and farmers' appeal. Further, closed pedigree and the necessity of reconstituting hybrids from their parents allow the private seed sector to protect intellectual property owing to investment in research and development. Both private and public sectors have been able to take the hybrid technology to farmers' fields in a complimentary way. Hybrid parental lines and segregating material suitable for developing male-sterile and restorer lines are shared with the private sector through Hybrid Parents Research Consortium of ICRISAT (Mula et al., 2007). Public-sector hybrids are made available to the private sector for seed production and marketing through mutually agreed upon terms and conditions.

The most extensive and productive partnership of the Indian public sector with the private seed sector has been in the area of multi-location evaluation of hybrids. Private seed companies not only evaluate their hybrids in the All India Coordinated Trial System but also offer their test locations in varied environments, thus permitting very extensive testing before the hybrids are identified for release. Government policy support to the seed industry is a must to deliver the product of improved genetics. One of the most important decisions of government of India that made seed industry viable and competitive is protection of their hybrids and their parental lines through the Protection of Plant Variety and Famers' Right Act by establishing an independent authority in 2001. The government also provides permission to private seed company to develop their own product and deploy it in target regions through Truthfully Labeled Seed. Presently, there is a market of more than 20,000 t of hybrid seed in India, and the private companies have major share in it.

Genetic Gain in Productivity

The trends in pearl millet production in different regions present very interesting trends. The increased production in SSA has happened largely due to increase in area, with a little contribution from increased productivity. In contrast, pearl millet productivity in India has increased from 303 kg/ha during 1950–1954 to 1239 kg/ha during 2015–2019, which translates to an increase of >300% owing to the widespread use of high-yielding and disease-resistant cultivars with improved production technology.

A critical analysis of the genetic improvement has been recently done in which seven decades of breeding were divided into four unique phases (Yadav et al., 2019) (Table 6.4). During Phase I (1950–1966), when genetic improvement largely concentrated on the enhancement of yield in locally adapted materials, the rate of productivity improvement was 4.5 kg/ha per year. Discovery and use of CMS in hybrid development marked Phase II (1967–1983), in which an annual increase of 6.6 kg/ha in productivity was realized despite the large-scale cultivation of a few hybrids. A large number of genetically diverse CMS lines were developed and used in hybrid breeding during Phase III (1984–2000), and the productivity increase was 19.0 kg/ha per year. During Phase IV (2001 onward), when genetic improvement put much greater emphasis on genetic diversification of hybrids and adaptation to niche areas of cultivation, the rate of improvement in grain productivity further increased to 31.1 kg/ha per year, which is 470% of the productivity gain achieved during Phase I.

The annual rate of gains in productivity in India is the combined outcome of improved genetics and management. This high quantum of productivity increase

Table 6.4 Different Phases of The Hybrid Development Program of India, Their Uniqueness, and Productivity Gain Achieved

Phase	Duration of phase	Uniqueness of phase	Gain in productivity
			kg/ha/yr
I	1950–1966	Enhancement of yield in locally adapted materials	4.5
II	1967–1983	Discovery and utilization of CMS in hybrid development	6.6
III	1984–2000	Utilizing genetically diverse CMS lines	19.0
IV	2001 onward	Genetic diversification of hybrids	31.1

Abbreviations: CMS, cytoplasmic male sterility.

in pearl millet assumes greater significance in two ways. First, >90% of pearl millet is grown as rainfed and often on marginal lands. Second, pearl millet has attracted much less infrastructure and human resources in comparison to other food crops. It also affirms the correctness of priorities set in the breeding programs and simultaneously demonstrates the role of hybrid technology in raising crop productivity in marginal drylands.

The development of genomic resources, the creation of genetic linkage maps, and identification of the genomic regions have created new opportunities in genomics-assisted breeding, which is poised to improve the precision and efficiency of the hybrid breeding for higher genetic gains. Genome sequencing of pearl millet (Varshney et al., 2017) has laid a solid foundation for carrying out trait discovery, mapping, and deployment of QTLs/alleles/candidate genes linked to traits of economic interest. Further, the whole-genome resequencing of the Pearl Millet Inbred Germplasm Association Panel as well as mapping population parents and elite hybrid parental lines have helped to develop a huge (>32 million) repository of genome-wide SNPs. These developments offer opportunities to rapidly map and deploy genes of agronomic importance and to rapidly re-sequence lines to mine and map genes of interest. The sequencing-based mapping strategy can also help us identify superior haplotypes for different traits to form the basis of haplotype-based breeding (Sinha et al., 2020).

Many traits of agronomic importance related to diseases, terminal drought, and grain and fodder quality, combining ability loci and heterotic gene pools, have been mapped with simple sequence repeat (SSR) markers. These SSR intervals can be mined for the presence of linked SNPs. The available QTLs can also be re-mapped using the currently available SNP-based high-throughput genotyping (HTPG) systems. This will allow integration into the modern breeding pipelines using HTPG platforms available currently in pearl millet (Srivastava et al., 2020a). Furthermore, with the availability of reference genome sequencing, large-scale whole-genome resequencing data, a cost-effective genotyping platform, and precise phenotyping platforms, it has become possible to quickly map breeding-related traits in hybrid development to achieve greater genetic gains.

Hybrid Seed Production

A robust pearl millet seed production program is central to providing high-quality seeds of improved hybrids. "High quality" in the case of seed refers to high genetic purity (true to type), high physical purity (freedom from objectionable weeds, other crops' seeds, inert matter, etc.), and high seed vigor and germination, in addition to freedom from seed-borne diseases. The use of improved seed has made a tremendous impact on pearl millet productivity.

Production Planning

Planning is the most critical component of seed production and essentially involves need assessment and carry-over strategy.

Need Assessment

Need assessment is the first most important step in production planning to produce adequate quantity of all classes of seed. An example of the method for calculating quantity of different classes of seed and land requirement is given for pearl millet seed production in Table 6.5.

Yield potential of parental lines of hybrids and OPVs is the most critical factor in determining the requirement of area and seed, although yield levels are also influenced by climate, soil conditions, and crop management.

Carry-over Strategy

The quantity of seed of any class produced in excess of the actual requirement for the season or the year is known as carry-over seed, which is produced as a buffer stock to ensure against sudden demand or unforeseen shortfalls due to the

Table 6.5 Area and Seed Requirement for Various Seed Classes to Produce 22,000 t (Enough to Plant 5.5 million ha) of Certified Pearl Millet Hybrid Seed

	Year 1, Season I		Year 1, Season II		Year 2, Season I	
	For BS production		**For FS production**		**For CS production**	
Parental line	**Area**	**Nucleus seed quantity**	**Area**	**BS quantity**	**Area**	**FS quantity**
	ha	kg	ha	kg	ha	kg
A-line	0.352	1.408	88.0	352.0	22,000	88,000
B-line	0.088	0.352	22.0	88.0	–	–
R-line	0.088	0.352	22.0	88.0	5,500	22,000
Total	0.528	2.112	132.0	528.0	27,500	110,000

Note: The calculations assume 1,000 kg/ha of seed yield in production plots, 4 kg/ha of seed rate and female: male row ratio of 4:1 (the direction of arrow indicates that planning starts from estimation of certified seed requirements). Adapted from Yadav et al. (2012b).

Abbreviations: BS, breeder seed; CS, certified seed; FS, foundation seed.

vagaries of nature. Because cost is involved for maintenance of carry-over seed, seed-producing agencies keep only a critical carry-over quantity, which depends on the past experience of carry-over requirement, essentially based on the product and market demand. A general approximate limit of carry-over seed for different classes of seed for future use is given below:

- Nucleus seed—100%
- Breeder seed—100%
- Foundation seed—50%
- Certified seed—20%

Seed Multiplication

The CMS system provides a genetic mechanism to produce pure single-cross hybrid seed on a commercial scale from cross between A-line and R-line under open pollination in isolation. The counterparts of A-lines with the same nuclear gene constitution but fertile cytoplasm are maintainer lines, commonly referred to as B-lines. OPVs are random mating populations. These are maintained and multiplied by random-mating their representative bulk in isolation. The seed multiplication process involves multiplication of various classes of seeds at different stages (Figure 6.2) and linkages among different departments and agencies.

Nucleus Seed Multiplication
The nucleus seed production should be undertaken with excellent crop management and under expert supervision to maintain the highest genetic purity. Because seed requirement for nucleus seed is substantially less, it is advised that multiplication of A- and B-lines should be done by hand pollination, preferably in isolation, to avoid contamination. Pollen from the selected B-line plants is used for hand pollination of selected A-line plants. Ear-to-row progenies of the A-line and B-line are planted side by side in paired row fashion. The seed of the A-line produced by hand pollination with the respective B-line is harvested and bulked to constitute the nucleus seed of the A-line. Seed from selfed B-line plants is harvested and bulked as B-line seed. B-line seed is multiplied by selfing/sib-mating, although it should be remembered that continued selfing for several generations may lead to degeneration (inbreeding depression) of the B-line. The R-line is also multiplied by selfing/sibmating the same way as the B-line but in a separate isolation.

Breeder Seed Multiplication
Many breeders in the private sector prefer to multiply breeder seed by selfing and hand pollination the same way as nucleus seed multiplication. However, if breeder seed requirement is greater, it is done by planting of A-line and B-line seed in isolation. R-line multiplication is also done through the isolation method. The

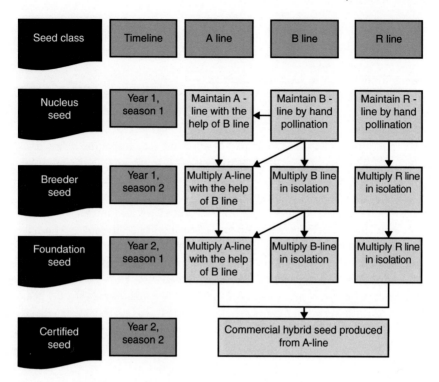

Figure 6.2 Procedure and timeline for multiplication of various classes of seed in pearl millet. R. S. Mahala

preference for maintenance under open-pollination in isolation is to avoid unnecessary inbreeding and the consequent loss of vigor. Isolation distance for breeder seed production is 1,000 m for A-, B-, and R-lines.

Foundation Seed Multiplication

Foundation seed is multiplied in isolation. The recommended isolation distance for pearl millet foundation seed production is 1,000 m for A-, B-, and R-lines and 400 m for OPVs. For multiplication of A-line, A- and B-lines are planted in ratios of 4:2 or 6:2. Generally four to six rows of B-line are planted all around the plot to ensure good A-line seed setting. It also acts as a barrier for foreign pollen. Generally, A-line starts flowering 2–3 days earlier than its counterpart B-line. Therefore, it is advised to plant one row of B-line 3–4 days earlier than planting of A-line; the rest of rows of A- and B-lines are planted the same day. This approach results in better seed set due to adequate pollen load during flowering. To avoid the risk of possible chance mixture, B-line rows are harvested immediately after pollination. Foundation seed stocks of B-lines, R-lines, and OPVs are multiplied under bulk planting in isolation following the specified isolation distance and other prescribed field standards.

Certified/Truthful Labeled Seed Multiplication

The recommended isolation distance for certified hybrid seed production is 200 m. The minimum field standards for certified seed production are followed to ensure good seed quality. The pattern of planting of hybrid parents is the same as for the breeder and foundation seed production of an A-line except that the B-line rows are replaced by R-line rows. Different row ratios (4:1, 8:2, or 6:2) are being followed in production areas. Row ratio depends on several factors, such as duration of stigma receptivity of the A-line, pollen supply ability of the R-line, etc. To avoid the risk of possible chance mixture, R-line rows are harvested immediately after pollination. The production of certified seed of an OPV is like breeder and foundation seed production in isolation following prescribed field standards.

The most critical component in certified seed production is to ensure adequate isolation distance. This requirement is easily fulfilled through "seed village" concept to undertake a seed production program in a non-traditional area for pearl millet, such as in Telangana and Karnataka states in India. The "seed village" approach is adopted where most farmers in a village agree to produce seed of a single hybrid.

Harvesting and Drying

The appropriate time of harvest to ensure maximum seed yield and quality is of great significance. Fully mature seed is easily harvested and cleaned with minimal harvest losses. Delayed harvesting may result in increased losses due to lodging and seed shattering. Sun drying of seeds on a clean threshing floor may be necessary to reduce moisture content, preserve viability and vigor, and improve storage quality. Drying of seed to the recommended moisture level of 12% is necessary to preserve its viability and vigor.

Seed Processing

The process of removing undesirable material from the desired seed lot and various steps in preparation of seed for marketing is called seed processing. The objectives of seed processing include ensuring high germination/seedling vigor, maintaining good appearance of seed, ensuring high physical purity, and minimizing deterioration during storage. Types of materials removed during seed processing include inert materials, common weed seeds, noxious weed seeds, deteriorated seeds, damaged seeds, other crop seeds, other variety seeds, and unacceptably small-sized seeds.

Commercialization

Marketing is the management process responsible for identifying, anticipating, and satisfying customer requirements profitably. The pricing strategy of seed from private companies largely involves the brand value, the benefit farmers' gain from growing pearl millet hybrid, and the cost of producing hybrid seed. On the other hand, the government agencies sell the seed that is cheaper due to subsidies provided on improved seed.

Maintenance of the Genetic Purity

Genetic purity is essential for the maintenance of seed purity and production of hybrids and inbred lines. Further, it is required for timely delivery of high-quality certified/truthfully labeled seed to the growers. The field-based grow-out test (GOT) and the use of phenotypic markers were common practices followed by the seed industry in India. However, GOT is time consuming, resource intensive, and influenced by environmental factors. The pearl millet seed production window in India is from January to April. Therefore, conducting GOT is challenging due to a shortage of time. To overcome these challenges, the use of reliable DNA markers for genetic purity assessment is becoming quite popular due to their high accuracy, quick results, and cost effectiveness. The most popular one is the SNP markers. The elite lines are genotyped using ~400 genome-wide SNP markers to identify a set of 16–20 informative purity markers. This set of purity markers is used for determining the purity of parental lines, which is essential for producing pure seeds.

Agronomic Interventions

Seed production of hybrids is full of innovations and has evolved rapidly and very extensively in India. The "seed village" concept involving multiplication of specific cultivars in particular villages has been effectively used to undertake a seed production program in nontraditional areas for pearl millet, largely in southern India. Seed production by small farmers under contractual arrangement with seed corporations and private seed companies is a unique example of the willingness of Indian farmers to embrace new and intricate technology to produce genetically pure seed with high seed vigor and quality. During the last three decades, yield levels in seed production plots of hybrids have doubled, mainly due to the development of high-yielding parental lines and improved crop management skills of farmers. The seed processing period has also been shortened considerably, reducing the time from harvesting to delivery of seed in target regions within 1 month.

Figure 6.3   Right synchronization in flowering of male-sterile line (central rows) and restorer line (outer rows) in pearl millet seed production.

This circumvents the need for storage of seed over a season, resulting in reduced overhead cost.

Earlier synchronization of flowering in male and female parents of hybrids has been a serious challenge. Synchronization of flowering of the A-line with the R-line in certified seed production plots is essential to ensure pollen availability in the R-line when stigmas emerge in the A-line. Synchronized flowering results in good seeds in the A-line and higher yields in production plots (Figure 6.3). The A-line and R-line may differ for flowering due to their inherent genetic differences for this trait and their differential photo-thermal sensitivity. Information is needed on flowering time of parental lines of hybrids in actual seed production areas rather than in experimental areas. Synchronization of flowering of A-lines and R-lines can also be enhanced by management interventions. Hastening flowering time by 6–8 days can be achieved by three to four sprays of 4% urea at 2- to 3-day intervals at boot leaf stage. Application of micronutrients also helps in fast growth. Light-selective irrigation to late-flowering parents also helps in hastening flowering. If the difference in flowering time of parental lines is more, a practice of staggered sowing is followed. Removal of main shoots in early-flowering parental lines also helps in the synchronization of flowering of the parental lines. Non-synchronization results in poor seed set, low yield, and thus practically a failure of seed production (Figure 6.4).

Roguing, which is the process of removing pollen shedders from A-lines and off-types from parental lines in seed production field, represents another past challenge that has been overcome effectively by the training of seed-producing farmers. Pollen shedders are male fertile plants in an A-line with similar morphology. Pollen shedders in an A-line are results of mutations or mechanical mixtures. Off-types are the plants distinctly different in morphological characteristics from those with typical characteristics of the line under production. The off-type plants

Figure 6.4 Wrong synchronization in flowering of male-sterile line (central rows) and restorer line (outer rows) in seed production of pearl millet.

may arise through mechanical mixture or outcrossing and rarely as mutants. Adequate and timely roguing constitutes the most important operation in seed production. Rogues differing from normal plants in phenotypes should be pulled out and discarded at the earliest possible stage of plant growth, before flowering, to avoid genetic contamination. In the case of pollen shedders, the best time for roguing is morning, when the A-line will have the least exposure to outcrossing and there is little wind movement.

Hybrid Seed Production Profitability

During the last 25 years, seed yield levels in production plots of hybrids have increased considerably, mainly due to the development of high-yielding parental lines and improved crop management skills of farmers. In the early 1980s, average A-lines seed yields were 0.8–1.0 t/ha, whereas current yield levels have been reported to be ~1.5–2.0 t/ha (personal communication with seed industry experts). As a result, the seed production of pearl millet hybrids has become as economical as rice cultivation in southern India, and fields of both of these crops can be found side by side (Figure 6.5). The margin of profit can be as high as US$500–1,000 per hectare because pearl millet seed production requires fewer resources (e.g., less water), leading to lower cultivation costs. The risk in hybrid seed production due to diseases is minimized because hybrid parental lines possess a high level of disease resistance and because there is assured procurement of seed by the production agencies.

Figure 6.5 Certified seed production of pearl millet hybrid (in the background field) and commercial crop of rice (in front) being grown in Nizamabad area of the Telangana state.

Government Policies to Promote Hybrids

During the last more than six decades, enormous progress has been made in the genetic improvement and agronomic management of pearl millet, and it is often referred to as a success story in Indian agriculture. The substantial yield improvement in pearl millet under rainfed conditions is a successful demonstration of technology-led development and highlights the role of hybrid technology in raising crop productivity even in marginal drylands. Government policies have contributed significantly to this success. Government policies have been very favorable for increased investment in research and development of pearl millet by the private sector. Protection of hybrids through the Protection of Plant Variety and Farmers' Rights Authority has promoted research investments in hybrid development by the private seed industry (Ravi, 2004).

Future Outlook

Pearl millet hybrids have a great potential in contributing to achieving food, fodder, and nutritional security. The continuing genetic gains in pearl millet suggest that, with the greater use of the hybrid technology and modern tools and wider inter-institutional and inter-sectoral participation, this productivity momentum can not only be maintained but can even be further accelerated through integrating advanced molecular technologies. Although the national productivity of pearl millet in India is 1.2 t/ha in the rainy season, full adoption of hybrids and improved crop management in the summer season have demonstrated 4–5 t/ha of grain yield in 85 days of crop growth, indicating enormous yield potential and management responsiveness of pearl millet. The availability of extra-early–maturing hybrids with tolerance to multiple environmental constraints would make pearl millet an ideal cereal for multiple cropping.

However, pearl millet hybrids have yet to take off fully in Africa despite the fact that several innovations in hybrid production technologies have been demonstrated in similar agricultural production environments of India. Pearl millet research and development are a mandate of several national and international organizations, with a common goal of making an impact on the communities. With this shared goal, the partnership needs to be pursued systematically, with specific focus on Africa. Partnership with the private sector has been critical in delivering the products of improved genetics to the farming community. The role of the private sector needs to be enhanced through appropriate policy intervention.

Pearl millet is still a relatively new grain crop, and no readily available markets are available in the United States. In other countries, pearl millet is milled both dry and wet and consumed as a food grain. Although pearl millet grain is well suited for both human and animal consumption, producers in the United States should secure a market prior to adoption to new dryland cropping and livestock systems. Because the grain is gluten free, marketing opportunities in the United States are slowly gaining in health-food outlets. Currently, research efforts on the development of a mapping population, parental lines, and hybrids for grain and forage with improved nutritional value integrating both classical and molecular breeding approaches are being continuously focused on only at the Kansas State University Agricultural Research Center, Hays, Kansas.

The productivity of pearl millet is still much lower than other food crops (e.g., rice, wheat, and maize), although it has shown impressive genetic gains in India for the past seven decades (Yadav et al., 2019). The magnitude of heterosis needs to be taken to next level to further increase yield potential. Heterotic grouping of hybrid parental lines and heterosis prediction in hybrids are the two most effective means to achieve this objective. Some attempts have been made in pearl millet to define heterotic groups (Pucher et al., 2016; Ramya et al., 2018; Singh et al., 2018). Recently, in a study involving 320 R-lines and 260 B-lines derived from pearl millet breeding programs in India, two B-line and two R-line heterotic groups were identified based on the heterotic performance and combining ability (Gupta et al., 2020). Hybrids from these identified B×R heterotic groups showed grain yield heterosis of >10% over the best commercial hybrid checks. This study also indicated that distinct parental groups can be formed based on molecular markers, which can help in assigning hybrid parental lines to heterotic groups to develop high-yielding hybrids.

The global germplasm collection is an excellent resource for discovering new sources of key traits of abiotic stress tolerance, disease and pest resistance, and improved nutritional quality. So far, a meager percentage of germplasm has been exploited, owing to the large number of germplasms and the prevalence of undesirable traits in the unadapted genetic background. These two issues have been mostly avoided. The development of core and mini-core collections is encouraging breeders to use desired germplasm in extending the genetic base of commercial cultivars, which is critical for reducing disease/pest epidemics and mitigating the effects of climate change (Upadhyaya et al., 2011). The sequencing of 1,000 genomes has been a significant step forward for pearl millet (Varshney et al., 2017). The available genotyping-by-sequencing data of 398 pearl millet accessions (Serba et al., 2019) from KSU-ARC and CRIPSR-Cas9 gene editing in plants (Zhu et al., 2020) provide an opportunity for molecular-assisted breeding and gene identification for pearl millet trait improvement. To increase pearl millet adaptation in the United States, the integration of genomic tools is imperative to accelerate the breeding process on the targeted traits. Development of robust mapping population (nested association mapping and multi-parent advanced generation intercrosses) by stacking and introgressing many complex traits in the developed recombinant inbred lines would help to broaden the genetic base and to develop many heterotic groups with increased genetic gain and yield potential in the new diversified hybrid combinations.

Identification of promising hybrids requires generating and testing large numbers of hybrids under different field conditions. Performing crosses to generate F_1 hybrids is labor intensive, and further evaluating hybrids across field trials under several environments requires significant time and resources. Hence, methods for selecting parental inbred lines and determining which crosses are likely to yield the best hybrids is a critical part of crop improvement. Traditionally, mid-parental values have been a common way to predict the performance of hybrids on the basis of inbred values, combined with estimates of GCA. Genetic markers across the whole genome can be used to implement breeding programs based on estimated breeding values from genome-wide sets of markers. In a study on pearl millet, Liang et al. (2018), based on the data on inbreds and its hybrids, reported prediction accuracy ranging from 0.48 to 0.89 for various traits including grain yield, which suggested that even small numbers of selected SNPs can achieve relatively high prediction accuracy in pearl millet. The implementation of a hybrid genomic selection/genomic prediction–guided pearl millet breeding program has the potential to significantly improve the efficiency of breeding efforts. Genomics-assisted breeding methods like genomic selection, molecular marker discovery, gene expression, and validation analyses are integral to accelerating the breeding process. The innovative results from genomic selection and prediction can be used for further advanced gene expression studies, particularly for food grain resilient cereal crops like pearl millet, to increase the genetic gain for yield and grain quality traits under stress environments.

HarvestPlus, a CGIAR Challenge Program, started breeding for micronutrients and vitamins. ICRISAT established a database for Fe and Zn in partnership with national partners. As part of the ICRISAT product profile, Fe and Zn concentrations >60 ppm were sought for breeding. This is a significant step toward the mainstreaming of elite breeding lineages. The dissemination and use of these mainstreamed breeding lines and hybrid parents (both public and private sectors) would make biofortified hybrid development a regular activity to improve genetic gains for micronutrient traits over time (Yadav et al., 2021). Higher genetic improvements in pearl millet have been attained over the past few decades in quantitative genetics and wide application of data analytics in multilocational trials. To achieve greater genetic advances, pearl millet breeding must take advantage of the current era's ultrahigh-speed computers, big-data analytics, crop simulations, machine learning, and artificial intelligence (Yadav et al., 2021). Taken together, integration of high-throughput genomic tools, speed breeding, and precision phenotyping protocols can be efficiently used to exploit the unexplored genetic resources available in pearl millet to develop more diversified heterotic groups for hybrid development for stress environments with increased yield potential and nutritional value.

References

Amarender, R., A., Yadav, O. P., Malik, D. P., Ardeshna, N., Kundu, K., Gupta, S. K., Sharma, R., Moses Shyam, S. G., & Sammi Reddy, K. (2013). *Utilization pattern, demand and supply of pearl millet grain and fodder in western India.* ICRISAT.

Andrews, D. J., & Kumar, K. A. (1996). Use of the West African pearl millet landrace Iniadi in cultivar development. *Plant Genetic Resources Newsletter, 105*, 15–22.

Athwal, D. S. (1961). Recent developments in the breeding and improvement of bajra (pearl millet) in the Punjab. *Madras Agricultural Journal, 48*, 18–19.

Athwal, D. S. (1966). Current plant breeding research with special reference to *Pennisetum. Indian Journal of Genetics and Plant Breeding, 26A*, 73–85.

Ayyangar, G. N. R., Vijiaraghavan, C., & Pillai, V. G. (1933). Studies on *Pennisetum typhoideum* (Rich.), the pearl millet, part I: Anthesis. *Indian Journal of Agricultural Sciences, 3*, 688–694.

Bhattarai, B., Singh, S., West, C. P., & Saini, R. (2019). Forage potential of pearl millet and forage sorghum alternatives to corn under the water-limiting conditions of the Texas High Plains: A review. *Crop, Forage & Turfgrass Management, 5*(1), 1–12. https://doi.org/10.2134/cftm2019.08.0058

Burton, G. W. (1948). The performance of various mixtures of hybrid and parent inbred pearl millet, *Pennisetum glaucum* (L.) R. BR. *Agronomy Journal, 40*(10), 908–915. https://doi.org/10.2134/agronj1948.00021962004000100007x

Burton, G. W. (1958). Cytoplasmic male-sterility in pearl millet (*Pennisetum glaucum*) (L.) R. Br. *Agronomy Journal, 50*(4), 230–230. https://doi.org/10.2134/agronj195 8.00021962005000040018x

Burton, G. W. (1965). Pearl millet Tift 23A released. *Crops & Soils, 17*(5), 19.

Chapman, S. R., & Carter, L. P. (1976). *Crop production, principle and practices.* WH Freeman.

Cherney, D. J., Patterson, J. A., & Johnson, K. D. (1990). Digestibility and feeding value of pearl millet as influenced by the brown-midrib, low-lignin trait. *Journal of Animal Science, 68*, 4345–4351. https://doi.org/10.2527/1990.68124345x

Crossa, J., & Gardner, C. O. (1987). Introgression of an exotic germplasm for improving an adapted maize population. *Crop Science, 27*(2), 187–190. https://doi .org/10.2135/cropsci1987.0011183X002700020008x

Darwin, C. R. (1876). *The effects of cross- and self-fertilisation in the vegetable kingdom.* John Murray.

Dave, H. R. (1986). Pearl millet hybrids. In J. R. Witcombe & S. R. Beckerman (Eds.), *Proceedings of the international pearl millet workshop* (pp. 7–11). ICRISAT.

Davis, A. J., Dale, N., & Ferreira, F. J. (2003). Pearl millet as an alternative feed ingredient in broiler diets. *Journal of Applied Poultry Research, 12*, 137–144. https://doi.org/10.1093/japr/12.2.137

de Lima, G. S. (1998). *Estudo comparativo da resistencia a' seca no sorgo forrageiro (Sorghum bicolor (L.) Moench) em differentes estadios de desenvolvimento.* UFRPE.

Djanaguiraman, M., Perumal, R., Ciampitti, I. A., Gupta, S. K., & Prasad, P. V. V. (2018). Quantifying pearl millet response to high temperature stress: Thresholds, sensitive stages, genetic variability and relative sensitivity of pollen and pistil. *Plant, Cell & Environment, 41*(5), 993–1007. https://doi.org/10.1111/ pce.12931

East, E. M. (1908). Inbreeding in corn. In *Report of connecticut agricultural experiment station* (pp. 419–428). CAES.

FAOSTAT. (2023). FAOSTAT. https://www.fao.org/faostat/en/#data/QCL/metadata

Gill, K. S. (1991). Pearlmillet and its improvement. Indian Council of Agricultural Research, New Delhi. 305 pp.

Godbole, S. V. (1925). *Pennisetum typhoideum.* Studies on the Bajri crop. I. The morphology of *Pennisetum typhoideum.* In *Memoirs of the Department of Agriculture in India* (pp. 247–268). India Department of Agriculture.

Govindaraj, M., Rai, K. N., Cherian, B., Pfeiffer, W. H., Kanatti, A., & Shivade, H. (2019a). Breeding biforti fied pearl millet varieties and hybrids to enhance millet markets for human nutrition. *Agriculture, 9*(5), 106. https://doi.org/10.3390/ agriculture9050106

Govindaraj, M., Rai, K. N., Shanmugasundaram, P., Dwivedi, S. L., Sahrawat, K. L., Muthaiah, A. R., & Rao, A. S. (2013). Combining ability and heterosis for grain iron and zinc densities in pearl millet. *Crop Science, 53*, 507–517.

Govindaraj, M., Yadav, O. P., Srivastava, R. K., & Gupta, S. K. (2019b). Conventional and molecular breeding approaches for biofortification of pearl millet. In Z. A. Dar & S. H. Wani (Eds.), *Quality breeding in field crops Qureshi AMI* (pp. 85–107). Springer International Publishing.

Govintharaj, P., Maheswaran, M., Blümmel, M., Sumathi, P., Vemula, A. K., Rathore, A., Sivasubramani, S., Kale, S. M., Varshney, R. K., & Gupta, S. K. (2021). Understanding heterosis, genetic effects, and genome wide associations for forage quantity and quality traits in multi-cut pearl millet. *Frontiers in Plant Science, 12*, 687859. https://doi.org/10.3389/fpls.2021.687859

Green, P. H. R., Lebwohl, B., & Greywoode, R. (2015). Celiac disease. *The Journal of Allergy and Clinical Immunology, 135*(5), 1099–1106. https://doi.org/10.1016/j.jaci.2015.01.044

Gupta, S. K., Govintharaj, P., & Bhardwaj, R. (2022). Three-way top-cross hybrids to enhance production of forage with improved quality in pearl millet (*Pennisetum glaucum* (L.) R. Br.). *Agriculture, 12*(9), 1–16.

Gupta, S. K., Patil, K. S., Rathore, A., Yadav, D. V., Sharma, L. D., Mungra, K. D., Patil, H. T., Gupta, S. K., Kumar, R., Chaudhary, V., & Das, R. R. (2020). Identification of heterotic groups in South-Asian-bred hybrid parents of pearl millet. *Theoretical and Applied Genetics, 133*, 873–888. https://doi.org/10.1007/s00122-019-03512-z

Gupta, S. K., Rai, K. N., Govindaraj, M., & Rao, A. S. (2012). Genetics of fertility restoration of the A4 cytoplasmic-nuclear male sterility system in pearl millet. *Czech Journal of Genetics, 48*(2), 87–92. http://10.0.67.69/164/2011-CJGPB

Gupta, S. K., Rai, K. N., Singh, P., Ameta, V. L., Gupta, S. K., Jayalekha, A. K., Mahala, R. S., Pareek, S., Swami, M. L., & Verma, Y. S. (2015). Seed set variability under high temperatures during flowering period in pearl millet (*Pennisetum glaucum* L.(R.) Br.). *Field Crops Research, 171*, 41–53. https://doi.org/10.1016/j.fcr.2014.11.005

Gupta, S. K., Rathore, A., Yadav, O. P., Rai, K. N., Khairwal, I. S., Rajpurohit, B. S., & Das, R. R. (2013). Identifying mega-environments and essential test locations for pearl millet cultivar selection in India. *Crop Science, 53*(6), 2444–2453. https://doi.org/10.2135/cropsci2013.01.0053

Gupta, S. K., Yadav, D. V., Govindaraj, M., Boratkar, M., Kulkarni, V. N., & Rai, K. N. (2018). Inheritance of fertility restoration of A5 cytoplasmic-nuclear male sterility system in pearl millet [*Pennisetum glaucum* (L.) R. Br.]. *Indian Journal of Genetics and Plant Breeding, 78*(2), 228–232. https://doi.org/10.5958/0975-6906.2018.00029.9

Hanna, W. W. (1989). Characteristics and stability of a new cytoplasmic-nuclear male-sterile source in pearl millet. *Crop Science, 29*(6), 1457–1459. https://doi.org/10.2135/cropsci1989.0011183X002900060026x

Harinarayana, G., Melkania, N. P., Reddy, B. V. S., Gupta, S. K., Rai, K. N. & Kumar, P. S. (2005). Forage potential of sorghum and pearl millet. In *Sustainable development and management of drylands in the twenty-first century: Proceedings of the Seventh International Conference on the Development of Dryland, September 14–17, 2003.* ICARDA.

Hash, C. T., Sharma, A., Kolesnikova-Allen, M. A., Singh, S. D., Thakur, R. P., Bhasker Raj, A. G., Ratnaji Rao, M. N. V., Nijhawan, D. C., Beniwal, C. R., Sagar, P., Yadav, H. P., Yadav, Y. P., Srikant Bhatnagar, S. K., Khairwal, I. S., Howarth, C. J., Cavan, G. P., Gale, M. D., Liu, C., . . . Witcombe, J. R. (2006). Teamwork delivers biotechnology products to Indian small-holder crop-livestock producers: Pearl millet hybrid "HHB 67 improved" enters seed delivery pipeline. *Journal of SAT Agricultural Research, 2*(1), 1–3.

Hassan, Z. M., Sebola, N. A., & Mabelebele, M. (2021). The nutritional use of millet grain for food and feed: A review. *Agriculture & Food Security, 10*, 16. https://doi .org/10.1186/s40066-020-00282-6

Hassanat, F. M. (2007). *Evaluation of pearl millet forage* [Doctoral dissertation, McGill University]. https://escholarship.mcgill.ca/downloads/q811kn460.pdf

Jukanti, A. K., Gowda, C. L., Rai, K. N., Manga, V. K., & Bhatt, R. K. (2016). Crops that feed the world 11. Pearl millet (*Pennisetum glaucum* L.): An important source of food security, nutrition and health in the arid and semi-arid tropics. *Food Security, 8*, 307–329. https://doi.org/10.1007/s12571-016-0557-y

Kanatti, A., Rai, K. N., Radhika, K., Govindaraj, M., Sahrawat, K. L., & Rao, A. S. (2014a). *Grain iron and zinc density in pearl millet: Combining ability, heterosis and association with grain yield and grain size* (Vol. 3, pp. 763). *Springerplus.*

Kanatti, A., Rai, K. N., Radhika, K., Govindaraj, M., Sahrawat, K. L., Srinivasu, K., & Shivade, H. (2014b). Relationship of grain iron and zinc content with grain yield in pearl millet hybrids. *Crop Improvement, 41*, 91–96.

Khairwal, I. S. (1999). *Pearl millet breeding.* Science Publications.

Kichel, A. N., Miranda, C. H. B., & da Silva, J. M. (1999). Pearl millet (Pennisetum americanum L. Leek) as a forage plant. In A. L. F. Neto, R. F. Amabile, D. A. M. Netto, T. Yamashita, & H. Gocho (Eds.), *Proceedings of International Pearl Millet Workshop, Planaltina, Brazil, 9–10 June 1999* (pp. 93–98). Embrapa Cerrados.

Kumar, K. A. (1989). Pearl millet: Current status and future potential. *Outlook on Agriculture, 18*, 46–53. https://doi.org/10.1177/003072708901800201

Leuck, D. B., & Burton, G. W. (1966). Pollination of pearl millet by insects. *Journal of Economic Entomology, 59*(5), 1308–1309.

Liang, Z., Gupta, S. K., Yeh, C. T., Zhang, Y., Ngu, D. W., Kumar, R., Patil, H. T., Mungra, K. D., Yadav, D. V., Rathore, A., & Srivastava, R. K. (2018). Phenotypic data from inbred parents can improve genomic prediction in pearl millet hybrids. *G3: Genes|Genomes|Genetics, 8*(7), 2513–2522. https://doi.org/10.1534/g3.118 .200242

Machicek, J. (2018). *Evaluating forage sorghum and pearl millet for forage production and quality in the Texas high plains* [Master's thesis, West Texas A&M University, Department of Agricultural Sciences] https://wtamu-ir.tdl.org/items/ c723a428-e574-4254-9a76-42f7f512965f.

Mula, R. P., Rai, K. N., Kulkarni, V. N., & Singh, A. K. (2007). Public-private partnership and impact of ICRISAT's pearl millet hybrid parents research. *Journal of SAT Agricultural Research, 5*(1), 1–5.

Muthamilarasan, M., Dhaka, A., Yadav, R., & Prasad, M. (2016). Exploration of millet models for developing nutrient rich graminaceous crops. *Plant Science, 242,* 89–97. https://doi.org/10.1016/j. plantsci.2015.08.0230

Ndoye, M. & Gahukar, R. (1987). *Proceedings of the International Pearl Millet Workshop.* Cambridge University Press.

Nedumaran, S., Bantilan, M., Gupta, S., Irshad, A. & Davis, J. (2014). *Potential welfare benefit of millets improvement research at ICRISAT: Multi country-economic surplus model approach, socioeconomics discussion paper series number 15.* ICRISAT.

Ouendeba, B., Ejeta, G., Nyquist, W. E., Hanna, W. W., & Kumar, A. (1993). Heterosis and combining ability among African pearl millet landraces. *Crop Science, 33*(4), 735–739. https://doi.org/10.2135/cropsci1993.0011183X003300040020x

Patil, K. S., Gupta, S. K., Marathi, B., Danam, S., Thatikunta, R., Rathore, A., Das, R. R., Dangi, K. S., & Yadav, O. P. (2020). African and Asian origin pearl millet populations: Genetic diversity pattern and its association with yield heterosis. *Crop Science, 60*(6), 3035–3048. https://doi.org/10.1002/csc2.20245

Pucher, A., Sy, O., Sanogo, M. D., Angarawai, I. I., Zangre, R., Ouedraogo, M., Boureima, S., Hash, C. T., & Haussmann, B. I. (2016). Combining ability patterns among West African pearl millet landraces and prospects for pearl millet hybrid breeding. *Field Crops Research, 195,* 9–20. https://doi.org/10.1016/j.fcr.2016.04.035

Rai, K. N. (1995). A new cytoplasmic-nuclear male sterility system in pearl millet. *Plant Breeding, 114*(5), 445–447. https://doi.org/10.1111/j.1439-0523.1995.tb00829.x

Rai, K. N., Govindaraj, M., Kanatti, A., Rao, A. S., & Shivade, H. (2017). Inbreeding effects on grain iron and zinc concentrations in pearl millet. *Crop Science, 57,* 1–8.

Rai, K. N., Govindaraj, M., & Rao, A. S. (2012). Genetic enhancement of grain iron and zinc content in pearl millet. *Quality Assurance and Safety of Crops and Food, 4,* 119–125.

Rai, K. N., & Hanna, W. W. (1990). Morphological characteristics of tall and dwarf pearl millet isolines. *Crop Science, 30*(1), 23–25. https://doi.org/10.2135/ cropsci1990.0011183X003000010005x

Rai, K. N., Khairwal, I. S., Dangaria, C. J., Singh, A. K., & Rao, A. S. (2009). Seed parent breeding efficiency of three diverse cytoplasmic-nuclear male-sterility systems in pearl millet. *Euphytica, 165,* 495–507. https://doi.org/10.1007/s10681-008-9765-7

Rai, K. N., & Kumar, K. A. (1994). Pearl millet improvement at ICRISAT-an update. *International Sorghum and Millets Newsletter, 35*(1), 1–29.

Rai, K. N., Patil, H. T., Yadav, O. P., Govindaraj, M., Khairwal, I. S., Cherian, B., Rajpurohit, B. S., Rao, A. S., & Kulkarni, M. P. (2014). Dhanashakti: A high-iron pearl millet variety. *Indian Farming*, *64*, 32–34.

Rai, K. N., & Rao, A. S. (1991). Effect of d_2 dwarfing gene on grain yield and yield components in pearl millet near-isogenic lines. *Euphytica*, *52*, 25–31. https://doi .org/10.1007/BF00037853

Rai, K. N., Virk, D. S., Harinarayana, G., & Rao, A. S. (1996). Stability of male-sterile sources and fertility restoration of their hybrids in pearl millet. *Plant Breeding*, *115*(6), 494–500. https://doi.org/10.1111/j.1439-0523.1996.tb00964.x

Ramya, A. R., Ahamed, M. L., Satyavathi, C. T., Rathore, A., Katiyar, P., Raj, A. B., Kumar, S., Gupta, R., Mahendrakar, M. D., Yadav, R. S., & Srivastava, R. K. (2018). Towards defining heterotic gene pools in pearl millet [*Pennisetum glaucum* (L.) R. Br.]. *Frontiers in Plant Science*, *8*, 1934. https://doi.org/10.3389/fpls.2017.01934

Ravi, S. B. (2004). Effectiveness of Indian sui generis law on plant variety protection and its potential to attract private investment in crop improvement. *Journal of Intellectual Property Rights*, *9*, 533–548.

Sattler, F. T., & Haussmann, B. I. (2020). A unified strategy for West African pearl millet hybrid and heterotic group development. *Crop Science*, *60*(1), 1–13. https://doi.org/10.1002/csc2.20033

Sattler, F. T., Pucher, A., Kassari Ango, I., Sy, O., Ahmadou, I., Hash, C. T., & Haussmann, B. I. (2019). Identification of combining ability patterns for pearl millet hybrid breeding in West Africa. *Crop Science*, *59*(4), 1590–1603. https://doi .org/10.2135/cropsci2018.12.0727

Sattler, S. E., Saballos, A., Xin, Z., Funnell-Harris, D. L., Vermerris, W., & Pedersen, J. F. (2014). Characterization of novel sorghum brown midrib mutants from an EMS-mutagenized population. *G3: Genes|Genomes|Genetics*, *4*, 2115–2124. https://doi.org/10.1534/g3.114.014001

Serba, D. D., Muleta, K. T., St. Amand, P., Bernardo, A., Bai, G., Perumal, R., & Bashir, E. (2019). Genetic diversity, population structure, and linkage disequilibrium of pearl millet. *Plant Genome*, *12*, 180091. https://doi.org/10.3835/plantgenome2018 .11.0091

Shull, G. H. (1908). The composition of a field of maize. *The Journal of Heredity*, *1*, 296–301. https://doi.org/10.1093/jhered/os-4.1.296

Singh, F., Singh, R. M., Singh, R. B., & Singh, R. K. (1980). Genetic studies of downy mildew resistance in pearl millet. In V. P. Gupta & J. L. Minocha (Eds.), *Trends in genetical research on pennisetums* (pp. 171–172)). Punjab Agricultural University

Singh, S., Gupta, S. K., Thudi, M., Das, R. R., Vemula, A., Garg, V., Varshney, R. K., Rathore, A., Pahuja, S. K., & Yadav, D. V. (2018). Genetic diversity patterns and heterosis prediction based on SSRs and SNPs in hybrid parents of pearl millet. *Crop Science*, *58*(6), 2379–2390. https://doi.org/10.2135/cropsci2018.03.0163

Singh, S. D. (1995). Downy mildew of pearl millet. *Plant Disease*, *79*(6), 545–550.

Sinha, P., Singh, V. K., Saxena, R. K., Khan, A. W., Abbai, R., Chitikineni, A., Desai, A., Molla, J., Upadhyaya, H. D., Kumar, A., & Varshney, R. K. (2020). Superior haplotypes for haplotype-based breeding for drought tolerance in pigeonpea (*Cajanus cajan* L.). *Plant Biotechnology Journal, 18*, 2482–2490. https://doi.org/10.1111/pbi.13422

Srivastava, R. K., Bollam, S., Pujarula, V., Pusuluri, M., Singh, R. B., Potupureddi, G., & Gupta, R. (2020a). Exploitation of heterosis in pearl millet: A review. *Plants, 9*(7), 807. https://doi.org/10.3390/plants9070807

Srivastava, R. K., Singh, R. B., Srikanth, B., Satyavathi, C. T., Yadav, R., & Gupta, R. (2020b). Genome-wide association studies (GWAS) and genomic selection (GS) in pearl millet: Advances and prospects. *Frontiers in Genetics, 10*, 1389.

Tang, H. M., Liu, S., Hill-Skinner, S., Wu, W., Reed, D., Yeh, Y. C. T., Nettleton, D., & Schnable, P. S. (2014). The maize brown midrib2 (*bm2*) gene encodes a methylenetetrahydrofolate reductase that contributes to lignin accumulation. *The Plant Journal, 77*, 380–392. https://doi.org/10.1111/tpj.12394

Tewodros, M., & Mohammed, A. (2016). Heterotic response in major cereals and vegetable crops. *International Journal of Plant Breeding and Genetics, 10*(2), 69–78. https://doi.org/10.3923/ijpbg.2016.69.78

Thakur, R. P., Shetty, H. S., & Khairwal, I. S. (2006). Pearl millet downy mildew research in India: Progress and perspectives. *International Sorghum and Millets Newsletter, 47*, 125–130.

Upadhyaya, H. D., Yadav, D., Reddy, K. N., Gowda, C. L. L., & Singh, S. (2011). Development of pearl millet minicore collection for enhanced utilization of germplasm. *Crop Science, 51*(1), 217–223. https://doi.org/10.2135/cropsci2010.06.0336

Van Oosterom, E. J., Bidinger, F. R., & Weltzien, E. R. (2003). A yield architecture framework to explain adaptation of pearl millet to environmental stress. *Field Crops Research, 80*(1), 33–56. https://doi.org/0.1016/S0378-4290(02)00153-3

Varshney, R. K., Shi, C., Thudi, M., Mariac, C., Wallace, J., Qi, P., Zhang, H., Zhao, Y., Wang, X., Rathore, A., Srivastava, R. K., Chitikineni, A., Fan, G., Bajaj, P., Punnuri, S., Gupta, S. K., Wang, H., Jiang, Y., Couderc, M., . . . Xu, X. (2017). Pearl millet genome sequence provides a resource to improve agronomic traits in arid environments. *Nature Biotechnology, 35*, 969–974.

Velu, G., Rai, K. N., Muralidharan, V., Longvah, T., & Crossa, J. (2011). Gene effects and heterosis for grain iron and zinc density in pearl millet (*Pennisetum glaucum* (L.) R. Br). *Euphytica, 180*, 251–259.

Weltzien, E., & Witcombe, J. R. (1989). *Pearl millet improvement for the Thar desert, Rajasthan, India.* Vortraege fuer Pflanzenzuechtung.

Yadav, O. P. (1996). Downy mildew incidence of pearl millet hybrids with different male-sterility inducing cytoplasms. *Theoretical and Applied Genetics, 92*, 278–280. https://doi.org/10.1007/BF00223386

Yadav, O. P., & Bidinger, F. R. (2007). Utilization, diversification and improvement of landraces for enhancing pearl millet productivity in arid environments. *Annals of Arid Zone, 46*(1), 49–57.

Yadav, O. P., & Bidinger, F. R. (2008). Dual-purpose landraces of pearl millet (*Pennisetum glaucum*) as sources of high stover and grain yield for arid zone environments. *Plant Genetic Resources, 6*(2), 73–78. https://doi.org/10.1017/S1479262108993084

Yadav, O. P., Gupta, S. K., Govindaraj, M., Sharma, R., Varshney, R. K., Srivastava, R. K., Rathore, A., & Mahala, R. S. (2021). Genetic gains in pearl millet in India: Insights into historic breeding strategies and future perspective. *Frontiers in Plant Science, 12*, 396. https://doi.org/10.3389/fpls.2021.645038

Yadav, O. P., Manga, V. K., & Gupta, G. K. (1993). Influence of A 1 cytoplasmic substitution on the downy-mildew incidence of pearl millet. *Theoretical and Applied Genetics, 87*, 558–560. https://doi.org/10.1007/BF00221878

Yadav, O. P., & Rai, K. N. (2011). Hybridization of Indian landraces and African elite composites of pearl millet results in biomass and stover yield improvement under arid zone conditions. *Crop Science, 51*(5), 1980–1987. https://doi.org/10.2135/cropsci2010.12.0731

Yadav O. P., Rai K. N., Rajpurohit B. S., Hash C. T., Mahala R. S., Gupta S. K., Shetty H. S., Bishnoi H. R., Rathore M. S., Kumar A., Sehgal S. and Raghvani K. L. 2012a. Twenty-five Years of Pearl Millet Improvement in India. All India Coordinated Pearl Millet Improvement Project, Jodhpur, India. 122 pp.

Yadav, O. P., & Rai, K. N. (2013). Genetic improvement of pearl millet in India. *Agricultural Research, 2*, 275–292. https://doi.org/10.1007/s40003-013-0089-z

Yadav, O. P., Rai, K. N., Bidinger, F. R., Gupta, S. K., Rajpurohit, B. S., & Bhatnagar, S. K. (2012a). Pearl millet (*Pennisetum glaucum*) restorer lines for breeding dual-purpose hybrids adapted to arid environments. *Indian Journal of Agricultural Sciences, 82*(11), 922–927.

Yadav, O. P., Rai, K. N., Rajpurohit, B. S., Hash, C. T., Mahala, R. S., Gupta, S. K., Shetty, H. S., Bishnoi, H. R., Rathore, M. S., Kumar, A., & Sehgal, S. (2012b). *Twenty-five years of pearl millet improvement in India.* ICAR.

Yadav, O. P., Singh, D. V., Dhillon, B. S., & Mohapatra, T. (2019). India's evergreen revolution in cereals. *Current Science, 116*(11), 1805–1808.

Yadav, O. P., Upadhyaya, H. D., Reddy, K. N., Jukanti, A. K., Pandey, S., & Tyagi, R. K. (2017). Genetic resources of pearl millet: Status and utilization. *The Indian Journal of Plant Genetic Resources, 30*(1), 31–47. https://doi.org/10.5958/0976-1926.2017.00004.3

Yadav, O. P., & Weltzien Rattunde, E. (2000). Differential response of landrace-based populations and high yielding varieties of pearl millet in contrasting environments. *Annals of Arid Zone, 39*(1), 39–45.

Yadav, R. S., Sehgal, D., & Vadez, V. (2011). Using genetic mapping and genomics approaches in understanding and improving drought tolerance in pearl millet. *Journal of Experimental Botany*, *62*(2), 397–408. https://doi.org/10.1093/jxb/erq265

Yadava, D. K., Choudhury, P. R., Hossain, F., & Kumar, D. (2020). *Biofortified varieties: Sustainable way to alleviate malnutrition*. Indian Council of Agricultural Research.

Zhu, H., Li, C., & Gao, C. (2020). Applications of CRISPR–Cas in agriculture and plant biotechnology. *Nature Reviews Molecular Cell Biology*, *21*, 661–677. https://doi.org/10.1038/s41580-020-00288-9

7

Challenges and Opportunities of Pearl Millet Hybrid Development and Seed Production in West Africa

Prakash I. Gangashetty, Drabo Inoussa, Ghislain Kanfany, P. Rakshith, Mohammed Riyazaddin, Omar Diack, Desalegn D. Serba, and Ramasamy Perumal

Chapter Overview

Enhancing agricultural productivity in response to the growing population and food demand in West Africa is imperative. Pearl millet [*Pennisetum glaucum* (L.) R. Br.], which is well suited to the region's challenging conditions, holds significant potential for strengthening food security. However, current yields are low, and the adoption of improved hybrid varieties lags. Organizations like International Crop Research Institute for the Semi-Arid Tropics and national agricultural research institutes are working to establish hybrid breeding programs, emphasizing harnessing heterosis for higher yields and stress resilience, but challenges include finding suitable cytoplasmic male sterility sources and developing restorer lines. Initiatives by the Alliance for Green Revolution in Africa and Harvest Plus support hybrid development, as evidenced by Burkina Faso's release of Nufagnon, a high-yield, downy mildew–resistant hybrid adapted to local conditions. Key to success is addressing agronomic and socio-economic challenges through collaboration with farmers, optimizing planting density, and ensuring proper fertilizer use. Public and private sectors play vital roles, requiring training for local seed companies, incentivizing hybrid seed production, and implementing policies to prevent malpractice. Ongoing genetic research, including identifying heterotic groups and utilizing advanced molecular techniques, is crucial for shaping hybrid development. The outlook is promising, with collaborative efforts expected to yield improved hybrid varieties. The adoption of "speed breeding" and precision breeding techniques holds potential for quicker adaptation to changing agricultural landscapes. In conclusion, success hinges on continued collaboration, favorable policies, and concerted efforts to address challenges, ultimately enhancing food security and fostering economic growth in the region.

Pearl Millet: A Resilient Cereal Crop for Food, Nutrition, and Climate Security, First Edition.
Edited by Ramasamy Perumal, P. V. Vara Prasad, C. Tara Satyavathi, Mahalingam Govindaraj, and Abdou Tenkouano.
© 2025 American Society of Agronomy, Inc., Crop Science Society of America, Inc., Soil Science Society of America, Inc. Published 2025 by John Wiley & Sons, Inc.

Introduction

The world population is projected to reach over nine billion by 2050. Western Africa (WA) has the highest population growth, and the population of WA increased by over 400 million from 1950 to 2020 (Kaba, 2020). This trend is expected to continue in the foreseeable future. In this context, food demand will increase dramatically in the coming years in this part of the world. Pearl millet [*Pennisetum glaucum* (L.) R. Br.], which is grown by subsistence farmers, is expected to play an important role in achieving food security. This cereal crop can be grown in drought-prone and high-temperature areas like the Sahel of WA where other cereals will face greater yield reduction or total crop failure (Serba et al., 2017). Pearl millet additionally has a high potential for fighting malnutrition due to its nutritional benefits. It is more nutritious than maize (*Zea mays* L.), rice (*Oryza sativa* L.), wheat (*Triticum aestivum* L.), and sorghum [*Sorghum bicolor* (L.) Moench] because its grain contains higher levels of protein, vitamins, and essential micronutrients like Fe and Zn (Ghatak et al., 2021; Saleh et al., 2013; Serba et al., 2017). The inherent ability of pearl millet to grow in the harsh environments of African regions provides great opportunity for hassle-free cultivation, and this has made pearl millet an ideal crop of the WA cropping system, wherein it is widely grown as intercrop with cowpea [*Vigna unguiculata* (L.) Walp] (Reddy et al., 1992), groundnut (*Arachis hypogaea* L.), grain sorghum, or maize and often in agroforestry systems (Riej & Smaling, 2008). Apart from these practices, pearl millet is also grown as a sole crop.

Of the 30 million ha that are planted for pearl millet worldwide, more than 18 million ha are in Africa. Africa consists of two major regions: West/Central Africa (WCA) (Nigeria, Niger, Chad, Mali, and Senegal) and East/Southern Africa (Sudan, Ethiopia, and Tanzania). In sub-Saharan Africa, pearl millet is the third most important crop, with Nigeria, Senegal, Chad, Mali, Niger, and Burkina Faso being the main producers. With 95% of its total area dedicated to the production of pearl millet, WCA is the greatest pearl millet–producing region in Africa and the globe (Jukanti et al., 2016). Being the largest producer of pearl millet, WCA constitutes the top five countries within it, led by Nigeria (41%), Niger (16%), Burkina Faso (7%), Mali (6.4%), and Senegal and Sudan (4.8%). Overall, 28 countries in Africa produce 11.36 million t of yield over 18 million ha covering 30% of the continent having diverse agroecologies. This makes up 49% of the world's millet acreage (FAO, 2019).

Despite the importance of this cereal in WA, productivity is still very low (648 kg/ha) compared with its yield performance in India (FAO, 2019). Undeniably, farmers in WA have not adopted large-scale production of improved open-pollinated and hybrid varieties (Christinck et al., 2014). In addition, pearl millet in Africa is mostly cultivated on fields with very low soil fertility, which is associated with poor soil resources, soil degradation, extensive management practices, and a low level of external inputs (Bekunda et al., 1997; Somda et al., 2002). This is made more difficult by the fact that many smallholder farmers in WA are prevented from using fertilizers due to a lack of capital, high costs, risk aversion, and inadequate infrastructure. Particularly low P input will become a bigger limitation because there are limited and non-renewable supplies of P fertilizer (Pucher, 2018). In contrast, in India, most of the fields use improved hybrids, and yield significantly increased from 305 kg/ha during 1951–1955 to 998 kg/ha during 2008–2012 (Yadav & Rai, 2013). Development of single-cross hybrid and cultivation is prevalent in most of the pearl millet–growing regions in India, facilitated by the presence of essential expertise, resources, a robust seed industry, and a well-implemented cytoplasmic male sterility (CMS) system within adapted germplasm and elite lines (Kumara Charyulu et al., 2014; Sattler & Haussmann, 2020; Serba et al., 2017).

Therefore, enhancement of pearl millet production in WA can mostly be reached through the development and dissemination of adapted pearl millet hybrid varieties. Even though the majority of WA's national pearl millet breeding programs are still in their infancy, the International Crop Research Institute for the Semi-Arid Tropics (ICRISAT) and national agricultural research institutes (NARS) have jointly demonstrated the prevalence of high levels of heterosis and the potential to breed for hybrids in WA environments. Several hybrids, such as Nufagnon and Toronio TCH, have been developed and released for commercial cultivation.

Heterosis and Genetic Gain

The use of the open-pollinated variety (OPV) option leads to limited hybrid vigor utilization. Although crosses involving similar genotypes enable rapid OPV generation with acceptable uniformity, they lack the genetic potential needed for substantial yield gains. Then, achieving lasting grain production improvements through OPVs poses a significant challenge. In contrast, the hybrid approach offers a chance to harness diverse parental lines, maximizing heterosis and uniformity. In pearl millet breeding, the prevalence of economic heterosis, and availability of CMS make hybrids economically viable. Hybrid cultivars in WA's arid regions require high heterosis to compensate for higher seed costs and yield stability in challenging environmental conditions (Sattler & Haussmann, 2020).

Even though certain nations, such as Senegal, initiated a minor hybrid breeding program in 1982 (Ndoye & Gupta, 1987), pearl millet breeding efforts in WCA have focused on intra-population enhancement for grain output in recent years. Inter-population hybrids were made using African germplasm and evaluated in Niger in 1989 and 1990. Significant expression of better parent, standard OPV, and standard hybrid heterosis was observed in different types of pearl millet hybrids, giving evidence of hybrid superiority over open-pollinated varieties (Drabo, 2016; Kanfany et al., 2018; Ouendeba et al., 1993). The first generation of hybrids developed at ICRISAT India and tested in Africa were high performing (>50% yield increase), but they were not adapted to the production constraints (Ndoye & Gupta, 1987). Despite this performance, the hybrid was not disseminated to farmers because of its susceptibility to downy mildew (DM) and short panicle length. ICRISAT spearheaded several projects in the area to set up hybrid breeding programs using locally adapted germplasm in WA. Four national institutes, including Burkina Faso, Nigeria, Mali, and Niger, were supported by the Alliance for Green Revolution in Africa from 2008 to 2012 to develop pearl millet hybrid. The BMZ-heterotic project, which involved Nigeria, Senegal, and Niger, was another endeavor. In WA, HarvestPlus also provided funding for hybrid development (www.icrisat.org). These supports have contributed to establishing hybrid breeding programs in WA. ICRISAT-West Africa has been focused on hybrid development and testing across NARS in WCA. Several hybrids were identified by NARS as promising in their environment. In 2021, Burkina Faso released its first pearl millet hybrid, Nufagnon, developed by ICRISAT and released by the Institute of Environment and Agricultural Research (INERA)- Burkina Faso. Nufagnon has a high yield, is resistant to DM, and has adapted to the local growing conditions. In view of the urgent need for a rapid increase in productivity to achieve food security, the pearl millet breeding program in WA has dedicated at least 25% of their efforts for hybrid development. Thus, several combining ability studies using populations or inbred lines have clearly showed the prevalence of high levels of heterosis in West African environments (Kanfany et al., 2018; Pucher et al., 2016; Sattler et al., 2019). Currently, several hybrids, including an inter-population hybrid (Mali), a top-cross hybrid (Senegal), and a single-cross hybrid (Burkina Faso), were released by NARS in WA (www.icrisat.org). Adoption of hybrids is increasing especially in Burkina Faso, where seed companies are producing and commercializing the seed.

Available Grain and Forage Hybrids and Yield Potential Compared with Landraces

Since the establishment of pearl millet breeding activities by different programs in WCA, several hybrids have been developed and released, and some others are in the pipeline with NARS (Burkina Faso, Senegal, Niger, and Mali) and ICRISAT. One inter-population hybrid (TORONIOU C1 arsite HYBRIDE TOP CROSS) was released and registered in the regional seed catalog by Mali (CEDEAO-UEMOA-CILSS, 2016). However, they are heterogeneous, and there is a lack of synchronization between parental lines. The R-line is Maïwa, a highly photosensitive landrace from Nigeria. One top-cross hybrid (Taaw) was registered by Senegal and registered in the ECOWAS regional seed catalog (CEDEAO-UEMOA-CILSS, 2021). The first African single-cross pearl millet hybrid (Nafagnon) developed by ICRISAT was released in Burkina Faso by INERA in 2019 and registered in the ECOWAS regional seed catalog in 2021 (CEDEAO-UEMOA-CILSS, 2021). It is a dual-purpose hybrid with a stay-green trait. Current and future hybrid breeding efforts will be based on identification of a pearl millet heterotic group to maximize heterosis. Elite germplasm, including dual purpose lines identified in previous studies (Serba et al., 2020), will be useful in developing new parental lines.

Constraints in Hybrids Seed Production Systems

Better-yielding hybrids have been produced in the area as a result of multiple hybrid development initiatives. However, the primary obstacles to the production of hybrid pearl millet seed remain the absence of multinational companies operating in the area, the inadequate training of regional seed companies, the absence of a formal seed system, the government's lack of interest in promoting hybrid pearl millet, and the inability of farmers to obtain seed at the appropriate time. There is also a need for research on seed production of released hybrids. Researchers are putting effort in the region to breed better hybrids but are lacking in the research on seed production of the hybrids. In addition, considering the current situation, millet value chains (starting from production and reaching up to consumers) have disparate quality requirements and enforcements. However, the methods for improving quality (e.g., varietal purity and post-harvest techniques) are seldom used during the production stage (Kaminski et al., 2013). Furthermore, grain size, product homogeneity, purity, criteria for taste and odor, and water composition appear to be primarily within the control of buyers, particularly institutional or industrial large-scale customers (such as state agencies or the World Food Program). However, in general, quality assurance is not yet developed in local or even urban markets, giving the final consumer the task of upgrading the quality of the purchased cereal crops. This is a serious hindrance to the development of a millet value chain and a significant demand-suppressing factor.

It takes certification and standardization to encourage the growth of downstream millet value chains. The goal of standard setting should be to meet global production and sales norms (such as those pertaining to aflatoxin). Time, money, human resources, physical investments, and the participation of all direct and indirect value chain players are necessary for compliance with quality requirements. In the domains of standardization, quality assurance, and the marketing of agricultural goods, capacity building is crucial. To encourage high quality standards, a supportive industry and marketplace environment is crucial (Kaminski et al., 2013).

In order to properly plan and commercialize future hybrids in the region, as well as to identify specialized zones for hybrid seed production in each country, a systematic approach to hybrid seed production research should be established. In Burkina Faso, four seed companies were trained under the AVISA project, and a hybrid parent consortium was established between those seed companies (NAFASO, FAGRI, EPAM, and EPC-SAC), INERA, and ICRISAT. Hybrid parent research was conducted to optimize the seed production in 2019 and 2020, and a planting ratio of 2:6 rows of male/female with plot size of 20×80 cm (within and between row spacing) was identified as the best design for hybrid NAFAGNON seed production. Similar efforts are needed for the hybrids under consideration for release in the region.

Challenges in Using Alternative CMS Sources

The CMS system is a maternally inherited trait that is characterized by anthers that are not able to produce viable pollen and thus facilitate large-scale hybrid seed production. The A_1 CMS system, which is extensively used in the pearl millet hybrid development program, was the first CMS system ever reported in pearl millet and is based on the Tift 23 A_1 cytoplasm source (Burton, 1958). Subsequently, several other CMS sources, including A_2, A_3, A_v, A_4, and A_5, were discovered in different genetic stocks (Appadurai et al., 1982; Burton & Athwal, 1967). Although the A_1 CMS system has a higher frequency of restorer lines than the A_4 CMS system, its sterility is less stable. Thus, the A_1 CMS is most extensively used as a seed parent, and many male sterile lines developed with this source are used in Indian and African pearl millet breeding programs. The two released hybrid varieties in WA were developed using the A_1 CMS system from ICRISAT-Niger.

To avoid cytoplasmic uniformity, cytoplasmic diversification studies were conducted, and several alternative sources to the A_1 system were discovered (Hanna, 1989; Sujata et al., 1994). However, only A_4 and A_5 CMS systems were identified as commercially viable (Rai et al., 2001, 2009). In WA, current activities include mainly the identification of promising hybrid parents and conversion of adapted materials into the CMS line. The breeding program of the INERA of Burkina Faso has succeeded in converting one popular and widely adapted variety in WCA (SOSAT-C88) in the CMS line, A_4 system, named MS-SOSAT or IKMA18002/IKMB18002. The conversion was done by crossing SOSAT-C88 to 38A from ICRISAT, followed by successive backcrosses. Several versions of CMS lines were also developed from the CIVAREX population from Mali. The DM-resistant

B-line called MS-CIVAREXB-S3-45 is resistant to three diverse DM pathogen biotypes from Burkina Faso and Niger (Drabo, 2016). MS-CIVAREXB bristled was also developed. The SOSAT-C88 version of A_1 was also developed by ICRISAT, named ICMA 177111/ICMB 177111. Since then, many West African genetic backgrounds were converted to CMS lines. Similarly, restorer lines were identified from West African germplasm. Hybrids developed with the new parental lines are more adapted to the environmental conditions in WCA, are resistant to DM, and are adapted to the crop cycle. There is a good synchronization between restorer and seed parental lines. Another limit to the use of Indian origin CMS-lines was their extra-earliness compared with restorer lines from WCA. It was very difficult to get restorer lines in West African germplasm with synchronized flowering period to the Indian source of CMS lines. Although a hybrid program is established in WCA and parental development is progressing, there are no heterotic groups identified or developed yet. Harriet et al. (2020) studied combining ability, fertility restoration, and sterility maintenance of West African released and popular pearl millet varieties on A_1 and A_4 CMS lines. Five genotypes were categorized as strong fertility restorers, with ICMA 177001 × ICMV IS 99001 and ICMA 177001 × Exborno crosses out of the A_1 and A_3 (ICMA 177002 × Ankoutess, ICMA 177002 × ICMV 167002 and ICMA 177002 × Exborno) in the A_4 CMS system. Exborno, ICMV 167002, and ICMV IS 99001 restore fertility on both A_1 and A_4. Most of the crosses for A_1 and A_4 genotypes from WA fell under sterility maintainer. The comparative study between the two CMS systems showed that partial restoration is possible for both cytoplasms. This study will help to prioritize developing pearl millet hybrids and identification of restorers and maintenance sources for future heterotic pooling and hybrid development in WA.

Agronomic and Socio-economic Challenges

Agronomic packages for each hybrid, along with the seed production optimization for each hybrid, are a must. Breeders should work with agronomists to optimize crop and seed production. Seed technologists must be involved in optimizing the planting density and date of planting male and female parents of the hybrids. An agronomic study has been conducted for hybrid seed production in Burkina Faso over two years (Drabo Innousa [INERA] and Gangashetty Prakash [ICRISAT]). The objective was to determine the optimum seed production by varying male/female rows (2:6, and 2:8), different fertilizer rates (75, 100, and 120 kg/ha of NPK + 50 kg/ha of urea), and variable plant spacing to suit ideal plant density (20 × 80 cm and 40 × 80 cm). This highly effective strategy will maximize productivity in terms of yield in the hybrid grain production and in the seed production of hybrids.

Farmers in the region are deprived of the improved production packages and newer innovative technologies. Gaoh et al. (2023) conducted a survey on the uptake of hybrid pearl millet in the Dosso region of Niger. Results revealed that pearl millet is a primary staple crop (98% of respondents) that is grown and consumed on a daily basis as food and also used as feed for animals. The majority of farmers preferred a long panicle (50.7%) and a good seed set (45.3%). For grain traits, a white color (50%) and larger size (100%) were preferred, which fetches them higher prices in the market, where they compete with sorghum grains. These traits should be considered while making the selection in the hybrid pearl millet. The adoption of enhanced high-yielding cultivars presents the bigger challenge for pearl millet seed production in Africa. The regions under improved pearl millet types are extremely few, given the current situation. Furthermore, farmers in Africa are undoubtedly limited in their ability to use high-yielding modern varieties due to the lack of a well-established seed infrastructure. These varieties primarily rely on the availability of breeder seed, the evaluation of seed quality, and an efficient seed distribution system. The example of Niger's pearl millet adoption pattern shows this quite effectively. The government's meager resources for producing breeder seeds in Niger are made more difficult by the absence of procedures for distribution, demand assessment, and quality control. Because of this, there is a relatively low acceptance rate of improved new varieties of pearl millet (~5%), which results in low yield (Ndjeunga, 1997).

The region's ability to produce seeds has been severely hampered by the absence of an accurate demand assessment mechanism. Accurate demand assessment is the key to successful seed production; otherwise, the entire seed chain—including seed distribution, seed needed for extension use, and wholesale and retail sales—would be adversely affected. However, a number of factors, such as the cost of seed available from alternate sources, farmers' earnings, farmers' preferences, and the type of crop, influence the demand for seed. A crucial factor in providing farmers with seeds at the appropriate time is seed distribution. The involvement of several channels—extension services, national, and regional development initiatives, non-governmental organizations, and merchants—allows this to be done effectively. With few exceptions, most African nations lack these amenities and frequently have relatively few wholesale and retail locations that are far from the widely dispersed farming villages in remote places with inadequate road connectivity. This will undoubtedly result in high shipping costs and may limit the number of markets that seeds can reach (Ndjeunga, 1997).

Research Strategies and Traits of Focus

Grain yield is the main interest of all breeding programs. In WA, it has been unanimously considered as the most important trait by farmers when selecting a variety to grow in their environment (Kanfany et al., 2020). Many varieties have been developed and released that out-yield farmers' landraces. However, the adoption rate of improved pearl millet varieties remains the lowest among the major crops in WCA (<13%) (Vabi et al., 2020). Participatory varietal selection has been implemented to increase adoption rate (Omanya et al., 2007). The current strategy consists of identification of market segments and target product profiles by the country product design team, a multidisciplinary team composed of representatives of pearl millet value chain actors. Breeding pipelines are developed for each target product profile, and the product design team decides the advancement of the breeding lines and the release date when they meet the target product profile standard for a specific market segment. Thus, the use of improved cultivars like hybrids could be an effective way to considerably increase grain yield and meet farmers' needs. In addition to grain yield, priority can be given to resistance to DM, panicle length, larger grain size, resistance to striga, and high Fe and Zn levels (Drabo et al., 2019; Gaoh et al., 2023; Kanfany et al., 2020; Rouamba et al., 2021). A sustainable hybrid strategy is needed. The priority will be the identification and development of heterotic groups, use of molecular tools, and rapid generation advancement.

The future improvement of crop yields using hybrid technology will be contingent on having enough genetic diversity among the hybrid parents and on accurately identifying the specific cross combinations that yield high production, thus enabling the creation of even more productive hybrid varieties. Thus, available germplasm needs to be organized into heterotic groups to increase the efficiency of the hybrid breeding program. Heterotic patterns have a strong impact in crop improvement because they can provide information about the type of germplasm that should be used in a hybrid breeding program over a longer period of time (Reif et al., 2005).

The origin of pearl millet in WA has led to significant phenotypic and genotypic diversity among cultivated and wild varieties due to ongoing gene flow (Brunken et al., 1977; Burgarella et al., 2018; Manning et al., 2011). Core collections representing genetic variations have been developed, revealing a mixture of Sahelian and West African pearl millet (Bashir et al., 2014; Padilla, 2007; Pucher et al., 2015; Sattler et al., 2018). Genotyping techniques like simple sequence repeat and Diversity Arrays Technology markers have identified subgroups based on geography and traits shaped by local practices (Bashir et al., 2014). Senegalese cultivars display unique allele frequencies linked to flowering time genes (Diack et al., 2017). Despite these findings, there is no clear association between genetic distance and hybrid performance, highlighting the need for markers linked to yield-related QTL (Bashir et al., 2014; Charcosset et al., 1991; Jordan et al., 2003; Zhang et al., 1994, 1995). The publication of the reference genome for pearl millet by Varshney et al. (2017) has further advanced the understanding of its genetic diversity and population structure.

Although studies have identified some clusters of West African pearl millet varieties, naturally existing heterotic groups are absent. Large-scale combining ability studies have been conducted, involving diallel (Dutta, 2019) and factorial mating designs (Pucher et al., 2016; Sattler et al., 2019), to identify potential starting points for heterotic group development. Despite the absence of distinct groups, recommendations for initial heterotic groups were made based on combining ability patterns of specific cultivars from Niger and Senegal targeted at developing long-panicle pearl millet hybrids (Sattler et al., 2019). Further research is needed because the superiority of specific varieties might be influenced by adaptation to local conditions. The performance of parents strongly affects hybrid performance, with a significant impact on general combining ability (GCA) variance relative to specific combining ability (SCA) variance. Increasing genetic distance is believed to enhance not only heterosis but also hybrid prediction accuracy. Hybrids from parents belonging to distinct heterotic groups express a larger GCA/SCA variance ratio. This trend is observed in combining ability studies of West African pearl millet (Dutta, 2019; Ouendeba et al., 1993; Pucher et al., 2016; Sattler et al., 2019). This highlights the significance of comprehensively grasping combining ability patterns in the absence of naturally existing heterotic groups because such understanding can streamline the methodical establishment of heterotic groups and foster the feasibility of sustainable hybrid breeding. Challenges in identifying ideal hybrid parent combinations and selecting top hybrids have led to the development of genomic selection. This approach combines molecular markers with phenotypic data, offering promise in predicting hybrid performance (Bernardo, 1998). High-density genome-wide markers enable this technique, although it requires substantial funding. Despite limitations in the private sector, genomic selection can expedite heterotic grouping in pearl millet. Similar strategies applied in wheat have shown potential, although genetic diversity and heterosis within each pool warrant exploration (Zhao et al., 2015).

In the context of pearl millet, a staple crop in many African regions, speed breeding techniques offer potential solutions to challenges like climate change, pest and disease pressures, and food security needs. Through controlled environments, advanced genetics, and optimized agronomic practices, breeders can hasten the selection of desired traits, resulting in improved pearl millet varieties better suited to local conditions. Speed breeding methods often involve creating ideal growth conditions, such as controlled photoperiods, temperature, and humidity, to accelerate plant growth and development. This allows for multiple generations to be completed in a shorter timeframe. Moreover, modern molecular techniques, like marker-assisted selection and genomic selection, can be integrated into speed breeding programs to enhance the precision and efficiency of trait selection. The adoption of speed breeding facilities, exemplified by the RapidGen facility developed at ICRISAT in India (Yadav et al., 2021), will gain traction in African pearl millet breeding efforts. This approach has the potential to significantly reducing the time needed to develop and release new cultivars that better suit changing environmental and market conditions.

Public- and Private-Sector Efforts for Hybrid Promotion

Most NARs and ICRISAT have developed promising hybrid technologies in the region. The hybrid systems need more systematic seed systems involving private seed industries. The seed system should also be supported by local governments and authorities. The benefits of heterosis can only be realized through holistic seed systems supported by hybrid research and development. West African countries, mainly in the Sahel and Sudanian regions, are the dominant pearl millet–growing regions. Due to the regions' soil fertility, moisture availability, soil type, and distribution of rainfall, hybrid technology introduction should initially concentrate on the Sudanian region. However, the Sahel region has extremely variable rainfall patterns, sandy soils, and very low soil fertility. Hence, the OPV should focus on early maturity, long heads, and higher biomass production in the given rainfall and soil conditions. ICRISAT and NARs focus according to the regional agro-ecological conditions. To fully reap the benefits of heterosis, local private seed companies must be trained, in addition to creating hybrids and promoting them locally. In contrast to India, there are no multinational companies operating in the area that are involved in the hybrid pearl millet. Hence, focus should be given to train the whole value chain of hybrid pearl millet development to the local seed companies in the region.

After the release of the first hybrid 'IKMH 18001' (recommended for the Sudanian Zone of Burkina Faso), a partnership was established between INERA and the NAFASO seed company in Burkina Faso (ICRISAT, 2020) to produce and commercialize certified seed of the hybrid in Burkina Faso and in the region. Thus far, 1,200 kg of hybrid-certified seed has been produced and sold. In 2022, 500 kg of foundation seed of the parental lines was produced in partnership with NAFASO. As the demand from the farmers for hybrid seeds increases, with the support of governments, private seed industries can take steps to produce more of these hybrid seeds and sell them regionally. This has been a great success story for the reason of adopting improved varieties/hybrids that indeed will enhance productivity. Nevertheless, IKMH 18001 yields 35%–40% more in comparison to the local and enhanced OPV Misari-1 in Burkina Faso. This hybrid is distinguished by its strong producing potential of up to 3.2 tons/ha, lengthy greening period, dual function, and medium maturation rate.

Similarly, the high Fe–biofortified variety 'Chakti' released in the year 2020 (ICRISAT, 2020) in Niger promises great potential to expand the cultivated area and hence to scale up seed production as it is gaining popularity among the farmers due to its early maturity, high production potential, resistance to DM, and enhanced Fe and Zn content. As reported by the Niger government's Ministry of Agriculture, owing to demand for the variety, 118 t of respective certified seeds were produced in the country during the year 2020. Considering its success, in the Niger Falwel region, Mooriben, an association of farmers has additionally advocated for the cultivation of Chakti. This in turn will create greater demand for Chakti, and, in collaboration with either public or private entities or both, the demand can be met.

Hybrid Cost Profit Ratio and Government Policies

The cost of producing seed as well as the viability of seed distribution and multiplication initiatives are determined by seed price, which is a crucial aspect of seed marketing. The average sale price must exceed the average cost of production in order to support sustainable seed multiplication and distribution. The type of product determines the seed price. Grain is an excellent alternative for seed, so one cannot expect the price of seed to be significantly higher than the price of grain for varieties or composites. Nevertheless, margins over the cost of seed production ultimately decide the financial efficiency and hence the sustainability of the seed system. However, hybrid technology for pearl millet in WCA is still relatively new, and the local government should support the development of hybrid pearl millet by providing incentives for the production of hybrid seeds. Policies should be established to curb the malpractice of selling the wrong hybrids or OPV seeds as hybrids, and proper labeling should be provided. A systematic study of the cost-benefit ratio for production and marketing is needed. Hybrid pearl millet production is cost-effective because of the availability of high-performance seed parents. The ratio between the male and female can be up to eight rows of female and two rows of male plants for the hybrid ICPMH 177111 developed by ICRISAT-Niger. Hybrid yields obtained from the field are up to 1.2 t/ha. Hybrid seed yield obtained from Burkina Faso (Drabo & Gangashetty, unpublished data) varied from 650 to 1,150 kg/ha, with an average of 800 kg/ha. A simulation study based on the yield and the production cost conducted shows that hybrid seed can be sold at 1,500–2,000 CFA (US$2.5–3.4) per kilogram and still provide a profit. The seed rate is 4–5 kg/ha. Pearl millet hybrids in the region can be effectively produced and marketed.

Summary and Outlook

The significance of pearl millet as a potential solution to food security challenges in WA is highlighted due to its ability to thrive in tough environmental conditions such as drought and high temperatures. Its nutritional richness, with high levels of protein, vitamins, and micronutrients, makes it an effective tool against malnutrition. Despite these advantages, pearl millet yields in the region are low compared with other areas, and the adoption of improved hybrid varieties has been limited. Challenges and possibilities surround hybrid development and seed production of pearl millet in Africa, such as the importance of harnessing hybrid technology to maximize heterosis, resulting in higher yields, and increased resilience against environmental pressures. The efforts of organizations like ICRISAT, NARS, and private seed companies are highlighted in establishing hybrid breeding programs. These endeavors include identifying suitable CMS sources, creating restorer lines, and developing hybrid combinations suitable for local conditions. Furthermore, the use of genomic tools like molecular markers and genomic selection is emphasized to enhance the efficiency and precision of hybrid breeding programs. The identification of heterotic groups within the diverse pearl millet germplasm is considered crucial for generating effective hybrid combinations and enhancing breeding outcomes. The paper also touches upon the concept of speed breeding techniques, which can expedite the development of pearl millet varieties that can adapt to changing environmental dynamics.

The outlook for pearl millet hybrid development and seed production in Africa is promising, with collaborative efforts between public and private sectors expected to yield positive results. The establishment of hybrid breeding programs and partnerships between research institutions and seed companies are anticipated to lead to the release of improved and higher-yielding hybrid varieties. Genetic research, including the identification of heterotic groups and advanced molecular techniques, will likely continue to shape the development of hybrids that are suited to local conditions. The adoption of speed breeding techniques and precision breeding methods holds the potential to expedite the creation of new pearl millet cultivars, enabling quicker adaptation to changing agricultural landscapes and market demands. The increasing involvement of private seed companies in hybrid seed production and distribution suggests a potential rise in the availability and accessibility of hybrid seeds for farmers. This trend could contribute to enhanced food security and economic growth in the region. Government policies and support play a pivotal role in the success of pearl millet hybrid development and seed production. Favorable policies that encourage research, development, and adoption of improved pearl millet varieties, alongside investments in infrastructure and training, are critical to amplifying pearl millet's impact on food security and livelihoods in West Africa. The recent adoption of Chakti by large-scale farmers in Niger, which is now advocated to extend to the Falwel region, may encourage the private sector to scale up seed production. The recent release of the first-ever African pearl millet hybrid IKMH 18001 developed in collaboration between the public and private sectors would lay a strong platform to take up such hybrid breeding in WCA.

References

Appadurai, R., Raveendran, T. S., & Nagarajan, C. (1982). A new male-sterility system in pearl millet. *Indian Journal of Agricultural Sciences, 52*, 832–834.

Bashir, E. M. A., Ali, A. M., Ali, A. M., Melchinger, A. E., Parzies, H. K., & Haussmann, B. I. G. (2014). Characterization of Sudanese pearl millet germplasm for agro-morphological traits and grain nutritional values. *Plant Genetic Resources, 12*, 35–47. https://doi.org/10.1017/S1479262113000233

Bekunda, M. A., Bationo, A., & Ssali, H. (1997). Soil fertility management in Africa: A review of selected research trials. In R. J. Buresh, P. A. Sanchez, & F. Calhoun (Eds.), *Replenishing soil fertility in Africa* (pp. 63–79). SSSA and ASA.

Bernardo, R. (1998). A model for marker-assisted selection among single crosses with multiple genetic markers. *Theoretical and Applied Genetics, 97*, 473–478. https://doi.org/10.1007/s001220050919

Brunken, J., De Wet, J. M. J., & Harlan, J. R. (1977). The morphology and domestication of pearl millet. *Economic Botany*, 31, 163–31, 174.

Burgarella, C., Cubry, P., Kane, N. A., Varshney, R. K., Mariac, C., Liu, X., Shi, C., Thudi, M., Couderc, M., Xu, X., Chitikineni, A., Scarcelli, N., Barnaud, A., Rhoné, B., Dupuy, C., François, O., Berthouly-Salazar, C., & Vigouroux, Y. (2018). A western Sahara centre of domestication inferred from pearl millet genomes. *Nature Ecology & Evolution, 2*, 1377–1380. https://doi.org/10.1038/s41559-018-0643-y

Burton, G. W. (1958). Cytoplasmic male-sterility in pearl millet (*Pennisetum glaucum*) (L.) R. Br.1. *Agronomy Journal, 50*(4), 230–230. https://doi.org/10.2134/agronj1958.00021962005000040018x

Burton, G. W., & Athwal, D. S. (1967). Two additional sources of cytoplasmic male-sterility in pearl millet and their relationship to Tift 23A 1. *Crop Science, 7*(3), 209–211. https://doi.org/10.2135/cropsci1967.0011183X000700030011x

CEDEAO-UEMOA-CILSS. (2016). *Catalogue régional des espèces et variétés végétales.* 548–109. http://www.coraf.org/paired/wp-content/uploads/2019/11/Catalogue-Re%CC%81gional-des-Espe%CC%80ces.pdf

Charcosset, A., Lefort-Buson, M., & Gallais, A. (1991). Relationship between heterosis and heterozygosity at marker loci: A theoretical computation. *Theoretical and Applied Genetics, 81*, 571–575. https://doi.org/10.1007/BF00226720

Christinck, B. A., Diarra, M., & Horneber, G. (2014). *Innovations in seed systems— Lessons from the CCRP funded project Sustaining farmer-managed seed initiatives in Mali, Niger and Burkina Faso.* FAO.

Diack, O., Kane, N. A., Berthouly-Salazar, C., Gueye, M. C., Diop, B. M., Fofana, A., Sy, O., Tall, H., Zekraoui, L., Piquet, M., Couderc, M., Vigouroux, Y., Diouf, D., & Barnaud, A. (2017). New genetic insights into pearl millet diversity as revealed by characterization of early- and late-flowering landraces from Senegal. *Frontiers in Plant Science, 8*, 818. https://doi.org/10.3389/fpls.2017.00818

Drabo, I., Zangre, R. G., Danquah, E. Y., Ofori, K., Witcombe, J. &. R., & Hash, C. T. (2019). Identifying famers' preferences and constraints to pearl millet production in the Sahel and North-Sudan zones of Burkina Faso. *Experimental Agriculture, 55*(5), 765–775.

Dutta, S. (2019). *Combining ability patterns among West and Central African pearl millet populations.* University of Hohenheim.

FAO. (2019). *FAO database. Statistiques agricoles.* https://www.fao.org/3/ca6030en/ca6030en.pdf

Gaoh, B. S. B., Gangashetty, P. I., Mohammed, R., Govindaraj, M., Dzidzienyo, D. K., & Tongoona, P. (2023). Establishing breeding priorities for developing biofortified high-yielding pearl millet (*Pennisetum glaucum* (L.) R. Br.) varieties and hybrids in Dosso region of Niger. *Agronomy, 13*(1), 166. https://doi.org/10.3390/agronomy13010166

Ghatak, A., Chaturvedi, P., Bachmann, G., Valledor, L., Ramšak, Ž., Bazargani, M. M., Bajaj, P., Jayadevan, S., Li, W., Sun, X., Gruden, K., Varshney, R. K., & Weckwerth, W. (2021). Physiological and proteomic signatures reveal mechanisms of superior drought resilience in pearl millet compared to wheat. *Frontiers in Plant Science, 11*, 600278. https://doi.org/10.3389/fpls.2020.600278

Harriet et al. (2020) Studies on combining ability, restoration and maintenance of West African released and popular pearl millet varieties on A1 and A4 cytoplasm of pearl millet (Pennisetum glaucum L). Master thesis. PAN African University, Ibadan, Nigeria.

Hanna, W. W. (1989). Characteristics and stability of a new cytoplasmic-nuclear male-sterile source in pearl millet. *Crop Science, 29*(6), 1457–1459. https://doi.org/10.2135/cropsci1989.0011183X002900060026x

ICRISAT. (2020). *Research highlights 2020. Doing science, planting hope.* ICRISAT.

Jordan, D. R., Tao, Y., Godwin, I. D., Henzell, R. G., Cooper, M., & McIntyre, C. L. (2003). Prediction of hybrid performance in grain sorghum using RFLP markers. *Theoretical and Applied Genetics, 106*, 559–567. https://doi.org/10.1007/s00122-002-1144-5

Jukanti, A. K., Gowda, C. L. L., Rai, K. N., Manga, V. K., & Bhatt, R. K. (2016). Crops that feed the world 11. Pearl millet (*Pennisetum glaucum* L.): An important source of food security, nutrition and health in the arid and semi-arid tropics. *Food Security, 8*, 307–329. https://doi.org/10.1007/s12571-016-0557-y

Kaba, A. J. (2020). Explaining Africa's rapid population growth, 1950 to 2020: Trends, factors, implications, and recommendations. *Sociology Mind, 10*(04), 226–268. https://doi.org/10.4236/sm.2020.104015

Kaminski, J., Elbehri, A., & Samake, M. (2013). An assessment of sorghum and millet in Mali and implications for competitive and inclusive value chains. In A. Elbehri (Ed.), *Rebuilding West Africa's food potential* (pp. 481–501). FAO/IFAD.

Kanfany, G., Diack, O., Kane, N. A., & Gangashetty, P. (2020). Implications of farmer perceived production constraints and varietal preferences to pearl millet breeding in Senegal. *African Crop Science Journal, 28*(October), 411–420.

Kanfany, G., Fofana, A., Tongoona, P., Danquah, A., Offei, S., Danquah, E., & Cisse, N. (2018). Estimates of combining ability and heterosis for yield and its related traits in pearl millet inbred lines under downy mildew prevalent areas of Senegal. *International Journal of Agronomy*, *2018*, 3439090.

Kumara Charyulu, D., Bantilan, M. C. S., Rajalaxmi, A., Rai, K. N., Yadav, O. P., Gupta, S. K., Singh, N. P., & Shyam, D. M. (2014). *Development and diffusion of pearl millet improved cultivars in India: Impact on growth and yield stability.* ICRISAT.

Manning, K., Pelling, R., Higham, T., Schwenniger, J. L., & Fuller, D. Q. (2011). 4500-year old domesticated pearl millet (*Pennisetum glaucum*) from the Tilemsi Valley, Mali: New insights into an alternative cereal domestication pathway. *Journal of Archaeological Science*, *38*(2), 312–322.

Ndjeunga, J. (1997). Constraints to variety release, seed multiplication, and distribution of sorghum, pearl millet, and groundnut in Western and Central Africa. In D. D. Rohrbach, Z. Bishaw, & A. J. G. van Gastel (Eds.), *Alternative strategies for smallholder seed supply: proceedings of an International Conference on Options for Strengthening National and Regional Seed Systems in Africa and West Asia* (pp. 34–46). ICRISAT.

Ndoye, A. T., & Gupta, S. K. (1987). Research on pearl millet hybrids in Senegal. In *Proceedings of the International Pearl Millet Workshop*. ICRISAT.

Omanya, G. O., Weltzien-Rattunde, E., Sogodogo, D., Sanogo, M., Hanssens, N., Guero, Y., & Zangre, R. (2007). Participatory varietal selection with improved pearl millet in West Africa. *Experimental Agriculture*, *43*(01), 5–19. https://doi.org/10.1017/S0014479706004248

Ouendeba, B., Ejeta, G., Nyquist, W. E., Hanna, W. W., & Kumar, A. (1993). Heterosis and combining ability among African pearl millet landraces. *Crop Science*, *33*, 735–739. https://doi.org/10.2135/cropsci1993.0011183X003300040020x

Padilla, J. C. (2007). *Phenotypic characterization of a pearl millet [*Pennisetum glaucum *(L.) R. Br.] core collection under field conditions in Niger.* University of Hohenheim.

Pucher, A., Sy, O., Angarawai, I. I., Gondah, J., Zangre, R., Ouedraogo, M., Sanogo, M. D., Boureima, S., Hash, C. T., & Haussmann, B. I. G. (2015). Agro-morphological characterization of West and Central African pearl millet accessions. *Crop Science*, *55*, 737–748. https://doi.org/10.2135/cropsci2014.06.04

Pucher, A., Sy, O., Sanogo, M. D., Angarawai, I. I., Zangre, R., Ouedraogo, M., Boureima, S., Hash, C. T., & Haussmann, B. I. G. (2016). Combining ability patterns among West African pearl millet landraces and prospects for pearl millet hybrid breeding. *Field Crops Research*, *195*, 9–20. https://doi.org/10.1016/j.fcr.2016.04.035

Pucher, A. I. (2018). *Pearl millet breeding in West Africa –Steps towards higher productivity and nutritional value* [Doctoral dissertation]. Institute of Plant Breeding, Seed Science and Population Genetics, University of Hohenheim.

https://opus.uni-hohenheim.de/volltexte/2018/1477/pdf/Dissertation_A. Pucher_13.02.18_print.pdf

Rai, K. N., Anand Kumar, K., Andrews, D. J., & Rao, A. S. (2001). Commercial viability of alternative cytoplasmic-nuclear male-sterility systems in pearl millet. *Euphytica*, *121*(1), 107–114. https://doi.org/10.1023/A:1012039720538

Rai, K. N., Khairwal, I. S., Dangaria, C. J., Singh, A. K., & Rao, A. S. (2009). Seed parent breeding efficiency of three diverse cytoplasmic-nuclear male-sterility systems in pearl millet. *Euphytica*, *165*(3), 495. https://doi.org/10.1007/s10681-008-9765-7

Reddy, K. C., Visser, P., & Buckner, P. (1992). Pearl millet and cowpea yields in sole and intercrop systems, and their after-effects on soil and crop productivity. *Field Crops Research*, *28*, 315–326.

Reif, J. C., Hallauer, A. R., & Melchinger, A. E. (2005). Heterosis and heterotic patterns in maize. *Maydica*, *50*(3–4), 215–223.

Riej, C. P., & Smaling, E. M. A. (2008). Analyzing successes in agriculture and land management in Sub-Saharan Africa: Is macro-level gloom obscuring positive micro-level change? *Land Use Policy*, *25*, 410–420.

Rouamba, A., Shimelis, H., Drabo, I., Laing, M., Gangashetty, P., Mathew, I., Mrema, E., & Shayanowako, A. I. T. (2021). Constraints to pearl millet (*Pennisetum glaucum*) production and farmers' approaches to Striga hermonthica management in Burkina Faso. *Sustainability*, *13*(15), 8460. https://doi.org/10.3390/su13158460

Sattler, F. T., & Haussmann, B. I. (2020). A unified strategy for West African pearl millet hybrid and heterotic group development. *Crop Science*, *60*(1), 1–13.

Sattler, F. T., Pucher, A., Kassari Ango, I., Sy, O., Isaaka, A., Hash, C. T., & Haussmann, B. I. G. (2019). Identification of combining ability patterns for pearl millet hybrid breeding in West Africa. *Crop Science*, *59*, 1590–1603. https://doi.org/10.2135/cropsci2018. 12.0727

Sattler, F. T., Sanogo, M. D., Kassari Ango, I., Angarawai, I. I., Gwadi, K. W., Dodo, H., & Haussmann, B. I. G. (2018). Characterization of West and Central African accessions from a pearl millet reference collection for agro-morphological traits and Striga resistance. *Plant Genetic Resources*, *16*, 260–272. https://doi.org/10.1017/S1479262117000272

Serba, D. D., Perumal, R., Tesso, T. T., & Min, D. (2017). Status of global pearl millet breeding programs and the way forward. *Crop Science*, *57*(6), 2891–2905. https://doi.org/10.2135/cropsci2016.11.0936

Somda, J., Nianogo, A. J., Nassa, S., & Sanou, S. (2002). Soil fertility management and socio-economic factors in crop-livestock systems in Burkina Faso: A case study of composting technology. *Ecological Economics*, *43*, 175–183. http://dx.doi.org/10.1016/S0921-8009(02)00208-2

Sujata, V., Sivaramakrishnan, S., Rai, K. N., & Seetha, K. (1994). A new source of cytoplasmic male sterility in pearl millet: RFLP analysis of mitochondrial DNA. *Genome*, *37*(3), 482–486. https://doi.org/10.1139/g94-067

Vabi, M. B., Abdulqudus, I. A., Angarawai, I. I., Adogoba, D. S. Kamara, A. Y., Ajeigbe, H. A. & Ojiewo, C. (2020). Adoption and welfare impacts of pearl millet technologies in Nigeria. ICRISAT.

Varshney, R. K., Shi, C., Thudi, M., Mariac, C., Wallace, J., Qi, P., Zhang, H., Zhao, Y., Wang, X., Rathore, A., Srivastava, R. K., Chitikineni, A., Fan, G., Bajaj, P., Punnuri, S., Gupta, S. K., Wang, H., Jiang, Y., Couderc, M., . . . Xu, X. (2017). Pearl millet genome sequence provides a resource to improve agronomic traits in arid environments. *Nature Biotechnology*, *35*, 969–976. [erratum *36*, 368] https://doi .org/10.1038/nbt.3943

Yadav, O. P., Gupta, S. K., Govindaraj, M., Sharma, R., Varshney, R. K., Srivastava, R. K., Rathore, A., & Mahala, R. S. (2021). Genetic gains in pearl millet in India: Insights into historic breeding strategies and future perspective. *Frontiers in Plant Science*, *12*, 396.

Yadav, O. P., & Rai, K. N. (2013). Genetic improvement of pearl millet in India. *Agricultural Research*, *2*(4), 275–292. https://doi.org/10.1007/s40003-013-0089-z

Zhang, Q., Gao, Y. J., Maroof, M. A. S., Yang, S. H., & Li, J. X. (1995). Molecular divergence and hybrid performance in rice. *Molecular Breeding*, *1*, 133–142. https://doi.org/10.1007/BF01249698

Zhang, Q., Gao, Y. J., Yang, S. H., Ragab, R. A., Maroof, M. A. S., & Li, Z. B. (1994). A diallel analysis of heterosis in elite hybrid rice based on RFLPs and microsatellites. *Theoretical and Applied Genetics*, *89*, 185–192. https://doi.org/10.1007/BF00225139

Zhao, Y., Li, Z., Liu, G., Jiang, Y., Maurer, H. P., Würschum, T., Mock, H.-P., Matros, A., Ebmeyer, E., Schachschneider, R., Kazman, E., Schacht, J., Gowda, M., Longin, C. F. H., & Reif, J. C. (2015). Genome-based establishment of a high-yielding het erotic pattern for hybrid wheat breeding. *Proceedings of the National Academy of Sciences of the United States of America*, *112*, 15624–15629. https://doi.org/10.1073/pnas.1514547112

Saleh, A. S. M., Zhang, Q., Chen, J., & Shen, Q. (2013). Millet Grains: Nutritional Quality, Processing, and Potential Health Benefits. *Comprehensive Reviews in Food Science and Food Safety*, *12*, 281–295. https://doi.org/10.1111/1541-4337.12012

Serba, D. D., Yadav, R. S., Varshney, R. K., Gupta, S. K., Mahalingam, G., Srivastava, R. K., Gupta, R., Ramasamy, P., & Tesfaye, T. T. (2020). Genomic designing of pearl millet: a resilient crop for arid and semi-arid environments. In C. Kole (Ed.), *Genomic Designing of Climate-Smart Cereal Crops* (pp. 221–286). Springer. https:// doi.org/10.1007/978-3-319-93381-8_6

Drabo I. (2016). Breeding pearl millet (*Pennisetum glaucum* [L.] R. BR.) for downy mildew resistance and improved yield in Burkina Faso. PhD Thesis, University of Ghana, pp. 154. http://ugspace.ug.edu.gh.

CEDEAO UEMOA & CILSS. (2021). *Catalogue Régional des Espèces et Variétés Végétales CEDEAO–UEMOA–CILSS Variétés homologuées 2018–2021*. CEDEAO-UEMOA-CILSS.

8

Growth and Development of Pearl Millet

Maduraimuthu Djanaguiraman, Parvaze A. Sofi, Arun K. Shanker,
Ignacio A. Ciampitti, Ramasamy Perumal, and P. V. Vara Prasad

Chapter Overview

Pearl millet [*Pennisetum glaucum* (L.) R. Br.] is the second most widely grown millet crop after sorghum, partly due to its wider adaptability. It originated in Western Africa, was subsequently introduced into India and other parts of Africa, and is now grown in many parts of the semi-arid tropics. Being a C_4 (photosynthetic pathway) plant, it possesses improved photosynthetic efficiency and accumulates greater biomass under low-input agricultural production systems. Globally, pearl millet ranks sixth after rice, wheat, maize, barley, and sorghum and is cultivated over 30 million ha. It has desirable nutritional and nutraceutical properties with higher concentrations of micronutrients, particularly Fe and Zn. It is a good source of energy, carbohydrates, crude fibers (including resistant starch and soluble and insoluble dietary fibers), soluble and insoluble fat, proteins, ash, and antioxidants. The genus *Pennisetum* comprises about 140 species that are widely distributed in the tropics and subtropics. Pearl millet originated from its wild progenitor *Pennisetum violaceum*, and its domestication dates back to the agricultural civilizations in West Africa. Pearl millet is an annual herbaceous grass; plants are tall and erect, with profuse tillering, bearing grains on terminal spikes that are erect and candle shaped. Pearl millet has wide developmental plasticity in growth and grows under various ecological conditions. It is generally grown in areas with annual precipitation ranging from 350 to 500 mm, indicating its low water requirement and wider adaptability to rainfed conditions, with optimum growth at 20–28 °C. Pearl millet can be grown on a wide variety of soils and grows best in sandy loam to loamy soils with good drainage and less salinity and lower soil pH. In this chapter, details about growth and development and taxonomy of pearl millet are discussed.

Pearl Millet: A Resilient Cereal Crop for Food, Nutrition, and Climate Security, First Edition.
Edited by Ramasamy Perumal, P. V. Vara Prasad, C. Tara Satyavathi, Mahalingam Govindaraj, and Abdou Tenkouano.
© 2025 American Society of Agronomy, Inc., Crop Science Society of America, Inc., Soil Science Society of America, Inc. Published 2025 by John Wiley & Sons, Inc.

Introduction

Millet is a common term for small-seeded grasses, including sorghum [*Sorghum bicolor* (L.) Moench], pearl millet [*Pennisetum glaucum* (L.) R. Br.], and other minor millets, like finger millet (*Eleusine coracana* L.), little millet (*Panicum sumatrense* L.), foxtail millet (*Setaria italica* L.), proso millet (*Panicum miliaceum* L.), barnyard millet (*Echinochloa esculenta* L.), kodo millet (*Paspalum scrobiculatum* L.), and teff (*Eragrostis tef* L.). Due to their nutritional properties and climate resilience, millets are also known as nutri-cereals or dryland cereals. Pearl millet is the world's second most widely grown millet after sorghum, partly due to its low water requirement and its drought and heat tolerance among all domesticated cereals (Prasad et al., 2008; Stevens & Fuller, 2018). Pearl millet is also known as bulrush, cattail, or spiked millet (English); bajra (Hindi); dukhn (Arabic); mil à chandelles or petit mil (French); and mhunga or mahango (African languages) (Arun & Faraday, 2020). It originated in Western Africa, followed by subsequent introduction into India. It is now widely distributed in the semi-arid tropics of Asia and Africa. Pearl millet has relatively higher ceiling temperatures ($T_{max} \sim 42\,°C$) for yield (Krishnan & Meera, 2018; Prasad et al., 2008) compared with other cereal grains. Given its low water requirement, temperature tolerance, and greater resilience to climate change, pearl millet can be a viable option for sustainable food systems in marginal farming systems under rainfed conditions. Millet can produce significant grain yields even in severe water-limited areas with <250 mm annual rainfall (Nambiar et al., 2011). In India, pearl millet is grown under irrigated and extremely hot summers in northern Gujarat and eastern Uttar Pradesh and southern India due to its ability to tolerate high temperatures. Being a plant with C_4 pathway, it possesses higher photosynthetic efficiency, resulting in greater biomass accumulation and survival under low-input agricultural systems (Nambiar et al., 2011). Pearl millet and other C_4 crops, such as maize ˋand sorghum, account for about 30% of global terrestrial C fixation (Choudhary et al., 2020). Across the globe, it is grown between 15 °W and 90 °E long and 5 °S to 40 °N lat in areas with annual precipitation of 350–500 cm, indicating its wider adaptability to diverse environmental conditions (Pandya & Lunagariya, 2022).

Globally, pearl millet ranks sixth after rice (*Oryza sativa* L.), wheat (*Triticum aestivum* L.), maize (*Zea mays* L.), barley (*Hordeum vulgare* L.), and sorghum and is cultivated over an area of 30 million ha, with production and productivity of 48.12 million tonnes and 1,110 kg ha^{-1}, respectively (Garin et al., 2023). Major producers of millet include India, China, Nepal, Pakistan, and Myanmar in Asia and Nigeria, Niger, Senegal, Cameroon, Burkina Faso, Mali, Uganda, Kenya, Namibia, Tanzania, Togo, Senegal, Chad, and Zimbabwe in sub-Saharan Africa. Pearl millet is the staple food for 90 million poor and marginal people in Asia and Africa.

About 50% of pearl millet production is used as feed and fodder (Srivastava et al., 2020). India is the largest producer of pearl millet in the world; during 2021–2022, the pearl millet area in India was 7.65 million ha, with a total production of 18.02 million tonnes (APEDA, 2024). In India, pearl millet is the fourth most extensively grown cereal crop after rice, wheat, and maize.

Pearl millet has desirable nutritional and nutraceutical properties (Singhal et al., 2018) because it supplies 30%–40% of inorganic nutrients and adequate amounts of Fe and Zn (Rao et al., 2006). It is a good source of energy, carbohydrates, and crude fibers, including resistant starch, soluble, and insoluble fat, proteins (8%–19%), ash, dietary fibers (1.2 g 100 g^{-1}), antioxidants, and fat (3%–8%) with better fat digestibility, Fe, and Zn (Uppal et al., 2015). Pearl millet is also rich in vitamins (riboflavin, niacin, and thiamine) and minerals such as K, P, Mg, Fe, Zn, Cu, and Mn (Weckwerth et al., 2020). It has a comparatively better essential amino acid profile than rice and maize and is low in prolamins, resulting in higher protein digestibility. Pearl millet fat has 74% polyunsaturated fatty acids and is rich in omega-3 fatty acids, such as oleic acid (25%), linoleic acid (45%), and linolenic acid (4%) (Singh et al., 2018). The gluten-free grain retains alkaline properties under cooking and is an ideal alternative for people allergic to gluten and has a low glycemic index (Satyavathi et al., 2021). The antioxidant potential of pearl millet is attributed to higher amounts of polyphenols, such as coumaric and ferulic acids (Goswami et al., 2020). Regarding overall health benefits, pearl millet consumption can protect us from various cancers, cardiovascular diseases, and certain age-related diseases. Because of its broader health-promoting potential, pearl millet has been classified as a nutri-cereal.

Taxonomy, Origin, Spread, and Adaptation

The genus *Pennisetum* is the largest in the tribe Paniceae, with five sections and approximately 140 species widely distributed in the tropics and subtropics (Clayton, 1972). Taxonomically, it is placed close to the genus Cenchrus (Bor, 1960) based on the arrangement of spikelets in groups surrounded by involucre. *Pennisetum* includes two reproductively isolated species: *Pennisetum purpureum*, a tetraploid (2n = 28) perennial species, and *P. glaucum* (syn. *Pennisetum americanum, Pennisetum typhoides*), an annual diploid species (2n = 14). *Pennisetum purpureum* is cultivated throughout the wet tropics of the world, whereas *P. glaucum* is grown in the semi-arid tropics of Africa and India (Brunken, 1977). *Pennisetum glaucum* comprises three subspecies: ssp. *glaucum*, ssp. *violaceum (monodii)*, and ssp., *stenostachyum*, with ssp. *glaucum* being the only cultivated species (Clayton, 1972).

Pearl millet originated from its wild progenitor *Pennisetum violaceum* (syn. *P. americanum* ssp. Monodii) (Brunken et al., 1977; D'Andrea & Casey, 2002). The domestication of pearl millet dates back to the development of agricultural civilizations in West Africa and has been extended to Ethiopia (Vavilov, 1950) and Niger (Murdock, 1959). Upon domestication, it was integrated into the pastoral system and provided livelihood to subsistence farmers. The oldest records of pastoral domestication of pearl millet are available for Lower Tilemsi Valley (Mali), around 2500 BCE (Manning & Fuller, 2014). Subsequently, it spread across the Sahara, including eastern Sudan, around 1850 BCE and southern India by 1700 BCE (Pokharia et al., 2014). Recent evidence showed that domestication occurred between Niger and Mauritania (Fuller & Hildebrand, 2013) and was supported by modeling studies using modern genomic data (Burgarella et al., 2018). Recently, Fuller et al. (2021) provided new insights about the domestication of pearl millet dating back to the middle Holocene (~5000 BCE) from northeast Mali. Based on available long-term archeological evidence in western Africa, the post-domestication evolution includes reduced husk and bristle length, leading to free-threshing pearl millet (Brunken et al., 1977). Three important domestication syndromes in pearl millet are (a) the evolution of non-shattering spikes, making seed dispersal reliant on human harvesting, threshing, and sowing (Fuller et al., 2021); (b) an increase in seed size and volume by increasing breadth and thickness, resulting in club-shaped seeds of domesticated pearl millet (Zach & Klee, 2003); and (c) floret duplication, resulting in two spikelets in involucres compared with the single-spikelet involucre in *P. violaceum* (Fuller & MacDonald, 2007). Clotault et al. (2012) estimated the domestication date of pearl millet around 4,800 years ago, using a set of 20 random genes in 33 cultivated and 13 wild pearl millet accessions. The study showed that a set of flowering pathway genes was preferentially selected during the domestication process of pearl millet.

Botany

Pearl millet is an annual herbaceous grass; plants are tall and erect, attaining a height up to 3.5 m. It is profusely tillering, with tropical C_4 (photosynthetic pathway) cereal-bearing grains on the surface of terminal spikes that are erect and candle-shaped. The seed size ranges from 0.5 to 2.0 g 100 seeds^{-1}. Each ear head may bear 300–2,500 seeds. It is a diploid plant (2n = 14) and is cross-pollinated. The pearl millet plant is invariably stoloniferous, and the stem or culm is herbaceous but stout. The stem orientation varies from erect to ascending to decumbent or mat-forming. It may be glabrous or pubescent with solid or hollow internodes. Leaves are usually not well differentiated (basal and cauline), are not distinctly distichous, and are ligulate with a ciliate membrane. Leaf blades are flat with a prominently pubescent midrib.

Inflorescence in pearl millet is a false spike with spikelets on contracted axes bearing a single-seeded caryopsis. Inflorescence of pearl millet is a contracted or fusiform panicle that consists of a central rachis covered with short hairs and bears fascicles on rachillae. The density of fascicles and the length of rachillae determine the compaction of the panicle. Variability in panicle length and thickness exists among different varieties, typically ranging from 20 to 25 cm long and from 7 to 9 cm thick. The panicle can exhibit various shapes, including cylindrical, conical, spindle, candle, lanceolate, dumbbell, club, oblanceolate, or globose.

The main rachis supports numerous spirally arranged rachillae, each capable of holding around 25 spikelets. Within each spikelet, there are two florets. The lower floret is staminate, featuring an oblong pointed lemma enveloping three stamens, and lodicules are usually absent, occasionally appearing sterile. The upper floret is bisexual, characterized by a broad, pointed, leathery lemma (possibly hairy or hairless at the tip) and a thin oval palea. The upper bisexual floret contains three stamens with long filaments and bi-lobed, dorsifixed, or versatile anthers. The ovary possesses two styles jointed at the base of the fruit.

The emergence of the style begins approximately 2 or 3 days after the panicle appears. Initially, the style protrudes from the florets in the panicle's upper-middle section and extends both upward and downward. In hermaphrodite flowers, the stigmas emerge ahead of the anthers, allowing them to receive pollen from the inflorescence of other plants. The complete process of stigma emergence takes about 2–3 days, after which they remain receptive for an additional 2 or 3 days.

As anther emergence initiates, the stigma has already emerged and undergone pollination, preventing self-pollination. Typically, the first anther emerges approximately 3–4 days after the initial stigma emergence. The process of anther emergence unfolds in two phases: The first phase involves only hermaphrodite flowers, while the second phase includes staminate flowers. The first phase progresses until the basal spikelets, and the second phase begins as staminate flowers become functional in the upper part of the panicle.

Pollen shedding from the panicle lasts about 3 days, and anther emergence occurs continuously, both day and night. The anthesis occurs between 8 a.m. and 2 p.m., peaking at about 10 a.m. Environmental conditions influence anther emergence; increased humidity and decreased temperature tend to inhibit it, whereas decreased humidity and elevated temperature promote anthesis. Thus, pearl millet is a highly cross-pollinated crop because of its protogynous nature.

Based on grain shape, pearl millet is grouped into four races: *Typhoides* (obovate caryopsis and the cross sections are obtuse and terete, inflorescences shape is cylindrical and has more diverse morphology among the four races and grown in all African countries and in India); *Nigritarium* (the cross-section in caryopsis is angular with three and six facets in each grain, inflorescence shape is candle-like and mature grain is longer and protrudes beyond the floral bracts compared with other groups of races and found in western Sudan and Nigeria); *Globosum* (spherical caryopsis with candle shape inflorescence, and mainly found in central Nigeria, Niger, Ghana, Togo, and Benin); *Leonis* (acute and terete caryopsis, a unique character of the acute apex, which is ended by the remnants of the stylar base, candle-like inflorescence shape and mainly grown in Sierra Leone, Senegal, and Mauritania) (https://iastate.pressbooks.pub/cropimprovement/chapter/millet-breeding). A description of various botanical aspects of pearl millet is given in Table 8.1.

Table 8.1 Botanical Description of Pearl Millet.

Plant part	Description
Growth habit	Annual
Root system	Typical monocot root system consisting of primary or seminal, adventitious roots, and collar or crown roots
Plant height	1.2–3.5 m
Stem diameter	1–3 cm
Tillering	Basal or nodal tillering habit with 2–20 culms
Nodes	Slightly swollen, long, white cilia pointing upward with adventitious roots on the basal side
Internodes	Light green to purple
Leaves	Upright or drooping with a membranous ligule, long (90–100 cm) and slender (5–8 cm), scabrous, and rather slender, varying levels of pubescence, with lanceolate or cordate shapes
Ear head	15–45 cm long and 2–5 cm thick, varies from loose to compact, with cylindrical, conical, or spindle shape
Rachis	Straight, cylindrical, solid, often 8–9 mm thick, and unbranched
Spikelets	Spikelets are with or without bristles
Seeds	Seeds are grey, brown, white, purple or yellowish and weight of 0.03–0.14 g
Endosperm	Endosperm color range from white to yellow and gray; comprised of outer hard or vitreous layer and an inner opaque and soft layer; inner endosperm does not contain protein bodies
Aleurone layer	One celled thick

Growth and Development

The growth period of pearl millet can be broadly divided into three major developmental stages: (a) growth phase (GP)-1 (vegetative phase), (b) GP-2 (panicle development phase), and (c) GP-3 (grain-filling phase). The details of these phases are given below.

GP-1 (vegetative phase)

This phase starts with the emergence of the seedling and continues up to panicle initiation. Seedlings establish their primary root system during the vegetative stage and produce adventitious roots. In late varieties, leaves are initiated during GP-1, but for early varieties, leaves are initiated and fully expanded during this phase. The tiller buds are formed, and their leaf primordia are initiated, resulting in the emergence of many side tillers by the end of GP-1, but there is very little elongation of internodes. During this phase, dry matter accumulation is almost entirely confined to leaves and roots. At the end of this stage, panicle initiation is marked by the elongation of the apical dome and the formation of a constriction at the base of the apex.

GP-2 (panicle development phase)

This stage spans from the panicle initiation to flowering of the main stem. During this phase, all leaves are fully expanded, and the earliest emerged (oldest) leaves begin to senesce. The sequential elongation of internodes from the base of the stem causes progressive stem elongation. There is profuse tillering that undergoes leaf expansion as well as floral initiation. The earliest emerged tillers follow the main stem in their development, whereas the later emerged tillers frequently cease due to competition. During GP-2, dry matter accumulates in roots, leaves, and stems. The panicle undergoes a series of distinct morphological and developmental changes, including the development of spikelets, florets, glumes, stigmas, anthers, and finally, stigma emergence and pollination, which heralds the end of the GP-2.

GP-3 (grain-filling phase)

This stage spans from the fertilization of florets in the panicle to the physiological maturity of the main stem and tillers. The bulk of dry matter increases in the grain, but tillers may also continue to elongate and flower in many varieties. Senescence of the lower leaves continues, and by the end of GP-3, only a few (two to four) upper leaves remain green. Some tillers emerge from upper nodes but usually have a shorter developmental cycle than the basal tillers, with just a few leaves and a small panicle. The development of a small dark layer of tissue in the hilar region of the grain marks the end of GP-3.

Growth Stages

The various distinct growth stages that span across these three developmental phases are listed below.

Stage 0 (emergence)
Emergence begins with the coleoptile emergence from the soil and is preceded by water imbibition by the seed, which activates metabolism. The radicle emerges near the hilar region within about 16 hours after germination initiation, followed by the plumule approximately 2 hours later. The radicle and plumule grow downward and upward, producing roots and coleoptile, respectively. The duration from germination to emergence depends on the planting depth, soil moisture, and soil temperature, generally taking 2–3 days (Figure 8.1).

Stage 1 (three-leaf stage)
Stage 1 occurs approximately 5 days after emergence of the coleoptile. The lamina of the third leaf is visible in the whorl of the second leaf. The first leaf is fully expanded, and the second leaf is slightly rolled at the base. During this stage, the seminal root grows rapidly and develops branching on it (Figure 8.2).

Stage 2 (five-leaf stage)
Stage 2 occurs 15 days after emergence, with a visible lamina of the fifth leaf. The first and second leaves are fully expanded, and the third leaf is slightly rolled. The growing point remains below the soil and surrounded by leaf primordia. The seminal root is well developed and has a number of branches, with evidence of adventitious roots. The tiller leaves are emerging from inside of the sheath of the basal leaves. The stem is sturdy and leaves dark green (Figure 8.3).

Figure 8.1   Emergence stage.

Figure 8.2   Three-leaf stage.

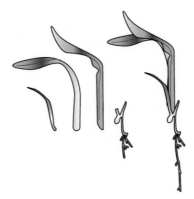

Figure 8.3   Five-leaf stage.

Stage 3 (panicle initiation)

In Stage 3, the plant transitions from vegetative to reproductive phase, marked by the development of spikelet primordia. The apex develops a constriction at its base and presents a shape as dome-like. Almost all leaves have emerged at this stage, and six to seven leaves are fully expanded. At this point, the growing point is above the soil surface. The root system has proliferated as a network of seminal roots and their lateral branches with the rapid growth of adventitious roots. Tillers are now conspicuous and undergo a similar developmental pattern to the main shoot (Figure 8.4).

Stage 4 (flag leaf stage)

At Stage 4, the lamina of the final leaf is visible and is easily distinguished from the preceding leaves. Internodes elongate in sequence from the base, with each successive internode longer than the preceding one. Similarly, other organs, such as branches, spikelet, and floral primordia, emerge in sequence from the base. At this point, the panicle is enclosed by the sheaths of the flag and penultimate leaves (Figure 8.5).

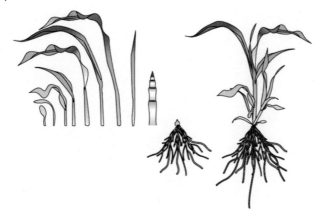

Figure 8.4 Panicle initiation stage.

Figure 8.5 Flag leaf stage.

Stage 5 (boot stage)
In Stage 5, the panicle is enclosed within the flag leaf sheath but has not yet emerged from the collar. The panicle rapidly increases in length and width, and its development is completed by the end of this stage. Finally, the panicle emerges from the collar of the flag leaf as the uppermost internode (peduncle) begins to elongate (Figure 8.6).

Stage 6 (half bloom)
Pearl millet is a cross-pollinated crop with protogynous (i.e., the stigmas appear first within 3–5 days after panicle emergence). The stigma emerges in the florets a

Figure 8.6   Boot stage.

few centimeters below the tip of the panicle and proceeds upward and downward simultaneously. Half bloom (50% flowering) coincides with stigma emergence in the middle region of the panicle. Upon pollination, the stigmas may shrivel within a few hours. The anthers emerge in two flushes (2–3 days after stigma emerges), the first in perfect flowers and the second in the male flowers extending the period over 5–6 days. Different heads on a plant may flower at the same time or sequentially (Figure 8.7).

Figure 8.7   Half-bloom stage.

Stage 7 (milking stage)
Approximately 6–7 days after fertilization, seeds start becoming sufficiently visible within the floret and consist of the seed coat filled with a watery and milky liquid. This is followed by rapid starch deposition in the endosperm cells and a concomitant increase in seed dry weight (Figure 8.8).

Stage 8 (dough stage)
The dough stage is marked by the sequential change in the endosperm from liquid milk to a solid state via a rapid increase in the starch content in the endosperm and a gradual decrease in seed moisture. This process leads to a reasonably hard dough as grain filling nears completion. The lower leaves start showing signs of senescence (Figure 8.9).

Stage 9 (physiological maturity)
Stage 9 involves the formation of a small black layer in the hilar region of the seed, coinciding with the cessation of remobilization of materials into the grain, leading to the termination of seed size increase. The black layer formation begins on the upper part of the panicle and proceeds downward. As this stage is near completion, seeds achieve their maximum dry weight, and the endosperm is fairly hard. Most of the leaves of the plant are senesced (Figure 8.10; Table 8.2). The growth stages of pearlmillet was shown in Figure 8.11.

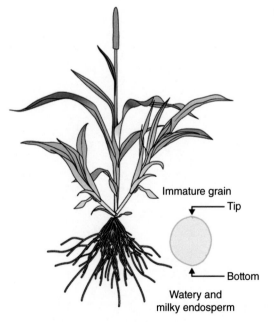

Figure 8.8 Milking stage.

Immature grain

Tip

Bottom

Watery and milky endosperm

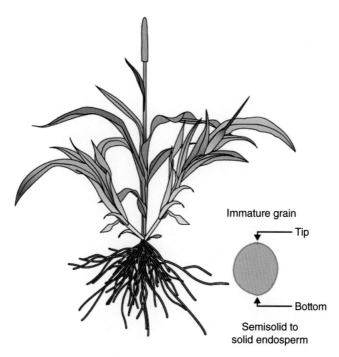

Figure 8.9   Dough stage.

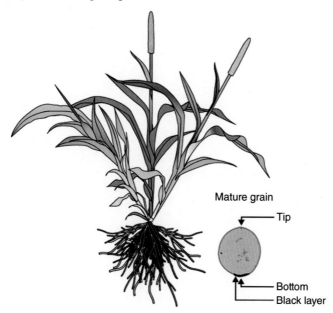

Figure 8.10   Physiological maturity stage.

Table 8.2   Three Major Growth Phases and Distinct Morphological Development Stages of Pearl Millet.

Major growth phase code	Component stage	Growth stage (days after emergence)	Salient features
GP1 (vegetative) 0–21 days	Emergence	2–4	Upon germination, the radicle starts emerging from the hilar region within 16 hours of the initiation of the germination, followed by the development of the plumule and coleoptile sheath ~2 hours later. Radicle grows downward rapidly and produces fine root hairs. Coleoptile grows upward slowly through the soil until it emerges from the soil surface. The emergence of coleoptile from the soil surface depends upon the sowing depth, soil texture, soil moisture, and soil temperature. It takes 2–3 days under favorable conditions.
	Three leaf	3–7	
	Five leaf	7–14	
	Panicle initiation	14–21	
GP2 (reproductive / panicle development) 21–42 days	Flag leaf	21–28	A change from vegetative to reproductive phase is marked by the formation of an apical dome-like structure and a constriction at the base of the shoot meristem, leading to a change from leaf primordia to spikelet primordia, development of spikelets, florets, glumes, stigma, and anther. The period between panicle initiation to anthesis is critical in determining the grain number.
	Booting	28–35	
	Half bloom	35–42	
GP3 (grain filling) 42–77 days	Milking	42–49	Pearl millet seed is a caryopsis, and it is highly variable, ranging from globular to conical shape. The seed color varies from ivory to purplish black, with light to deep gray being the most common seed color. A small embryo is present on the depressed or flat surface at the tapering end of the seed. The size of the grain depends on its position in the panicle, being largest at the base, medium in the middle, and smallest at the apex. Variations exist in grain size among varieties generally ranging from 4 to 12 g per 1,000 grains.
	Dough stage	49–56	
	Black layer formation or physiological maturity	56–63	

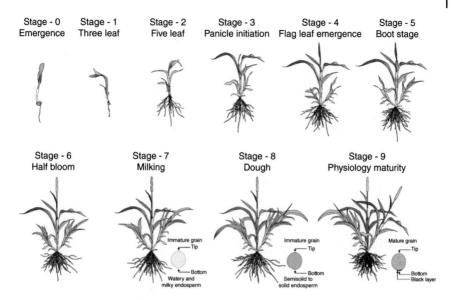

Figure 8.11   Overview of pearl millet growth stages.

Genetic Variation

The developmental stages of pearl millet have wide genetic variation for yield and other component traits with distinct differences between grain and forage types (Tables 8.3 and 8.4). These unique genetic variation resources are the foundation for grain and forage breeding programs to develop biotic and abiotic stress-tolerant, high-yielding grain with improved nutritional traits and high-biomass cultivars with superior forage quality traits.

Table 8.3   Genetic Variability for Grain Traits in Pearl Millet.

Trait	Range	References
Days to 50% flowering	34.0–137.0	Gunguniya et al. (2023), Maiti et al. (1989), Patil & Gupta (2022), Singh & Chhabra (2018)
Plant height, cm	63.7–250.0	
Days to maturity	79.0–89.0	Gunguniya et al. (2023)
Leaf number	4.0–28.0	Maiti et al. (1989), Shah et al. (2012), Singh & Singh (2016)
Leaf blade length, cm	24.4–63.3	Maiti et al. (1989)
Leaf blade width, cm	1.7–4.5	

(continued overleaf)

Table 8.3 (Continued)

Trait	Range	References
Leaf area, cm^2	22.0–256.0	Shah et al. (2012), Singh & Chhabra (2018)
Flag leaf length, cm	17.0–37.0	Reddy et al. (2023)
Flag leaf width, cm	1.7–3.6	
Flag leaf area, cm^2	14.0–109.0	Shah et al. (2012)
Sheath length, cm	9.4–26.9	Maiti et al. (1989)
Stem thickness, mm	5.5–17.3	Maiti et al. (1989), Singh & Chhabra (2018)
Head diameter, cm	1.03–4.3	Maiti et al. (1989), Patil & Gupta (2022), Gunguniya et al. (2023)
Panicle length, cm	7.6–77.4	
Productive tillers plant^{-1}, numbers	0.1–11.1	Gunguniya et al. (2023), Maiti et al. (1989), Patil & Gupta (2022), Singh & Chhabra (2018).
Peduncle exertion, cm	0–32.0	Maiti et al. (1989), Shah et al. (2012)
Internode length, cm	12.0–23.0	Shah et al. (2012)
Number of grains per panicle	202.0–8405.0	Singh & Chhabra (2018)
1,000 grain weight, g	3.07–19.10	Gunguniya et al. (2023), Maiti et al. (1989), Patil & Gupta (2022)
Grain yield plant^{-1}, g	0.6–142.1	Gunguniya et al. (2023), Singh & Chhabra, (2018), Singh & Singh (2016)
Protein, %	8.3–20.1	Gunguniya et al. (2023)
Lipid, %	2.72–6.95	
Fe, ppm	19.7–145	Bashir et al. (2014), Govindaraj et al. (2022), Gunguniya et al. (2023)
Zn, ppm	13.5–96.0	
Na, ppm	3.0–41.0	Govindaraj et al. (2022)
Mg, ppm	940.0–1890.0	
Mn, ppm	7.2–23.0	Govindaraj et al. (2022), Gunguniya et al. (2023)
Ca, ppm	42.0–400.0	
Cu, ppm	4.9–22.6	
K, ppm	1,800.0–7,250.0	
P, ppm	2,258.0–4,950.0	
Phytate, mg 100 g^{-1}	201.5–542.5	Gunguniya et al. (2023)
Total phenolic acid, mg 100 g^{-1}	75.1–44.4	

Table 8.4   Genetic Variability for Forage Traits in Pearl Millet.

Trait	Range	References
Plant height, cm	49.0–442.3	Arya et al. (2009, 2010),
Tillers plant^{-1}, *numbers*	1.0–9.3	Khairwal et al. (2007),
Stem girth, mm	5.0–31.2	Kulkarni et al. (2000)
Leaves, *n*	4.3–37.0	
Leaf length, cm	19.3–130.0	
Leaf width, cm	1.1–8.6	
Biomass yield, t ha^{-1}	13.7–106.7	

Climatic Requirements

Growth Conditions

Pearl millet has wide developmental plasticity in growth and grows in a broad range of ecological conditions. It is generally grown between 15°W and 90°E long. and 5°S to 40°N lat., indicating its wide adaptability (Pandya & Lunagariya, 2022). Despite being a warm-weather crop, it can grow best at 20–28°C. Among cultivated cereals, pearl millet has the greatest tolerance to higher temperatures. The optimum temperature for the germination of pearl millet seed is 23–32°C, and in fact, its seed does not germinate and grow well under cool soil conditions. Temperatures >40°C decrease the germination percentage and rate of emergence. A temperature range of 30–35°C is optimum for growth and development.

The optimum rainfall requirement for pearl millet is 35–50cm, and it can be grown in areas with <35cm of annual rainfall, such as sandy soils of Rajasthan, India, where the average annual rainfall is just 150–200mm. Pearl millet is successfully grown with moisture conservation practices. Prolonged spells of drought and high temperatures may be detrimental and may lead to significant yield reduction. Dry, warm weather is most suitable at harvest time. Despite being one of the most stress-tolerant crops, pearl millet can optimally respond to adequate water during its growth. A well-distributed and light rainfall with few cloudy days is ideal for the growth and development of pearl millet. However, there should be no rains during flowering and grain formation stages. High humidity and low temperature at flowering increase the incidence of ergot disease, leading to low yield.

Djanaguiraman et al. (2017) conducted a detailed study on the effect of temperature on various growth and reproductive traits and observed that high-temperature stress (≥36/26°C) imposed at different stages and durations caused a decrease in the number of seeds, individual seed weight, and seed yield. Two sensitive phases, 10–12 days and 2–0 days before anthesis, were identified as most vulnerable to short episodes of temperature stress, recording maximum decreases in pollen germination percentage and seeds numbers. A higher temperature of $\geq 36/26\,°C$ results in floret sterility. However, pistils were relatively more sensitive than pollen grains, causing a decreased number of seeds and seed yield. Higher temperature stress also increased the reactive oxygen species contents and decreased the activity of the antioxidant enzymes in both pollen and pistils. The increase of reactive oxygen species under stress was relatively higher in pistils, whereas antioxidant enzyme activity was lower than in the pollen grains, accounting for the greater susceptibility of pistils. Similarly, the critical photoperiod for pearl millet ranges from 13 to 13.35 hours, and the photoperiod sensitivity ranges from 142 to 6,184 growing degree days per hour of photoperiod extension (Sanon et al., 2014).

The ability of pearl millet to grow in harsh drier environments is due to several physiological and morphological characteristics, including (a) rapid and deep root penetration (root depths of 3.6 m have been recorded), (b) well-developed and specialized cell walls in the root system that prevent desiccation, (c) compensation of reduced yield by profuse tillering capacity, and (d) short day warm weather crop and greater drought tolerance than sorghum.

Soil Requirement

Pearl millet can be grown on a wide variety of soils, including the sandy loams of Punjab and Utter Pradesh; the light sandy light soils of Rajasthan and north Gujarat; the heavy clays of Andhra Pradesh and Tamil Nadu; and the shallow black, red, and light soils of Deccan and southern India. However, sandy loam to loamy soils with good drainage and less salinity or alkalinity are best suited for pearl millet cultivation. The crop can perform better on alkaline soils but not on highly saline soils. Compared with other cereals, pearl millet can grow well at a lower soil pH (5.5–6.5). Pearl millet is invariably grown in low-input farming systems; however, high yields can be achieved with a more balanced fertilization. The optimum fertilizer rates (per hectare) is 80 kg N, 40 kg P, and 40 kg K. Slightly lower fertilizer rates are recommended for arid [40–20–40; nitrogen:phosphorus:potassium (N–P–K)] and semiarid regions (60–30–40; N–P–K).

Conclusions and Future Needs

Pearl millet is increasingly recognized as a crop with climate resilience, disease, and pest resilience as well as a balanced nutritional profile (Satyavathi et al., 2021).

Pearl millet continues to be an invaluable asset due to its hardiness and adaptability to abiotic stress conditions. Its inherent resilience to harsh environments, coupled with its efficient C_4 photosynthetic pathway, offers ample opportunities for understanding the complex physiological, biochemical, and molecular plant responses to climate change–induced stresses. The insights gleaned from studying pearl millet's adaptive mechanisms can significantly contribute to the development of climate-resilient strategies for other less tolerant crop species, potentially revolutionizing our approaches to ensure global food security in the face of increasingly volatile climatic conditions. Pearl millet a tolerant crop that can provide fair yields under stress conditions, yet it has a better yield potential under more favorable conditions. In order to promote and increase pearl millet cultivation, appropriate policy support to create an enabling environment for promoting pearl millet cultivation is required. Adequate genetic and genomic resources are now available in pearl millet, as are new technologies (e.g., gene editing) for improved plant breeding and precision tools for better crop management. These new tools will enhance the development of pearl millet cultivars with broader environmental spectrum that can help with area expansion, the development of high-yielding cultivars, refinement of production technology, as well as price support by the government. This will promote the profits of small and marginal farmers, enhance demand of pearl millet, create value addition chains, and promote new profitable markets for millet.

References

APEDA, Agricultural and Processed Food Products Export Development Authority. (2024). *Indian Millets.* https://apeda.gov.in/apedawebsite/SubHead_Products/Indian_Millets.htm. Visited on 07.04.2024.

Arun, A., & Faraday, M. K. (2020). A perceptive study to endorse the nutritional aspects of pearl millet (*Pennisetum glaucum* L) and formulated recipes. *Research Journal of Pharmacy and Technology, 13*(2), 911–914.

Arya, R. K., Yadav, H. P., Yadav, A. K., & Deshraj. (2009). *Forage symposium on emerging trends in Forage research and livestock production.* CAZRI, RRS.

Arya, R. K., Yadav, H. P., Yadav, A. K., & Singh, M. K. (2010). Effect of environment on yield and its contributing traits in pearl millet. *Forage Research, 36*, 176–180.

Bashir, E., Ali, A., Ali, A., Melchinger, A., Parzies, H., & Haussmann, B. (2014). Characterization of Sudanese pearl millet germplasm for agro-morphological traits and grain nutritional values. *Plant Genetic Resources, 12*(1), 35–47. https://doi.org/10.1017/S1479262113000233

Bor, N. L. (1960). *The grasses of Burma, Ceylon, India and Pakistan.* Pergamon Press.

Brunken, J., De Wet, J. M. J., & Harlan, J. R. (1977). The morphology and domestication of pearl millet. *Economic Botany*, *31*, 163–174.

Brunken, J. N. (1977). A systematic study of Pennisetum sect. Pennisetum (Gramineae). *American Journal of Botany*, *64*, 161–176.

Burgarella, C., Cubry, P., Kane, N. A., Varshney, R. K., Mariac, C., Liu, X., Shi, C., Thudi, M., Couderc, M., Xu, X., Chitikineni, A., Scarcelli, N., Barnaud, A., Rhoné, B., Dupuy, C., François, O., Berthouly-Salazar, C., & Vigouroux, Y. (2018). A western Sahara centre of domestication inferred from pearl millet genomes. *Nature Ecology and Evolution*, *2*(9), 1377–1380.

Choudhary, S., Guha, A., Kholova, J., Pandravada, A., Messina, C. D., Cooper, M., & Vadez, V. (2020). Maize, sorghum, and pearl millet have highly contrasting species strategies to adapt to water stress and climate change-like conditions. *Plant Science*, *295*, 110297.

Clayton, W. D. (1972). Gramineae. In F. N. Hepper (Ed.), *Flora of west tropical Africa* (pp. 170–465). Crown Agents.

Clotault, J., Thuillet, A. C., Buiron, M., De Mita, S., Couderc, M., Haussmann, B. I., Mariac, C., & Vigouroux, Y. (2012). Evolutionary history of pearl millet [*Pennisetum glaucum* (L.) R. Br.] and selection on flowering genes since its domestication. *Molecular Biology and Evolution*, *29*(4), 1199–1212.

D'Andrea, A. C., & Casey, J. (2002). Pearl millet and Kintampo subsistence. *African Archaeological Review*, *19*, 147–173.

Djanaguiraman, M., Perumal, R., Ciampitti, I. A., Gupta, S. K., & Prasad, P. V. V. (2017). Quantifying pearl millet response to high temperature stress: Thresholds, sensitive stages, genetic variability and relative sensitivity of pollen and pistil. *Plant, Cell and Environment*, *41*, 993–1007.

Fuller, D. Q., Barron, A., Champion, L., Dupuy, C., Commelin, D., Raimbault, M., & Denham, T. (2021). Transition from wild to domesticated pearl millet (*Pennisetum glaucum*) revealed in ceramic temper at three middle Holocene sites in northern Mali. *African Archaeological Review*, *38*, 211–230.

Fuller, D. Q., & Hildebrand, L. (2013). Domesticating plants in Africa. In P. Mitchell & P. Lane (Eds.), *The Oxford handbook of African archaeology* (pp. 507–525). Oxford University.

Fuller, D. Q., & MacDonald, K. (2007). Early domesticated pearl millet in Dhar Nema (Mauritania): Evidence of crop processing waste as ceramic temper. In R. Cappers (Ed.), *Fields of change: Progress in African archaeobotany* (pp. 71–76). Groningen Institute of Archeology.

Garin, V., Choudhary, S., Murugesan, T., Kaliamoorthy, S., Diancumba, M., Hajjarpoor, A., Chellapilla, T. S., Gupta, S. K., & Kholovà, J. (2023). Characterization of the pearl millet cultivation environments in india: status and perspectives enabled by expanded data analytics and digital tools. *Agronomy*, *13*, 1607. https://doi.org/10.3390/agronomy13061607

Goswami, S., Asrani, P., Ansheef Ali, T. P., Kumar, R. D., Vinutha, T., Veda, K., Kumari, S., Archana, A., Singh, S. P., Satyavathi, C. T., Kumar, R. R., & Praveen, S. (2020). Rancidity matrix: Development of biochemical indicators for analyzing the keeping quality of pearl millet flour. *Food Analytical Methods, 13,* 2147–2164.

Govindaraj, M., Kanatti, A., Rai, K. N., Pfeiffer, W. H., & Shivade, H. (2022). Association of grain iron and zinc content with other nutrients in pearl millet germplasm, breeding lines, and hybrids. *Frontiers in Nutrition, 8,* 746625.

Gunguniya, D. F., Kumar, S., Patel, M. P., Sakure, A. A., Patel, R., Kumar, D., & Khandelwal, V. (2023). Morpho-biochemical characterization and molecular marker based genetic diversity of pearl millet (*Pennisetum glaucum* (L.) R. Br.). *Peer Journal, 11,* e15403. https://doi.org/10.7717/peerj.15403

Khairwal, I. S., Yadav, S. K., Rai, K. N., Upadhyaya, H. D., Kachhawa, D., Nirwan, B., Bhattacharjee, R., Rajpurohit, B. S., & Dangaria, C. J. (2007). Evaluation and identification of promising pearl millet germplasm for grain and fodder traits. *Journal of SAT Agricultural Research, 5*(1), 1–6.

Krishnan, R., & Meera, M. S. (2018). Pearl millet minerals: Effect of processing on bioaccessibility. *Journal of Food Science and Technology, 55,* 3362–3372.

Kulkarni, V. M., Navale, P. A., & Harinarayana, G. (2000). Variability and path analysis in white grain pearl millet [*Pennisetum glaucum* (L.) R. Br.]. *Tropical Agriculture, 77*(2), 130–132.

Maiti, R. K., Gonzalez, H., & Landa, H. (1989). Evaluation of ninety international pearl millet germplasm collections for morpho-physiological characters in Nuevo Leon, Mexico. *Turrialba, 39*(1), 34–39.

Manning, K., & Fuller, D. (2014). Early millet farmers in the lower Tilemsi Valley, northeastern Mali. In C. J. Stevens, S. Nixon, M. A. Murray, & D. Q. Fuller (Eds.), *The archaeology of African plant use* (pp. 73–81). Walnut Creek.

Murdock, G. P. (1959). *Africa: Its peoples and their culture history.* McGraw-Hill.

Nambiar, V. S., Dhaduk, J. J., Sareen, N., Shahu, T., & Desai, R. (2011). Potential functional implications of pearl millet (*Pennisetum glaucum*) in health and disease. *Journal of Applied Pharmaceutical Science, 1,* 62–67.

Pandya, H. R., & Lunagariya, M. M. (2022). Effect of rainfall on soil moisture content in pearl millet crop. *The Pharma Innovation Journal, 11,* 878–880.

Patil, K. S., & Gupta, S. K. (2022). Geographic patterns of genetic diversity and fertility restoration ability of Asian and African origin pearl millet populations. *The Crop Journal, 10*(2), 468–477. https://doi.org/10.1016/j.cj.(2021).04.013

Pokharia, A. K., Kharakwal, J. S., & Srivastava, A. (2014). Archaeobotanical evidence of millets in the Indian subcontinent with some observations on their role in the Indus civilization. *Journal of Archaeological Science, 42,* 442–455.

Prasad, P. V. V., Staggenborg, S., & Ristic, Z. (2008). Impacts of drought and/or heat stress on physiological, developmental, growth, and yield processes of crop plants. In L. R. Ahuja, V. R. Reddy, S. A. Saseendran, & Q. Yu (Eds.), *Response of crops to limited water: Understanding and modeling water stress effects on plant growth processes*, Advances in Agricultural Systems Modeling Series (Vol. 1, pp. 301–355). American Society of Agronomy, Inc. https://doi.org/10.2134/advagricsystmodel1.c11

Rao, P. P., Birthal, P. S., Reddy, B. V., Rai, K. N., & Ramesh, S. (2006). Diagnostics of sorghum and pearl millet grains-based nutrition in India. *International Sorghum and Millets Newsletter, 47*, 93–96.

Reddy, P. S., Srividya, S., Khandelwal, V., & Satyavathi, C. T. (2023). Association of photosynthesis of flag leaves with grain yield in pearl millet (*Pennisetum glaucum* (L.) R. Br.). *Annals of Arid Zone, 62*(1), 91–96.

Sanon, M., Hoogenboom, G., Traoré, S. B., Sarr, B., Garcia, A. G. Y., Somé, L., & Roncoli, C. (2014). Photoperiod sensitivity of local millet and sorghum varieties in West Africa. *NJAS: Wageningen Journal of Life Sciences, 68*, 29–39.

Satyavathi, C. T., Ambawat, S., Khandelwal, V., & Srivastava, R. K. (2021). Pearl millet: A climate-resilient nutricereal for mitigating hidden hunger and provide nutritional security. *Frontiers in Plant Science, 12*, 659938.

Shah, I. A., Rahman, H., Shah, S. M. A., Shah, Z., Rahman, S., Interamullah, & Noori, M. (2012). Caracterization of pearl millet germplasm for various morphological and fodder yield parameters. *Pakistan Journal of Botany, 44*(1), 273–279.

Singh, D., Raghuvanshi, K., Chaurasiya, A., Dutta, S. K., & Dubey, S. K. (2018). Enhancing the nutrient uptake and quality of pearlmillet (*Pennisetum glaucum* L.) through use of biofertilizers. *International Journal of Current Microbiology and Applied Science, 7*, 3296–3306.

Singh, J., & Chhabra, A. K. (2018). Genetic variability and character association in advance inbred lines of pearl millet under optimal and drought condition. *Ekin Journal of Crop Breeding and Genetics, 4*(2), 45–51.

Singh, O. V., & Singh, A. K. (2016). Analysis of genetic variability and correlation among traits in exotic germplasm of pearl millet (*Pennisetum glaucum* (L.) R. Br.). *Indian Journal of Agricultural Research, 50*(1), 76–79.

Singhal, T., Satyavathi, C. T., Kumar, A., Sankar, S. M., Singh, S. P., Bharadwaj, C., Aravind, J., Anuradha, N., Meena, M. C., & Singh, N. (2018). Genotype × environment interaction and genetic association of grain iron and zinc content with other agronomic traits in RIL population of pearl millet. *Crop and Pasture Science, 69*, 1092–1102.

Srivastava, R. K., Singh, R. B., Pujarula, V. L., Bollam, S., Pusuluri, M., Chellapilla, T. S., Aravind, J., Anuradha, N., Meena, M. C., Yadav, R. S., & Gupta, R. (2020). Genome-wide association studies and genomic selection in pearl millet: Advances and prospects. *Frontiers in Genetics, 10*, 1389.

Stevens, C., & Fuller, D. (2018). Sorghum and pearl millet. In S. L. L. Varela (Ed.), *The encyclopedia of archaeological sciences* (pp. 1–4). Wiley.

Uppal, R. K., Wani, S. P., Garg, K. K., & Alagarswamy, G. (2015). Balanced nutrition increases yield of pearl millet under drought. *Field Crops Research*, *177*, 86–97.

Vavilov, N. I. (1950). The origin, variation, immunity and breeding of cultivated plants. *Chronica Botanica*, *13*, 1–366.

Weckwerth, W., Ghatak, A., Bellaire, A., Chaturvedi, P., & Varshney, R. K. (2020). PANOMICS meets germplasm. *Plant Biotechnology Journal*, *18*, 1507–1525.

Zach, B., & Klee, M. (2003). Four thousand years of plant exploitation in the Chad Basin of NE Nigeria II: Discussion on the morphology of caryopses of domesticated Pennisetum and complete catalogue of the fruits and seeds of Kursakata. *Vegetation History and Archaeobotany*, *12*, 187–204.

9

Impact of Drought and High-temperature Stresses on Growth and Development Stages, Physiological, Reproductive, and Yield Traits on Pearl Millet

Maduraimuthu Djanaguiraman, A. S. Priyanka, S. J. Vaishnavi, Ramasamy Perumal, Ignacio A. Ciampitti, and P. V. Vara Prasad

Chapter Overview

Pearl millet [*Pennisetum glaucum* (L.) R. Br.] is one of the important staple food crops grown in arid and semi-arid regions of the world for food and nutritional security. Generally, pearl millet is better adapted to drought and high-temperature stress conditions than other cereal crops. However, it is susceptible to drought and high-temperature stress, particularly during early seedling and reproductive stages of crop development. In the past, pearl millet was grown during the rainy season, but in recent years it has been grown under hot-dry seasons for yield and quality advantages. Hence, the drought and high-temperature stress tolerance during sensitive stages of pearl millet must be improved, and this will require better understanding of the impacts and mechanisms of tolerance or susceptibility. Pearl millet is sensitive to drought stress during the early seedling, gametogenesis, anthesis, and grain-filling stages. Similarly, it is susceptible to high-temperature stress during gametogenesis, anthesis, and grain-filling stages. The primary effects of drought stress are decreased seed germination or emergence and early seedling vigor due to limited soil moisture, decreased photosynthesis due to osmotic imbalance, reactive oxygen species (ROS) production, and damage to the ultrastructure of cell organelles. High-temperature stress causes membrane damage to the leaf, resulting in decreased photosynthesis, increased ROS, and membrane damage to gametes (pistil and pollen), leading to decreased seed-set percentage and final grain number. Drought tolerance in pearl millet is associated with shoot growth characteristics, a deep root system, restricted/limited transpiration, and stay-green trait and improved partitioning to grain for post-flowering drought tolerance mechanisms, whereas high-temperature stress tolerance is associated with an antioxidant defence system and heat shock proteins, greater gamete viability, early-morning flowering (escape), and improved percent seed-set.

Pearl Millet: A Resilient Cereal Crop for Food, Nutrition, and Climate Security, First Edition.
Edited by Ramasamy Perumal, P. V. Vara Prasad, C. Tara Satyavathi, Mahalingam Govindaraj, and Abdou Tenkouano.
© 2025 American Society of Agronomy, Inc., Crop Science Society of America, Inc., Soil Science Society of America, Inc. Published 2025 by John Wiley & Sons, Inc.

Introduction

Millets are neglected and underutilized crops that are nutritionally rich and gluten free. Cereal grain crops like wheat (*Triticum aestivum* L.), rice (*Oryza sativa* L.), maize (*Zea mays* L.), barley (*Hordeum vulgare* L.), sorghum [*Sorghum bicolor* L. (Moench)], and pearl millet [*Pennisetum glaucum* (L.) R. Br.] are the major staple food grain crops around the world. Millets account for only 2% of global cereal production, of which 95% is from Asia and Africa. Among millets, pearl millet is one of the most important and widely grown after sorghum. The major producers of pearl millet include India, China, Nepal, Pakistan, and Myanmar in Asia and Nigeria, Niger, Senegal, Cameroon, Burkina Faso, Mali, Uganda, Kenya, Namibia, Tanzania, Togo, Senegal, Chad, and Zimbabwe in sub-Saharan Africa. Pearl millet can grow in regions with <400 mm of rainfall and when the air temperature is ~42 °C (Krishnan & Meera, 2018). Millet can produce significant grain yields even in severe water-limited areas with <250 mm annual rainfall (Nambiar et al., 2011). However, different climate projection models have indicated that the global millet grain yield will decrease by 7.0%–8.5% by 2050 due to the changing climate, occurrence of extreme events, and rising mean temperatures (Nelson et al., 2009).

By the end of this century, the global population is projected to increase by 11.2 billion. Across the globe, despite decades of steady decline in global hunger, after the COVID-19 pandemic and many global crises, the number of people suffering from hunger increased to over 820 million (FAO, IFAD, UNICEF, WFP, & WHO, 2023). The observed and projected climate variability with high probability of occurrences of extreme events, including long-periods of dry spells and increased intensity and duration of extreme high temperatures, challenges current and future crop production (IPCC, 2022). The duration and intensity of the drought or high-temperature stress significantly affect crop productivity (Toker et al., 2007). The year-to-year variability in crop yield is primarily associated with rainfall and air temperatures during the production period and critical stages of crop development (Kukal & Irmak, 2018). Compared with the baseline, increasing global surface temperature up to 1.5 °C will cause multiple climate hazards and risks to the ecosystem (IPCC, 2022). Among the various environmental variables, drought and high temperature are the most critical factors that negatively affect major food grain crops, including pearl millet (Saleem et al., 2021). Hence, the current global food production systems must be climate resilient to achieve food and nutritional security. Understanding the impacts of drought and high-temperature stress at physiological, biochemical, and whole plant/crop levels will help to improve the productivity of pearl millet by developing stress-tolerant cultivars and improved crop, soil, and water management strategies. The challenges of abiotic stresses can be succeeded by growing crops like pearl millet, sorghum, and other minor millets using stress-tolerant cultivars with improved agronomic practices, which can ensure yield stability and minimize environmental risks. Plants develop several morphological, physiological, and biochemical mechanisms to withstand and tolerate drought and high-temperature stress during their growth and development. In this chapter, we provide an overview of the major impacts of drought and high-temperature stress on pearl millet and discuss the mechanisms associated with tolerance or susceptibility.

Drought Stress

Drought stress negatively affects crop growth, development, and yield (Mir et al., 2012; Prasad et al., 2008). Drought is one of the major environmental stresses affecting ~35% of the Earth's surface, affecting crop production (Tripathy et al., 2023). The effects of drought stress on the crop depend on the duration, intensity, and timing (Mir et al., 2012; Prasad et al., 2008). The primary effect of drought stress at the time of planting is inhibition of seed germination, poor emergence, and seedling establishment (Farooq et al., 2009). Drought impairs the overall plant growth and development, leaf area, root biomass, shoot biomass, partitioning, and grain yield (Eziz et al., 2017; Farooq et al., 2009; Hossain et al., 2021; Hussain et al., 2008; Prasad et al., 2008; Zhu, 2002).

Drought tolerance in plants can be classified into three types: escape, avoidance, and tolerance (Levitt, 1980). Drought escape is a mechanism by which plants complete the life cycle before the onset of drought, without experiencing this stress. The plants possessing the drought escape mechanism will control the duration of the vegetative and reproductive growth stages. The plant growth stages will be completed based on the water availability by rapidly accomplishing the phenological phases or adopting developmental plasticity (Jones et al., 1981). Plants complete their phenological phase quickly by speeding plant growth, producing lower yield and fewer seeds in the dry season but producing high biomass and yield in the wet season. In drought avoidance, plants maintain greater cell turgidity even when the soil moisture is substantially reduced (Levitt, 1980). Cell turgidity is maintained under drought stress by minimizing water loss (water savers) or by taking more water from the soil (water spenders). Water savers minimize water loss under drought by reducing leaf area, transpiration, and interception of solar radiation. Water spenders take more water from the soil by spending energy for water absorption or increased rooting depth and/or root length. Under drought tolerance, the plants tolerate lower plant tissue water content by expressing traits like osmotic adjustment, cellular elasticity, protoplasmic resistance (Morgan, 1984), maintaining reproductive fertility through production of dehydrins, and maintaining greener canopies (stay-green) to meet the demands for photosynthates from sink (particularly grains).

Pearl millet is one of the most important drought-tolerant millets grown in regions characterized by hot climates and low rainfall. Pearl millet was domesticated from its wild ancestor *Cenchrus americanus* ssp. *violaceum* from Africa, where the annual rainfall was very low (~200–400 mm) (Bidinger et al., 1987; Burgarella et al., 2018; Oumar et al., 2008). Studies also showed that some pearl millet germplasm can withstand drought in regions receiving <50 mm of annual rainfall (Harlan, 1975; Pucher et al., 2015). However, pearl millet is susceptible to drought stress at its critical growth stages, particularly during seedling emergence and establishment and reproductive development. Studies showed that leaf rolling, reduced canopy leaf area, and decreased transpiration rate are the adaptive traits of pearl millet under drought stress (Srivastava et al., 2022). In pearl millet, short-term drought responses include stomatal closure, osmotic adjustment, accumulation of ROS, antioxidant synthesis, and increased antioxidant enzyme activity. Long-term drought responses include changes in rooting patterns, asynchronous tiller development, and flowering plasticity (Shivhare et al., 2020; Shrestha et al., 2023).

In general, pearl millet is usually planted at the onset of the rainy season. Due to the current rainfall pattern and unpredictable start or delayed onset, pearl millet will face drought stress during the seedling emergence and establishment stages if the initial rain events of the season are distant from each other. Apart from the vegetative stage, the flowering and grain development stages are also sensitive to drought stress (Bidinger et al., 1987; Kholova et al., 2010; Serba & Yadav, 2016; Yadav et al., 2002). Although drought stress can coincide with any time of crop growth, late-season and terminal drought stresses (from flowering to grain filling) are more damaging to the yield than stresses at the vegetative stage (Mahalakshmi et al., 1987). The increased sensitivity during the terminal stage was associated with pearl millet's asynchronous tillering ability and rapid growth rate, allowing it to recover rapidly from intermittent drought, but was limited under terminal stress stage (Mahalakshmi et al., 1987). Drought during the mid- or late vegetative stage or early reproductive stages can cause increased tillering to compensate for limited grain yield in the main tiller (Bidinger et al., 1987; Mahalakshmi et al., 1987; van Oosterom et al., 2006). In contrast, drought stress at the flowering and post-flowering stages causes drastic reduction in grain and stover yield without the possibility of a compensation effect (Bidinger & Hash, 2004; Kholova et al., 2009; Kholova & Vadez, 2013; Mahalakshmi et al., 1987). Debieu et al. (2018) reported that drought stress during the seedling stage at 28 days decreased the plant height, number of leaves, and seed yield, with an overall yield reduction between 40% and 50%. Similarly, drought stress during the terminal stage of crop development decreased yield from 27% to 61% over the irrigated control (Yadav et al., 2002). The study clearly indicates that drought stress in the reproductive stage was more deleterious to yield than in the vegetative stage.

Impacts of Drought Stress on Morphological and Growth Traits

In general, leaf characteristics (e.g., glossy, profuse trichomes, thick cuticles, compact palisade cells, and thick collenchyma) are associated with drought tolerance in crops (Maiti, 2013), and these leaf traits are altered under drought stress (Do et al., 1996). Pearl millet mainly depends on morphological and phenological characteristics rather than physiological traits like osmotic adjustment or tissue elasticity (Do et al., 1996) to tolerate drought stress. Pearl millet leaves with narrow leaf angles exhibit better tolerance to drought by effectively reducing the evaporative surface area of the canopy (Kusaka et al., 2005a). Leaf rolling during the early vegetative stage can recover after 12 days of drought stress, indicating the ability to withstand drought stress severity (Kusaka et al., 2005b). Among the morphological traits, reduced leaf area, increased leaf rolling, and the development of waxy cuticles confer drought tolerance through decreased light interception, thereby reducing the transpiration rate in pearl millet (Do et al., 1996; Srivastava et al., 2022). However, reduced transpiration rate caused decreased photosynthesis, light interception, and CO_2 diffusion because of decreased stomatal conductance (Kusaka et al., 2005a).

In pearl millet, the presence of a waxy leaf cuticle is an important drought tolerance mechanism (Bi et al., 2017). Apart from drought, the lines with an increased wax content and thicker cuticle show increased resistance to downy mildew fungal disease. Teowalde et al. (1986) observed a wide genetic variability among the pearl millet genotypes for the stomatal frequencies. As an adaptive mechanism, the stomata size and number are reduced during drought stress. In pearl millet, stomatal density increased under mild stress; however, it decreased during severe drought stress (Xu & Zhou, 2008). It was also observed that stomatal density did not vary significantly between the tolerant and susceptible parental lines and among the near-isogenic lines under normal or drought-stress conditions (Kholova et al., 2010). Another important morphological trait associated with drought stress was leaf senescence. In pearl millet, premature leaf senescence is a strategy to decrease the leaf area under severe drought stress, and it usually occurs during the grain-filling stage (Do et al., 1996). Studies have indicated that the QTL associated with leaf rolling was colocalized with the QTL of water conservation at the vegetative stage (Kholova et al., 2012; Yadav et al., 2002, 2004).

Pearl millet is a monocot plant with a fibrous root system. The primary root length and lateral root density are associated with drought tolerance (Iwuala et al., 2022). Apart from its role in drought tolerance, the root system is responsible for nutrient acquisition. A study showed that the root length density in pearl millet was higher at 30 and 100 cm of soil depth than in other soil horizons (Zegada-Lizarazu et al., 2006) due to rapid emergence and vertical growth of the primary root (Passot et al., 2016). The primary root grows at the rate of 7 cm/d, compared with 3 cm/d in maize and wheat (Muller et al., 1998; Pahlavanian & Silk, 1988; Pritchard et al., 1987). Passot et al. (2016) showed that the pearl millet root system had three distinct lateral root types based on the diameter and radial anatomy. It was hypothesized that type 1 might be involved in the exploration of soil near the root, type 3 may be involved in branching of the root system and in exploration of new soil, and the role of type 2 is unclear. They hypothesize that the increased root length at the deeper soil horizon will help pearl millet to survive drought stress during seedling stage. The root developmental plasticity is also known to change due to water availability, resulting in altered root system architecture (Fromm, 2019).

A deep, wide-spreading, branching root structure is an important trait for drought tolerance. Pearl millet exhibits both elongated and dense root growth systems and affects the ability of plants to absorb water from moist subsoil layers by increasing lateral root density when drought stress occurs in the topsoil (Zegada-Lizarazu & Iijima, 2005). In semiarid environments, root elongation in deep soil layers occurs in pearl millet (Sivakumar & Salaam, 1994). To maintain high leaf water potential during drought stress conditions without changing the physiological and biochemical status, plants must develop elongated deep roots (Zegada-Lizarazu & Iijima, 2005). In addition, pearl millet roots can penetrate the compact subsoil horizon (in heavy soil) for water under drought-stress conditions, and a negative relationship between total root length and leaf osmotic adjustment was reported (Kusaka et al., 2005b). Significant genotypic differences were observed for total root length and root number in all soil moisture conditions, and the reduction in total root length under long-term drought stress conditions may be due to greater variation in root number (Kusaka et al., 2005a).

Impact of Drought Stress on Physiological Traits

Many physiological traits, including stomatal distribution in the leaf, may affect the gas exchange parameters (Xu & Zhou, 2008) and stomatal closure, a trait important for the physiological adaptation of pearl millet (Xiong & Zhu, 2002) that reduces photosynthesis and carbohydrate content under drought stress (Henson et al., 1981). The opened stomata maintain higher photosynthetic and transpiration rates. The genotypes with greater stomatal conductance are more susceptible to drought stress because of the rapid water loss through opened stomata. Hence, the plants tend to close the stomata partially to reduce water loss, causing less CO_2 fixed under drought stress. Yadav et al. (2004) reported that terminal drought-tolerant lines had a lower stomatal conductance than the drought-sensitive line. In addition, studies indicated that vapor pressure deficit (VPD) influences stomatal conductance under fully irrigated conditions (Kholova et al., 2010) The genotypes with reduced stomatal conductance under high soil moisture and high VPD conditions will conserve the soil moisture during early phases of growth, and the stored water will be available during later stages, which is termed as "limited transpiration" (Vadez et al., 2012). The above strategy would not be helpful if the pearl millet were grown in the lighter-texture soil because of deep drainage and low water holding capacity. Water use efficiency is negatively associated with evaporative demand, so maintaining photosynthetic activity under high evaporative demand is critical for improving the water use efficeincy. In pearl millet, a rapid change in transpiration rate was observed upon reaching VPD thresholds, indicating that hydraulic signals are needed to induce the stomatal closure to avoid loss of turgor. A wide genetic variability for restricting the transpiration rate under high VPD was observed for pearl millet (Kholova et al., 2010) and sorghum (Gholipoor et al., 2012). Similar variability was observed in maize, wheat and other crops, like soybean [*Glycine max* (L.) Moench], chickpea (*Cicer arietinum* L.), peanut (*Arachis hypogaea* L.), and cowpea (*Vigna unguiculata* L.). The mapping study showed that QTLs for transpiration rate co-localize with major terminal drought-tolerant QTL (root dry weight and grain yield) on linkage group (LG)2 (Kholova et al., 2012; Yadav et al., 2002), indicating that reduced/limited transpiration through decreased stomatal conductance is also a drought avoidance mechanism in pearl millet.

Stomatal closure during drought stress leads to excessive accumulation of ROS, leading to lipid peroxidation and membrane damage (Bhatt et al., 2011). Abscisic acid (ABA) accumulation was noticed upon the quick development of drought stress, which is transported to the leaves to cause stomatal closure; thus, water loss under drought stress is decreased. ABA serves as the first signal for a drought response to reduce water loss and increase water use efficiency (Kholova et al., 2010). Drought stress induces stomatal closure, leading to increased accumulation of ROS and oxidative stress (Bhatt et al., 2011). Drought stress in pearl millet activates antioxidant defense systems due to an imbalance in cellular redox homeostasis (Choudhury et al., 2022). The imbalance in cellular redox homeostasis occurs due to increased production and accumulation of ROS such as singlet oxygen ($^1O_2$), superoxide radical (O_2^-), hydroxyl radical (HO^-), and hydrogen peroxide (H_2O_2) (Choudhury et al., 2021). The produced ROS can react with cellular components and cause interference in normal cellular function. Apart from the damage, ROS are involved in cellular signalling to regulate various plant processes (Choudhury et al., 2013). Varshney et al. (2017) reported that ROS will accumulate in the pearl millet root tip under drought stress, and nicotinamide adenine dinucleotide hydrogenase (NADH) controls the meristematic activity through the

ABA pathway, indicating that ROS is involved in cell signalling processes during drought stress. Similarly, Steinkellner et al. (2007) and Ghatak et al. (2022) reported that enhanced production of flavonoids from the pearl millet root was observed under drought stress, and the flavonoids can function as antioxidants and in the signalling process. ROS controls the growth of plant roots along with hormones and other signal molecules. To balance root meristem growth, ROS and auxin signalling are antagonistically regulated in the root apical meristem (Tognetti et al., 2017). High glucose levels cause an imbalance of various ROS species or an accumulation of ROS that oxidizes active indole acetic acid (IAA), leading to its breakdown, affecting root meristem activity. This demonstrates that constitutive autophagy regulates the production of ROS and IAA by the peroxisome to modulate root meristem activity (Huang et al., 2019). Similarly, in the hydrotropic response, H_2O_2 also accumulates in the root curvature at a similar rate as the autophagosomes under drought stress (Jimenez-Nopala et al., 2018), suggesting that oxidative stress was induced during the hydrotropic response.

Previous studies showed that the primary site of ROS generation in a cell are in the electron transport chains of mitochondria and chloroplasts. In chloroplasts, the production of ROS occurs in thylakoids because of the over-reduction of both photosystems, which are no longer capable of accepting excess excitation energy from the light-harvesting protein complexes (Demmig-Adams & Adams Iii, 1992; Schmid, 2008). In chloroplasts, the production of ROS can be avoided by (a) degrading chlorophyll (Keiper et al., 1998; Maslova & Popova, 1993) and (b) maintaining chlorophyll content by activating the antioxidative scavenging systems (Farrant et al., 2003). In addition, under drought stress, the respiration rate is decreased due to the reduction in the availability of reductants like ATP, NADH, and tricarboxylic acid cycle intermediates, resulting in an increased production of ROS in the roots of pearl millet (Zhang et al., 2021). In contrast, activation of energy-intensive processes and increased respiratory rates were observed under drought stress (Florez-Sarasa et al., 2007). Hence, under drought treatment, ROS are generated due to the metabolic perturbation of cells, and these molecules cause cell damage and death (Nxele et al., 2017).

In pearl millet, drought stress for 21 days resulted in increased production of ROS and malondialdehyde (MDA) content in leaves, causing oxidative damage (Choudhury et al., 2022). The same authors reported similar damage in root tissues during drought stress (Choudhury et al., 2022). In another study, drought stress for 1 hour (short term) increased the ROS content in the root, which can damage the proteins and lipids; however, ROS content decreased after 3 and 7 hours of drought stress, indicating that the ROS produced was scavenged by the root's antioxidant systems and/or antioxidant enzyme activity (Zhang et al., 2021). Drought-tolerant plants produce enzymatic and non-enzymatic antioxidants to detoxify oxidative stress by scavenging the produced ROS (Mates, 2000). According to Yu et al. (2013), glutathione maintains ROS homeostasis in the root apical meristem by controlling the ROS content in the cells. Xanthophylls account for 50% of total carotenoid content and contribute to maintaining cellular redox homeostasis. Studies emphasized the function of carotenoids (particularly xanthophylls) in the direct deactivation of ROS (Demmig-Adams & Adams Iii, 1992).

In pearl millet, studies on the expression profiles of antioxidant genes under drought stress showed significant increases in genes encoding superoxide dismutase (SOD), ascorbate peroxide (APX), catalase (CAT), and peroxidase (POX) (Shivhare & Lata, 2017; Sun et al., 2020; Zhang et al., 2021). The enhanced expression of these genes under drought stress indicates that these enzymes are involved in scavenging ROS in pearl millet (Shivhare & Lata, 2017; Sun et al., 2020). In contrast, Kholova and Vadez (2013) observed that the antioxidant enzyme activity does not differ among the genotypes, except the APX under terminal drought stress, indicating that the antioxidant defense system may not be the only mechanism involved in drought tolerance in pearl millet.

Osmotic adjustment is an important drought tolerance mechanism to maintain cell turgor because cell expansion, photosynthesis, and stomatal conductance are affected under flaccid conditions (Henson et al., 1982). Osmotic adjustment is achieved by osmoregulation; the accumulation of osmolytes decreases the water potential by reducing osmotic potential, and therefore the water enters from the soil to the root cell under drought stress (Bajji et al., 2001). Pearl millet is known to adjust the osmotic potential under drought stress (Henson et al., 1982). A similar response has been reported in other drought-tolerant C_3 and C_4 crops like sorghum (Blum & Ebercon, 1976), wheat (Hong-Bo et al., 2006), and upland rice (Lum et al., 2014).

In pearl millet, proline content increased with drought intensity (Yadav et al., 2020). Under drought stress there is a reduction in leaf osmotic potential in pearl millet due to accumulation of osmolytes (Ali & Golombek, 2016). A significant reduction in leaf osmotic potential and water potential under drought stress and the resulting increase under well-watered conditions in pearl millet cultivars indicate the importance of osmotic adjustment (Ashraf et al., 2001).

Root tissues first sense drought stress (Fitter, 2002) and have the developmental plasticity to change the related architectural, biochemical, and physiological traits on the basis of water availability (Fromm, 2019; Gupta et al., 2020; Kang et al., 2022; Siddiqui et al., 2021). Drought stress inhibits the lateral root meristem, significantly affecting lateral root growth (Spollen & Sharp, 1991). A high-performance liquid chromatography ultraviolet study of root exudates of pearl millet showed that drought stress increased organic acids (e.g., succinic acid, oxalic acid, lactic acid, fumaric acid, malonic acid, and citric acid), which might have enhanced the osmotic adjustment, ultimately increasing cell turgor (Ghatak et al., 2022).

The significance of various osmotic constituents differs between the genotype and stage of plant development. During the vegetative stage, plants tend to accumulate relatively greater quantities of inorganic elements like potassium (K^+) and nitrate ions (NO_3^-) when compared to organic compounds such as proline, amino acids, and soluble sugars (Kusaka et al., 2005b). Osmotic adjustment can differ substantially among species and genotypes, and the adjustment capacity depends on the types of solutes accumulated and their relative contribution in lowering osmotic potential (Chen & Jiang, 2010). Drought-stress–related proteins and their functional relevance to drought tolerance in agricultural crops have been reviewed by Priya et al. (2019).

Impact of Drought Stress on Reproductive and Yield Traits

An increase in the frequency of drought stress during flowering stage affects plant reproduction (Srivastava et al., 2012). Pearl millet has large genetic diversity, with different levels of stress tolerance and yield potentials (Burgarella et al., 2018). Drought stress induces early flowering due to rapid crop phenological development. In pearl millet, early and asynchronous development of panicles and flowers could spread flowering over a long period, thereby maintaining yield potential even when the main stem is affected by drought stress (Mahalakshmi et al., 1987). Primary tillers develop in the axils of leaves in the main culm and secondary tillers from the buds on the primary tillers (Vadez et al., 2012). The transition of tillers to flowering at different times is called "asynchronous tillering." Maximum yield reduction due to drought stress occurs during the flowering and grain filling stages, with no possibility of recovering. However, drought stress during late vegetative stage affects flowering, but increased tillering (Mahalakshmi et al., 1987) can affect panicle size, the number of fertile florets per panicle, and grain size in pearl millet. The drought-tolerant genotypes alter their crop phenology by developing late-season tillers that flower and produce grains after the dry spell or when stress is relieved (Craufurd & Bidinger, 1988).

Developmental plasticity is linked to growth plasticity (i.e., a plant produces large amount of leaves and stems when water and nutrients are not limiting). Pearl millet can store up to 70% of its biomass in tillers (Azam-Ali et al., 1984), and the carbohydrates stored in the tillers are readily available during the cropping cycle and serve as a carbon buffer during drought stress. In addition, under drought, the plants reduce the leaf area through senescence (Soegaard & Boegh, 1995). Early floral initiation is an adaptive trait for late-season drought stress. Hence, early-maturing pearl millet varieties would adapt better to drought stress in arid environments. However, late-maturing pearl millets would adapt better to humid environments

(Diack et al., 2020). Pearl millet is a short-day plant, delaying flowering when the day length increases. However, there is large genetic variability in terms of photoperiod sensitivity. Almost 54.4% of total cultivated pearl millet germplasms were found to flower irrespective of day length, although the majority are photoperiod sensitive, with a delay in flowering time with increasing day length (Rai & Yadav, 2013). Pearl millet flowers either early or late as a drought escape mechanism. Pearl millet grown under long-day conditions with mid-season drought stress resulted in delayed flowering, but there was no significant effect on grain yield (Mahalakshmi et al., 1987).

Flower or seed abortion is the major issue affecting crop yield under drought (Boyer & McLaughlin, 2007; Patrick & Stoddard, 2010; Prasad et al., 2008; Ruan et al., 2010). Drought at the flowering stage causes seed abortion by reduced carbon allocation to the newly formed seeds due to decreases in photosynthesis or partitioning (McLaughlin & Boyer, 2004), leading to fewer mature seeds. If drought occurs after seed-set, grain size is decreased. Grain-filling in cereals is the process of converting simple sugars to starch, which is performed by enzymes such as sucrose synthase, adenosine diphosphate-glucose-pyro phosphorylase, starch synthase, and starch branching enzyme (Taiz & Zeiger, 2006). Under drought stress, the grain growth rate was decreased by reduced sucrose synthase activity, and grain growth was stopped by inhibiting adenosine diphosphate-glucose-pyro phosphorylase activity. Grain filling is affected due to the imbalanced allocation of resources to grains by reduced activities of sucrose and starch synthesizing enzymes (Farooq et al., 2009). During the grain-filling stage, mature leaves and flag leaves synthesize the carbohydrates and transport the photosynthates to the seeds for their development and growth. Drought stress during this stage will have a significant impact on seed size and seed weight (Begcy & Walia, 2015; Fougereux et al., 1997). An important QTL responsible for drought tolerance in pearl millet is LG2, associated with a QTL for change in flowering time under drought stress. Other co-localized genes included are transcription factors belonging to known flowering time gene families, including zinc finger CCCH type and MADS-box gene families (Sehgal et al., 2012). The genomic regions associated with 1,000-seed weight and seed yield per plant were mapped on LG1, LG2, and LG6. Kumar et al. (2021) reported QTL for 1,000-seed weight on LG3 and LG7. Bidinger et al. (2007) mapped the QTL for grain yield on LG2, LG3, and LG4. The co-localized QTLs had pleiotropic effects with these associated traits, indicating difficulties in developing an ideal plant ideotype for drought stress. With the identified co-localized QTLs, common genes controlling the different process may be identified to validate the developed QTLs in diverse genetic backgrounds to develop reliable molecular markers for parental line development with enhanced pearl millet hybrid grain yield under terminal drought stress environments.

High-temperature Stress

Pearl millet has relatively greater tolerance to high temperatures, as evidenced by the regions where it is grown. However, when high temperature coincides with critical growth stages, plant growth is decreased due to the negative impacts on physiological and biochemical changes (Ong & Monteith, 1985). Seed germination is the first sensitive stage to high-temperature stress because high temperatures decrease the germination and emergence percentages. Ong and Monteith (1985) reported that germination percentage of pearl millet increased linearly from 10 to 12 °C; maximum germination percentage was observed at 33–34 °C, and it declined to 0% at 45–47 °C. The growth and development of the leaf, panicle initiation, and tillering were similar to the seed germination process. The leaf extension rate showed a linear response up to 34 °C (Ong & Monteith, 1985), and the canopy cover was faster up to an air temperature of 34 °C in pearl millet. Pearl millet growth was slow at 19.5 °C and fastest at 31.0 °C (Maiti & Wesche-Ebeling, 1997). Pearl millet is relatively more tolerant to high temperatures than other cereals with higher ceiling temperatures (Gupta et al., 2015; Prasad et al., 2017).

Impact of High-temperature Stress on Morphological and Growth Traits

The initial visible symptom of high-temperature stress in pearl millet was leaf rolling and folding of leaves, followed by leaf senescence (Kadioglu & Terzi, 2007), resulting in decreased leaf transpiration rate via reduced leaf area. In pearl millet, the number of tillers was significantly decreased, along with a lower leaf area, under high-temperature stress (Ong & Monteith, 1985). Generally, it is thought that a genotype with increased tillering and small grains will outyield a genotype with fewer tillers and large grains under stressful or marginal environments. However, under optimal growth conditions, the opposite result occurred (van Oosterom et al., 2006). Plasticity in tiller development is a characteristic of pearl millet under abiotic stress. Primary tillers develop from the axils of leaves of the main culm, and secondary tillers develop from the buds in the axils of leaves on the primary tillers. Based on the emergence pattern, tillers can be either basal or nodal, which are crucial characteristics that allow pearl millet to withstand high-temperature stress. This developmental plasticity is responsible for yield compensation (i.e., if the primary tiller fails to flower and set the seeds during high temperatures, then the secondary tiller develops to compensate for the yield loss on the main and primary tiller) (Vadez et al., 2012). Pearl millet biomass production and leaf area remained unaffected under high-temperature stress (38 °C). However, the relative growth rate and net assimilation rate were significantly increased (Ashraf & Hafeez, 2004), indicating that the carbohydrates formed will be stored in the stem, and during recovery, it may be mobilized to produce grains under high-temperature stress. Apart from this, under high temperatures, pearl millet accumulates higher concentrations of P and K in the leaves (Ashraf & Hafeez, 2004), and it was hypothesized that the high-temperature stress tolerance in pearl millet is associated with nutrient accumulation (Marschner, 1995).

Impact of High-temperature Stress on Physiological Traits

High-temperature stress negatively affects many physiological traits, including photosynthesis and respiration, driven by loss of chlorophyll and membrane damage. Chlorophyll loss is related to environmental stress, and changes in chlorophyll content under high-temperature stress indicate the negative impacts on the integrity of the thylakoid membrane because the chlorophyll molecule is localized in the thylakoid membranes (Djanaguiraman et al., 2018; Fisher & Maurer, 1978; Meena et al., 2021). In addition to the pigment loss, high-temperature stress decreased the photosynthetic rate due to the downregulation of linear electron transport in photosystem (PS) II (Shanker et al., 2022) in pearl millet. This finding indicates that reduced carbon availability under high-temperature stress requires fewer electrons from PSII for carbon assimilation; the additional electron can lead to photoinhibition at PSII, as documented by Shanker et al. (2022) in pearl millet. The reduction in the F_v/F_m ratio under high-temperature stress, a trait associated with photoinhibition, could be due to the physical disconnection of the antenna from PSII (Shanker et al., 2022). It was observed that the photosynthetic rate in some pearl millet lines was not affected up to 40/30 °C (daytime/nighttime temperature) (Djanaguiraman et al., 2018). However, at 44/34 °C, the photosynthetic rate was decreased by 10% (Djanaguiraman et al., 2018). There was variation in the photosynthetic rates in response to temperature stress.

The integrity of the thylakoid membrane was highly susceptible to high-temperature stress. Within chloroplasts, significant changes (e.g., alterations in the structural arrangement of thylakoids, granum stacking breakdown, and grana swelling) occur under high-temperature stress (Rivas & Barber, 1997). At high temperatures, damage to the oxygen-evolving complex occurs, resulting in an uneven distribution of electrons to the PSII acceptor site. Additionally, it leads to the denaturation of the D1 and D2 proteins (Telfer, 2005). An increased transpiration rate in pearl millet under high-temperature stress (Shanker et al., 2022) can be a mechanism of leaf surface cooling under high VPD condition.

In most field crops, high-temperature stress adversely affects photosynthesis, respiration, water relations, and membrane stability. High-temperature stress causes a change in hormone levels and primary and secondary metabolites in pearl millet (Arya et al., 2014). In plants, stomatal conductance estimates the rate of gas exchange (H_2O, CO_2, and O_2) through the stomata. High-temperature stress affects the water status of the leaf by modulating the stomatal conductance (Greer & Weedon, 2012). Stomatal closure under high temperatures is a distinct cause of impaired photosynthesis that could have affected the intercellular CO_2 concentration (Yang et al., 2012). Increased stomatal conductance because of increased stomatal opening was associated with the leaf-cooling effect, which can contribute to high-temperature stress tolerance (Narayanan, 2018; Shanker et al., 2022). During high-temperature stress, the leaves are folded and rolled or senesced in pearl millet, leading to a decreased transpiration rate as a result of the reduced leaf area (Kadioglu & Terzi, 2007).

In pearl millet, high-temperature stress induces the production of ROS in leaves (Sankar et al., 2021), which can strongly react with organic molecules and disrupts fundamental metabolic pathways and cell structures (Choudhury et al., 2022). The ROS, such as H_2O_2, singlet oxygen $^1O_2$, and hydroxyl radical OH, are produced during high-temperature stress (Dudhate et al., 2018). Like drought stress, lipid peroxidation of the cell membrane (Sun et al., 2020), protein inactivation, DNA damage, and cell apoptosis occur under high-temperature stress in pearl millet (Huang et al., 2021).

The effect of high temperature during seedling stage indicated that genes encoding amine oxidases and polyamine oxidases were upregulated after 24 hours. However, at 48 hours, the genes involved in ROS production were inhibited due to upregulation of genes of antioxidant enzymes (Sun et al., 2020). In pearl millet, high-temperature stress (43 °C) during early vegetative stage resulted in increased ROS, MDA, ascorbate peroxidase (APX), and glycine betaine (GB) levels (Jacob et al., 2022). From this study, it was inferred that high temperature damages the membrane through enhanced production of ROS, and to protect the membrane from damage, enhanced activity of APX and GB was observed, because APX and GB are enzymatic and non-enzymatic antioxidants, respectively. Similarly, high-temperature stress during reproductive stages resulted in increased ROS (O_2^- and H_2O_2) production in pollen and pistil compared with the optimum temperature (Djanaguiraman et al., 2018). Antioxidant enzyme (SOD, CAT, POX) activity was also decreased under high temperature stress, resulting in enhanced oxidative damage.

In pearl millet, understanding the effect of high temperature at the molecular level began by examining the role of small heat shock proteins HSP70, HSP90, and purple acid phosphatases (Reddy et al., 2017). About 529 transcription factors (TFs) were differentially expressed in pearl millet leaves under high-temperature stress, and these TFs were distributed in 50 transcription families. The most significant TFs are heat shock factor (HSF), basic leucine zipper, ethylene responsive factor, and N-acetyl cysteine transcription factor (Huang et al., 2021). Short-term high-temperature stress has resulted in changes in genes involved in protein folding, indicating that protein homeostasis is another key process for plants to adapt to high-temperature stress (Huang et al., 2021). For example, HSP70 and HSP90 can bind to unfolded peptide chains or denatured proteins to promote the correct folding or degradation of peptide chains and proteins, thereby playing an important role in high-temperature stress tolerance. Mid-term high-temperature stress induces the expression of antioxidant enzyme genes like SOD, CAT, and APX. Under long-term stress, SOD was upregulated, and there were no changes in other antioxidant enzyme activity, indicating that, in pearl millet, antioxidant defense is only for short-term and not for long-term stress conditions (Huang et al., 2021; Sun et al., 2020). In contrast, exposing pearl millet to high temperatures decreased the genes associated with flavonoid biosynthesis, which indicates that pearl millet conserves energy to resist high temperatures by inhibiting energy-intensive processes like flavonoid biosynthesis. At the transcription level, the pearl millet quickly responds to prevent temperature-induced protein misfolding by co-regulation of the transcription factor family gene (*RWP-RK*) with heat shock factor genes and endoplasmic reticulum-related genes (Yan et al., 2023). Similarly, protein synthesis elongation factor and molecular chaperones such as elongation factor thermo unstable (EF-Tu) are known to impart heat tolerance in many plant species (Fu et al., 2012).

Impact of High-temperature Stress on Reproductive and Yield Traits

Generally, the reproductive stage (anthesis, flowering, and grain filling) is more sensitive to high-temperature stress than other crop growth stages (Prasad et al., 2008, 2017). The viability and receptivity of male and female reproductive tissues respectively determines the success of fertilization, and failure of this process occurs if the viability of gametes is lost. The development of reproductive tissues in different species indicates that high temperature during meiosis causes loss of viability of male and female gametes (Barnabas et al., 2008; Hedley et al., 2009; Prasad et al., 2017). Brief exposure to temperature stress disrupts pollen development during meiosis. Specifically, the uninucleate stage is susceptible to high-temperature stress during pollen development (Sage et al., 2015). In addition, high temperature during the anther development results in poor anther dehiscence, which leads to poor release of mature pollen grains from the compartments within the anther (Jiang et al., 2019).

Generally, male reproductive tissue is more vulnerable to high-temperature stress than female reproductive tissue (Jagadish, 2020), causing sterility of pollen grains by reducing the pollen germination to almost close to zero when the temperature was around 40 °C in pearl millet. Djanaguiraman et al. (2018) reported that high-temperature stress affects the pistil more than the pollen grain in pearl millet. The failure of pollen grains and pistil function is due to the greater membrane damage caused by ROS. Pearl millet seed yield becomes almost zero, and no seed formation occurs when the temperature is around 42 °C. The seed-set percentage was decreased by 75% if the pistil alone was stressed under high temperature (Djanaguiraman et al., 2018; Gupta et al., 2015). Similar observations were made in a field study, where the critical temperature for the seed set was 42 °C (Gupta et al., 2015). Seed size was also decreased by 50% when air temperature ranged from 32 to 42 °C (Djanaguiraman et al., 2018). Genetic variability was observed in pearl millet (Djanaguiraman et al., 2018; Gupta et al., 2015) and other cereals, like sorghum (Djanaguiraman et al., 2014; Sunoj et al., 2017), finger millet (Opole et al., 2018), and wheat (Bheemanahalli et al., 2019), for gamete viability or seed-set under high-temperature stress.

Plants possess various mechanisms to cope with high-temperature stress during the flowering stage. An effective strategy is to keep the canopy cooler by enhancing transpiration (Lin et al., 2017). This approach is particularly valuable for crops to adapt to high temperatures, especially when soil moisture is available or under irrigated conditions. When plants keep the canopy cooler during peak flowering, they can escape high-temperature stress. Conceptually, this has been demonstrated in cereal crops like rice (Bheemanahalli et al., 2017), wheat (Aiqing et al., 2018), sorghum (Chiluwal et al., 2019), and pearl millet (Jagadish, 2020). The modification in the flower opening time is a promising technique to protect the reproductive organs from high-temperature stress, thereby reducing crop yield losses. Some of the crop species, including pearl millet, that flower early in the morning can also escape the high temperatures that typically occur during mid-day or afternoons. The early-morning flowering trait has been identified in some wild species and varieties of rice (Jagadish et al., 2015; Prasad et al., 2006) and wheat (Aiqing et al., 2018) that escape high-temperature stress and continue to set seeds without affecting the yield potential. This trait needs to be evaluated in pearl millet to determine its importance.

Conclusion

Pearl millet is an important coarse-grain cereal cultivated in arid and semiarid regions of the world. Compared with other cereals, pearl millet is more tolerant to drought and high-temperature stresses. However, pearl millet is sensitive to drought and high-temperature stress during some critical crop growth stages, such as emergence, flowering, and seed development. Traditionally, pearl millet is grown during the rainy season, particularly in India; however, in recent years it has been grown during the hot-dry post-rainy season. Similarly, in Africa, pearl millet is grown during the hot-dry season. Hence, the tolerance to drought and high-temperature stress in pearl millet during sensitive stages of crop growth and development will be a benefit and increase the production area and productivity. Drought tolerance of pearl millet is associated with the deep root system, restricted transpiration, and stay-green trait during post-flowering stages. Research focus on the effects of high-temperature stress in pearl is limited when compared to drought stress. The high-temperature stress tolerance in pearl millet is associated with an antioxidant defense system and HSPs as well as improved reproductive fertility, seed-set, and partitioning of carbohydrates to seeds. Although our understanding of drought and high-temperature tolerance mechanisms in pearl millet has improved on yield traits, the impacts of drought and high-temperature stress on grain nutritional qualities, including micronutrients (e.g., grain Zn and Fe) and vitamins, and the associated trade-off are not well understood. In addition, there is a need to systematically screen larger germplasm collections of pearl millet for drought and tolerance to high temperatures to identify tolerant lines and to use in targeted breeding programs that focus on improving stress tolerance, yields, and grain quality. Further, to improve the nutritional quality of locally adapted high-yielding pearl millet cultivars, genes associated with nutrition uptake, mobility, and translocation must be identified by integrating classical and genomics-assisted breeding tools. Mainstreaming the biofortification in pearl millet for Fe and Zn is necessary in all breeding programs. There is also a need to focus on value addition of grain to increase existing market opportunities and to create new markets for food products of pearl millet. Such efforts are critical for the economic security of small and marginal farmers who are dependent on pearl millet for their food, nutrition, and climate security.

References

Aiqing, S., Somayanda, I., Sebastian, S. V., Singh, K., Gill, K., Prasad, P. V. V., & Jagadish, S. V. K. (2018). Heat stress during flowering affects time of day of flowering, seed set and grain quality in spring wheat. *Crop Science, 58*, 380–392.

Ali, Z. I., & Golombek, S. D. (2016). Effect of drought and nitrogen availability on osmotic adjustment of five pearl millet cultivars in the vegetative growth stage. *Journal of Agronomy and Crop Science, 202*, 433–444.

Arya, R. K., Singh, M. K., Yadav, A. K., Kumar, A., & Kumar, S. (2014). Advances in pearl millet to mitigate adverse environmental conditions emerged due to global warming. *Forage Research, 40*, 57–70.

Ashraf, M., Ahmad, A., & McNeilly, T. (2001). Growth and photosynthetic characteristics in pearl millet under water stress and different potassium supply. *Photosynthetica, 39*, 389–394.

Ashraf, M., & Hafeez, M. (2004). Thermotolerance of pearl millet and maize at early growth stages: Growth and nutrient relations. *Journal of Plant Biology, 48*, 81–86.

Azam-Ali, S. N., Gregory, P. J., & Monteith, J. L. (1984). Effects of planting density in water use and productivity of pearl millet (*Pennisetum typhoides*) grown on stored water. II. Water use, light interception, and dry matter production. *Experimental Agriculture, 20*, 215–224.

Bajji, M., Lutts, S., & Kinet, J. M. (2001). Water deficit effects on solute contribution to osmotic adjustment as a function of leaf ageing in three durum wheat (*Triticum durum* Desf.) cultivars performing differently in arid conditions. *Plant Science, 160*, 669–681.

Barnabas, B., Jager, K., & Feher, A. (2008). The effect of drought and heat stress on reproductive processes in cereals. *Plant Cell & Environment, 31*, 11–38.

Begcy, K., & Walia, H. (2015). Drought stress delays endosperm development and misregulates genes associated with cytoskeleton organization and grain quality proteins in developing wheat seeds. *Plant Science, 240*, 109–119.

Bhatt, D., Negi, M., Sharma, P., Saxena, S. C., Dobriyal, A. K., & Arora, S. (2011). Responses to drought induced oxidative stress in five finger millet varieties differing in their geographical distribution. *Physiology and Molecular Biology of Plants, 17*, 347–353.

Bheemanahalli, R., Sathishraj, R., Manoharan, M., Sumanth, H. N., Muthurajan, R., Ishimaru, T., & Jagadish, S. V. K. (2017). Is early morning flowering an effective trait to minimize heat stress damage during flowering in rice? *Field Crops Research, 203*, 238–242.

Bheemanahalli, R., Sunoj, V. S. J., Saripalli, G., Prasad, P. V. V., Balyan, H. S., Gupta, P. K., Grant, N., Gill, K. S., & Jagadish, S. V. K. (2019). Quantifying the impact of heat stress on pollen germination, seed set and grain filling in spring wheat. *Crop Science, 59*, 684–696.

Bi, H., Kovalchuk, N., Langridge, P., Tricker, P. J., Lopato, S., & Borisjuk, N. (2017). The impact of drought on wheat leaf cuticle properties. *BMC Plant Biology, 17*, 1–13.

Bidinger, F. R., & Hash, C. T. (2004). Pearl millet. In H. T. Nguyen & A. Blum (Eds.), *Physiology and biotechnology integration for plant breeding* (pp. 225–270). Dekker.

Bidinger, F. R., Mahalakshmi, V., & Rao, G. D. P. (1987). Assessment of drought resistance in pearl millet [*Pennisetum amencanum* (L.) Leeke]: I. Factors affecting yield under stress. *Australian Journal of Agricultural Research, 38*, 37–48.

Bidinger, F. R., Nepolean, T., Hash, C. T., Yadav, R. S., & Howarth, C. J. (2007). Quantitative trait loci for grain yield in pearl millet under variable post flowering moisture conditions. *Crop Science, 47*(3), 969–980.

Blum, A., & Ebercon, A. (1976). Genotypic responses in sorghum to drought stress. III. Free proline accumulation and drought resistance. *Crop Science, 16,* 428–431.

Boyer, J. S., & McLaughlin, J. E. (2007). Functional reversion to identify controlling genes in multigenic responses: Analysis of floral abortion. *Journal of Experimental Botany, 58,* 267–277.

Burgarella, C., Cubry, P., Kane, N. A., Varshney, R. K., Mariac, C., Liu, X., Shi, C., Thudi, M., Couderc, M., Xu, X., Chitikineni, A., Scarcelli, N., Barnaud, A., Rhoné, B., Dupuy, C., François, O., Berthouly-Salazar, C., & Vigouroux, Y. (2018). A western Sahara centre of domestication inferred from pearl millet genomes. *Nature Ecology & Evolution, 2,* 1377–1380.

Chen, H., & Jiang, J. (2010). Osmotic adjustment and plant adaptation to environmental changes related to drought and salinity. *Environmental Review, 18,* 309–319.

Chiluwal, A., Bheemanahalli, R., Kanaganahalli, V., Boyle, D., Perumal, R., Pokharel, M., Oumarou, H., & Jagadish, S. V. K. (2019). Deterioration of ovary plays a key role in heat stress-induced spikelet sterility in sorghum. *Plant, Cell & Environment, 43,* 448–462.

Choudhury, S., Moulick, D., Ghosh, D., Soliman, M., Alkhedaide, A., Gaber, A., & Hossain, A. (2022). Drought-induced oxidative stress in pearl millet (*Cenchrus americanus* L.) at seedling stage: Survival mechanisms through alteration of morphophysiological and antioxidants activity. *Life, 12,* 1171.

Choudhury, S., Moulick, D., & Mazumder, M. K. (2021). Secondary metabolites protect against metal and metalloid stress in rice: An in-silico investigation using dehydroascorbate reductase. *Acta Physiologiae Plantarum, 43,* 1–10.

Choudhury, S., Panda, P., Sahoo, L., & Panda, S. K. (2013). Reactive oxygen species signaling in plants under abiotic stress. *Plant Signaling & Behavior, 8,* e23681.

Craufurd, P., & Bidinger, F. (1988). Effect of the duration of the vegetative phase on crop growth, development, and yield in two contrasting pearl millet hybrids. *Journal of Agricultural Sciences, 110,* 71–79.

Debieu, M., Sine, B., Passot, S., Grondin, A., Akata, E., Gangashetty, P., Vadez, V., Gantet, P., Fonceka, D., Cournac, L., Hash, C. T., Kane, N. A., Vigouroux, Y., & Laplaze, L. (2018). Response to early drought stress and identification of QTLs controlling biomass production under drought in pearl millet. *PLoS One, 13,* e0201635.

Demmig-Adams, B., & Adams Iii, W. W. (1992). Photoprotection and other responses of plants to high light stress. *Annual Review of Plant Biology, 43,* 599–626.

Diack, O., Kanfany, G., Gueye, M. C., Sy, O., Fofana, A., Tall, H., Serba, D. D., Zekraoui, L., Berthouly-Salazar, C., Vigouroux, Y., Diouf, D., & Kane, N. A. (2020). GWAS unveils features between early- and late-flowering pearl millets. *BMC Genomics, 21,* 777.

Djanaguiraman, M., Perumal, R., Ciampitti, I. A., Gupta, S. K., & Prasad, P. V. V. (2018). Quantifying pearl millet response to high temperature stress: Thresholds, sensitive stages, genetic variability and relative sensitivity of pollen and pistil. *Plant, Cell & Environment, 41*, 993–1007.

Djanaguiraman, M., Prasad, P. V., Murugan, M., Perumal, R., & Reddy, U. K. (2014). Physiological differences among sorghum (*Sorghum bicolor* L. Moench) genotypes under high temperature stress. *Environmental and Experimental Botany, 100*, 43–54.

Do, F., Winkel, T., Cournac, L., & Louguet, P. (1996). Impact of late-season drought on water relations in a sparse canopy of millet (*Pennisetum glaucum* (L.) R. Br.). *Field Crops Research, 48*, 103–113.

Dudhate, A., Shinde, H., Tsugama, D., Liu, S., & Takano, T. (2018). Transcriptomic analysis reveals the differentially expressed genes and pathways involved in drought tolerance in pearl millet [*Pennisetum glaucum* (L.) R. Br]. *PLoS One, 13*, e0195908.

Eziz, A., Yan, Z., Tian, D., Han, W., Tang, Z., & Fang, J. (2017). Drought effect on plant biomass allocation: A meta-analysis. *Ecology and Evolution, 7*, 11002–11010.

FAO, IFAD, UNICEF, WFP, & WHO. (2023). *The state of food security and nutrition in the world 2023. Urbanization, agrifood systems transformation and healthy diets across the rural–urban continuum.* FAO.

Farooq, M., Wahid, A., Kobayashi, N., Fujita, D., & Basra, S. M. A. (2009). Plant drought stress: Effects, mechanisms, and management. In E. Lichtfouse, M. Navarrete, P. Debaeke, S. Véronique, & C. Alberola (Eds.), *Sustainable agriculture* (pp. 153–188)). Springer.

Farrant, J. M., Vander Willigen, C., Loffell, D. A., Bartsch, S., & Whittaker, A. (2003). An investigation into the role of light during desiccation of three angiosperm resurrection plants. *Plant, Cell & Environment, 26*, 1275–1286.

Fisher, R. A., & Maurer, R. (1978). Drought resistance in spring wheat cultivars. I. Grain yield responses in spring wheat. *Australian Journal of Agriculture, 29*, 892–912.

Fitter, A. (2002). Characteristics and functions of root systems. In Y. Waisel, A. Eshel, & U. Kafkafi (Eds.), *The hidden half* (3rd ed., pp. 15–32). Marcel Dekker.

Florez-Sarasa, I. D., Bouma, T. J., Medrano, H., Azcon-Bieto, J., & Ribas-Carbo, M. (2007). Contribution of the cytochrome and alternative pathways to growth respiration and maintenance respiration in *Arabidopsis thaliana. Physiologia Plantarum, 129*, 143–151.

Fougereux, J. A., Doré, T., Ladonne, F., & Fleury, A. (1997). Water stress during reproductive stages affects seed quality and yield of pea (*Pisum sativum* L.). *Crop Science, 37*, 1247–1252.

Fromm, H. (2019). Root plasticity in the pursuit of water. *Plants, 8*, 236.

Fu, J., Momcilovic, I., & Prasad, P. V. V. (2012). Roles of protein synthesis elongation factor EF-Tu in heat tolerance in plants. *Journal of Botany, 2012*, 835836.

Ghatak, A., Schindler, F., Bachmann, G., Engelmeier, D., Bajaj, P., Brenner, M. L., Varshney, R. K., Subbarao, G. V., Chaturvedi, P., & Weckwerth, W. (2022). Root exudation of contrasting drought-stressed pearl millet genotypes conveys varying biological nitrification inhibition (BNI) activity. *Biology and Fertility of Soils, 58,* 291–306.

Gholipoor, M., Sinclair, T. R., & Prasad, P. V. V. (2012). Genotypic variation within sorghum for transpiration response to drying soil. *Plant and Soil, 357,* 35–40.

Greer, D. H., & Weedon, M. M. (2012). Modelling photosynthetic responses to temperature of grapevine (*Vitis vinifera* cv. Semillon) leaves on vines grown in a hot climate. *Plant, Cell & Environment, 35,* 1050–1064.

Gupta, A., Rico-Medina, A., & Cano-Delgado, A. I. (2020). The physiology of plant responses to drought. *Science, 368,* 266–269.

Gupta, S. K., Rai, K. N., Singh, P., Ameta, V. L., Gupta, S. K., Jayalekha, A. K., Mahala, R. S., Pareek, S., Swami, M. L., & Verma, Y. S. (2015). Seed set variability under high temperatures during flowering period in pearl millet (*Pennisetum glaucum* L. (R.) Br.). *Field Crops Research, 171,* 41–53.

Harlan, J. (1975). *Crops and man.* ASA and CSSA.

Hedley, A., Hormaza, J. I., & Herrero, M. (2009). Global warming and sexual plant reproduction. *Trends in Plant Science, 14,* 30–36.

Henson, I. E., Mahalakhmi, V., Bidinger, F. R., & Algarswamy, G. (1981). Stomatal responses of pearl millet (*Pennisetum americanum* (*L.*) *Leeke*) genotypes, in relation to abscisic acid and water stress. *Journal of Experimental Botany, 32,* 1211–1221.

Henson, I. E., Mahalakshmi, V., Bidinger, F. R., & Alagarswamy, G. (1982). Osmotic adjustment to water stress in pearl millet (*Pennisetum americanum* [L.] *Leeke*) under field conditions. *Plant, Cell & Environment, 5,* 147–154.

Hong-Bo, S., Xiao-Yan, C., Li-Ye, C., Xi-Ning, Z., Gang, W., Yong-Bing, Y., Zhao Chang-Xing, Z., & Zan-Min, H. (2006). Investigation on the relationship of proline with wheat anti-drought under soil water deficits. *Colloids and Surfaces B: Biointerfaces, 53,* 113–119.

Hossain, A., Pramanick, B., Bhutia, K. L., Ahmad, Z., Moulick, D., Maitra, S., Ahmad, A., & Aftab, T. (2021). Emerging roles of osmoprotectant glycine betaine against salt-induced oxidative stress in plants: A major outlook of maize (*Zea mays* L.). In T. Aftab & K. R. Hakeem (Eds.), *Frontiers in plant-soil interaction* (pp. 567–587). Academic Press.

Huang, D., Sun, M., Zhang, A., Chen, J., Zhang, J., Lin, C., Zhang, H., Lu, X., Wang, X., Yan, H., Tang, J., & Huang, L. (2021). Transcriptional changes in pearl millet leaves under heat stress. *Genes, 12,* 1716.

Huang, L., Yu, L. J., Zhang, X., Fan, B., Wang, F. Z., Dai, Y. S., Qi, H., Zhou, Y., Xie, L. J., & Xiao, S. (2019). Autophagy regulates glucose-mediated root meristem activity by modulating ROS production in *Arabidopsis. Autophagy, 15,* 407–422.

Hussain, M., Malik, M. A., Farooq, M., Ashraf, M. Y., & Cheema, M. A. (2008). Improving drought tolerance by exogenous application of glycine betaine and salicylic acid in sunflower. *Journal of Agronomy and Crop Science, 194*, 193–199.

IPCC. (2022). Summary for policymakers. In H.-O. Pörtner, D. C. Roberts, E. S. Poloczanska, K. Mintenbeck, M. Tignor, A. Alegría, M. Craig, S. Langsdorf, S. Löschke, V. Möller, & A. Okem (Eds.), *Climate change 2022: Impacts, adaptation and vulnerability. Contribution of working group II to the sixth assessment report of the intergovernmental panel on climate change* (pp. 3–33). Cambridge University Press.

Iwuala, E., Adu, M. O., Odjegba, V., Odiong Unung, O. O., Ajiboye, A., Opoku, V. A., Umebese, C., & Alam, A. (2022). Mechanisms underlying root system architecture and gene expression pattern in pearl millet (*Pennisetum glaucum*). *Gesunde Pflanzen, 74*, 983–996.

Jacob, J., Sanjana, P., Visarada, K. B. R. S., Shobha, E., Ratnavathi, C. V., & Sooganna, D. (2022). Seedling stage heat tolerance mechanisms in pearl millet (*Pennisetum glaucum* (L.) R. Br.). *Russian. Journal of Plant Physiology, 69*, 128.

Jagadish, S. K. (2020). Heat stress during flowering in cereals–effects and adaptation strategies. *New Phytologist, 226*, 1567–1572.

Jagadish, S. V. K., Murty, M. V. R., & Quick, W. P. (2015). Rice responses to rising temperatures–challenges, perspectives and future directions. *Plant, Cell & Environment, 38*(9), 1686–1698.

Jiang, Y., Lahlali, R., Karunakaran, C., Warkentin, T. D., Davis, A. R., & Bueckert, R. A. (2019). Pollen, ovules, and pollination in pea: Success, failure, and resilience in heat. *Plant, Cell & Environment, 42*, 354–372.

Jimenez-Nopala, G., Salgado-Escobar, A. E., Cevallos-Porta, D., Cardenas, L., Sepulveda-Jimenez, G., Cassab, G., & Porta, H. (2018). Autophagy mediates hydrotropic response in *Arabidopsis thaliana* roots. *Plant Science, 272*, 1–13.

Jones, M. M., Turne, N. C., & Osmond, C. B. (1981). Mechanisms of drought resistance. In E. G. Paleg & D. Aspinall (Eds.), *Physiology and biochemistry of drought resistance in plants* (pp. 15–37). Academic Press.

Kadioglu, A., & Terzi, R. (2007). A dehydration avoidance mechanism: Leaf rolling. *Botanical Review, 73*, 290–302.

Kang, J., Peng, Y.,., & Xu, W. (2022). Crop root responses to drought stress: Molecular mechanisms, nutrient regulations, and interactions with microorganisms in the rhizosphere. *International Journal of Molecular Sciences, 23*, 9310.

Keiper, F. J., Chen, D. M., & De Filippis, L. F. (1998). Respiratory, photosynthetic, and ultrastructural changes accompanying salt adaptation in culture of Eucalyptus microcorys. *Journal of Plant Physiology, 152*, 564–573.

Kholova, J., Hash, C. T., Kakkera, A., Kocova, M., & Vadez, V. (2009). Constitutive water-conserving mechanisms are correlated with the terminal drought tolerance of pearl millet [*Pennisetum glaucum* (L.) R. Br.]. *Journal of Experimental Botany, 61*, 369–377.

Kholova, J., Hash, C. T., Kumar, P. L., Yadav, R. S., Kocova, M., & Vadez, V. (2010). Terminal drought tolerant pearl millet [*Pennisetum glaucum* (L.) R. Br.] have high leaf ABA and limit transpiration at high vapour pressure deficit. *Journal of Experimental Botany, 61*, 1431–1440.

Kholova, J., Nepolean, T., Hash, C. T., Supriya, A., Rajaram, V., Senthilvel, S., & Vadez, V. (2012). Water saving traits co-map with a major terminal drought tolerance quantitative trait locus in pearl millet [*Pennisetum glaucum* (L.) R. Br.]. *Molecular Breeding, 30*, 1337–1353.

Kholova, J., & Vadez, V. (2013). Water extraction under terminal drought explains the genotypic differences in yield, not the antioxidant changes in leaves of pearl millet (*Pennisetum glaucum*). *Functional Plant Biology, 40*, 44–53.

Krishnan, R., & Meera, M. S. (2018). Pearl millet minerals: Effect of processing on bio accessibility. *Journal of Food Science and Technology, 55*, 3362–3372.

Kukal, M. S., & Irmak, S. (2018). Climate-driven crop yield and yield variability and climate change impacts on the US Great Plains agricultural production. *Scientific Reports, 8*, 1–18.

Kumar, S., Hash, C. T., Singh, G., Nepolean, T., & Srivastava, R. K. (2021). Mapping QTLs for important agronomic traits in an Iniadi-derived immortal population of pearl millet. *Biotechnology Notes, 2*, 26–32.

Kusaka, M., Lalusin, A. G., & Fugimura, T. (2005a). The maintenance of growth and turgor in pearl millet (*Pennisetum glaucum L.*) cultivars with different root structures and osmoregulation under drought stress. *Plant Science, 168*, 1–14.

Kusaka, M., Ohta, M., & Fujimura, T. (2005b). Contribution of inorganic components to osmotic adjustment and leaf folding for drought tolerance in pearl millet. *Physiologia Plantarum, 125*, 474–489.

Levitt, J. (1980). *Responses of plants to environmental stresses. Volume II. Water, radiation, salt, and other stresses.* Academic Press.

Lin, H., Chen, Y., Zhang, H., Fu, P., & Fan, Z. (2017). Stronger cooling effects of transpiration and leaf physical traits of plants from a hot dry habitat than from a hot wet habitat. *Functional Ecology, 31*, 2202–2211.

Lum, M. S., Hanafi, M. M., Rafii, Y. M., & Akmar, A. S. N. (2014). Effect of drought stress on growth, proline and antioxidant enzyme activities of upland rice. *JAPS: Journal of Animal and Plant Sciences, 24*(5), 1487–1493.

Mahalakshmi, V., Bidinger, F. R., & Raju, D. S. (1987). Effect of timing of water deficit on pearl millet (*Pennisetum americanum*). *Field Crops Research, 15*, 327–339.

Maiti, R. (2013). Application of botany in abiotic and biotic stress resistances in crops: A synthesis. *International Journal of Bio-resource and Stress Management, 4.* https://ojs.pphouse.org/index.php/IJBSM/article/view/336

Maiti, R., & Wesche-Ebeling, P. (1997). *Pearl millet science.* Oxford & IBH Publishing.

Marschner, H. (1995). *Mineral nutrition of higher plants* (Vol. 2). Academic Press.

Maslova, T. G., & Popova, I. A. (1993). Adaptive properties of the plant pigment systems. *Photosynthetica, 29*, 195–203.

Mates, J. M. (2000). Effects of antioxidant enzymes in the molecular control of reactive oxygen species toxicology. *Toxicology*, *153*, 83–104.

McLaughlin, J. E., & Boyer, J. S. (2004). Sugar-responsive gene expression, invertase activity, and senescence in aborting maize ovaries at low water potentials. *Annals of Botany*, *94*, 675–689.

Meena, R. C., Ram, M., Ambawat, S., Tara Satyavathi, C., Khandelwal, V., Meena, R. B., Kumar, M., Bishnoi, J. P., & Mungra, K. D. (2021). Screening of pearl millet genotypes for high temperature and drought tolerance based on morpho-physiological characters. *International Journal of Environment and Climate Change*, *11*, 74–80.

Mir, R. R., Zaman-Allah, M., Sreenivasulu, N., Trethowan, R., & Varshney, R. K. (2012). Integrated genomics, physiology, and breeding approaches for improving drought tolerance in crops. *Theoretical and Applied Genetics*, *125*, 625–645.

Morgan, J. M. (1984). Osmoregulation and water stress in higher plants. *Annual Review of Plant Physiology*, *35*, 299–319.

Muller, B., Stosser, M., & Tardieu, F. (1998). Spatial distributions of tissue expansion and cell division rates are related to irradiance and to sugar content in the growing zone of maize roots. *Plant Cell & Environment*, *21*, 149–158.

Nambiar, V. S., Dhaduk, J. J., Sareen, N., Shahu, T., & Desai, R. (2011). Potential functional implications of pearl millet (*Pennisetum glaucum*) in health and disease. *Journal of Applied Pharmaceutical Science*, *1*, 62–67.

Narayanan, S. (2018). Effects of high temperature stress and traits associated with tolerance in wheat. *Open Access Journal of Science*, *2*, 177–186.

Nelson, G. C., Rosegrant, M. W., Koo, J., Robertson, R., Sulser, T., Zhu, T., Ringler, C., Msangi, S., Palazzo, A., Batka, M., Magalhaes, M., Valmonte-Santos, R., Ewing, M., & Lee, D. (2009) *Climate change: Impact on agriculture and costs of adaptation.* Food Policy Report. IFPRI.

Nxele, X., Klein, A., & Ndimba, B. K. (2017). Drought and salinity stress alters ROS accumulation, water retention, and osmolyte content in sorghum plants. *South African Journal of Botany*, *108*, 261–266.

Ong, C. K., & Monteith, J. L. (1985). Response of pearl millet to light and temperature. *Field Crops Research*, *11*, 141–160.

Opole, R. A., Prasad, P. V. V., Djanaguiraman, M., Vimala, K., Kirkham, M. B., & Upadhyaya, H. D. (2018). Thresholds, sensitive stages and genetic variability of finger millet to high temperature stress. *Journal of Agronomy and Crop Science*, *204*, 477–492.

Oumar, I., Mariac, C., Pham, J.-L., & Vigouroux, Y. (2008). Phylogeny and origin of pearl millet (*Pennisetum glaucum* [L.] R. Br) as revealed by microsatellite loci. *Theoretical and Applied Genetics*, *117*, 489–497.

Pahlavanian, A. M., & Silk, W. K. (1988). Effect of temperature on spatial and temporal aspects of growth in the primary maize root. *Plant Physiology*, *87*, 529–532.

Passot, S., Gnacko, F., Moukouanga, D., Lucas, M., Guyomarc, H. S., Ortega, B. M., Atkinson, J. A., Belko, M. N., Bennett, M. J., Gantet, P., Wells, D. M., Guédon, Y., Vigouroux, Y., Verdeil, J.-L., Muller, B., & Laplaze, L. (2016). Characterization of pearl millet root architecture and anatomy reveals three types of lateral roots. *Frontiers in Plant Science, 7*, 829.

Patrick, J. W., & Stoddard, F. L. (2010). Physiology of flowering and grain filling in faba bean. *Field Crops Research, 115*, 234–242.

Prasad, P. V. V., Bheemanahalli, R., & Jagadish, S. V. K. (2017). Field crops and the fear of heat stress – Opportunities, challenges, and future directions. *Field Crops Research, 200*, 114–121.

Prasad, P. V. V., Boote, K. J., Allen, L. H., Jr., Sheehy, J. E., & Thomas, J. M. G. (2006). Species, ecotype and cultivar differences in spikelet fertility and harvest index of rice in response to high temperature stress. *Field Crops Research, 95*, 398–411.

Prasad, P. V. V., Staggenborg, S. A., & Ristic, Z. (2008). Impacts of drought and/or heat stress on physiological, developmental, growth, and yield processes of crop plants. In *Advances in Agricultural Systems Modeling* (Vol. *1*, pp. 301–355).

Pritchard, J., Tomos, A. D., & Wyn Jones, R. G. (1987). Control of wheat root elongation growth. *Journal of Experimental Botany, 38*, 948–959.

Priya, M., Dhanker, O. P., Siddique, K. H. M., Hanumanta Rao, B., Nair, R. M., Pandy, S., Singh, S., Varshney, R. K., Prasad, P. V. V., & Nayyar, H. (2019). Drought and heat stress-related proteins: An update about their functional relevance in imparting stress tolerance in agricultural crops. *Theoretical and Applied Genetics, 132*, 1607–1638.

Pucher, A., Sy, O., Angarawai, I. I., Gondah, J., Zangre, R., Ouedraogo, M., Sanogo, M. D., Boureima, S., Hash, C. T., & Haussmann, B. I. G. (2015). Agro-morphological characterization of west and central African pearl millet accessions. *Crop Science, 55*, 737–748.

Rai, K. N., & Yadav, O. P. (2013). Genetic improvement of pearl millet in India. *Agricultural Research, 2*, 275–292.

Reddy, C. S., Kim, K. M., James, D., Varakumar, P., & Reddy, M. K. (2017). PgPAP18, A heat-inducible novel purple acid phosphatase 18-like gene (PgPAP18-like) from *Pennisetum glaucum*, may play a crucial role in environmental stress adaptation. *Acta Physiologia Plantarum, 39*, 54.

Rivas, D. L. J., & Barber, J. (1997). Structure and thermal stability of photosystem II reaction centers studied by infrared spectroscopy. *Biochemistry, 36*, 8897–8903.

Ruan, Y. L., Jin, Y., Yang, Y. J., Li, G. J., & Boyer, J. S. (2010). Sugar input, metabolism, and signaling mediated by invertase: Roles in development, yield potential, and response to drought and heat. *Molecular Plant, 3*, 942–955.

Sage, T. L., Bagha, S., Lundsgaard-Nielsen, V., Branch, H. A., Sultmanis, S., & Sage, R. F. (2015). The effect of high temperature stress on male and female reproduction in plants. *Field Crops Research, 182*, 30–42.

Saleem, S., Mushtaq, N. U., Shah, W. H. A., Hakeem, K. R., & Rehman, R. U. (2021). Morpho-physiological, biochemical, and molecular adaptation of millets to abiotic stresses: A review. *International Journal of Experimental Botany*, *90*, 1363–1385.

Sankar, M. S., Tara Satyavathi, C., Barthakur, S., Singh, S. P., Bharadwaj, C., & Soumya, S. L. (2021). Differential modulation of heat-inducible genes across diverse genotypes and molecular cloning of a sHSP from pearl millet [*Pennisetum glaucum* (L.) R. Br.]. *Frontiers in Plant Science*, *12*, 659893.

Schmid, V. H. R. (2008). Light-harvesting complexes of vascular plants. *Cellular and Molecular Life Sciences*, *65*, 3619–3639.

Sehgal, D., Rajaram, V., Armstead, I. P., Vadez, V., Yadav, Y. P., Hash, C. T., & Yadav, R. S. (2012). Integration of gene-based markers in a pearl millet genetic map for identification of candidate genes underlying drought tolerance quantitative trait loci. *BMC Plant Biology*, *12*, 9.

Serba, D. D., & Yadav, R. S. (2016). Genomic tools in pearl millet breeding for drought tolerance: Status and prospects. *Frontiers in Plant Science*, *7*, 1–10.

Shanker, A. K., Amirineni, S., Bhanu, D., Yadav, S. K., Jyothilakshmi, N., Vanaja, M., Singh, J., Sarkar, B., Maheswari, M., & Singh, V. K. (2022). High-resolution dissection of photosystem II electron transport reveals differential response to water deficit and heat stress in isolation and combination in pearl millet [*Pennisetum glaucum* (L.) R. Br.]. *Frontiers in Plant Science*, *13*, 892676.

Shivhare, R., Asif, M. H., & Lata, C. (2020). Comparative transcriptome analysis reveals the genes and pathways involved in terminal drought tolerance in pearl millet. *Plant Molecular Biology*, *103*, 639–652.

Shivhare, R., & Lata, C. (2017). Exploration of genetic and genomic resources for abiotic and biotic stress tolerance in pearl millet. *Frontiers in Plant Science*, *7*, 2069.

Shrestha, N., Hu, H., Shrestha, K., & Doust, A. N. (2023). Pearl millet response to drought: A review. *Frontiers in Plant Science*, *14*, 1059574.

Siddiqui, M. N., Leon, J., Naz, A. A., & Ballvora, A. (2021). Genetics and genomics of root system variation in adaptation to drought stress in cereal crops. *Journal of Experimental Botany*, *72*, 1007–1019.

Sivakumar, M. V., & Salaam, S. A. (1994). A wet excavation method for root/shoot studies of pearl millet on the sandy soils of the Sahel. *Experimental Agriculture*, *30*, 329–336.

Soegaard, H., & Boegh, E. (1995). Estimation of evapotranspiration from a millet crop in the Sahel combining sap flow, leaf area index and eddy correlation technique. *Journal of Hydrology*, *166*, 265–282.

Spollen, W. G., & Sharp, R. E. (1991). Spatial distribution of turgor and root growth at low water potentials. *Plant Physiology*, *96*, 438–443.

Srivastava, K., Preethi, B., Niranjan Kumar, M., & Rajeevan. (2012). On the observed variability of monsoon droughts over India. *Weather and Climate Extremes*, *1*, 42–50.

Srivastava, R. K., Yadav, O. P., Kaliamoorthy, S., Gupta, S. K., Serba, D. D., Choudhary, S., & Varshney, R. K. (2022). Breeding drought-tolerant pearl millet using conventional and genomic approaches: Achievements and prospects. *Frontiers in Plant Science, 13,* 781524.

Steinkellner, S., Lendzemo, V., Langer, I., Schweiger, P., Khaosaad, T., Toussaint, J. P., & Vierheilig, H. (2007). Flavonoids and strigolactones in root exudates as signals in symbiotic and pathogenic plant-fungus interactions. *Molecules, 12,* 1290–1306.

Sun, M., Huang, D., Zhang, A., Khan, I., Yan, H., Wang, X., & Huang, L. (2020). Transcriptome analysis of heat stress and drought stress in pearl millet based on Pacbio full-length transcriptome sequencing. *BMC Plant Biology, 20,* 1–15.

Sunoj, J. V. S., Somayananda, I. M., Chiluwal, A., Perumal, R., Prasad, P. V. V., & Jagadish, S. V. K. (2017). Resilience of pollen and post-flowering response in diverse sorghum genotypes exposed to heat stress under field conditions. *Crop Science, 57,* 1658–1669.

Taiz, L., & Zeiger, E. (2006). *Plant physiology* (4th ed.). Sinauer Associates.

Telfer, A. (2005). Too much light ? How beta-carotene protects the photosystem II reaction centre. *Photochemical & Photobiological Sciences, 4,* 950–956.

Teowalde, H., Osman, M., Voigt, R., & Dobrenz, A. (1986). Stomata distribution of three pearl millet genotypes. In *Forage and grain: A College of Agriculture Report* (pp. 131). University of Arizona.

Tognetti, V. B., Bielach, A., & Hrtyan, M. (2017). Redox regulation at the site of primary growth: Auxin, cytokinin and ROS crosstalk. *Plant, Cell & Environment, 40,* 2586–2605.

Toker, C., Canci, H., & Yildirim, T. (2007). Evaluation of perennial wild *Cicer* species for drought resistance. *Genetic Resources and Crop Evolution, 54,* 1781–1786.

Tripathy, K. P., Mukherjee, S., Mishra, A. K., Mann, M. E., & Williams, A. (2023). Climate change will accelerate the high-end risk of compound drought and heatwave events. *Proceedings of the National Academy of Sciences, 120,* e2219825120.

Vadez, V., Hash, T., & Kholova, J. (2012). Phenotyping pearl millet for adaptation to drought. *Frontiers in Physiology, 3,* 386.

van Oosterom, E. J., Weltzien, E., Yadav, O. P., & Bidinger, F. R. (2006). Grain yield components of pearl millet under optimum conditions can be used to identify germplasm with adaptation to arid zones. *Field Crops Research, 96,* 407–421.

Varshney, R. K., Shi, C., Thudi, M., Mariac, C., Wallace, J., Qi, P., & Xu, X. (2017). Pearl millet genome sequence provides a resource to improve agronomic traits in arid environments. *Nature Biotechnology, 35,* 969–976.

Xiong, L., & Zhu, J. K. (2002). Molecular and genetic aspects of plant responses to osmotic stress. *Plant, Cell & Environment, 25*(2), 131–139.

Xu, Z., & Zhou, G. (2008). Responses of leaf stomatal density to water status and its relationship with photosynthesis in the grass. *Journal of Experimental Botany, 59,* 3317–3325.

Yadav, R., Hash, C., Bidinger, F., Devos, K., & Howarth, C. (2004). Genomic regions associated with grain yield and aspects of post-flowering drought tolerance in pearl millet across stress environments and tester background. *Euphytica, 136*, 265–277.

Yadav, R. S., Hash, C. T., Bidinger, F. R., Cavan, G. P., & Howarth, C. J. (2002). Quantitative trait loci associated with traits determining grain and stover yield in pearl millet under terminal drought-stress conditions. *Theoretical and Applied Genetics, 104*, 67–83.

Yadav, T., Kumar, A., Yadav, R. K., Yadav, G., Kumar, R., & Kushwaha, M. (2020). Salicylic acid and thiourea mitigate the salinity and drought stress on physiological traits governing yield in pearl millet-wheat. *Saudi Journal of Biological Sciences, 27*, 2010–2017.

Yan, H., Sun, M., Zhang, Z., Jin, Y., Zhang, A., Lin, C., Wu, B., He, M., Xu, B., Wang, J., Qin, P., Mendieta, J. P., Nie, G., Wang, J., Jones, C. S., Feng, G., Srivastava, R. K., Zhang, X., Bombarely, A., . . . Huang, L. (2023). Pangenomic analysis identifies structural variation associated with heat tolerance in pearl millet. *Nature Genetics, 55*(3), 507–518.

Yang, Z., Sinclair, T. R., Zhu, M., Messina, C. D., Cooper, M., & Hammer, G. L. (2012). Temperature effect on transpiration response of maize plants to vapour pressure deficit. *Environmental and Experimental Botany, 78*, 157–162.

Yu, X., Pasternak, T., Eiblmeier, M., Ditengou, F., Kochersperger, P., Sun, J., & Palme, K. (2013). Plastid-localized glutathione reductase–regulated glutathione redox status is essential for *Arabidopsis* root apical meristem maintenance. *The Plant Cell, 25*, 4451–4468.

Zegada-Lizarazu, W., & Iijima, M. (2005). Deep root water uptake ability and water use efficiency of pearl millet in comparison to other millet species. *Plant Production Science, 8*, 454–460.

Zegada-Lizarazu, W., Kanyomeka, L., Izumi, Y., & Iijima, M. (2006). Pearl millet developed deep roots and changed water sources by competition with intercropped cowpea in the semiarid environment of northern Namibia. *Plant Production Science, 9*, 355–363.

Zhang, A., Ji, Y., Sun, M., Lin, C., Zhou, P., Ren, J., Luo, D., Wang, X., Ma, C., Zhang, X., Feng, G., Nie, G., & Huang, L. (2021). Research on the drought tolerance mechanism of *Pennisetum glaucum* (L.) in the root during the seedling stage. *BMC Genomics, 22*, 568.

Zhu, J. K. (2002). Salt and drought stress signal transduction in plants. *Annual Review of Plant Biology, 53*, 247–273.

10

Weed Management in Pearl Millet: Challenges and Opportunities

Vipan Kumar, Midhat Zulafkar Tugoo, Prashant Jha, Antonio DiTommaso, and Kassim Al-Khatib

Chapter Overview

Pearl millet [*Pennisetum glaucum* (L.) R. Br.], commonly known as "bajra," is an important coarse cereal crop grown for food, feed, and nutritional security around the world (Mishra, 2015). A member of the "Poaceae" family, it is the sixth most important cereal crop globally after rice, wheat, maize, barley, and sorghum (Tara Satyavathi et al., 2021). Pearl millet is a C4 plant that cycles CO_2 into four-C sugar compounds to enter into the Calvin cycle C fixation pathway. Pearl millet is native to Africa and is very efficient in hot and dry climates; thus, it can be cultivated in areas that are often too hot and dry for other crops (Singh et al., 2017). Pearl millet is mostly grown in arid and semi-arid regions of the United States, Puerto Rico, Africa, India, and other Asian countries as a high-quality grain or silage crop (Pudelko et al., 1993; Sheahan, 2014). This crop has a higher protein content (12%–14%) and quality than sorghum (Andrews et al., 1993). Pearl millet grains are rich in oil, fiber, essential vitamins, and minerals. Its deep root system is capable of scavenging residual soil nutrients (Hannaway & Larson, 2004). Pearl millet is an energy-efficient crop and is a top choice crop for low-input sustainable agricultural production systems (Sheahan, 2014) (Figure 10.1).

Globally, pearl millet is grown on over 31 million ha, mainly in Africa and Asia (ICRISAT, 2021) and in some parts of the United States, where it is grown as a forage crop. In 2022, total millet (including all types) production was estimated around 30,923,000 t worldwide, with the greatest production recorded in India (contributing 39% of the total global production) (USDA, 2023) (Figure 10.1). India is the leading producer of pearl millet, followed by Niger, Burkina Faso, Mali, and Pakistan (FAO, 2019). These five countries contribute about 75% of global pearl millet production. Other significant pearl millet–producing countries include Nigeria, Sudan, Ethiopia,

Pearl Millet: A Resilient Cereal Crop for Food, Nutrition, and Climate Security, First Edition. Edited by Ramasamy Perumal, P. V. Vara Prasad, C. Tara Satyavathi, Mahalingam Govindaraj, and Abdou Tenkouano.

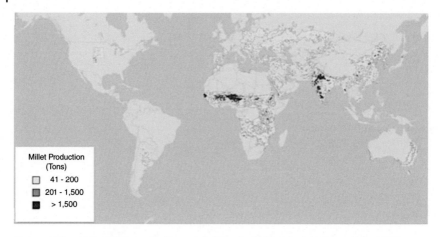

Figure 10.1   Global status of millet production in 2022. Adapted from USDA (2023).

Senegal, and Chad (USDA, 2023) (Figure 10.1). Total millet production in the United States was estimated around 349 t in 2021, with major production in Colorado, Nebraska, and South Dakota (USDA, 2023) (Figure 10.1). In addition to grain, pearl millet is a preferred choice for forage compared with other millets and used for grazing, silage, hay, and green chop (Newman et al., 2010; Sheahan, 2014).

Weeds pose a serious production challenge to pearl millet producers globally. Due to early slow growth (first few weeks), millets are relatively poor competitors against weeds (Mishra et al., 2016). Major grass and broadleaf weed species commonly found across the pearl millet–producing regions are summarized in Tables 10.1 and 10.2. Despite weeds being a major challenge for successful production of pearl millet, very few herbicide options are available for weed control in this crop. Grass weed control is especially challenging due to a lack of effective pre-emergence (PRE) or post-emergence (POST) herbicides labeled in pearl millet. Producers mainly rely on non-chemical weed control techniques in pearl millet. More integrated weed management (IWM) tools and technologies are needed to provide improved weed management options in pearl millet that can help to further expand the growing regions of this crop, especially under changing climatic conditions (i.e., high temperature or low and infrequent rainfall) across arid and semi-arid regions. This chapter provides an overview of major weed challenges in global pearl millet production, including the critical period of weed control (CPWC), grain yield losses by various weed species, and limited herbicide options and occurrence of herbicide-resistant (HR) weeds. We also highlight knowledge gaps and future opportunities in weed management, including development of herbicide-tolerant crop traits; the role of cultural, mechanical, and agronomic practices on weed populations and soil seed-bank dynamics; and precision weed control technologies for integration in pearl millet production.

Table 10.1   Major Grass Weed Species Commonly Found Across The Pearl Millet–Producing Regions Worldwide

Species	Common name
Brachiaria ramosa (L.) Stapf	Browntop millet
Cenchrus longispinus (Hack.) Fernald	Longspine sandbur
Chloris barbata Sw.	Peacock plumegrass
Cynodon dactylon Pers.	Bermuda grass
Cyperus rotundus L.	Purple nutsedge
Dactyloctenium aegyptium L.	Crowfoot grass
Digitaria insularis L.	Sourgrass
Digitaria sanguinalis (L.) Scop.	Large crabgrass
Dinebra retroflexa (Vahl) Panz.	Viper grass
Echinochloa crus-galli (L.) P. Beauv.	Barnyardgrass
Echinochloa colona (L.) Link	Junglerice
Eleusine indica (L.) Gaertn.	Goosegrass
Panicum repens L.	Torpedograss
Setaria faberi Herrm.	Giant foxtail
Setaria pumila (Poir.) Roem. & Schult.	Yellow foxtail
Setaria viridis (L.) P. Beauv.	Green foxtail
Sorghum bicolor (L.) Moench ssp. verticilliflorum (Steud.) de Wet ex Wiersema & J. Dahlb.	Shattercane
Sorghum halepense (L.) Pers.	Johnsongrass

Note: Adapted from Mishra, 2016.

Weed Management Challenges in Pearl Millet

Critical Period of Weed Control in Pearl Millet

The critical period of weed control (CPWC) refers to a time period in the crop life cycle when a weed-free environment is critically important to prevent economic yield losses (Zimdahl, 1988). Knowing the CPWC is crucial in making in-crop weed control decisions to limit crop–weed competition. Previous researchers have defined the CPWC in different ways. Zimdahl (1988) defined the CPWC as a "span of time between that period after seeding or emergence when weed competition does not reduce crop yield and the time after which weed competition will no longer reduce crop yield." Swanton and Wiese (1991) defined the CPWC as "the time interval when it is essential to maintain a weed-free environment to prevent

Table 10.2   Major Broadleaf Weed Species Commonly Found Across The Pearl Millet–Producing Regions Worldwide

Species	Common name
Acanthospermum hispidum DC.	Bristly starbur
Achyranthes japonica (Miq.) Nakai	Japanese chaff flower
Ageratum conyzoides L.	Billygoat weed
Amaranthus palmeri S. Watson	Palmer amaranth
Amaranthus retroflexus L.	Redroot pigweed
Amaranthus tuberculatus (Moq.) J. D. Sauer	Common waterhemp
Bassia scoparia (L.) A. J. Scott	Kochia
Celosia argentea L.	Silver cock's comb
Cleome hassleriana L.	Spider flower
Commelina benghalensis L.	Benghal dayflower
Convolvulus arvensis L.	Field bindweed
Corchorus acutangulus Lamk.	Jew's mallow
Cyperus rotundus L.	Purple nutsedge
Digera muricata (L.) Mart.	False amaranth
Eclipta alba Hassk.	False daisy
Erigeron canadensis L.	Horseweed
Euphorbia hirta L.	Pillpod spurge
Heracleum mantegazzianum Sommier & Levier	Giant hogweed
Ipomoea hederacea Jacq.	Ivyleaf morningglory
Phyllanthus niruri	Gale of the wind
Portulaca oleracea L.	Common purslane
Salsola tragus L.	Russian-thistle
Striga asiatica (L.) Kuntze	Witchweed
Trianthema portulacastrum L.	Carpetweed
Tribulus terrestris L.	Puncturevine
Tridax procumbens L.	Coat buttons
Xanthium strumarium L.	Common cocklebur

Note: Adapted from Mishra, 2016.

crop yield loss." This definition involves two separate components: (a) the critical duration of weed interference (a time period during which weeds can coexist with the crop and not affect yield) and (b) a critical weed-free period (a time period from planting to when later-emerging weeds will not affect crop yield) (Swanton

& Wiese, 1991). More recently, Knezevic et al. (2002) defined CPWC as a "window in the crop growth cycle during which weeds must be controlled to prevent unacceptable yield losses."

Millets are generally susceptible to weed competition early in their life cycle, and early-season weed control is crucial to prevent economic grain yield losses (Mishra & Dash, 2014). Due to slight differences in early growth patterns, the CPWC ranges from 15 to 30 days for pearl millets and from 25 to 42 days for finger millets (Labrada et al., 1994; Sundraseh et al., 1975). Weed species composition, density, emergence patterns, and periodicity and size of weed seedbanks are important factors that can influence the length and timing of CPWC in pearl millet (Knezevic et al., 2002; Martin et al., 2001). In addition, environmental conditions (e.g., temperature, soil moisture) and cultural practices (e.g., time of crop planting, crop row spacing, soil nutrient management, etc.) can influence crop–weed competitive relationships (Di Tomaso, 1995; Lindquist et al., 1999; Little et al., 2021; Martin et al., 2001), which may result in CPWC variations in pearl millet. The influence of environmental and cultural practices on CPWC in pearl millet warrants further investigation. After millets reach canopy closure (about 50 cm in height), weed control generally does not affect grain yields. However, it is equally important to manage late-emerging weed cohorts to prevent additions to the soil seedbank for infestations in subsequent years.

Pearl Millet Yield Losses due to Weeds

Weeds compete with pearl millet for limited soil moisture, nutrients, sunlight, CO_2, and space, resulting in grain yield and quality losses and increased production costs. Allelochemicals released by certain weeds such as field bindweed have also shown inhibitory effects on germination and growth of pearl millet (Fateh et al., 2012). Pearl millet is generally grown on marginal low-fertility and water-limited soils in arid and semi-arid regions (Sheahan, 2014). Rapid emergence and growth of weeds, especially during the early growth stages of pearl millet, deplete available soil nutrients and water needed by the crop (Mishra et al., 2016). In general, actively growing broadleaf weeds are known to capture more soil nutrients and soil water than a cash crop such as sorghum (Stahlman & Wicks, 2000). This can adversely affect the yields of pearl millet (Mishra & Dash, 2014). The degree of grain yield losses depends on pearl millet hybrids/cultivars, time of planting, row spacing, the nature and duration of weed infestations, environmental conditions, and production practices. Depending upon growing conditions, grain yield losses due to weed interference can range from 5% to 94% in pearl millet (Balyan et al., 1993; Banga et al., 2000; Rao et al., 2005; Sharma & Jain, 2003). For instance, grain yield reductions of up to 40% have been reported from season-long weed interference (Girase et al., 2017; Sharma & Jain, 2003). Samota et al. (2022a) documented a 47% yield reduction of pearl millet due to season-long competition from grass and broadleaf weeds in a field study conducted in Rajasthan, India. Furthermore, Kumar et al. (2022a, 2022b) reported a 43% grain yield reduction in pearl millet infested by a variety of weed species, including bermudagrass, signal grass, false amaranth, puncture vine, slender amaranth, congress grass, Indian sandbur, wild jute, nutsedge, hairy spurge, purple nutsedge, and gale of the wind. Banga et al. (2000) reported a 55% reduction in pearl millet grain yields with a season-long infestation of horse purslane and Junglerice. Balyan et al. (1993) found 16%–94% grain yield reductions in pearl millet when competing with carpetweed and barnyard grass. Season-long competition from witchweed alone can result in more than a 70% grain yield reduction in pearl millet crops grown in India and sub-Saharan Africa (Emechebe et al., 2004; Rao, 1987).

Herbicide-resistant Weeds

Evolution of HR weed populations pose additional challenges to weed control efforts in some pearl millet growing regions. Despite limited herbicide options labeled for use in pearl millet, the widespread evolution of HR weeds is evident in the semiarid US Great Plains. For instance, increasing cases of HR weed species, including kochia, horseweed, Palmer amaranth, Russian-thistle, green foxtail, shattercane, Johnsongrass, and barnyard grass, have been reported in the region (Heap, 2023; Kumar et al., 2023). This shift is mainly due to a repeated use of herbicides with the same site-of-action in rotational crops or fallow periods. Lack of effective herbicide options and prevalent no–till–based production systems complicate the management of HR weeds in millets across the US Great Plains region (Lyon et al., 2007). These reported cases of HR weeds further highlight the need for developing a holistic IWM approach for pearl millet production.

Weed Management Opportunities in Pearl Millet

Herbicide Options in Pearl Millet

Compared with several other row crops, few herbicides are labeled for use in pearl millet. Therefore, growers should carefully consider their weed management options before planting pearl millet. Several studies reported superior weed control in pearl millet when herbicides were integrated with other methods of weed control. Based on the timing of application, herbicides are generally categorized as PRE (applied after crop planting but prior to weed emergence) and POST (applied after crop and weed emergence). Application of a PRE herbicide followed by another timely application of a POST herbicide can provide effective control of broadleaf weeds in pearl millet (Mishra & Dash, 2014). However, no effective POST herbicide option is currently available for selective control of grass weed species in pearl millet.

Preemergence Herbicides

PRE herbicides are mainly applied to the soil surface, which results in a chemical barrier that prevents the emergence or weed seedlings. Several studies have evaluated PRE herbicides for weed control in pearl millet. Atrazine (at various rates) is the most used PRE herbicide for weed control in millets (Mishra, 2015). Dower et al. (1993) reported 88% grass control and 98% broadleaf control in pearl millet, respectively, with a single PRE application of atrazine at 2240 g active ingredient (ai)/ha. Nibhoria et al. (2021) reported 72% weed control using a PRE application of atrazine at 500 g ai/ha. A combination of PRE applied atrazine with manual hand weeding has also been evaluated for weed control in pearl millet. Das et al. (2013) reported that PRE applied atrazine at 1000 g ai/ha followed by one hand weeding at 30 days after pearl millet planting significantly reduced total weed density and total weed dry weight. Similarly, Girase et al. (2017) found that a PRE application of atrazine at 500 g ai/ha plus one hand weeding at 35 days after pearl millet planting resulted in the lowest weed index. Excellent control (>90%) of carpetweed and barnyard grass was reported with PRE applied atrazine (500 g ai/ha) or acetochlor + atrazine (750 + 375 g ai/ha) applied at 10 days after sowing of pearl millet (Banga et al., 2000). Furthermore, Choudhary et al. (2022) observed the lowest total weed density, weed dry weight, and percent visual control throughout the season with PRE applied atrazine at 750 g ai/ha plus one hand weeding at 21 days after pearl millet planting. Although atrazine has provided effective weed control in several crops, including millets, the likelihood of atrazine being labeled for use in pearl millets in the United States is very slim due to environmental concerns. Atrazine was originally labeled as a proso millet in the United States in the 1980s, but proso millet was removed from the atrazine label in the 1990s during re-registration. The USEPA is currently trying to adopt a new aquatic ecosystem concentration equivalent level of concern for atrazine at 3.4 parts per billion (ppb) down from 15 ppb, further indicating the slim chance of its registration for use in pearl millets in the United States. A PRE application of pendimethalin at 750–1000 g ai/ha effectively controlled a range of grass weeds in pearl millet (Ram et al., 2005). Dowler et al. (1995) found that a combination of PRE applied propachlor plus pendimethalin improved grass weed control in pearl millet compared with PRE applied atrazine alone. In that same study, pendimethalin-treated plots resulted in poor germination of pearl millet due to phytotoxic effects (Dowler et al., 1995). Asbil (2001) documented that PRE applied pendimethalin at 1680 and 3360 g ai/ha provided effective control of annual grasses with minor visual injury (5% or less) on pearl millet. Ram et al. (2005) documented effective control of grass and broadleaf weeds with a PRE applied oxadiazon at 1000 g ai/ha plus one hand-weeding at 45 days after pearl millet planting. Dhayal et al. (2020) found a 56% lower weed infestation with PRE applied oxyfluorfen at 200 g ai/ha and one hand-weeding at 20 days after pearl millet planting. In the same study, PRE applied alachlor at 1000 g ai/ha plus one hand-weeding at 60 days after pearl millet planting reduced total weed infestation by 68% (Dhayal et al., 2020). Similarly, Ramesh et al. (2019) found that PRE applied pretilachlor at 450 g ai/ha significantly reduced the total weed density in pearl millet. PRE applied saflufenacil at 50 g ai/ha provided broad-spectrum weed control in pearl millet (Reddy et al., 2014).

Postemergence Herbicides

POST or foliar herbicides are applied after the emergence of crop and weeds. Like PRE herbicides, POST options also are limited in pearl millet. Samota et al. (2022b) reported that POST applications of atrazine (100–300 g ai/ha) provided effective weed control and shoot dry weight reductions (1145–1273 g/m^2) and resulted in the highest grain yields (2150–2180 kg/ha) of pearl millet. Similarly, Girase et al. (2017) found that POST applied atrazine at 400 g ai/ha reduced total weed density by 20% in pearl millets when compared with the weedy check at 30 days after treatment. Dowler et al. (1995) also observed that an early POST application of atrazine provided the most effective control of broad-spectrum weeds in pearl millet. Singh et al. (2010) reported that POST (4–6 weeks after planting) application of 2,4-D at 500–750 g ai/ha controlled broad-leaved weeds in pearl millet. In the same study, pearl millet plants treated with POST applied 2,4-D at 500 g ai/ha had a maximum chlorophyll content (3.75 mg/g) and resulted in maximum grain and straw yield (Singh et al., 2010). Furthermore, POST applications of pendimethalin (840 and 1680 g ai/ha) combined with dicamba (140 and 280 g ai/ha) or atrazine (750 g ai/ha) were found to be safer on pearl millet and resulted in similar grain yields as a hand-weeded control (Cuerrier et al., 2010). A POST application of tembotrione at 80–100 g ai/ha resulted in significantly lower weed density, lower total weed dry weights, and higher grain yield of pearl millets (Choudhary et al., 2022). In contrast, a POST application of imazethapyr at 40 g ai/ha showed phytotoxic injury symptoms on pearl millet with reduced grain yields (Singh et al., 2016).

While selecting PRE or POST herbicide programs, producers should carefully consider the subsequent crop(s) to be planted after pearl millet. This is important because some herbicides, such as atrazine or atrazine-containing premixtures or tank mixtures, may require a long plant-back interval or rotational restriction based on their rates and timing. Producers should consult herbicide labels before use when planning crop rotations with pearl millets. Further research on additional herbicides for crop safety and weed control in pearl millets is warranted.

Herbicide-resistant Pearl Millet

The discovery and widespread adoption of genetically engineered (GE) crops resistant to glyphosate (e.g., corn [*Zea mays* L.], soybean [*Glycine max* (L.) Merr.], cotton [*Gossypium hirsutum* L.], and canola [*Bromus napus* L.]) have revolutionized agricultural production and weed management and have facilitated no-tillage–based production systems to conserve soil moisture and reduce soil erosion (Bagavathiannan & Davis, 2018). Since the discovery of glyphosate-resistant crops, several other GE crops resistant to other herbicides, including glufosinate, 2,4-D, dicamba, and 4-hydroxyphenylpyruvate dioxygenase (HPPD) inhibitors, have been developed and commercialized. The primary reason for widespread cultivation of these GE crops (corn, soybean, cotton, canola) is the availability of herbicide options that facilitate weed control. Development of pearl millet hybrids resistant to herbicides will help in providing season-long weed control options, thereby reducing crop yield losses and increasing productivity. New HR trait technologies in combination with drought- and heat-tolerant traits can facilitate rapid expansion of pearl millet production in arid or semi-arid regions amid a changing climate. Availability of HR pearl millet cultivars will help in improving the competitiveness of the pearl millet industry. Research is underway at Kansas State University Agricultural Research Center (KSU-ARC) near Hays, KS, in identifying and characterizing ALS-resistant, ACCase-resistant, and HPPD inhibitor–resistant traits from wide collection of pearl millet genotypes for subsequent introgression into elite pearl millet hybrids (V. Kumar et al., unpublished data). In addition, research is in progress at KSU-ARC exploring tolerance of various pearl millet hybrids to commonly used PRE and POST herbicides in grain sorghum (V. Kumar et al., unpublished data). Development and availability of these herbicide resistance traits and expansion of herbicide tools in pearl millet will offer broad-spectrum weed control options. Furthermore, the use of advanced CRISPR/Cas9 systems for gene editing may play a critical role in developing novel HR pearl millet lines. However, as mentioned previously, evolution of herbicide resistance has been reported in various weed species in pearl millet growing regions globally. Therefore, proper stewardship guidelines and the use of diversified multi-tactic strategies would be needed for the success of HR pearl millet technologies.

Nonchemical Options in Pearl Millet

Due to limited herbicide options available for use in pearl millets, growers can rely on various nonchemical weed control methods. Nonchemical weed control strategies include cultural practices (e.g., competitive crop rotations, competitive hybrids, optimum seeding rates, proper row spacing, proper fertilization, intercropping, and cover crops, etc.), mechanical practices (stale seedbed preparation, interrow cultivation, interrow mowing, electrocution, harvest weed seed control [HWSC], etc.), and precision agricultural tools (site-specific weed detection and management). In the following sections, we review the role of various non-chemical weed management strategies in pearl millet.

Cultural Practices

Competitive Crop Rotations, Hybrids, and Improved Agronomic Practices

The selection of competitive crops in rotation or crop sequence is an important cultural practice for disrupting weed life cycles and reducing their abundance (Barberi & Lo Cascio, 2001). Competitive rotational crop(s) with vigorous growth characteristics offer a competitive advantage over weeds and allow the use of different herbicide options and more diversified weed control tactics. For example, crop rotation can play an important role in managing green or giant foxtail in pearl millet–based production systems in the semiarid Central Great Plains of the United States, wherein the seedbank of these foxtail species can be depleted in rotational crops such as maize, soybean, or cotton using glyphosate or other effective herbicides. Rotation of broadleaf crops (e.g., canola, cotton, soybean, sunflower, etc.) with pearl millet can allow the use of graminicides for grass weed control. Because pearl millet is a warm-season crop, rotation with winter or cool-season crops (including winter wheat, triticale, winter peas, etc.) would also help with the management of the warm-season grass weeds that are the most problematic in pearl millet.

Selection and development of crop varieties/hybrids with weed-suppressive traits can also provide a competitive advantage over weeds (Vander Meulen & Chauhan, 2017). These crop varieties/hybrids can suppress weeds because of extended crop competition and/or release of allelochemicals. For instance, selection of pearl millet hybrids that emerge quickly and develop aggressive early growth/biomass can provide early-season competitive advantage over weeds (when most of the competition with weeds occurs). Similarly, sorghum (closely related to pearl millet) produces sorgoleone, an important allelochemical, which inhibits germination and growth of some weed species (Dayan & Duke, 2009; Dayan et al., 2003). Sorghum hybrids and environmental conditions have been found to influence the production of sorgoleone (Hess et al., 1992; Weston & Czarnota, 2001). Therefore, identification and development of weed-suppressive pearl millet hybrids can be a valuable cultural strategy and warrants further research.

Improved agronomic practices, such as optimum planting time, higher seeding rates, narrow row spacing, and early nutrient supply, can enhance crop canopy formation and improve the smothering effect on weeds, especially early in the growing season. For example, Nelson (1977) and Agdag (2001) reported a significant reduction in weed competition and improved grain yields of proso millets when planted in narrow rows (<30 cm) compared with a nontreated weedy check. Further research is needed on optimizing these agronomic practices for weed control in pearl millet hybrids/varieties grown in various regions of the world.

Cover Crops and Intercropping

The use of cover crops has emerged as one of the most important ecologically based practices for weed suppression and soil health benefits in various cropping systems (Fernando & Shrestha, 2023; Kumar et al., 2020; Obour et al., 2022; Osipitan et al., 2018). Replacement of fallow periods with fall- or spring-planted cover crops has shown weed suppression benefits in subsequent cash crops, including grain sorghum (closely related to pearl millet) in the semiarid Central Great Plains of the United States (Dhanda et al., 2022; Obour et al., 2022). Cover crops (e.g., cereal rye, triticale, oats, etc.) can produce sufficient aboveground biomass to suppress weed emergence and growth in subsequent cash crops (Teasdale, 1996). Like grain sorghum, pearl millet can possibly be planted into the standing/rolled cover crop residues using a no-till or strip-till planter. Cover crop biomass can presumably provide effective weed suppression during the early stages of pearl millet establishment. However, growers should be cautioned that growing cover crops in arid and semi-arid environments can result in negative effects (e.g., reduction in stored soil water after cover crops) on subsequent crop yields (Nielsen et al., 2015). Similar to cover crops, short-duration intercrops, such as cowpea [*Vigna unguiculata* (L.) Walp.], mungbean (*Vigna radiata* L.), or peanut (*Arachis hypogaea* L.), can possibly be established between pearl millet rows, acting as a living mulch for weed suppression (Islam et al., 2018; Mishra et al., 2016; Osman et al., 2021; Zhu et al., 2023). However, the optimum termination timing of these short-duration intercrops would be needed to avoid competition with pearl millet. Currently, there is a lack of research on the use of cover crops and intercrops for weed suppression in pearl millet production.

Mechanical Practices

Tillage, Mowing, and Manual Weeding

Tillage has historically been a key method of weed control in various crops. Tillage operations prior to or at the time of seedbed preparation can provide early-season weed control in pearl millet. Similarly, occasional and/or strategic tillage operations during the fallow phase can be effective in controlling perennial tumble windmill grass and HR kochia in the no-tillage semi-arid Central Great Plains (pearl millet growing region) (Obour et al., 2019). However, there is a lack of information on the productivity of pearl millet in tilled versus no-tilled soil in arid and semi-arid environments and warrants further investigation. In addition, weeds that emerge between pearl millet rows can be controlled using cultivation or mowing and hand weeding. For instance, a one-pass interrow cultivation reduced early-season weed infestation by 45% in grain sorghum compared with the weedy check plots (Thate et al., 2017). Interestingly, manual weeding is the most commonly used method in pearl millet in India and other millet-growing regions (Mishra et al., 2016). However, an increasing labor shortage and time requirements make manual weeding inefficient and costly. In contrast, mechanical interrow cultivation or mowing operations can be more cost-effective for early-season weed control in pearl millet (Figure 10.2). Recent advancements in camera-guided interrow cultivators can increase the effectiveness of this method of weed control and prevent crop injury in pearl millet (Figure 10.3). Further research efforts are needed to implement interrow cultivation and interrow mowing for weed control in pearl millet.

Figure 10.2   Tractor-operated interrow mower for mowing weeds in row crops.
Cambridge University Press.

Figure 10.3 Camera-guided interrow cultivator for weed control in row crops. Dr. Matt Ryan, Cornell University.

Harvest Weed Seed Control

HWSC strategies involve collection and/or destruction of weed seeds during crop harvest, thus minimizing replenishment of the soil seedbank from late-season weed escapes (Walsh et al., 2013, 2017, 2018). Weed seeds are intercepted by a combine and separated from the bulk crop residue and grain for subsequent destruction/management. The success of HWSC relies on the ability of annual weed species to retain seeds until crop harvest (Schwartz-Lazaro et al., 2017). Weed seed destruction and chaff lining are two important HWSC strategies. The weed seed destruction method of HWSC is comprised of a high-impact, cage mill–based chaff processing unit in which weed seeds in the grain chaff fraction are captured and destroyed at crop harvest (Walsh et al., 2012). Seed destructor technology has been widely adopted in major grain crops of Australia (wheat, barley, and lupin) to drastically reduce (>95%) seed inputs of multiple HR ryegrass (*Lolium* spp.) and other weed species (Walsh et al., 2012). In contrast, chaff lining involves making a simple chute to divert the weed seed–bearing chaff fraction (from the sieves) into a narrow windrow (15–20 cm), which is left to rot or mulch while the straw is chopped and spread (Walsh et al., 2017, 2018) (Figure 10.4). Weeds emerging in the chaff lines need to be effectively managed to prevent seed additions in the subsequent growing seasons. However, this method is very effective in concentrating >99% weed seeds in narrow windrows and can drastically reduce the spread of weed seeds to 5% of the field (Bennet et al., 2023). Preliminary field studies in western Kansas have shown that the use of Redekop Seed Destructor (Figure 10.5) at the time of grain sorghum harvest can destroy (physical damage) >90% of Palmer amaranth seeds and simulated sorghum chaff lines can prevent 30%–40% germination of summer annual weeds in the subsequent

Figure 10.4 Modified version of chaff lining kit on John Deere combine. Dr. Vipan Kumar, Cornell University.

year (V. Kumar, unpublished data) (Figures 10.4 and 10.5). These novel HWSC strategies hold a high potential for weed seed bank management and should be explored in pearl millet and other crops grown in rotation with pearl millet.

Weed Electrocution

Weed electrocution was conceived in 1970 and was first tested to control bolting beets in a sugar beet crop (Diprose et al., 1980). An effective control of bolting beets required a voltage of 5 kV that can be achieved in 20 seconds, thereby avoiding excessively long treatment times (Diprose et al., 1980). The mobile generating unit progressed into a tractor-driven electrocution system over time (Diprose et al., 1980; Schreier et al., 2022). Recent field studies on the effectiveness of a tractor-driven electrocution system (Weed Zapper) (Figure 10.6) in soybean have demonstrated that electrocution is effective in controlling waterhemp, cocklebur, giant ragweed (*Ambrosia trifida* L.), common ragweed (*Ambrosia artemisiifolia* L.), and horseweed at the later growth stages (Schreier et al., 2022). All weed species containing seed at the time of electrocution had 54%–80% reductions in seed viability. In a separate field study, electrocution applied at the reproductive growth stage of soybean not only controlled late-season water hemp by 51%–97% but also reduced soybean grain yields by 11%–26%. These findings indicate that proper timing (when

Figure 10.5 John Deere combine equipped with Redekop Seed Destructor. Dr. Vipan Kumar, Cornell University.

Figure 10.6 Tractor-operated Weed Zapper for weed electrocution. Cambridge University Press.

weeds are taller than the crop) of electrocution can be an important component of an integrated weed-management strategy in pearl millet, potentially reducing the viable weed seed returns to the soil seedbank. Future research needs to explore the efficacy of electrocution in controlling problem weed species in pearl millet.

Precision Agricultural Technologies

Advanced precision agricultural technologies, including weed sensing and mapping using various sensor technologies mounted on aerial or ground-driven systems, herbicide applications using precision sprayers, as well as various autonomous robotics, have shown great potential for site-specific weed control in different crops (Gerhards et al., 2022; Hunter III et al., 2020; Lamb & Brown, 2001). Similarly, advanced computer vision and machine learning approaches have demonstrated improved accuracy and precision in image-based detection and control of various weed species, including HR biotypes (Coleman et al., 2022; Nugent et al., 2018; Scherrer et al., 2019; Wu et al., 2021). Several studies have reported the use of various types of precision agricultural technologies for weed control in specialty or high-value cash crops (Fennimore et al., 2016; Lee et al., 2010). However, a paucity of published information exists on this topic in pearl millets, and future research is warranted for increased adoption and production of this crop globally.

Conclusion

Due to a lack of effective herbicide options, weeds pose a serious production challenge for pearl millet production. Future research efforts should focus in developing and implementing novel, cost-effective, and sustainable IWM programs and technologies in pearl millet. The use of advanced breeding and genetic tools for developing weed-competitive hybrids and herbicide-tolerant pearl millet hybrids will be particularly helpful. Integration of chemical and non–chemical-based weed control strategies can provide improved and sustainable options for both grass and broadleaf weed control, especially selective control of grass weeds in pearl millet. Interdisciplinary weed control research is imperative to offer cost-effective weed management options to further enhance the productivity and global adoption of this orphan crop.

References

Agdag, M., Nelson, L., Baltensperger, D., Lyon, D., & Kachman, S. (2001). Row spacing affects grain yield and other agronomic characters of proso millet. *Communications in Soil Science and Plant Analysis, 32*, 2021–2032.

Andrews, D. J., Rajewski, J. F., & Kumar, K. A. (1993). Pearl millet: New feed grain crop. In J. Janick & J. E. Simon (Eds.), *New crops* (pp. 198–208). Wiley.

Asbil, W. (2001). Optimizing pearl millet and sorghum for forage and grain production in eastern Ontario. In *Sorghum, pearl millet, and chickpea, annual report* (pp. 153–185). Agriculture Environmental Renewal Canada Inc.

Bagavathiannan, M. V., & Davis, A. S. (2018). An ecological perspective on managing weeds during the great selection for herbicide resistance. *Pest Management Science, 74*(10), 2277–2286.

Balyan, R. S., Kumar, S., Malik, R. K., & Panwar, R. S. (1993). Post-emergence efficacy of atrazine in controlling weeds in pearl-millet. *Indian Journal of Weed Science, 25*, 7–11.

Banga, R. S., Yadav, A., Malik, R. K., Pahwa, S. K., & Malik, R. S. (2000). Evaluation of tank mixture of acetochlor and atrazine or 2, 4-D against weeds in pearl millet. *Indian Journal of Weed Science, 32*, 194–198.

Barberi, P., & Lo Cascio, B. (2001). Long-term tillage and crop rotation effects on weed seedbank size and composition. *Weed Research, 41*(4), 325–340.

Bennet, A. J., Yadav, R., & Jha, P. (2023). Using soybean chaff lining to manage waterhemp (*Amaranthus tuberculatus*) in a soybean–corn rotation. *Weed Science, 71*(4), 395.

Choudhary, V. K., Naidu, D., & Dixit, A. (2022). Weed prevalence and productivity of transplanted rice influenced by varieties, weed management regimes and row spacing. *Archives of Agronomy and Soil Science, 68*(13), 1872–1889.

Coleman, G. R. Y., Bender, A., Hu, K., Sharpe, S. M., Schumann, A. W., Wang, Z., Bagavathiannan, M. V., Boyd, N. S., & Walsh, W. J. (2022). Weed detection to weed recognition: Reviewing 50 years of research to identify constraints and opportunities for large-scale cropping systems. *Weed Science, 36*(6), 741–757.

Cuerrier, M. E., Vanasse, A., & Leroux, G. D. (2010). Chemical and mechanical weed management strategies for grain pearl millet and forage pearl millet. *Canadian Journal of Plant Science, 90*(3), 371–381.

Das, J., Patel, B. D., Patel, V. J., & Patel, R. B. (2013). Comparative efficacy of different herbicides in summer pearl millet. *Indian Journal of Weed Science, 45*(3), 217–221.

Dayan, F. E., & Duke, S. O. (2009). Biological activity of allelochemicals. In A. E. Osbourn & V. Lanzotti (Eds.).

Dayan, F. E., Kagan, I. A., & Rimando, A. M. (2003). Elucidation of the biosynthetic pathway of the allelochemical sorgoleone using retrobiosynthetic NMR analysis. *Journal of Biological Chemistry, 278*(31), 28607–28611.

Dhanda, S., Kumar, V., Obour, A. K., Dille, A., & Holman, J. D. (2022). Fall-planted cover crops for weed suppression in western Kansas. *Kansas Agricultural Experiment Station Research Reports, 8*(4), 14.

Dhayal, B. C., Yadav, S. S., Jat, M. L., & Dhayal, L. (2020). Efficacy of herbicides for weed control in pearl millet [*Pennisetum glaucum* (L.) R. Br. Emend Stuntz]. *Environment and Ecology, 38*(3), 651–656.

Di Tomaso, J. (1995). Approaches for improving crop competitiveness through the manipulation of fertilization strategies. *Weed Science, 43*, 491–497.

Diprose, M. F., Benson, F. A., & Hackam, R. (1980). Electrothermal control of weed beet and bolting sugar beet. *Weed Research, 20*(5), 311–322.

Dowler, C. C., & Wright, D. L. (1995). Weed management systems for pearl millet in the southeastern United States. In I. D. Teare (Ed.), *Proceeding of the first national grain pearl millet symposium* (pp. 64–71). University of Georgia.

Emechebe, A. M., Ellis-Jones, J., Schulz, S., Chikoye, D., Douthwaite, B., Kureh, I., Tarawali, G., Hussaini, M. A., Kormawa, P., & Sanni, A. (2004). Farmers perception of the *striga* problem and its control in northern Nigeria. *Experimental Agriculture, 40,* 215–232.

FAOSTAT. (2019). http://www.fao.org/faostat/en/#data/QC/visualize.

Fateh, E., Sohrabi, S. S., & Gerami, F. (2012). Evaluation the allelopathic effect of bindweed (*Convolvulus arvensis* L.) on germination and seedling growth of millet and basil. *Advances in Environmental Biology, 6*(3), 940–950.

Fennimore, S. A., Slaughter, D. C., Siemens, M. C., Leon, R. G., & Saber, M. N. (2016). Technology for automation of weed control in specialty crops. *Weed Technology, 30,* 823–837.

Fernando, M., & Shrestha, A. (2023). The Potential of Cover Crops for Weed Management: A Sole Tool or Component of an Integrated Weed Management System? *Plants, 12*(4), 752.

Gerhards, R., Sanchez, D. A., Hamouz, P., Peteinatos, G. G., Christensen, S., & Fernandez-Quintanilla, C. (2022). Advances in site-specific weed management in agriculture–a review. *Weed Research, 62*(2), 123–133.

Girase, P. P., Suryavanshi, R. T., Pawar, P. P., & Wadile, S. C. (2017). Integrated weed management in pearl millet. *Indian Journal of Weed Science, 49*(1), 41–43.

Hannaway, D. B., & Larson, C. (2004). *Forage fact sheet: pearl millet (Pennisetum americanum).* Oregon State University. http://forages.oregonstate.edu/php/fact_sheet_print_grass.php?SpecID=34&use=Forage

Heap, I. (2023). The International Herbicide-Resistant Weed Database. (accessed on 17 February 2021).

Hess, D. E., Ejeta, G., & Butler, L. G. (1992). Selecting sorghum genotypes expressing a quantitative biosynthetic trait that confers resistance to striga. *Phytochemistry, 31*(2), 493–497.

Hunter J. E., III, Gannon, T. W., Richardson, R. J., Yelverton, F. H., & Leon, R. G. (2020). Integration of remote-weed mapping and an autonomous spraying unmanned aerial vehicle for site-specific weed management. *Pest Management Science, 76*(4), 1386–1392.

International Crops Research Institute for the Semi-Arid Tropics (ICRISAT) (2021). http://exploreit.icrisat.org/profile/Pearl%20Millet/178.

Islam, N., Zamir, M. S. I., Din, S. M. U., Farooq, U., Arshad, H., Bilal, A., & Sajjad, M. T. (2018). Evaluating the intercropping of millet with cowpea for forage yield and quality. *American Journal of Plant Sciences, 9*(9), 1781–1793.

Knezevic, S. Z., Evans, S. P., Blankenship, E. E., Van Acker, R. C., & Lindquist, J. L. (2002). Critical period for weed control: The concept and data analysis. *Weed Science, 50*(6), 773–786.

Kumar, M., Meena, R. C., & Satyvathi, C. T. (2022a). Integrated weed management in pearl millet [*Pennisetum glaucum* (L.) R. Br.]. *International Journal of Agricultural Science, 18*(1), 231–234.

Kumar, R. R., Singh, S. P., Rai, G. K., Krishnan, V., Berwal, M. K., Goswami, S., Vinutha, T., Mishra, G. P., Satyavathi, C. T., Singh, B., & Praveen, S. (2022b). Iron and zinc at a cross-road: A trade-off between micronutrients and anti-nutritional factors in pearl millet flour for enhancing the bioavailability. *Journal of Food Composition and Analysis, 111*, 104591.

Kumar, V., Liu, R., Chauhan, D., Perumal, R., Morran, S., Gaines, T., & Jha, P. (2023). Characterization of imazamox-resistant shattercane (*Sorghum bicolor*) populations from Kansas. *Weed Technology, 37*(4). https://doi.org/10.1017/wet.2023.55

Kumar, V., Obour, A., Jha, P., Liu, R., Manuchehri, M. R., Dille, J. A., Holman, J., & Stahlman, P. W. (2020). Integrating cover crops for weed management in the semiarid US Great Plains: opportunities and challenges. *Weed Science, 68*(4), 311–323.

Labrada, R., Caseley, J. C., & Parker, C. (1994). *Weed management for developing countries.* FAO.

Lamb, D. W., & Brown, R. B. (2001). PA—Precision agriculture: Remote-sensing and mapping of weeds in crops. *Journal of Agricultural Engineering Research, 78*(2), 117–125.

Lee, W. S., Alchanatis, V., Yang, C., Hirafuji, M., Moshou, D., & Li, C. (2010). Sensing technologies for precision specialty crop production. *Computers and Electronics in Agriculture, 74*(1), 2–33.

Lindquist, J. L., Mortensen, D. A., Westra, P., Lambert, W. J., Bauman, T. T., Fausey, J. C., Kells, J. J., Langton, S. J., Harvey, R. G., Bussler, B. H., Banken, K., Clay, S., & Forcella, F. (1999). Stability of corn (*Zea mays*)–foxtail (*Setaria spp.*) interference relationships. *Weed Science, 47*, 195–200.

Little, N., DiTommaso, A., Westbrook, A., Ketterings, Q., & Mohler, C. (2021). Effects of fertility amendments on weed growth and weed–crop competition: A review. *Weed Science, 69*, 132–146.

Lyon, D., Kniss, A., & Miller, S. (2007). Carfentrazone improves broadleaf weed control in proso and foxtail millets. *Weed Technology, 21*(1), 84–87.

Martin, S. G., Van Acker, R. C., & Friesen, L. F. (2001). Critical period of weed control in spring canola. *Weed Science, 49*, 326–333.

Mishra, J. S., Rao, A. N., Singh, V. P., & Kumar, R. (2016). Weed management in major field crops. In B. S. Chauhan & G. Mahajan (Eds.), *Advances in weed management.* Indian Society of Agronomy.

Mishra, J. S., Rao, A. N., Singh, V. P., & Kumar, R. (2017). Weed management in major field crops. *Advances in Weed Management, 4*, 1–21.

Mishra, P., & Dash, D. (2014). Rejuvenation of biofertilizer for sustainable agriculture and economic development. *Consilience, 11*, 41–61.

Mishra, S. (2015). Weed management in millets: Retrospect and prospects. *Indian Journal of Weed Science, 47*(3), 246–253.

Nelson, L. A. (1977). Influence of various row widths on yields and agronomic characteristics of proso millet. *Agronomy Journal, 69*(3), 351–353.

Newman, Y., Jennings, E., Vendramini, J., & Blount, A. (2010). *Pearl millet (Pennisetum glaucum): Overview and management.* Institute of Food and Agricultural Sciences, University of Florida.

Nibhoria, A., Singh, B., Kumar, J., Soni, J. K., Dehinwal, A. K., & Kaushik, N. (2021). Enhancing productivity and profitability of pearl millet through mechanized inter-culture, suitable crop geometry, and agrochemicals under rainfed conditions. *Agricultural Mechanization in Asia, Africa, and Latin America, 52*(01), 2819–2830.

Nielsen, D. C., Lyon, D. J., Hergert, G. W., Higgins, R. K., & Holman, J. D. (2015). Cover crop biomass production and water use in the central Great Plains. *Agronomy Journal, 107*(6), 2047–2058.

Nugent, P. W., Shaw, J. A., Jha, P., Scherrer, B., Donelick, A., & Kumar, V. (2018). Discrimination of herbicide-resistant kochia with hyperspectral imaging. *Journal of Applied Remote Sensing, 12*(1), 016037. https://doi.org/10.1117/1.JRS.12.016037

Obour, A. K., Dille, J., Holman, J., Simon, L. M., Sancewich, B., & Kumar, V. (2022). Spring-planted cover crop effects on weed suppression, crop yield, and net returns in no-tillage dryland crop production. *Crop Science, 62*(5), 1981–1996.

Obour, A. K., Holman, J. D., & Schlegel, A. J. (2019). Strategic tillage in dryland no-tillage crop production systems. *Kansas Agricultural Experiment Station Research Reports, 5*(4), 4.

Osipitan, O. A., Dille, J. A., Assefa, Y., & Knezevic, S. Z. (2018). Cover crop for early season weed suppression in crops: Systematic review and meta-analysis. *Agronomy Journal, 110*(6), 2211–2221.

Osman, Z., Farah, Y., Hassan, H. A., & Elsayed, S. (2021). Comparative physicochemical evaluation of starch extracted from pearl millet seeds grown in Sudan as a pharmaceutical excipient against maize and potato starch, using paracetamol as a model drug. *Annales Pharmaceutiques Françaises, 79*(1), 28–35.

Pudelko, J. A., Wright, D. L., & Teare, I. D. (1993). *Herbicide effects on pearl millet in relation to weed control and crop damage. Quincy NFREC Research Report,* 94–95. https://projects.sare.org/wp-content/uploads/893LS93-053.009.pdf

Ram, B., Chaudhary, G. R., Jat, A. S., & Jat, M. L. (2005). Effect of integrated weed management and inter-cropping systems on growth and yield of pearl millet. *Indian Journal of Agronomy, 50*(3), 254–258.

Ramesh, N., Kalamani, M., Baradhan, G., Kumar, S. M., & Ramesh, S. (2019). Influence of weed management practices on nutrient uptake and productivity of hybrid pearl millet under different herbicides application. *Plant Archives, 19*(2), 2893–2898.

Rao, S. A., Mengesha, M. H., Harinarayana, G., & Reddy, C. R. (1987). Collection and evaluation of pearl millet germplasm from Maharashtra. *Indian Journal of Genetics and Plant Breeding, 47*(2), 125–132.

Rao, V. P., Thakur, R. P., Rai, K. N., & Sharma, Y. K. (2005). Downy mildew incidence on pearl millet cultivars and pathogenic variability among isolates of *Sclerospora graminicola* in Rajasthan. *SAT e-Journal, 46*, 107–110.

Reddy, S. S., Stahlman, P. W., Geier, P. W., Charvat, L. D., Wilson, R. G., & Moechnig, M. J. (2014). Tolerance of foxtail, proso and pearl millets to saflufenacil. *Crop Protection*, *57*, 57–62.

Samota, S. R., Singh, S. P., & Shivran, H. (2022b). Performance of pearl millet (*Pennisetum glaucum* L.) as affected by weed control measures. *Journal of Cereal Research*, *14*(2), 211–214.

Samota, S. R., Singh, S. P., Shivran, H., Singh, R., & Godara, A. S. (2022a). Effect of weeds control measures on weeds and yield of pearl millet [*Pennisetum glaucum* L.]. *Indian Journal of Weed Science*, *54*(1), 95–97.

Scherrer, B., Sheppard, J., Jha, P., & Shaw, J. A. (2019). Hyperspectral imaging and neural networks to classify herbicide-resistant weeds. *Journal of Applied Remote Sensing*, *13*(4), 044516–044516.

Schreier, H., Bish, M., & Bradley, K. W. (2022). The impact of electrocution treatments on weed control and weed seed viability in soybean. *Weed Technology*, *36*(4), 481–489.

Schwartz-Lazaro, L. M., Green, J. K., & Norsworthy, J. K. (2017). Seed retention of palmer amaranth (*Amaranthus palmeri*) and barnyardgrass (*Echinochloa crus-galli*) in soybean. *Weed Technology*, *31*(4), 617–622.

Sharma, O. L., & Jain, N. K. (2003). Integrated weed management in pearl millet (*Pennisetum glaucum*). *Indian Journal of Weed Science*, *35*, 34–35.

Sheahan, C. M. (2014). *Plant guide for pearl millet (Pennisetum glaucum)*. USDA-NRCS.

Singh, P., Boote, K. J., Kadiyala, M. D. M., Nedumaran, S., Gupta, S. K., Srinivas, K., & Bantilan, M. C. S. (2017). An assessment of yield gains under climate change due to genetic modification of pearl millet. *Science of the Total Environment*, *601*, 1226–1237.

Singh, R. K., Chakraborty, D., Garg, R. N., Sharma, P. K., & Sharma, U. C. (2010). Effect of different water regimes and nitrogen application on growth, yield, water use, and nitrogen uptake by pearl millet (*Pennisetum glaucum*). *Indian Journal of Agricultural Sciences*, *80*(3), 213–216.

Singh, V., Jat, M. L., Ganie, Z. A., Chauhan, B. S., & Gupta, R. K. (2016). Herbicide options for effective weed management in dry direct-seeded rice under scented rice-wheat rotation of western Indo-Gangetic Plains. *Crop Protection*, *81*, 168–176.

Stahlman, P. W., & Wicks, G. A. (2000). Weeds and their control in grain sorghum. In C. W. Smith & R. A. Frederikson (Eds.), *Sorghum: Origin, history, technology, and production* (pp. 535–582). Wiley.

Sundraseh, H. N., Rajappa, M. G., Lingegowda, B. K., & Krishnashastry, K. S. (1975). Critical stages of crop-weed competition in ragi (*Eleusine coracana*) under rainfed conditions. *The Mysore Journal of Agricultural Sciences*, *9*, 582–585.

Swanton, C. J., & Wiese, S. F. (1991). Integrated weed management: The rationale and approach. *Weed Technology*, *5*, 657–663.

Tara Satyavathi, C., Ambawat, S., Sehgal, D., Lata, C., Tiwari, S., Srivastava, R. K., Kumar, S., & Chinnusamy, V. (2021). Genomic designing for abiotic stress tolerance in pearl millet [*Pennisetum glaucum* (L.) R. Br.]. In C. Kole (Ed.), *Genomic designing for abiotic stress resistant cereal crops* (pp. 223–253). Springer.

Teasdale, J. R. (1996). Contribution of cover crops to weed management in sustainable agricultural systems. *Journal of Production Agriculture, 9*(4), 475–479.

Thate, T., Samuelson, S., Bastos-Martins, M., & Bagavathiannan, M. (2017). Evaluating different non-chemical weed management options for grain sorghum. In *Proceedings of the Texas Plant Protection Association Annual Meeting* (pp. 5–6). Texas A&M University.

USDA. (2023). *India millet area, yield and production.* USDA Foreign Agricultural Service. https://ipad.fas.usda.gov/countrysummary/Default.aspx?id=IN&crop=Millet.

Vander Meulen, A., & Chauhan, B. S. (2017). A review of weed management in wheat using crop competition. *Crop Protection, 95*, 38–44.

Walsh, M., Newman, P., & Powles, S. (2013). Targeting weed seeds in-crop: A new weed control paradigm for global agriculture. *Weed Technology, 27*(3), 431–436.

Walsh, M., Ouzman, J., Newman, P., Powles, S., & Llewellyn, R. (2017). High levels of adoption indicate that harvest weed seed control is now an established weed control practice in Australian cropping. *Weed Technology, 31*, 341–347.

Walsh, M. J., Broster, J. C., Schwartz-Lazaro, L. M., Norsworthy, J. K., Davis, A. S., Tidemann, B. D., Beckie, H. J., Lyon, D. J., Soni, N., Neve, P., & Bagavathiannan, M. V. (2018) Opportunities and challenges for harvest weed seed control in global cropping systems. *Pest Management Science, 74*(10), 2235–2245.

Walsh, M. J., Harrington, R. B., & Powles, S. B. (2012). Harrington seed destructor: A new nonchemical weed control tool for global grain crops. *Crop Science, 52*, 1343–1347.

Weston, L. A., & Czarnota, M. A. (2001). Activity and persistence of sorgoleone, a long-chain hydroquinone produced by *Sorghum bicolor. Journal of Crop Production, 4*(2), 363–377.

Wu, Z., Chen, Y., Zhao, B., Kang, X., & Ding, Y. (2021). Review of weed detection methods based on computer vision. *Sensors, 21*(11), 3647. https://doi.org/10.3390/s21113647

Zhu, L. I. U., Nan, Z. W., Lin, S. M., Yu, H. Q., Xie, L. Y., Meng, W. W., Zhang, Z., & Wan, S. B. (2023). Millet/peanut intercropping at a moderate N rate increases crop productivity and N use efficiency, as well as economic benefits, under rain-fed conditions. *Journal of Integrative Agriculture, 22*(3), 738–751.

Zimdahl, R. L. (1988). The concept and application of the critical weed-free period. In M. A. Altieri & M. Liebman (Eds.), *Weed management in agroecosystems: Ecological approaches* (pp. 145–155). CRC Press.

11

Diseases of Pearl Millet

Christopher R. Little, Ramasamy Perumal, and Timothy C. Todd

Chapter Overview

Pearl millet is the sixth most important cereal crop produced worldwide. As an environmentally resilient crop, it exhibits high productivity and is dependable across various agroecological zones. However, numerous biotic stressors affect pearl millet production, including fungi and fungal-like organisms, parasitic plants, nematodes, viruses, and prokaryotes. This chapter describes pearl millet pathology across the spectrum of agents that cause disease. A combination of historical and modern literature provides context, insights, and highlights for many of the recorded pathogens of pearl millet. Some diseases, such as blast, downy mildew, ergot, rust, and smut, have been given greater emphasis because their recorded impacts have drawn attention for many years. Further, this chapter delves into other groups of pathogens, including viruses and nematodes, that may be understudied but certainly affect production.

Introduction

Pearl millet [*Pennisetum glaucum* (L.) R. Br., syn. *Cenchrus americanus* (L.) Morrone] is a large-grain cereal row crop grown throughout the world. According to the United Nations Food and Agriculture Organization (FAO), in 2021, a total of 31,835,039 ha (78,666,094.6 ac) of millets were harvested worldwide (FAO, 2023). The countries with the most harvested acres were India (9,764,817 ha), Niger (6,145,774 ha), and Sudan (2,800,000 ha). The United States was in 15th place, with 267,900 ha (661,995.3 ac) harvested (FAO, 2023). For 2021, the average global yield was 15,602.1 hg/ha (~23.2 bu/ac; calculated based on a 60 lb bushel of wheat) and varied between a low of 1,429 hg/ha (~2.1 bu/ac) in Mauritania and a high of 148,983 hg/ha (~221.5 bu/ac) in Azerbaijan. The United States produced 13,017 hg/ha (~19.4 bu/ac) (FAO, 2023).

Pearl Millet: A Resilient Cereal Crop for Food, Nutrition, and Climate Security, First Edition. Edited by Ramasamy Perumal, P. V. Vara Prasad, C. Tara Satyavathi, Mahalingam Govindaraj, and Abdou Tenkouano.

In the United States, pearl millet has been an option for fodder production and poultry and livestock feed for many years (Burton et al., 1972; Hill & Hanna, 1990; Smith et al., 1989). Nutritionally, pearl millet has energy (kCal 100/g) comparable to wheat, rice, and sorghum; contains more cysteine, lysine, methionine, threonine, and tryptophan than maize; has comparable digestibility to other cereals; and is a good source of Ca, Fe, Mg, P, and vitamins including niacin, riboflavin, thiamine, and vitamins A and E (Adeola & Orban 2005; Hulse et al., 1980; Rathi et al., 2004). It has been widely produced in India and Africa for food and fodder (Malik et al., 2022; Raj et al., 2003; Sharma et al., 2013; Wenndt et al., 2021).

Pearl millet is typically viewed as an environmentally resilient crop due to its ability to grow in various soil types and quality, in soils with low moisture, and in soils with low fertility. This is due to an extensive root system that is well adapted to marginal soil. Pearl millet is often considered a cereal crop that can be grown in agroecologies where other crops are less productive (Reddy et al., 2021). Thus, pearl millet is frequently grown under suboptimal conditions. When pearl millet is produced under optimal rainfall and with good soil fertility, biomass, and yields, it can be competitive with other grain and forage crops (Singh & Singh, 1995; van Oosterom et al., 2006).

Aside from abiotic stressors such as heat and drought, pearl millet is affected by a wide range of diseases caused by microbial pathogens (fungal-like organisms and true fungi, parasitic plants, nematodes, viruses, and prokaryotes) (Luttrell, 1954; Wilson, 2000, 2002a). Among the most important, widespread, and damaging diseases are those caused by fungal-like organisms and true fungi, including blast (caused by *Pyricularia grisea*), downy mildew (caused by *Sclerospora graminicola*), ergot (caused by *Claviceps fusiformis*), rust (caused by *Puccinia substriata*), and smut (caused by *Moesziomyces penicilliariae*) (Miedaner & Geiger, 2015; Singh et al., 1993; Thakur & King, 1988a, 1988b, 1988c; Yadav & Rai, 2013). Table 11.1 ranks the subjective relative importance of various biotic stressors of pearl millet based on the authors' view of estimated reported disease impacts, losses, needs for further research, and numbers of current and historical literature citations.

Diseases Caused by Fungal-like Organisms and True Fungi

Many fungal-like organisms (including Oomycetes) and true fungi (including Ascomycetes and Basidiomycetes) cause global losses in pearl millet production areas. Most data come from just before or a little after the turn of the twenty-first century (Table 11.2). Therefore, there is a need to conduct updated yield loss assessments for fungal pathogens in pearl millet.

Table 11.1   Relative Importance of Diseases Affecting Pearl Millet.

Relative importance[a]	Disease/pathogen
High	Blast, downy mildew, ergot, rust, smut, viruses, witchweed
Medium	Bipolaris leaf spots, Curvularia leaf spots, head molds, mycotoxins, nematodes, seedling diseases
Low	Bacterial spot, Cercospora leaf spot, charcoal rot, false mildew, Fusarium stalk rot, Pantoea leaf spot, Phyllosticta leaf spot, Rhizoctonia blight, southern blight, top rot
Very low	Bacterial leaf streak, bacterial stem rot, bacterial stripe, Bipolaris stalk rot, Exserohilum leaf spot, Drechslera leaf spot, Dactuliophora leaf spot, Gloeocercospora leaf spot, Myrothecium leaf spot, Phytoplasmas, Plasmopara downy mildew

[a] Based upon the authors subjective view of estimated reported disease impacts, losses, needs for further research, and historical and current numbers of literature citations.

For any crop, stand establishment is a crucial development period and is among the most significant predictors of yield potential. Seedling pathogens, whether soilborne or seedborne, affect germination, emergence, and seedling vigor (Lamichhane et al., 2017). Generally, host resistance is a poor control method for many seedling pathogens, and seed treatments or other cultural control methods that help to avoid disease are preferred.

Foliar diseases are of particular importance for two reasons. First, foliar leaf spots and blight diseases reduce photosynthates that can be transferred to root and stalk tissues and ultimately to the panicle (Tesso et al., 2012). Second, for forage millets, foliage is the main output. Reduced biomass and reduced quality of foliage are of concern.

Several panicle diseases can cause direct yield losses or affect quality, harvestability, and harvest index. In addition to direct losses associated with lost grains, poor-quality grains due to grain mold and contamination with mycotoxins can be problematic. Typically, the smut fungi (e.g., *Moesziomyces bullatus*) and ergot (caused by *C. fusiformis*) replace infected grains, causing a direct loss. Grain molding fungi, such as *Aspergillus* and *Fusarium* spp., may produce a range of mycotoxins that contaminate harvested grain. Other grain mold fungi degrade caryopsis tissues and reduce endosperm and germ quality.

Root and stalk diseases develop aboveground symptoms that may be limited to stunting, chlorosis, or poor vigor. However, many fungal "root nibblers" can affect plant vigor and yield and may not show themselves clearly. Research on the root disease complex of pearl millet has been limited. However, many seedling disease pathogens can persist and parasitize roots as the plant matures.

Table 11.2   Disease Losses Associated with Fungal-like Organisms and True Fungi Diseases of Pearl Millet.

Disease	Pathogen	Losses[a]	References
Cercospora leaf spot	*Cercospora penniseti*	20%–25% (United States)	Burton & Wells (1981)
Downy mildew	*Sclerospora graminicola*	≤20% (India)	Singh (1995)
		57% (India)	Gupta & Singh (1996)
		6%–72% (Nigeria), 3.5%–21% (Mali)	Zarafi (1997)
		21%–60% (Burkina Faso)	Marley et al. (2002)
		10%–60%	Khairwal et al. (2007)
		20%–40, ≤80% (India)	Raj et al. (2018)
		80% (India)	Chelpuri et al. (2019)
Pyricularia leaf blight (and other foliar spots)	*Pyricularia grisea* et al.	~19% digestible dry matter forage yield (United States)	Wilson & Gates (1993)
Ergot	*Claviceps fusiformis*	58%–75% in grain production, quality, germination, and emergence of saved seed (India)	Hash et al. (1999)
Rust	*Puccinia substriata*	76% in yield, 41% in 500-grain weight (United States)	Monson et al. (1986)
		51% reduction in dry matter (United States)	Wilson et al. (1996)
Smut	*Moesziomyces bullatus*	<10% (West and central Africa)	Marley et al. (2002)
		5%–20% (India)	Thakur et al. (2011)

[a] Expressed in terms of "percent reduction" for the diseases and locations indicated.

Seedling Diseases

Seedling diseases affect every agricultural crop to one degree or another, and pearl millet is no exception. Optimum seed germination and emergence rates are essential for pearl millet to escape infection by soilborne pathogens. Such optima largely depend on the soil environment (Figure 11.1). For example, Ong (1983) found that the quickest and highest level of 'BK 560' pearl millet seedling

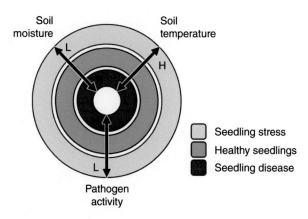

Figure 11.1 Interactions between soil moisture, soil temperature, and pathogen activity result in seedling disease, healthy seedling development, or seedling stress. H, high; L, low. Image by C. R. Little.

emergence occurred by ~2 days after sowing (DAS) when soil temperatures at 2.0 cm were 31.4°C. However, when soil temperatures were 17.9 or 19.4°C, the percent seedling emergence was approximately three-quarters of that at the high-temperature level, and emergence did not reach maximum until ~6 DAS.

Moisture is another aspect of the soil environment that is crucial for emergence. Soil moisture above 50% field capacity tends to reduce millet seed germination (Fawusi & Agboola, 1980). One explanation for reduced germination under high soil moisture conditions is anaerobiosis of the soil, which some seedling pathogens take advantage of as they can switch their metabolic activities from oxidative respiration to fermentation-type pathways. Alternatively, oxygen stress in the plant and high matric potentials cause root exudates to diffuse more quickly, thus stimulating the germination and growth of infection propagules of common seedling pathogens, including *Fusarium* and *Pythium* (Drew & Lynch, 1980).

Seedling disease may be characterized by pre- or post-emergence damping-off, slow emergence, reduced seedling vigor, and poor seasonal development of the adult plant (Khairwal et al., 2007). In pre-emergence damping-off, seeds either do not germinate or partially germinate, failing to break the soil surface (Lamichhane et al., 2018). In extreme cases, seeds rot in the ground, and no (or very little) coleoptile or radicle emerges from the pericarp. A pre-emergence damping-off often occurs due to a seed's inability to push the coleoptile through a wet, dense soil or break the soil surface if compaction or crusting occurs. An elongated period where seedling tissues remain beneath the soil provides longer periods where opportunistic pathogens may colonize and further weaken the seedling. Further, any mechanical damage that results from the seedling's attempt to emerge will provide passive entry points for fungal colonization.

In post-emergence damping-off, seedlings that have emerged from the soil line often wilt, become chlorotic, and die (i.e., seedling blight). Often, reddish or brown discoloration or sunken, watersoaked lesions may be observed at the base of the coleoptile. Furthermore, discoloration or necrosis of the limited root system may be observed. Seedlings that survive the initial emergence without damping-off often exhibit decreases in vigor, including seedling height, weight, and girth, compared with healthy plants. As the season progresses, host genotype and environmental conditions will determine if the diseased plant can compensate or if development and yield will be affected.

Classically, *Fusarium* spp., *Rhizoctonia* spp., and *Pythium* spp. are the primary seedling pathogens. However, other genera, such as *Alternaria, Curvularia, Epicoccum, Exserohilum,* and *Phyllosticta,* are also involved (Kumar et al., 2021b). These organisms represent a "seedling disease complex." Moreover, many pathogens discussed in the head mold section (see below) and leaf and sheath spots and blights section (see below) can cause disease in seedlings.

Akanmu et al. (2013) indicated that, among the millets, pearl millet seedlings showed the highest general resistance to *Fusarium* spp. In contrast, Guinea millet (*Brachiaria deflexa*), finger millet (*Eleusine coracana*), and black fonio (*Digitaria iburua*) were more susceptible. *Fusarium anthophilum, F. verticillioides, F. oxysporum,* and *F. scirpi* were pathogenic on all millet types in Nigeria. For pearl millet, *F. andiyazi, F. circinatum, F. compactum, F. dlamini, F. fujikuroi, F. napiforme, F. nygamai, F. oxysporum, F. phillophilum, F. proliferatum,* and *F. solani* significantly reduced stem girth in 14-day-old seedlings compared to control. *F. acutatum, F. andiyazi, F. compactum, F. nygamai, F. oxysporum,* and *F. solani* significantly reduced leaf numbers in 7-day-old seedlings, and *F. dlamini* significantly reduced leaf numbers in 14-day-old seedlings compared with control. Finally, *F. andiyazi* significantly reduced leaf area in 14-day-old seedlings compared with control.

Exserohilum rostratum and *Bipolaris setariae* are pathogens of millet seedlings when soil temperatures are <15 °C (59 °F) (Wells & Winstead, 1965). *Phyllosticta pennicillariae* causes seedling disease and, leaf blight in adult plants (Wilson & Burton, 1990).

Bacteria associated with the seedling rhizosphere show promise in reducing pearl millet seedling disease. Kumar et al. (2021a) report that several rhizosphere bacteria, including *Bacillus cereus, Bacillus licheniformis, Bacillus stearothermophilus, Bacillus subtilis, Chromobacterium violaceum, Pseudomonas chlororaphis,* and *Pseudomonas fluorescens* can suppress *F. oxysporum, Macrophomina phaseolina,* and *P. grisea.* These bacteria suppress fungal growth by producing degradative enzymes, including β-1,3-glucanases, chitinases, and proteases. They also produce siderophores and other secondary metabolites that impede spore germination and compromise fungal hyphae.

Kumar et al. (2021b) detail the seed endophytic bacteria (SEB) found in pearl millet caryopses and how these can be advantageous for seedling development and pathogen defense. These authors identified seven SEBs, including *Bacillus subtilis*, *B. tequilensis*, *B. velezensis*, *Kosakonia cowanii*, *Paenbacillus dendritiformis*, *Pantoea stewartii*, and *Pseudomonas aeruginosa*. Of these seven isolates, all produced auxin, five were phosphorus solubilizers, two were potassium solubilizers, and three produced siderophores. Seedling growth was reduced when pearl millet seeds were sterilized to remove the SEBs. When the SEBs were reinoculated into seeds, growth and development were improved. Further, most of the SEBs could inhibit pathogen growth in vitro. *Bacillus subtilis* was especially important because it produces a lipopeptide with antifungal activity against *Fusarium* sp.

Downy Mildews

Sclerospora graminicola

Downy mildew of pearl millet, or "green ear," is caused by the oomycete pathogen *S. graminicola* (Sacc.) J. Schröt (Oomycota, Oomycetes, Peronosporales, Peronosporaceae). As with all downy mildews, *S. graminicola* is an obligate biotroph that cannot be reared on artificial media and must be propagated in living host plants. *Sclerospora graminicola* is a long-studied pathogen, with original descriptions by Saccardo in 1876 on *Setaria verticillata*, Schroeder in 1879 in Germany, Farlow in 1884 in the United States on *Setaria viridis* (L.) Beauv., and in India by Butler (1907) and Kulkarni (1913). The disease was recognized as a problem related to low-lying, moisture-prone areas as early as 1918 (Butler, 1918) and 1930 (Mitter & Tandon, 1930). Subsequent disease outbreaks in India and Africa raised this pathogen to one of the most serious biotic stressors affecting pearl millet. It should be noted that even though the pathogen has been found on *S. viridis* in the United States, it has not been found on pearl millet (Figure 11.2).

Sclerospora graminicola has been found on a wide range of grass hosts in the family Poaceae besides pearl millet. In addition to *Setaria viridis*, Melhaus et al. (1927, 1928) found *S. graminicola* on *S. italica* and *Zea mays*. Further, Mitter and Mitra (1941) found *S. graminicola* on *S. verticillata* (hooked bristlegrass). Waterhouse (1964) identified *S. graminicola* on *Agrostis gigantea* (syn. *A. alba*; redtop), *Echinochloa crus-galli* (barnyardgrass), *E. frumentacea* (billion-dollar grass), *Panicum miliaceum* (proso millet), *Saccharum officinarum* (sugarcane), *Setaria pumila* (syn. *S. lutescens*; yellow foxtail), *Sorghum halepense* (Johnsongrass), *S. bicolor* subsp. *arundinaceum* (wild sorghum), *S. bicolor* subsp. *drummondii* (Sudangrass), and *Zea luxurians* (teosinte). Francis and Williams (1983) also identified this pathogen on *Eleusine indica* (Indian goosegrass) and *Z. mexicana* (Mexican teosinte).

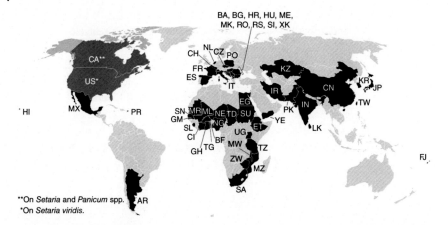

Figure 11.2 Global distribution of *Sclerospora graminicola*, the causal pathogen of pearl millet downy mildew. AR, Argentina; BA, Bosnia; BG, Bulgaria; BF, Burkina Faso; CA, Canada; CH, Switzerland; CI, Côte d'Ivoire; CN, China; CZ, Czech Republic; EG, Egypt; ES, Spain; ET, Ethiopia; FR, France; GH, Ghana; GM, The Gambia; HI, Hawaii (United States); HR, Croatia; HU, Hungary; IN, India; IR, Iran; IT, Italy; JP, Japan; KR, South Korea; KZ, Kazakhstan; LK, Sri Lanka; ME, Montenegro; MK, North Macedonia; ML, Mali; MW, Malawi; MX, Mexico; NE, Niger; NG, Nigeria; NL, Netherlands; PK, Pakistan; PO, Poland; PR, Puerto Rico (United States); RO, Romania; RS, Serbia; SA, South Africa; SI, Slovakia; SN, Senegal; SL, Sierra Leone; SU, Sudan; TD, Chad; TG, Togo; TW, Taiwan; TZ, Tanzania; UG, Uganda; US, United States; XK, Kosovo; YE, Yemen; ZW, Zimbabwe. References: N'Daiye (1995), Wilson (2000), Indira et al. (2002), Marley et al. (2002), Yadav and Rai (2013), CABI (2023). Image by C.R. Little.

Losses to the disease generally remain below 20% but can exceed this if a susceptible variety is planted and environmental conditions are favorable for an outbreak. However, numerous studies have indicated disease losses to this pathogen within and in excess of the abovementioned threshold, with up to ~80% yield loss. See Table 11.2 for a list of losses and relevant citations.

Disease losses due to phyllody or "green ear" occurring in the panicle (Figure 11.3). This symptom occurs when floret tissues are converted into deformed, leaf-like tissues instead of mature seeds (Singh et al., 1993). In addition to its direct effects on reproductive parts and seed development impacts due to phyllody, the pathogen appears to induce additional physiological and biochemical changes in the host that can lead to reduced yield. *Sclerospora graminicola* infection causes stomata to close resulting in reduced transpiration (Kumar & Gour, 1992). Mahatma et al. (2009) demonstrated that *S. graminicola* infection reduced total chlorophylls and carotenoids in susceptible plants compared with resistant plants. Also, amino acid content increased in resistant and susceptible genotypes after infection. Further, the pathogen's metabolic demand from the plant reduces photosynthates for developing grains and reroutes them to the pathogen sink.

Figure 11.3   Symptoms and signs of pearl millet infection by *Sclerospora graminicola*. (a) "Green ear" phyllody of pearl millet panicle. (b) Severe local lesion infection. (c) Lesion development and production of downy, white sporangia. "Larry E. Claflin Plant Disease Image Collection," Kansas State University Libraries, Manhattan, Kansas, USA / CC BY 1.0.

Sclerospora graminicola is a highly variable pathogen due to the development of a sexual stage via oosporogenesis (Thakur et al., 2009). Oosporogenesis, or the production of oospores, is characterized by the fusion of antheridia and oogonia, each containing post-meiotic haploid nuclei. The result of fertilization is the production of a diploid oospore (Figure 11.4). Genetic recombination leads to the

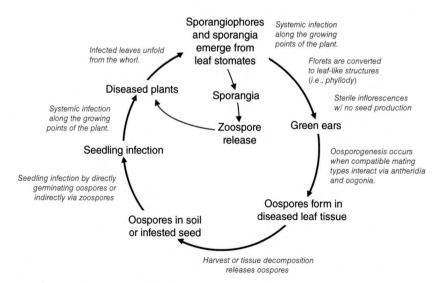

Figure 11.4   Disease cycle of pearl millet downy mildew caused by *Sclerospora graminicola*. Image by C. R. Little.

production of new virulent pathotypes in the pathogen population. In a review, Kumar et al. (2012) report on up to 15 pathotypes from multiple locations in India. Therefore, pathotypes must be considered when screening for resistance or developing resistance markers. For example, Thakur et al. (2009) examined 12 isolates when screening pearl millet germplasm in India, and each of these pathotypes had to be maintained on a genetically different germplasm. Likewise, Chelpuri et al. (2019) used three major pathotypes dominant in three Indian states to determine downy mildew resistance quantitative trait loci (QTL).

Pathotype 1 of *S. graminicola* has been sequenced (Nayaka et al., 2017). The result was a 40× genome coverage and a draft genome of 299,901,251 bp containing 65,404 genes. Gene Ontology (GO) annotation revealed many GO terms associated with pathogenicity functions. Further, among the GO terms identified, 699 were categorized in "DNA integration" (GO:0015074), far above any other set of GO terms.

Many pathogen-specific proteins exist on the mycelium and sporangial surfaces of oomycetes. Kumar et al. (1993a) used sporangia to produce *S. graminicola*-specific antibodies that would react with both sporangia and mycelium in in vitro tests. At the same time, Kumar et al. (1993b) determined that virulent races of *S. graminicola* carried a 90 kDa protein that specifically interacted with susceptible pearl millet cells in suspension. However, the protein did not bind to resistant pearl millet cell suspensions.

Control of downy mildew is achieved through resistant cultivars and hybrids. Field and greenhouse screening for disease resistance has been conducted by workers in India and other pearl millet-producing countries, and there are effective protocols for differentiating resistance and susceptibility (Singh, 1995; Singh & Gopinath, 1985; Singh et al., 1993; Williams et al., 1981). Additional sources of resistance and other genotypes with susceptible reactions are listed in Table 11.3.

The genetics of downy mildew resistance in pearl millet have been examined over the years. Work in the 1980s, despite methodological, environmental, pathogen population variability, and disease pressure shortcomings, showed that resistance to *S. graminicola* is quantitative (Basavaraju et al., 1981; Dass et al., 1984; Shinde et al., 1984). Further, numerous QTLs for downy mildew resistance have been identified and reviewed (Breese et al., 2002; Jones et al., 1995, 2002).

The ICMP 85410 (PI 597490)-resistant parent was used in the first pearl millet mapping population to be established, and the same population was used to identify QTLs for downy mildew resistance (Jones et al., 1995; Liu et al., 1994). Jones et al. (1995) screened the mapping population against *S. graminicola* isolates from India and several African locations to identify downy mildew resistance–associated QTLs. As a result, several QTLs were identified from this population, including one at a location designated M716 that appeared to be shared when the population

Table 11.3 Pearl Millet Entries and Their Reaction to Downy Mildew Caused by *Sclerospora graminicola.*

Plant ID	Reaction	References
Amanier	Susceptible	Shishupala et al. (1996)
BDP1	Susceptible	Maryam (2015)
BJ 104	Susceptible	Shishupala et al. (1996)
BK 560	Susceptible	Shishupala et al. (1996)
DMR15	Resistant	Maryam (2015)
GHB-719	Resistant	Dangaria et al. (2012)
HB3	Susceptible	Andrews et al. (1985a), Shishupala et al. (1996)
ICMA/B 841A	Resistant	Singh et al. (1990b, 1992)
ICML 12-16 (PI537577-581)	Resistant	Singh et al. (1990a)
ICM 423 (PI572306)	Resistant	Rai et al. (1994)
ICML 22 (GP-30; PI572474)	Resistant	Singh et al. (1994)
ICMP 312	Resistant	Witcombe et al. (1996)
ICMP 451	Resistant	Breese et al. (2002)
ICMP 85410 (PL-36; PI 597490)	Resistant	Talukdar et al. (1998)
ICMS 8019, -8282, -8283	Resistant	Andrews et al. (1985a)
IP18292, IP18293	Resistant	Thakur et al. (2004)
IP6147-2, IP6147-4	Resistant	Singh (1990), Jeger et al. (1998)
IP8695-1, IP8695-4	Resistant	Singh (1990), Jeger et al. (1998)
IP8749-1	Resistant	Singh (1990), Jeger et al. (1998)
IP18293, IP18294	Susceptible	Shishupala et al. (1996)
IP18292, IP18296		Shishupala et al. (1996)
IP8289, IP18295, IP18298	Resistant	Shishupala et al. (1996), Thakur et al. (2009)
IP8877-3	Resistant	Singh (1990), Jeger et al. (1998)
J104	Susceptible	Shishupala et al. (1996)
LCIC9702	Susceptible	Maryam (2015)
Makani	Susceptible	Shishupala et al. (1996)
MBH 110	Susceptible	Shishupala et al. (1996)
MOP1	Susceptible	Maryam (2015)
NHB3	Susceptible	Singh (1990), Jeger et al. (1998)
P7	Susceptible	Shishupala et al. (1996)

(Continued)

Table 11.3 (Continued)

Plant ID	Reaction	References
P7-4	Resistant	Shishupala et al. (1996)
P181-2	Resistant	Singh (1990), Jeger et al. (1998)
P310-17	Resistant	Shishupala et al. (1996), Thakur et al. (2004, 2007, 2008)
P462-4	Resistant	Singh (1990), Jeger et al. (1998)
P1449-2	Resistant	Singh (1990), Shishupala et al. (1996), Jeger et al. (1998), Thakur et al. (2009), Maryam (2015)
P-2895-3	Resistant	Singh (1990), Jeger et al. (1998)
P2910-2	Intermediate (?)	Singh (1990), Jeger et al. (1998)
P3281-1	Resistant	Singh (1990), Jeger et al. (1998)
PEO5532	Resistant	Maryam (2015)
PEO5984	Susceptible	Maryam (2015)
PNBM 85213	Resistant	Shishupala et al. (1996)
PMBV 9-21	Resistant	Shishupala et al. (1996)
PMCD 10	Resistant	Shishupala et al. (1996)
SOSATC88	Resistant	Maryam (2015)
Soyagon	Susceptible	Shishupala et al. (1996)
WC-C75	Resistant	Andrews et al. (1985a)
YL-18	Resistant	Thakur et al. (2009)
23BX	Susceptible	Shishupala et al. (1996)
75-3	Resistant	Singh (1990), Jeger et al. (1998)
841A/B	Resistant	Singh et al. (1990c), Shishupala et al. (1996)
843A	Susceptible	Shishupala et al. (1996)
843B	Susceptible	Shishupala et al. (1996)
5141A, 5141B	Susceptible	Singh et al. (1990a, 1990b, 1990c), Shishupala et al. (1996)
7042S	Susceptible	Shishupala et al. (1996), Thakur et al. (2007, 2008, 2009)
700481-5-3	Resistant	Singh (1990), Jeger et al. (1998)
700651	Susceptible	Shishupala et al. (1996), Thakur et al. (2004)

was screened by the Nigeria and Niger isolates. Each of the QTLs contributed greatly to the total phenotypic variation explained.

Many mechanistic studies have examined the physiological and molecular interaction between *S. graminicola* and pearl millet. These include (but are not limited to) (a) hypersensitive response (HR) and cell death; (b) accumulation of hydroxyproline-rich glycoproteins (HGRPs); (c) expression of polygalacturonase-inhibiting proteins (PGIPs); and (d) elicitation of defense-associated enzymes, including β-1,3-glucanases, chitinases, lipoxygenases, peroxidases, phenylalanine ammonia-lyase (PAL), ribonucleases, and superoxide dismutase.

Kumudini et al. (2005) found that HR could be induced in a downy mildew–resistant cultivar (IP18293) within 2 hours of inoculation. In contrast, it took 8 hours for the same response to be observed in the susceptible cultivar (HB3). Likewise, it took 2.9 hours for 50% localized cell death to occur at the inoculation site of the resistant cultivar and 6.5 hours in the susceptible cultivar. The time required to accumulate H_2O_2 in the periplasmic spaces was similar to that of 50% localized cell death. Kulkarni et al. (2016) found that HR-associated genes, including antioxidant enzymes, respiratory burst oxidase, superoxide dismutase, and HR-associated transcription factor families (MYB and WRKY) were upregulated in response to *S. graminicola* infection in a transcriptomics assay.

In 2-day-old downy mildew–resistant pearl millet seedlings, Shailasree et al. (2004) showed increased hydroxyproline (Hyp) accumulation 2 hours after inoculation by *S. graminicola*. This experiment used Hyp concentration in seedling tissues as a hydroxyproline-rich glycoprotein (HRGP) marker. HGRPs function as cell wall structural components accumulated in response to pathogens. In the case of the susceptible genotype, only a much slower and partial response was observed. Manjunatha et al. (2009) pre-treated pearl millet seedlings with sodium nitroprusside and S-nitrosoglutathione; both produce nitric oxide in planta and could induce HGRPs as well as hypersensitive cell death.

Prabhu et al. (2012) showed that polygalacturonase inhibiting proteins (PGIPs) were maximally induced within 24–48 hours after seedling infection with *S. graminicola*. Overall, PGIP accumulation was higher in the resistant cultivar (IP18296) versus the susceptible cultivar (7042S). Immunolocalization further showed that PGIPs accumulated in the epidermal layers of inoculated coleoptiles. PGIPs are host defense leucine-rich repeat proteins that inactivate pathogen polygalacturonases produced by facultative plant pathogens (Di Matteo et al., 2003). As a result, plant cell wall degradation leads to oligogalacturonide accumulation and activation of host defense responses.

Shivakumar et al. (2000) showed 10%–27% increased ribonuclease activity in pearl millet seedlings during incompatible (resistance) interactions, whereas compatible (susceptible) interactions showed 4%–13% reduced ribonuclease activity. Plant ribonucleases degrade RNA into nucleotides, which can occur in a sequence-specific or targeted manner. There are many classes, and most participate in the normal cellular processes of RNA degradation and phosphorylation. RNases are components of natural, induced, and engineered RNA interference processes.

Several studies report induction of resistance against downy mildew. Geetha and Shetty (2002) used benzathiadiazole (BTH), $CaCl_2$, and H_2O_2 to protect plants against *S. graminicola* infection. In their study, 0.75% BTH, 90 mM $CaCl_2$, and 1.0 mM H_2O_2 provided levels of protection between 59% and 78% against downy mildew in a susceptible host. Sudisha et al. (2011) used raw cow milk (RCM) and five amino acids to activate defense-associated enzymes in pearl millet to induce resistance to *S. graminicola*. By themselves, RCM (5%–30%, v/v), L-glutamic acid (Glu), L-isoleucine (Iso), L-phenylalanine (Phe), and L-proline (Pro), when used as a seed treatment, protected pearl millet from downy mildew infection at a significantly greater level than the distilled water control in both field and greenhouse conditions. However, these treatments gave less protection than the metalaxyl seed treatment, which was significantly better. Further, the RCM and amino acid treatments were able to induce the production of phenylalanine ammonia-lyase (PAL), which accumulated in pearl millet seedlings to its highest level by 72 hours post-inoculation (HPI). Among the seed treatments, Iso, Phe, and Pro induced the highest levels of PAL. In the case of peroxidases (POX), maximum expression of these enzymes appeared at ~24 HPI, and most of the treatments appeared to upregulate POX at a higher level than the distilled water control. Finally, maximal β-1,3-glucanase expression was observed in seedlings at ~36 HPI, with Glu, Iso, Phe, and Pro showing the highest level of enzyme induction.

Systemic acquired resistance (SAR) is a mechanism that underlies response to downy mildew in resistant plants but also facilitates induced resistance in susceptible or moderately susceptible plants. Devaiah et al. (2009) showed that the extract of *Datura metel* L. (Hindu datura, related to jimsonweed, *D. stromarium* L.) could induce SAR. In this study, treatment with *D. metel* extract induced salicylic acid by 4× and 10× in pearl millet seedling roots and shoots, respectively. Not only did *D. metel* extract induce SAR, but it also increased plant growth and yield. Kulkarni et al. (2016) explored the transcriptome in response to *S. graminicola*. They found that SAR-related genes, including phenylalanine ammonia-lyase and pathogenesis-related protein genes, were among those that responded in the host–pathogen interaction.

Plasmopara penniseti

The other downy mildew of pearl millet is *Plasmopara penniseti* R.G. Kenneth & J. Kranz (Oomycota, Oomycetes, Peronosporales, Peronosporaceae). Within the *Plasmopara* genus, this is the only species that infects monocot hosts

(Thines et al., 2007). Thus, due to host range, other morphological characters, and molecular phylogeny, the authors suggest that the pathogen genus name be changed to *Poakatesthia*.

The pathogen was first identified on pearl millet in Ethiopia (Kenneth & Kranz, 1973). Further, Kenneth (1976) asserted that the graminicolous *Plasmopara* downy mildews were only found in Africa and are their hypothesized place of origin. Here, the pathogens appeared to cause adaxial local lesions but did not result in systemic disease. Further, the lesions can follow the veins, streak, coalesce, and turn necrotic.

As in all downy mildews, the sporangiophores emerge through the stomata. Sporangiophores are branched dichotomously or pseudotrichotomously, and the papillate sporangia are azoosporic and function like conidia (Thines et al., 2007). No information exists in the literature to suggest that this pathogen undergoes a sexual cycle and produces oospores.

False Mildew

False mildew of pearl millet is caused by the ascomycete *Beniowskia sphaeroidea* (Kalchbr. & Cooke) E.W. Mason (Ascomycota; *Incertae sedis*). Currently, the taxonomic placement of this pathogen is unclear beyond inclusion in the Ascomycota. As far as is known, the host range for this pathogen is restricted to the Poaceae (de Milliano, 1992).

False mildew is characterized by white sporodochia (cushion-like masses of hyphae that support conidiophore and conidia development) and underlying local lesions with chlorosis and areas of pigment accumulation (Figure 11.5).

Figure 11.5 Symptoms and signs of false mildew caused by *Beniowskia sphaeroidea*. Note the numerous white sporodochia on the leaf surface. "Larry E. Claflin Plant Disease Image Collection," Kansas State University Libraries, Manhattan, Kansas, USA / CC BY 1.0.

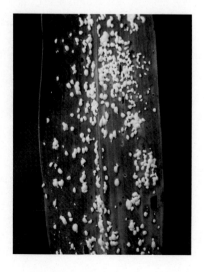

The degree to which these signs and symptoms affect yield is not clear. However, the incidence of the pathogen has raised alarms in various pearl millet–producing countries, such as Zimbabwe, from 1985 to 1989 (Mtisi & de Milliano, 1991, 1993). As of 1998, hot spots of false mildew in Zimbabwe remained (Monyo, 1998). However, in subsequent years after the initial epiphytotics, Mtisi and McLaren (2002) reported that incidence of the disease had decreased overall.

False mildew occurs in years with high moisture (de Milliano, 1992). Mtisi and de Milliano (1991) reported that false mildew occurred in Zimbabwe after ≥750 mm of rainfall and high humidity.

de Milliano (1992) reported that multiple plant introductions are resistant to *B. sphaeroidea*, including '700516', '7042 DMR', 'ICMPES 29', and 'ICMPES 1500-7-3-2'. This author indicated that entries in the Regional Cooperative Variety Trial in Zimbabwe, including 'IVS-A82' and 'IBVM 8402 SN', were resistant. In Zimbabwe, 'PMV 1' was susceptible (Mtisi, 1992). Monyo (1998) also indicated that '861B' and 'ICMB 87001' exhibited low severity, and hybrids derived from 'ICMA 87001 × SDPC 22' and 'ICMA 87001 × SDPC 40' were resistant.

In addition to pearl millet, *B. sphaeroidea* caused white mold disease on Napiergrass (syn. elephantgrass; *Ceratodon purpureus*) in Africa and Benowskia blight of knotroot bristlegrass [syn. marsh bristlegrass; *Setaria parviflora* (Poir.) Kerguélen] in the southern United States (Horst, 2013; Maher, 1936; Mtisi & de Milliano, 1993; Nattrass, 1941; Taber et al., 1978).

Ergot

Pearl millet ergot is caused by *C. fusiformis* Loveless (Ascomycota, Sordariomycetes, Hypocreales, Clavicipitaceae). *Claviceps fusiformis* is characterized by the production of fusiform or lunate-falcate conidia and is a pathogen of grasses in the Poaceae (Tribe Paniceae). Pažoutová et al. (2008) studied *Claviceps* and *Sphacelia* isolates from Texas, Africa, and Australia using DNA sequencing of the rDNA and β-tubulin genes, morphological characters, and alkaloid production. *Claviceps fusiformis* isolates that originated from Zimbabwe (Shamva) (CCC 525), French Central Africa (now Central African Republic, Chad, Gabon, and Republic of the Congo) (CCC 106), and Africa (CCC 110), and an isolate from India (Karnataka) (CCC 846) were grouped in a clade of *C. fusiformis* sensu Loveless (>98% confidence), which also included *Sphacelia eriochloae*, isolated from cup-grass (*Eriochloa sericea*) in Texas (Kingsville) (CCC 859). Most of the isolates from Zimbabwe (CCC 642, CCC 646, and CCC 647) were grouped in their own clade as *Sphacelia lovelessii* sp. nov., whereas the Australian isolates (Goondiwindi, Queensland) were grouped as *C. hirtella*, a sister species of *C. fusiformis*. Another set of Texas isolates (CCC 774, 776, and 778) isolated from pearl millet were grouped in a separate clade and were characterized as *Sphacelia texensis* sp. nov.

This disease reduces grain yield and quality (Thakur et al., 1984). *Claviceps fusiformis* induces the production of a sticky exudate called "honeydew" that disrupts clean harvesting, attracts insects, and provides a substrate on which other secondary fungi can colonize (Figure 11.6). Further, *C. fusiformis* produces toxic alkaloids in the sclerotia (or "ergots") that replace kernels. Toxins produced by *C. fusiformis* include the clavine alkaloids (Banks et al., 1974). The principal clavine alkaloid of *C. fusiformis* is agroclavine, with elymoclavine and chanoclavine produced at lower amounts (Pažoutová et al., 2008). Additionally, the fungus produces clavicipitic acid (King et al., 1977). The toxins of *C. fusiformis* have been associated with agalactia in sows (Loveless, 1967).

The ergot disease cycle is characterized by sexual and asexual stages (Figure 11.7). Infection occurs in unfertilized floret ovaries. Because pearl millet is a predominantly outcrossing crop, protogyny is essential to ensure good pollination (Miedaner & Geiger, 2015). In pearl millet, the pistil and stigmas emerge 2–3 days before the stamens. Therefore, reduced protogyny time leads to reduced ergot infection, a disease escape mechanism. Previous studies have shown that constriction of stigmas results in tissues that are not receptive to infection (Willingale & Mantle, 1985; Willingale et al., 1986). The introduction of cytoplasmic male sterility (CMS)-based hybrid production systems in pearl millet has increased the time between protogyny and fertilization. This gap leaves the plant vulnerable to floret infection.

About 1 week after successful floret infection, the asexual phase of the disease begins. A fungal mass, called a sphacelium, colonizes the base of the floret and exudes honeydew. Honeydew is composed of host exudates and fungal conidia. *Claviceps fusiformis* produces both macroconidia and microconidia, serving as asexual secondary inoculum. Both spore types are typically single celled.

Figure 11.6   Signs of pearl millet ergot caused by *Claviceps fusiformis*. (a) Honeydew exudate from florets. (b) Formation of black sclerotia replacing seed sites. "Larry E. Claflin Plant Disease Image Collection," Kansas State University Libraries, Manhattan, Kansas, USA / CC BY 1.0.

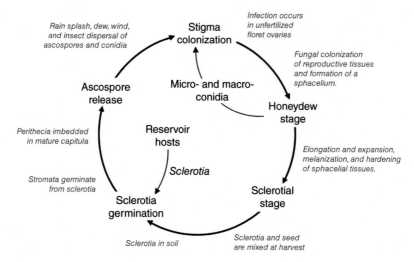

Figure 11.7 Disease cycle of pearl millet ergot caused by *Claviceps fusiformis*. Image by C. R. Little.

Claviceps fusiformis macroconidia can produce multiple germ tubes and typically produce multiple rounds of conidiation from a single conidium (i.e., a germinating conidium produces a germ tube and another conidium) (Thakur et al., 1984). When this occurs on the surface of the honeydew substrate, rain splash or dew, wind, and insects may disseminate conidia to uninfected florets.

Dakshinamoorthy and Sivaprakasam (1988) observed a strong relationship between rainfall, relative humidity, and disease incidence. Kumar et al. (1997) indicated that ergot disease could spread up to 18 m from the infection source and that disease occurrence in the field depended on wind direction. Dakshinamoorthy et al. (1988) observed several insect species visiting pearl millet panicles. Dwarf honeybees (*Apis florea* Fabricius), house flies (*Musca domestica* Linnaeus), and tachina flies (*Tachina fallax* Meigen) carried the most conidia and could potentially deposit these on other unfertilized florets.

The sphacelium, which is essentially a compact mycelial mass covered with minute conidiophores on the surface and interior, matures and dries as the host senesces and produces a hard, melanized structure that often grows beyond the length of the glume and may be exposed to the environment. Sclerotia may "shatter" from the panicle but are most often disturbed during harvest. Sclerotia may fall to the soil or be mixed with harvested seed.

The sclerotia can survive for at least 8–10 months. Thakur et al. (1984) could store sclerotia for a year at 15–30 °C, and these could germinate and produce infectious ascospores. Typically, sclerotia either remain in the field after harvest or are incorporated into harvested seed that is saved and used for replanting. Therefore, sclerotia are overseasoning structures that give rise to the primary ascospore inoculum. High moisture levels (>80% relative humidity, rain) are required to induce sclerotial germination, stipe, and capitulum formation, perithecium and ascospore maturity, and eventual dispersal of ascospores.

Sclerotia, which are typically much larger than the pearl millet grains, germinate stromatal stipes that produce capitula. Perithecia are embedded in the capitulum with necks that project from the capitulum surface stroma (Thakur et al., 1984). The asci formed within the perithecia produce long, thin ascospores, which are ejected. Like most ascomycetes, each ascus contains eight ascospores (Thakur et al., 1984).

Resistance is the most effective disease control measure for ergot, even though it is recessive and is controlled by multiple loci, which involves both nuclear and cytoplasmic inheritance (Thakur et al., 1983; Yadav, 1994; Yadav & Rai, 2013). Thakur and Williams (1980) developed a reliable field screen for ergot involving bagging male-sterile plants at the boot stage to prevent pollination and inoculating with a conidial suspension derived from infected panicle honeydew during the period when stigmas fully emerged prior to anthers, after which the bag is replaced.

Resistance screening is tedious, requires multiple environments, and is prone to escapes and uncertain disease responses due to variability in local pathogen populations. However, resistant sources have been identified using proper selection approaches under high pathogen pressure (Table 11.4). For example, Chalal et al. (1981) and Thakur et al. (1983) used recurrent and pedigree-type selection methods to identify and develop ergot resistance in millet. As a result, Thakur et al. (1985) were able to identify inbred lines (ICMPE) and "sib-bulk" populations (ICMPES) with 1%–7% ergot severity, compared with 30%–65% in BJ 104 (ergot susceptible check). These also had resistance to smut (*M. bullatus*) and downy mildew (*S. graminicola*). Further, Andrews et al. (1985a, 1985b) aimed to integrate ergot resistance with agronomic quality, especially yield. Much of this work was part of broad disease resistance screening programs and breeding nurseries carried out by ICRISAT in the 1980s.

Grain Molds and Storage Molds

Grain Molds

Grain molds of pearl millet are caused by a complex of fungi that colonize the developing floret, maturing grain, or the grain surface at or near harvest. Although the species associated with molded seeds are numerous, only a portion is considered pathogenic (Wilson, 2000). In addition, many species associated with seeds are also found in other tissues, such as leaves, seedlings, and stalks.

Grain mold typically affects the development and maturation of grain under wet or humid conditions. Also, grain left in the field due to delayed harvest may experience greater levels of mold and weathering.

Table 11.4   Selected Pearl Millet Entries and Their Reaction to Ergot Caused by *Claviceps fusiformis*

Plant I.D.	Reaction[a]	References
AHB 1200	Resistant	Kumari et al. (2022)
BJ 104	Susceptible	Thakur et al. (1985), Willingale et al. (1986)
BK 560	Susceptible	Willingale et al. (1986), Thakur et al. (1991)
Dhanskati	Resistant	Kumari et al. (2022)
ERBL 1, -2, -3, -7, -8, -9, -11	Resistant	Thakur et al. (1993)
ERH 4	Resistant	Thakur et al. (1993)
Geron Tsuntsu	Resistant	Abraham et al. (2019)
GHB 538, -558, -732, -744, -905	Resistant	Kumari et al. (2022)
HHB 67, -197, -223, -226, -272	Resistant	Kumari et al. (2022)
HHB 229	Susceptible	Kumari et al. (2022)
ICMA 91113,- 91115	Resistant	Rai & Thakur (1995)
ICMA 91333, -91444, -91555	Resistant	Hash et al. (1997)
ICMER 66, -92, -97, -98, -127	Resistant	Rai & Thakur (1995)
ICMPE 13-6-27	Resistant	Thakur et al. (1985), Willingale et al. (1986)
ICMPE 13-6-30	Resistant	Thakur et al. (1985), Willingale et al. (1986)
ICMPE 134-6-9, ICMPE-6-34	Resistant	Thakur et al. (1985)
ICMPE 134-6-11, ICMPE-6-25, ICMPE-6-41	Resistant	Thakur et al. (1985)
ICMPES 1, -2, -23, -27, -29, -28, -32, -34	Resistant	Thakur et al. (1985), Thakur et al. (1993)
ICMPES 29, 34	Resistant	Thakur et al. (1991)
ICMPES 8, 9	Moderately resistant	Thakur et al. (1991)
ICML 1, -2, -3, -4	Resistant	Thakur et al. (1993)
ICMR 356	Susceptible	Rai & Thakur (1995)
ICMS 7703	Susceptible	Thakur et al. (1993)
J 2210-2	Resistant	Thakur et al. (1982)
Kaveri Super Boss	Resistant	Kumari et al. (2022)
MPMH-17, -21	Resistant	Kumari et al. (2022)
Nirmal 4915	Resistant	Kumari et al. (2022)
PC 383	Susceptible	Kumari et al. (2022)
PC 443	Susceptible	Kumari et al. (2022)

Table 11.4   (Continued)

Plant I.D.	Reaction[a]	References
PEO 5842	Moderately resistant	Abraham et al. (2019)
PH-86M84	Resistant	Kumari et al. (2022)
Pusa 605	Resistant	Kumari et al. (2022)
Raj 171, -173, -177, -178	Susceptible	Kumari et al. (2022)
RHB 154, -173	Resistant	Kumari et al. (2022)
WC-C75	Susceptible	Thakur et al. (1991)
81A, 81B	Susceptible	Rai & Thakur (1995)
841A, B41B	Susceptible	Thakur et al. (1991)
842A, 842B	Susceptible	Thakur et al. (1991)
843A, 843B	Susceptible	Rai & Thakur (1995)
5141B	Susceptible	Thakur et al. (1982)
700599	Moderately resistant	Thakur et al. (1982)
86M01	Resistant	Kumari et al. (2022)

[a] Resistant = <10% disease severity; moderately resistant = 10%–20%; susceptible = >20%.

A comprehensive review of the pearl millet grain mold literature reveals that many of the fungal species associated with pearl millet caryopses are *Fusarium* spp. (30.5% of species*literature sources), *Aspergillus* spp. (12.0%), *Alternaria* spp. (11.5%), and *Curvularia* spp. (8.5%) (Figure 11.8). However, depending on genotype and production environment, fungal diversity on and within pearl millet caryopses differs but can be extensive.

Storage Molds

Mishra and Daradhiyar (1991) isolated 22 fungi from stored pearl millet samples in Bihar State, India. Fungal species included *Aspergillus flavus*, *Aspergillus parasiticus*, other species of *Aspergillus*, *Curvularia* spp., *Fusarium* spp., *Helminthosporium* (sensu lato) spp., and *Rhizopus* spp.

A review of the commonly isolated grain mold genera and specific species within these genera follows.

***Fusarium* spp.**

Many studies have found *Fusarium* spp. (Ascomycota, Sordariomycetes, Hypocreales, Nectriaceae) associated with pearl millet caryopses (Girisham et al., 1985; Ingle & Rout, 1994; Jurjevic et al., 2005; Konde et al., 1980; Luttrell, 1954; Luttrell et al., 1955; Marasas et al., 1988; Mathur et al., 1973; Mishra & Daradihiyar, 1991; Noor et al., 2023; Onyike et al., 1991; Wells & Winstead, 1965; Wilson et al., 1993) (Figure 11.8).

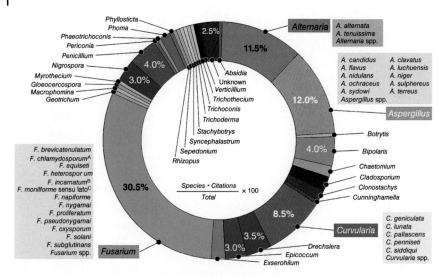

Figure 11.8 Distribution of head mold fungal species obtained from 15 literature sources. The top three most common genera and their species are listed in the inset boxes. Also, percentage frequencies of the top ten genera are indicated. Superscripts: (a), *Fusarium fusarioides* (synonym); (b), *F. semitectum* (synonym); (c) past nomenclature of members of the *Fusarium fujikuroi* species complex. Citations: Luttrell (1954), Luttrell et al. (1955), Wells and Winstead (1965), Mathur et al. (1973), Konde et al. (1980), Shetty et al. (1982), Girisham et al. (1985), Marasas et al. (1988), Mishra and Daradhiyar (1991), Onyike et al. (1991), Ahmed and Reddy (1993), Ingle and Rout (1994), Wilson et al. (1993), Jurjevic et al. (2005), Noor et al. (2023). Image by C. R. Little.

Onyike et al. (1991) examined pearl millet grain in Lesotho, Nigeria, and Zimbabwe. These authors found a predominance of *Fusarium* spp. In Lesotho, 100% of the *Fusarium* isolates recovered were *F. equiseti*. However, only nine *Fusarium* isolates were recovered from seeds obtained in one location. In Nigeria, averaged over five locations, *F. moniliforme* sensu lato (36.6%), *F. nygamai* (30.0%), *F. incarnatum* (syn. *F. semitectum*) (20.1%), *F. equiseti* (16.4%), *F. chlamydosporum* (7.0%), *F. subglutinans* (5.0%), and *F. napiforme* (1.9%) isolates were recovered from seed. In Zimbabwe, averaged over three locations, *F. moniliforme* sensu lato (47.6%), *F. equiseti* (47.1%), *F. incarnatum* (18.6%), *F. nygamai* (7.5%), *F. chlamydosporum* (2.4%), *F. oxysporum* (0.9%), and *F. napiforme* (0.8%) isolates were recovered.

Recent data, partially available in Noor et al. (2023), used morphological and polymerase chain reaction sequencing to identify *Fusarium* isolates from pearl millet seed derived from 54 parental lines and hybrids produced in Hays, Kansas, USA. These authors found several caryopsis-associated *Fusarium* spp. based on internal transcribed spacer region, primers for multiple loci, and sequencing analysis.

Table 11.5 outlines the potential *Fusarium* mycotoxins associated with pearl millet grain and cites many original studies that discovered these associations.

Table 11.5   Potential Mycotoxins Produced by Pearl Millet Seed–associated *Fusarium* spp

SPP	Beauvericin	Equisetin	Fumonisins	Fusapro liferin	Fusaric acid	Monili formin	Sambu toxin	Trichothe cenes	Zearale none
FBR	–	–	Tr	–	–	–	–	–	–
FCH	–	–	–	–	–	+	–	–	–
FEQ	+	+	–	–	–	–	–	+	+
FHE	–	–	–	–	+	–	–	–	–
FIN	+	+	–	–	–	+	+	+	+
FNA	–	–	+	–	+	+	–	–	–
FNY	–	–	+	–	–	+	–	–	–
FOX	+	–	+	–	+	+	+	–	–
FPS	–	–	–	+	–	+	–	–	–
FSO	–	–	–	–	+	+	–	–	–
FSU	–	–	–	+	+	+	–	–	–

Abbreviations: FBR, *F. brevicatenulatum*; FCH, *F. chlamydosporum*; FEQ, *F. equiseti*; FIN, *F. incarnatum*; FNA, *F. napiforme*; FNY, *F. nygamai*; FOX, *F. oxysporum*; FPS, *F. pseudonygamai*; FSO, *F. solani*; FSU, *F. subglutinans*; Tr, trace.

Citations: Amalfitano et al. (2002), Bacon et al. (1996), Capasso et al. (1996), Chelkowski et al. (1990), Eged (2002), Fotso et al. (2002), Hestbjerg et al. (2002), Leslie and Summerell (2006), Leslie et al. (2005), Lew et al. (1996), Logrieco et al. (1998), Marasas et al. (1984, 1986, 1991), Markell and Francl (2003), Moretti et al. (1995), Musser & Plattner (1997), Nelson et al. (1992), Rabie et al. (1978, 1982), Seo et al. (1999), Sewram et al. (1999, 2005), Shepherd et al. (1999), Thiel et al. (1991), Vismer et al. (2019), Wu et al. (2003).

It is important to note that not all *Fusarium* spp. are equally capable of producing mycotoxins, and not all isolates of mycotoxin-producing species produce myco-toxins at a level of concern. Host genotype and field and storage environment play a role in the level of fungal colonization of grain and subsequent mycotoxin elabo-ration. Further, many species of *Fusarium* carry the fumonisin gene cluster, although that does not always guarantee the ability of a particular isolate to pro-duce fumonisins.

Among the *Fusarium* spp. infecting pearl millet, caryopses, *F. pseudonygamai* produces mycotoxins, including fumonisin and moniliformin (Figure 11.9). Vismer et al. (2019) found that an *F. pseudonygamai* isolate (MRC 8723) produced fumoni-sin B_1, B_2, and B_3. However, the ratio of fumonisin B_1 production on pearl millet "patties" in vitro was three times higher for fumonisin B_1 compared with B_2 and B_3. Other isolates were tested in vitro on pearl millet patties and produced surprising levels of fumonisins. For example, isolates of *F. proliferatum* from maize (MRC 8742-8746) and sorghum (MRC 8726-8731) produced 2,200–3,800 mg/kg and 2,600–6,470 mg/kg total fumonisins, respectively. An average of these ranges is about one-third and one-half, respectively, of that produced by the *F. verticillioides* fumonisin positive control (MRC 826) at 9,090 mg/kg. The maximum tolerated daily intake suggested by the World Health Organization is 2 µg/kg of body weight (Voss & Riley, 2013). Although the above-mentioned fumonisin levels come from an experimental situation, there is a danger of exceeding this level if highly con-taminated foodstuffs are consumed.

In the same study, the *F. pseudonygamai* isolate produced 21,400 mg/kg of mon-iliformin on pearl millet patties, which is about two-thirds of the value produced by the *F. napiforme* moniliformin positive control at 33,400 mg/kg (Vismer et al., 2019). Also, three *F. proliferatum* isolates from maize (MRC 8744-8746) and seven *F. proliferatum* isolates from sorghum (MRC 8726-8732) produced 107–1,620 mg/kg and 853–8,890 mg/kg of moniliformin.

Aspergillus spp.

Multiple studies have found *Aspergillus* spp. (Ascomycota, Eurotiomycetes, Eurotiales, Aspergillaceae) associated with pearl millet caryopses (Girisham et al., 1985; Hussein et al., 2009; Konde et al., 1980; Luttrell et al., 1955; Mishra & Daradihiyar, 1991; Onyike et al., 1991; Wilson et al., 1993) (Figure 11.8).

Williams and McDonald (1983) report the association of *A. flavus* with mature seeds of various cereals. During wet seasons, Mishra and Daradhiyar (1991) indi-cated that *A. flavus* and *A. parasiticus* were common colonizers compared with other species. In Niger, *A. niger* was an important deteriorative pathogen of millet seed, resulting in seed decay and seedling damping-off (Makun et al., 2007). Hussein et al. (2009) found *Aspergillus alba*, *A. flavus*, and *A. niger* in association

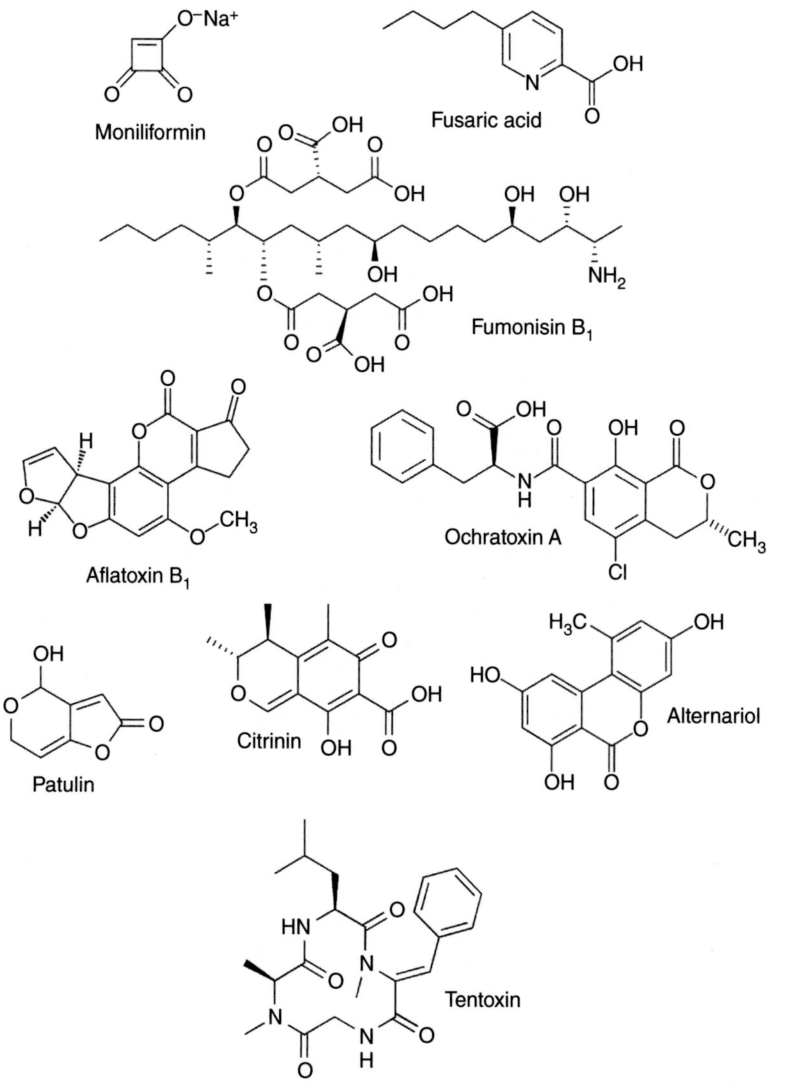

Figure 11.9 Common mycotoxins associated with *Fusarium* (moniliformin, fusaric acid, fumonisin B$_1$), *Aspergillus* (aflatoxin B$_1$, ochratoxin A), *Alternaria* (alternariol, tentoxin), and *Penicillium* spp. (patulin, citrinin). Citations: Bottalico and Logrieco (1994), Puel et al. (2010), Ismaiel and Papenbrock (2015), Pitt and Miller (2017), Perincherry et al. (2019), Kumar et al. (2022). Image by C. R. Little.

with pearl millet grain from several different cultivars in Pakistan. In these authors' study, the incidence of *Aspergillus* spp. was 6.5%, 14.0%, and 19.0%, respectively.

Aspergillus spp. produce a wide range of secondary metabolites, including aflatoxins (B1, B2, G1, and M1), agroclavine, kojic acid, malformins (A and C), ochratoxins (A and B) as well as aflatoxin precursors including averantin, averufin, sterigmatocystin, and versicolorins (A and C) (Houissa et al., 2019).

In India, Mishra and Daradhiyar (1991) found aflatoxin B1 (Figure 11.9) in both stored pearl millet grain (17–2,110 ppb) and cooked samples (18–549 ppb). In the Tunisian pearl millet mycotoxin study by Houissa et al. (2019), the median value for aflatoxin B1 across 220 samples was 0.296 mg/kg of sample (296 ppb), with a maximum of 1.046 mg/kg of sample (1,046 ppb).

Also in the Tunisian pearl millet mycotoxin study by Houissa et al. (2019), the median value for ochratoxin A across 220 samples was 0.036 mg/kg (36 ppb), with a maximum of 0.231 mg/kg (231 ppb). The World Health Organization Joint Expert Committee on Food Additives established a provisional weekly tolerance intake of 112 ng/kg (112 ppt) of body weight for ochratoxin A.

The most common reason for contamination of grain, including pearl millet, by mycotoxins derived from *Aspergillus* spp. is improper or suboptimal storage. *Aspergillus* spp. are xerophilic, meaning that these fungi can grow and reproduce on substrates with low levels of available water (Lasram et al., 2016). Therefore, even grain with low moisture levels stored at 25–35 °C can develop mycotoxin contamination issues by these fungi (Paterson & Lima, 2011).

Alternaria spp.

The most common species of *Alternaria* (Ascomycota, Dothideomycetes, Pleosporales, Pleosporaceae) associated with pearl millet grain is *Alternaria alternata* (Figure 11.8). Williams and McDonald (1983) reported *A. alternata* in association with pearl millet, resulting in grain deterioration and damaged seed sites in the panicle, affecting quantitative yield. Hussein et al. (2009) found *A. alternata* at an incidence of 35% in pearl millet grain from Pakistan. Noor et al. (2023) isolated and identified *A. alternaria* and *A. tenuissima* from pearl millet seed in Kansas, USA. Further, *A. alternata* can result in seed that shows reduced germination, pre-emergence damping-off, and seedling blight (Konde et al., 1980; Mishra & Prakash, 1975).

Alternaria spp. produce many phytotoxic metabolites that influence host range and pathogenicity, as well as those that may have allergic or toxic effects on animal cells. *Alternaria* produces toxic metabolites, including altersetin, altersolanol, alternariol (Figure 11.9), alternariol methylether, infectopyron, macrosporin, tentoxin (Figure 11.9), and tenuazonic acid (Bottalico & Logrieco, 1994; Houissa et al., 2019). In the Tunisian pearl millet multi-mycotoxin study by Houissa et al.

(2019), the median value for alternariol across 220 samples was 0.0269 mg/kg of the sample (26.9 ppb), with a maximum of 0.194 mg/kg of the sample (194 ppb). In the case of tentoxin, the median value across 220 samples was 0.0082 mg/kg (8.2 ppb), with a maximum value of 0.111 mg/kg (111 ppb).

Curvularia spp.

The most common species of *Curvularia* (Ascomycota, Dothideomycetes, Pleosporales, Pleosporaceae) associated with pearl millet grain is *C. lunata* (Figure 11.8). Hussain et al. (2009) found *C. lunata* at an incidence of 23.5% in pearl millet grain from Pakistan. Further, seeds were reduced in germination and emergence if infected by *C. lunata*. Sivanesan (1990) indicated that *C. penniseti*, another species associated with pearl millet, could result in seedling leaf lesions and stunted or blighted seedlings.

Penicillium spp.

Penicillium spp. (Ascomycota, Eurotiomycetes, Eurotiales, Aspergillaceae) are often considered storage molds and may be less of a grain mold or weathering fungus. Nevertheless, *Penicillium* spp. produce a range of metabolites, including atpenin A6, citreorosein, citrinin (Figure 11.9), curvularin, dihydrocitrinone, oxaline, and quinolactin A (Houissa et al., 2019; Pfohl-Leszkowicz & Manderville, 2007). Some *Penicillium* spp. produce ochratoxin A, and it is not known which *Penicillium* spp. associated with millet may be ochratoxin producers because ochratoxin in millet samples or food products is most likely elaborated by *Aspergillus* spp. (Houissa et al., 2019; Pfohl-Leszkowicz & Manderville, 2007).

Given the nephrotoxic nature of citrinin, the production of this metabolite by some species of *Penicillium*, including *P. citrinum*, is of concern. Doses of 10 mg/kg (10 ppm) of body weight can cause renal lesions in dogs, and 50 mg/kg (50 ppm) causes kidney damage in rats (Kogika et al., 1993; Lockard et al., 1980). Typically, citrinin and ochratoxin contamination of foodstuffs co-occurs (Pfohl-Leszkowicz & Manderville, 2007). In the Tunisian pearl millet multi-mycotoxin study by Houissa et al. (2019), the median value for citrinin across 220 samples was 2.47 mg/kg of sample (2.47 ppm), with a maximum of 25.66 mg/kg of sample (25.66 ppm).

Managing Head Molds and Mycotoxins

In addition to the production of mycotoxins, grain mold fungi affect seed germination if the seed is to be saved, stored, and replanted in subsequent cropping seasons. In Burkina Faso, Elisabeth et al. (2008) found that *A. alternata*, *C. lunata*, and *Fusarium* spp. were associated with deteriorated seeds that had poor germination. Hussain et al. (2009) found that pearl millet grain in Pakistan that was colonized by a range of grain mold fungi produced 59%–72% "abnormal" seedlings.

Chintapalli et al. (2006) used isolation rates from pearl millet caryopses to compare different genotypes for resistance. In the case of total fungi, fungi grew out of 'OP 303' caryopses, the grain mold susceptible line, at a higher rate (as measured by a significantly greater regression slope of time × kernels yielding fungal colonies) than that of '2034', 'OP 106', or 'TG 102'. Similarly, isolation rates from OP 303 for *Fusarium* spp. were significantly higher than those from the other three lines.

Leaf Blights, Sheath Blights, and Leaf Spots

Genus and species names of many leaf spot and blight fungi, especially those ascomycetes in the class Dothideomycetes, order Pleosporales, and family Pleosporaceae, have undergone revision over the past few decades. Here, the current taxonomic names, based upon phylogenetic taxonomic revisions, are presented, and synonyms are given as appropriate for *Bipolaris*, *Curvularia*, and *Drechslera* spp. (Manamgoda et al., 2012, 2014) (Figure 11.10).

Morphologically, many leaf spots coalesce to form blights, and it is difficult to distinguish between one leaf spot and another. Ultimately, it is necessary to determine the causal pathogen because the extent of chlorosis, necrosis, and other universal leaf spot symptoms varies depending on the cultivar, local pathogen population, and environment. In some cases, foliar symptoms are distinct enough to be diagnostic (Figure 11.11).

Bipolaris

Several species of *Bipolaris* Shoemaker (Ascomycota, Dothideomycetes, Pleosporales, Pleosporaceae) cause leaf spots in pearl millet. These include *B. bicolor* (Mitra) Shoemaker [syn. *Drechslera bicolor* (Mitra) Subram. & B.L. Jain], *B. colocasiae* (M.P. Tandon & V. Bhargava) Alcorn (syn. *Drechslera colocasiae* M.P. Tandon & V. Bhargava, or "*D. colocaseae*"), *B. cynodontis* (Marignoni) Shoemaker [syn. *Drechslera cynodontis* (Marignoni) Subram. & B.L. Jain], *B. papendorfii* (Aa) Alcorn (syn. *Curvularia papendorfii* Aa), *B. sacchari* (E.J. Butler) Shoemaker [syn. *Drechslera sacchari* (E.J. Butler) Subram. & B.L. Jain], *B. setariae* Shoemaker [syn. *Drechslera setariae* (Shoemaker) Subram. & B.L. Jain], and *B. urochloae* (V.A. Putterill) Shoemaker [syn. *Drechslera urochloae* (V.A. Putterill) Subram. & B.L. Jain] (Figure 11.12) (Singh et al., 1990a; Wilson, 2000). Of the *Bipolaris* spp. found on pearl millet, *B. bicolor*, *B. cynodontis*, and *B. setariae* have known *Cochliobolus* teleomorphs, which are *C. bicolor* A.R. Paul & Parbery, *C. cynodontis* R.R. Nelson, and *C. setariae* (S. Ito & Kurib.) Drechsler ex Dastur, respectively.

Bipolaris setariae is the most widespread fungus causing significant Bipolaris leaf spot disease. In addition to leaf spots on pearl millet, *B. setariae* is seedborne and causes seedling blight (Bhowmik, 1972; Shetty et al., 1982; Wells & Winstead, 1965; Wells, 1967). *Bipolaris setariae* infects a wide range of plants

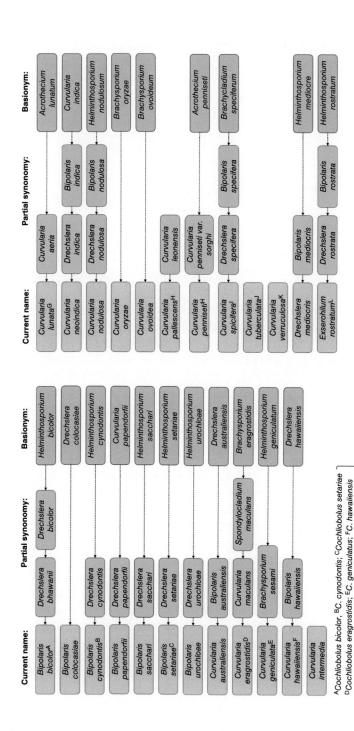

Figure 11.10 Current names, synonymous names, and basionyms for members of the Dothideomycetes causing leaf spots and blights on pearl millet. See Index Fungorum (indexfungorum.org) and Manamgoda et al. (2012, 2014), for more information. Image by C. R. Little.

Figure 11.11 Foliar symptoms of several minor leaf spots of pearl millet including (a) Drechslera leaf spot, (b) Dactuliophora leaf spot, (c) Curvularia leaf spot, (d) Bipolaris leaf spot, (e) Phyllosticta leaf spot, and (f) zonate leaf spot. "Larry E. Claflin Plant Disease Image Collection," Kansas State University Libraries, Manhattan, Kansas, USA / CC BY 1.0.

besides pearl millet, including browntop millet [*Brachiaria ramosa* (L.) Stapf.], cassava (*Manihot esculenta* Crantz), coconut seedlings (*Cocos nucifera* L.), maize (*Zea mays* L.), rubber trees (*Hevea brasiliensis* Muell. Arg.), causes brown stripe diseases of sugarcane (Li et al., 2022; Liu et al., 2016; Niu et al., 2014; Ramesh et al., 2021).

Balasubramanian (1980) reported that *B. setariae* also infected pearl millet plants susceptible to downy mildew, but healthy plants did not exhibit leaf spot. This suggests that downy mildew and leaf spot resistance are either closely linked or use similar basal resistance mechanisms.

A recent survey of Bipolaris leaf spot caused by *B. setariae* was conducted in Rajasthan, India, by Kardam et al. (2023). These authors found widespread incidence and severity of Bipolaris leaf spot throughout the state. Disease incidence ranged from 45.4% to 64.9%, and severity ranged from 11.3% to 32.0% in the locations surveyed. Further, local varieties (deshi) exhibited higher levels of susceptibility than improved varieties. Also, the authors examined other cultural factors, including soil type and cropping rotation, and their influence on the severity of Bipolaris leaf spot disease. Higher disease severity was found when plants were cultivated in clay soil versus loamy or sandy soil. Millet produced in a rotation with mustard also showed higher rates of leaf spot than in a wheat or chickpea rotation.

Wells and Hanna (1988) examined the complex inheritance of resistance to *B. setariae*. They found that resistance is controlled by four independent genes, consisting of two dominant resistance genes (Bp_1 and Bp_2), an inhibitory gene (Bp_3), and an anti-inhibitory gene (Bp_4).

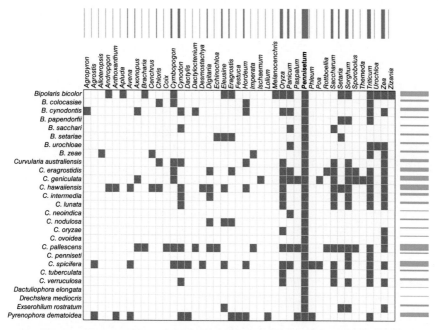

Figure 11.12 Monocot hosts of the leaf spot and blight pathogens of millet. References: Bhomik (1972), Sivanesan (1987), Wilson (2000), Manamgoda et al. (2012, 2014), Xiao et al. (2019), Ramesh et al. (2021), Li et al. (2022), CABI (2023). Image by C.R. Little.

Cercospora

The most important species of *Cercospora* (Ascomycota, Dothideomycetes, Capnodiales, Mycosphaerellaceae) affecting pearl millet is *C. penniseti* Chupp, which causes Cercospora leaf spot. *Cercospora penniseti* causes the most important damage to pearl millet varieties used for forage biomass. Narayanaswami and Veerraju (1970) first identified the disease in India.

A study in Georgia, USA, by Burton and Wells (1981) tested F_2 and F_3 populations from a Tift 13 × Tift 18 cross. As a result, 245 of 999 F_2 plants showed moderate, moderately severe, and severe Cercospora leaf spot. In the susceptible F_3 near-isogenic populations, Cercospora leaf spot reduced dry matter biomass yield (kg/ha) by an average of 13.5% compared with the resistant near-isogenic populations.

In Uganda, Lubadde et al. (2014) found that Cercospora leaf spot was among the significant diseases of pearl millet. In eastern Uganda, Cercospora leaf spot incidence and severity were less than that observed in northern Uganda, at an average of 2.5% and 1.0% versus 10.8% and 1.1%, respectively.

Curvularia

Several species of *Curvularia* also cause leaf spots. These include *C. australiensis* (Bugnic. ex M.B. Ellis) Manamgoda, L. Cal & K.D. Hyde (syn. *Bipolaris australiensis* Bugnic. ex M.B. Ellis) (Tsuda & Ueyama), *C. eragrostidis* (Henn.) J.A. Mey, *C. geniculata* R.R. Nelson, *C. hawaiiensis* [syn. *Bipolaris hawaiiensis* (Bugnic. ex M.B. Ellis) J.Y. Uchida & Aragaki], *C. intermedia* Boedijn, *C. lunata* (Wakker) Boedijn, *C. neoindica* Manamgoda, Rossman & K.D. Hyde (syn. *Bipolaris indica* J.N. Rai, Wadhwani & J.P. Tewari), *C. nodulosa* (Sacc.) Manamgodam Rossman & K.D. Hyde [syn. *B. nodulosa* (Sacc.) Shoemaker], *C. oryzae* Bugnic., *C. ovoidea* (Hiroë) Munt.-Cvetk., *C. pallescens* Boedijn, *C. penniseti* (Mitra) Boedijn, *C. spicifera* (Bainier) Boedijn [syn. *Bipolaris spicifera* (Bainier) Subram.], *C. tuberculata* B.L. Jain, and *C. verruculosa* Tandon & Bilgrami ex M.B. Ellis (Figure 11.12) (Wilson, 2000). *Curvularia eragrostidis, C. geniculatus, C. nodulosa, C. pallescens, C. tuberculata,* and *C. verruculosa* have *Cochliobolus eragrostidis* (Tsuda & Ueyama) Sivan., *C. geniculatus* R.R. Nelson, *C. nodulosus* Luttr., *C. pallescens* (Tsuda & Uemaya) Sivan., *C. tuberculatus* Sivan., and *C. verruculosus* (Tsuda & Uemaya) Sivan. teleomorphs, respectively. *Curvularia hawaiiensis, C. lunata,* and *C. spicifera* have *Pseudocochliobolus hawaiiensis* (Alcorn) Tsuda & Ueyama, *P. lunatus* (R.R. Nelson & F.A. Haasis) Tsuda, Ueyama & Nishih., *P. specifer* (R.R. Nelson) Tsuda, Ueyama & Nishih. teleomorphs, respectively.

Mahmood and Qureshi (1984) indicated that Curvularia leaf spot occurred on all pearl millet cultivars tested in field experiments in Pakistan but did not indicate which species was the causal pathogen.

Curvularia penniseti causes brown leaf spot disease in pearl millet. Singh et al. (2008) showed that seedborne *C. penniseti* could be transmitted from infected seed to seedlings. Further, seedborne inoculum may be responsible for the foliar leaf spots and blights caused by this fungus. *Curvularia eragrostidis* causes pearl millet midrib spot (Zarafi et al., 2004). *Curvularia lunata* is a seedborne fungus that causes leaf spot disease on pearl millet.

Nitharwal et al. (1991) showed that a *C. lunata* isolate causing leaf blight on pearl millet produced cellulases, which they hypothesized as a pathogenicity factor for the fungus. In addition to the production of cellulases, *C. lunata* produces a host and cultivar-specific phytotoxin (Gour et al., 1992).

Dactuliophora

Dactuliophora leaf spot is caused by *Dactuliophora elongata* C.L. Leakey (Ascomycota, Dothideomycetes, Pleosporales; Family, *Insertae sedis*), an ascomycetous fungus that remains taxonomically under-described. The fungus was first reported on pearl millet in Nigeria (Leakey, 1964; Tyagi, 1985). However, it is a common leaf spot pathogen in other West African pearl millet production areas. Wilson et al. (2000) reported damaging levels of Dactuliophora leaf spot in Niger and Mali. Here, lesions coalesced and caused entire leaves to become blighted. Like zonate leaf spot (see below), the fungus produces lesions characterized by concentric rings. In addition, the fungus produces sclerotia that develop in older lesions.

Drechslera

The causal agent of what has classically been considered Drechslera leaf spot is *Pyrenophora dematioidea* (Bubák & Wróbl.) Rossman & K.D. Hyde [syn. *Drechslera dematioidea* (Bubák & Wróbl.) Scharif] (Ascomycota, Dothideomycetes, Pleosporales, Pleosporaceae). Drechslera leaf spot may be an outmoded disease name because most of the most common fungal species historically associated with this disease have been renamed. However, for consistency, the historical disease name will be used here.

In addition to causing a leaf spot disease, *P. dematioidea* has been associated with the stalk rot complex of fungi in pearl millet (Wilson, 2002b), although at a very low frequency (<1% of stalk nodes).

Exserohilum

Exserohilum leaf blight of pearl millet is caused by *E. rostratum* (Drechs.) K.J. Leonard & E.G. Suggs (Ascomycota, Dothideomycetes, Pleosporales, Pleosporaceae). *Exserohilum rostratum* was identified on pearl millet by Young et al. (1947). Wilson and Hanna (1992) isolated *E. rostratum* from foliar lesions in a set of "Tift 23" derivatives at a rate of 0.4% in Georgia. Wilson and Gates (1993) isolated *Exserohilum* from 4.4% of isolations from "Tifleaf 1" leaf spots in Georgia. Wilson et al. (1999) reported that Exserohilum leaf blight was positively and significantly correlated to plant stand density. Further, Exserohilum leaf blight was negatively and significantly correlated to yield.

In Uganda, Lubadde et al. (2014) indicated that Exserohilum leaf blight showed the highest incidence and severity in the Katakwi district in the eastern part of the country.

Frederickson et al. (1999) found that *E. prolatum* K.J. Leonard & Suggs could be isolated from bacterial spot (*Pseudomonas syringae*) lesions in Zimbabwe. Because these authors proved Koch's postulates for *P. syringae* as the causal agent of bacterial spot, it may be hypothesized that *E. prolatum* is a facultative necrotroph or saprophyte that colonizes leaf tissues that a primary pathogen has already killed.

Gloeocercospora

Zonate leaf spot of pearl millet is caused by *Gloeocercospora sorghi* D.C. Bain & Edgerton ex Deighton (Ascomycota, Sordariomycetes, Xylariales; Family, *Insertae sedis*). Tyagi (1985) indicated confusion between Dactuliophora leaf spot (*D. elongata*) and zonate leaf spot on pearl millet. The main difference between the diseases is how microsclerotia are produced. *Gloeocercospora sorghi* microsclerotia are produced under the epidermis, whereas *D. elongata* microsclerotia become erumpent (i.e., break through the surface). Otherwise, both diseases can produce a bullseye-like lesion on the leaf.

Mahmood and Qureshi (1984) reported zonate leaf spot for the first time, although at a low incidence of 2%–6%, in Pakistan on cultivars C-47 and DB-2.

Myrothecium

Paramyrothecium roridum (Tode) L. Lombard & Crous (syn. *Myrothecium roridum* Tode) (Ascomycota, Sordariomycetes, Hypocreales, Stachybotryaceae) is the causal agent of Myrothecium leaf spot in pearl millet. In addition to pearl millet, *P. roridum* causes leaf spots in other hosts, including cluster bean, cotton, peanut (groundnut), and water hyacinth (Chauhan et al., 1997; Okunowo et al., 2013; Pal et al., 2014; Shivanna & Shetty, 1986).

Paramyrothecium roridum is considered a toxigenic fungus and produces the macrocyclic trichothecenes, verrucarin A, and roridin E under various experimental conditions (Bosio et al., 2017; Mantle, 1991).

Phyllosticta

Phyllosticta penicilliariae Speg. (Ascomycota, Dothideomycetes, Botryosphaeriales, Botryosphaeriaceae) causes Phyllosticta leaf blight. Wilson and Burton (1990) identified *P. penicilliariae* in Georgia, USA, causing leaf lesions. Individual lesions coalesced into a blight phase of the disease. The diagnostic feature of Phyllosticta leaf blight that differentiates it from other pearl millet leaf spots and blights is the production of pycnidia within the centers of necrotic lesions and the production of numerous, hyaline, single-celled, biguttulate, and ovoid conidia.

Pyricularia

Pyricularia leaf blight and blast is caused by the ascomycetous fungal pathogen *P. grisea* Cooke ex Sacc. [Teleomorph & synonym: *Magnaporthe grisea* (T.T. Herbert) M.E. Barr] (Ascomycota, Sordariomycetes, Magnaporthales, Magnaporthaceae) (Thakur et al., 2009). *Pyricularia grisea* can infect a range of hosts other than pearl millet, including finger millet, foxtail millet, rice, wheat, and other grasses (Prabhu et al., 1992; Sharma et al., 2013; Takan et al., 2012). The disease has been found to occur in India (Andhra Pradesh, Gujarat, Haryana, Karnataka, Maharashtra, Rajasthan, and Uttar Pradesh), Africa (Burkina Faso, Chad, Mali, Niger, Nigeria, and Senegal), and the southeastern United States (primarily Georgia and Florida). In India, the disease was first observed in Kanpur, Uttar Pradesh (Mehta et al., 1953). Likewise, increased disease incidence and losses to this disease have made it a particularly critical issue in India (Lakose et al., 2007).

Isolates from various hosts are highly variable and show a strong degree of host specialization. For example, isolates of *P. grisea* from pearl millet tend only to infect pearl millet and do not infect rice or finger millet (Sharma et al., 2013; Yadav &

Rai, 2013). Isolates from rice, finger millet, foxtail millet, wheat, and grassy weeds show wide pathogenic variation (Prabhu et al., 1992; Takan et al., 2012). In rice resistance to P. *grisea* is short-lived due to the complexity of the pathogen population and its ability to "break" host resistance due to pathogenicity changes and race specificity shifts (Srinivasachary et al., 2002; Suh et al., 2009). The underlying mechanisms for such variability and pathogen shifts are sexual recombination, heterokaryosis, parasexual recombination, and aneuploidy, which are characteristic of the pathogen (Sharma et al., 2013). All these factors suggest that the pearl millet P. *grisea* population is also highly variable due to many, if not all, the same reasons.

An isolate of the pathogen (RMg-DI) from India has been sequenced (Kumar et al., 2017) and assembled (Reddy et al., 2022). Using Illumina and PacBio sequencing, hybrid de novo genome assembly, and functional genome annotation, Reddy et al. (2022) found cell wall–degrading enzymes, carbohydrate-active enzymes, effectors, and virulence factors; transposons; carbohydrate metabolism, signal transduction, and secondary metabolism pathways; and orthologous proteins, protein families, and protein families, identified based upon a range of bioinformatics analyses tools. The genome of this isolate includes 51 virulence factors, 539 carbohydrate-active enzymes, 163 peptidases, 871 secretome proteins, and 594 effector proteins. These authors also identified over 18k transposable elements in the genome.

Dense canopies and high relative humidity increase the incidence of blast disease. Lesions begin as watersoaked areas on the leaf, become necrotic, and take on an "eyespot" or spindle-like appearance (Wlison et al., 1989) (Figure 11.13). Extensive lesion formation results in chlorosis and premature leaf senescence (Sharma et al., 2013). Ultimately, the disease reduces grain yield and results in forage losses, including reduced green forage, dry matter, and digestible dry matter yields (Timper & Wilson, 2006; Wilson & Gates, 1993).

Thakur et al. (2009) used a spray inoculation method at pre-tillering and flowering to screen for blast resistance in the field using a suspension of 1×10^5 conidia/ml. They maintained high humidity using "perfo-irrigation" twice daily for 30 minutes per irrigation event. Those pearl millet entries rated as resistant in the field were then subsequently screened at the 15-day seedling stage in the greenhouse using the same conidia concentration in a spray inoculum under a >90% relative humidity environment. The field and greenhouse screening methodologies used the 1–9 foliar rice blast rating scale developed at the International Rice Research Institute, Philippines. Here, ratings of 1, 2, and 3 are characterized by no lesions or minute specks, larger brown specks, and small (<2mm) necrotic spots, respectively, as resistant to blast. Ratings of 7, 8, and 9, characterized by spindle-shaped blast lesions that appear on 25%–50% and 51%–75% with many dead leaves and all dead leaves, respectively, are susceptible. Many resistance studies in India and elsewhere have adopted this approach for pearl millet genotype screening to blast disease Table 11.6.

Figure 11.13 Foliar lesions produced by *Pyricularia grisea*, the causal agent of pearl millet leaf blight and blast. Note the spindle or "eyespot" shape of the lesions. Image by C. R. Little.

Table 11.6 Some Pearl Millet Entries and Their Reaction to Blast Caused by *Pyricularia grisea*

Plant ID	Reaction[a]	References
ICMB 02111, 07111, 06444, 09333, 09999, 88004, 92444, 93333, 97222	Resistant	Goud et al. (2016), Singh et al. (2018)
ICMB 03333, 89111, 95444, 99666	Susceptible	Thakur et al. (2009), Goud et al. (2016), Sharma et al. (2018), Singh et al. (2018)
ICMB 97222-P1	Resistant	
ICMR 06222, 06444, 11003	Resistant	Thakur et al. (2009), Singh et al. (2018)
ICMR 08111, 10888	Resistant	Pujar et al. (2023)
IP 21187-P1	Resistant	Singh et al. (2018)
K-08-18400, K-08-18404	Resistant	Thakur et al. (2009)
Tift 85DB	Resistant	Wilson et al. (1989)
81B	Resistant	Goud et al. (2016)

[a] Using the modified rice-blast disease severity scale (1–9) developed by IRRI, where <3.0 is resistant and ≥7.0 is susceptible.

In *Pennisetum violaceum* (syn. *P. monodii*), a wild relative of pearl millet, Hanna and Wells (1989) found three independent resistance genes. Wilson et al. (1989) found a single *R* gene in Tift 85DB inherited dominantly. Gupta et al. (2012) used ICMR 06222 (*iniari* type, open-pollinated variety developed in southern Africa by ICRISAT) and ICMR 07555 (non-*iniari* type synthetic also developed by ICRISAT), each possessing a common *R* gene, to produce F_2 plants resistant to the Patancheru isolate of *P. grisea.*

Sharma et al. (2013) tested 25 *P. grisea* isolates against 10 pearl millet genotypes. ICMB 06444, ICMB 97222, and 863B exhibited decreased severity (<3.0 on the IRRI scale) across the isolates, whereas ICMB 95444, ICMB 96666, and ICMB 02777 showed high disease severity (>7 on the IRRI scale) across the isolates. As for pathogen isolates, Pg007 and Pg039 showed lower severity across the genotypes, whereas Pg056 and Pg118 showed the highest severity. In this study, the authors identified five pathogenic subgroups based on using 10 pearl millet genotypes as differentials. Sharma et al. (2020) found resistance in 17 accessions of the pearl millet minicore collection to five Indian pathotypes of *P. grisea.* Subsequently, Sharma et al. (2021) found 14 pathogenic groups of *P. grisea* based on 10 host differentials.

Due to the variability of the pathogen and the fact that many hybrids lack blast resistance, fungicides are another tool in the IPM kit for disease management. In addition to being present in the soil and in crop residue in the form of chlamydospores and mycelium, *P. grisea* may also be seedborne. All these modes may contribute as primary inoculum in the disease cycle (Singh & Pavgi, 1997).

Sharma et al. (2018) tested combinations of seed treatments and foliar sprays for control of blast. These authors found that seed treatment and two or three seasonal sprays of tebuconazole (50%) + trifloxystrobin (25% WG) ('Nativo') and propiconazole (25% EC) ('Tilt') treatments reduced foliar blast area under the disease progress curve (AUDPC) in blast-susceptible ICMB 95444 plants compared with little or no impact of the other fungicides tested, including carbendazim and tricyclazole. Bhardawaj et al. (2022) also evaluated tebuconazole (50%) + trifloxystrobin (25% WP) and tricyclazole (75% WP) at four different locations in India and found that, like the previous study, the tebuconazole + trifloxystrobin treatment reduced AUDPC compared with the control treatments and other fungicides. However, these authors also indicated that tricyclazole was effective if combined as a seed treatment and two seasonal foliar sprays.

Kaurav et al. (2018) compared several botanical-based products and found that neem kernel, *Aloe vera*, and *Lantana camara* extracts reduced disease incidence compared with control, although disease incidence remained higher than when tebuconazole + trifloxystrobin, propiconazole, tricyclazole, iprobenphos, hexaconazole, and azoxystrobin were used. Bhardawaj et al. (2022) tested neem oil (1% azadirachtin) and chitosan in the form of a seed treatment with or without two foliar sprays. Both neem oil and chitosan reduced blast AUDPC compared with control.

Stalk, Crown, and Root Diseases

A range of fungi are associated with the stalks of pearl millet (Wilson, 2000, 2002b). However, not all species cause disease, and none is associated with significant lodging. Among the common pathogens of stalks, crowns, and upper root systems are *Sclerotium rolfsii*, *Rhizoctonia* spp., *Fusarium* spp., and *M. phaseolina*.

Bipolaris Stalk Rot

In addition to causing leaf spots, blight, and infecting seeds, *Bipolaris* spp. can be associated with pearl millet stalks and cause a stalk rot disease. Navi et al. (1997) found that *B. panici-miliacei* was associated with pearl millet stalks in Rajasthan, India. Wilson et al. (1999) and Wilson (2002a, 2002b) found *B. setariae* associated with stalks. In fact, in Georgia, Wilson (2002b) reported that *B. setariae* was the most predominant fungus associated with discolored nodes of stalks, primarily when isolations occurred at later plant stages (> hard dough stage). Mofini et al. (2022) found that *Bipolaris* spp. were "hub taxa" in the mycobiome of pearl millet roots. Therefore, frequent association of this genus with all plant parts seems logical and may represent keystone species on and within the host plant. Stressors or physiological disturbances to the host may force the transition of *Bipolaris* from epiphytic or endophytic growth to pathogenic growth. However, this switch in lifestyle roles has yet to be established for *Bipolaris* spp. associated with pearl millet plant tissues.

Charcoal Rot

Charcoal rot is caused by the soilborne fungal pathogen *Macrophomina. phaseolina* (Tassi)Goid.(Ascomycota,Dothidomycetes,Botryosphaeriales,Botryosphaeriaceae). The pathogen is plurivorous and capable of causing disease in a wide range of hosts (>500 spp.). de Milliano (1992) reported its occurrence on pearl millet in southern Africa, specifically in Botswana, Malawi, South Africa, and Zimbabwe. The Botswana Ministry of Agriculture identified charcoal rot in pearl millet much earlier in 1987.

Singh et al. (1997) identified this pathogen on pearl millet (presumably) in India and named it "dry stalk rot." These authors observed that the disease occurred in soils with a more alkaline pH (8.5–9.5), sandy textures, and relatively low organic matter content. When rainfall fell below 350–450 mm, relative humidity was <85%, and temperatures were between 32.7 and 37.3 °C during the late August flowering period, dry stalk disease was induced. The disease appeared to be quite aggressive under these conditions, where charcoal rot occurred on the lower nodes, and plants died before flowering.

In India, the isolate(s) causing charcoal rot on pearl millet appeared to produce numerous pycnidia (Singh et al., 1997).

Fusarium Stalk Rot

A complex of *Fusarium* spp. is associated with stalk rots in pearl millet. Wilson (2002b) described *Fusarium incarnatum* (syn. *F. semitectum*) and *F. fujikuroi* species complex (*F. moniliforme* sensu lato) as being the most frequently isolated *Fusarium* spp. from pearl millet stalks in Georgia. These species complexes were isolated along the length of the stalk at multiple nodes and generally at <25% incidence at a particular node. Wilson et al. (1999) indicated that stalk and neck rot of pearl millet was also caused by *F. graminearum*, which was more frequent if millet followed canola. This study isolated *F. incarnatum*, *F. moniliforme* sensu lato, *F. graminearum*, and *F. equiseti* from stalks. *Fusarium madaense* C.N. Ezekiel, Sand.-Den., Houbraken & Crous has been isolated from pearl millet stalks, and Koch's postulates prove that it is a pathogen (Costa et al., 2022).

On the other hand, many *Fusarium* spp. are endophytic in pearl millet and may have beneficial activities. Nandhini et al. (2018) isolated *F. chlamydosporum*, *F. equiseti*, *F. oxysporum*, *F. redolens*, *F. solani*, and *F. thapsinum* from shoot and root tissues. In this study, an *F. oxysporum* isolate could suppress downy mildew incidence under greenhouse conditions.

Rhizoctonia Blight

Both *Rhizoctonia solani* J.G. Kühn and *R. zeae* Vorhees [syn. *Waitea zeae* (Vorhees) J.A. Crouch & Cubeta] (Basidiomycota Agaricomycetes, Ceratobasidiales, Ceratobasidiaceae) are pathogens causing Rhizoctonia blight on pearl millet. Luttrell (1954) reported the disease, and both associated pathogens, in pearl millet in Georgia. Saksena (1979) indicated that an "aerial strain" of *R. solani* found on rice caused banded blight disease on pearl millet, maize, and sorghum. Likewise, Mbwaga and Mdolwa (1995) reported a leaf and sheath blight in pearl millet in Tanzania.

Interestingly, cerebroside B, a glycosphingolipid elicitor, was isolated from *R. solani* and used to induce resistance to downy mildew in pearl millet (Deepak et al., 2003).

Southern Blight

Sclerotium rolfsii Sacc. [telomorph *Athelia rolfsii* (Curzi) C.C. Tu & Kimbr.] (Basidiomycota, Agaricomycetes, Atheliales, Atheliaceae) is the fungus causing southern blight in pearl millet. Weber (1963) observed this pathogen on pearl millet, causing root, crown, and stalk disease in Florida. *Sclerotium rolfsii* is a necrotroph with a broad host range (Aycock, 1966; Okabe et al., 1998; Sarma et al., 2002). Besides pearl millet, *S. rolfsii* causes foot rot and wilt of finger millet (*E. coracana* L. Gaertn.) and southern sclerotial rot on sorghum (Nagara & Reddy, 2009; Odvody, 2000). The main pathogenicity factor that allows *S. rolfsii* to be successful on so many hosts is the production of oxalic acid (Ferrar & Walker, 1993; Kritzman et al., 1977; Punja et al., 1985). Watersoaking and tissue necrosis are the primary symptoms. Large, brown sclerotia that are well defined by a rind, cortex, and pseudoparenchymatous medullary region are the diagnostic signs. These are generally produced in clusters and may be fused.

Under daily rain and prolonged high humidity conditions, *S. rolfsii* can colonize pearl millet root, crown, and lower leaf tissues. Weber (1963) observed mycelium growing along leaf sheaths and lower leaves. Additionally, lower leaves were easily killed by the pathogen. Sclerotia were formed behind advancing hyphae on nearly every tissue that was colonized.

Top Rot

Top rot is a disease caused by *Fusarium fujikuroi* Nirenberg [syn. *Fusarium subglutinans* (Wollenweber & Reinking) P.E. Nelson, Tousson & Marasas]. Not only does this disease occur in pearl millet, but it also occurs in sorghum (Das et al., 2019). In both cases, the other names given this disease are "pokkah boeng" or "twisted top" (Ramakrishnan, 1941; Wells, 1956). Such disease names are due to the malformation of the peduncle and panicle region of the plant and the failure of these tissues to develop fully. Immature panicles and leaf tissues will rot before exsertion from the leaf whorl (Wilson 2002a, 2002b). Original observations of this disease in Georgia were associated with the pathogen name *F. moniliforme* sensu lato and may be seen as such in earlier literature (Tchala, 1989).

Mathur et al. (1994) indicated that top rot can be seedborne, which makes sense because many fusaria associated with stalk rots and related diseases may be found in pearl millet seed.

Rust

The major rust of pearl millet is caused by *P. substriata* Ellis & Barthol. (Basidiomycota, Pucciniomycetes, Pucciniales, Pucciniaceae). At one time, several varieties had been described for this pathogen (including var. *decrospora*, *imposita*, *indica*, *penicilliare*, and *zimmermannii*), but such subspecies-level designations are no longer used and should be avoided.

Puccinia substriata occurs worldwide, wherever pearl millet is grown. After initial urediniospore infection, which occurs at optimum temperatures of 15–20 °C at night, chlorotic flecks appear on the leaf surface. Individual uredinia develop from these initial infection points. They may grow to coalesce and take up a large area on the leaf, usually in the direction from the leaf tip to the stalk. When mature, uredinia appear brick-red to dark orange, which gives rust diseases their chief characteristic and is the source of their name (Figure 11.14). In very susceptible plants, uredinia may form on the culms, sheaths, and stalks in addition to the leaves (Thakur et al., 2011). Maturing uredinia may convert into darker telia and teliospores, which are characteristically stalked and two-celled, as is typical for the Pucciniaceae.

The rust is heteroecious where *Solanum* spp. function as the aecial host (Figure 11.15). Wilson and Williamson (1997) identified naturally infected eggplant seedlings with aecia in Georgia. They subsequently infected pearl millet

Figure 11.14 Pearl millet rust pustules (uredinia) on the leaf blade and midrib produced by *Puccinia substriata.* "Larry E. Claflin Plant Disease Image Collection," Kansas State University Libraries, Manhattan, Kansas, USA / CC BY 1.0.

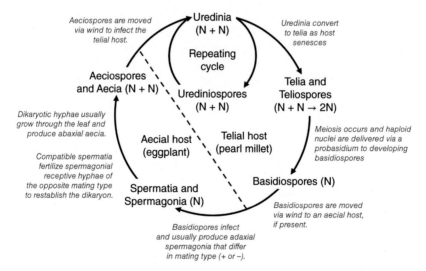

Figure 11.15 Disease cycle of pearl millet rust caused by *Puccinia substriata.* Image by C. R. Little.

seedlings in the greenhouse with aeciospores obtained from eggplant seedling leaves and obtained rust infection. Further, they observed that a large proportion of the aecial isolates could infect pearl millet with single gene resistance to

P. substriata (Rr_1) and that basidiospores developing from teliospores must be able to travel at least 9.7 km (the closest pearl millet rust source to the eggplant greenhouse) to infect eggplant seedlings.

Many studies and observations have reinforced that rust can be a significant limiting factor to yield potential. Rust is especially problematic on "multi-cut" forage hybrids. Wilson et al. (1991) showed that low levels of rust disease severity could significantly reduce dry matter digestible yield in such materials.

Wilson et al. (1996) evaluated 'Tift 23DA$_{1,E}$ × Tift 8677' hybrids combined with foliar fungicide treatments in Georgia. These authors found that rust severity increased when hybrids were planted late, and yield was reduced when >50% disease severity was observed. This can be even more problematic when rust occurs at the seedling or pre-flowering stages. When the disease occurs later in the season (or on more mature plants due to earlier planting), rust becomes a much more minor disease.

The search for rust resistance has been a long-term effort in many breeding programs worldwide and has found accessions with resistance (Anand Kumar et al., 1995; Singh et al., 1990c) (Table 11.7). Andrews et al. (1985a, 1985b) found that resistance to *P. substriata* is conditioned by a single, dominant gene. However, Wilson and Gates (1999) found partial resistance to rust in hybrids developed from a cross of several Tift 383 × ICMP 501 inbreds with Tift 23DA4. Tapsoba and Wilson (1999) evaluated physical mixtures, random-mated populations, and two- and three-way crosses from Tift 23DB (no resistance gene), Tift 85DB (Rr_1), PS748BC (unknown inheritance of rust resistance), and Tift 89D2 inbreds (non-Rr_1 type of resistance). They found that as random mating among inbreds progressed, the frequency of resistance to *P. substriata* races PS93-5 and PS93-6 increased. By the fourth random mating ("M4"), resistance frequencies to the two races were 68.6% and 71.6%, compared with 50.7% and 50.5% in the initial inbred mixture. In two

Table 11.7   Some Pearl Millet Entries and Their Reaction to Rust Caused by *Puccinia substriata*

Plant ID	Reaction	References
ICMB 96222	Resistant	Sharma et al. (2009)
ICML 11, ICML 17, ICML 18, ICML 19, ICML 20, ICML 21	Resistant	Singh et al. (1987, 1990a, 1990b, 1990c)
ICMP 451-P6, ICMP 451-P8	Resistant	Sharma et al. (2009)
ICMR 06999	Resistant	Sharma et al. (2009)
Tift #3, Tift #4	Resistant	Wilson et al. (1989) Wilson & Burton (1991)
P. glaucum ssp. *monodii* (wild relative)	Immune to some races	Hanna et al. (1982) Tapsoba & Wilson (1999)

years of their study, AUDPC values were significantly lower for the mixtures, random-mated populations, and two- and three-way crosses compared with the Tift 85DB stand-alone, which only had the single Rr_1 resistance gene. In most cases, dry matter yield, an important forage quantity measure, was significantly higher than in the single-gene resistance stand.

In a subsequent study, Wilson et al. (2001) examined "dynamic multilines" of pearl millet backcross derivatives from rust-resistant Burkino Faso landraces, *P. glaucum* subsp. *monodii* accessions, and Indian (ICMP 83506) and US elite (Tift 8677) inbreds. The male-sterile 'DMP-A_4' and female-fertile 'DMP-B' populations showed seedlings with significantly higher rust resistance than DMP-A_4 ×Tift 383 hybrids. According to Wilson (1994), Tift 383 has a low level of non–race-specific, partial resistance to rust.

Smuts

Several basidiomycetes can cause smut diseases in pearl millet, although not all are equally important. The primary smut pathogen that has gained the most attention over the years is *Moesziomyces bullatus* (J. Schröt.) Vanky (Basidiomycota Ustilaginomycetes, Ustilaginales, Ustilaginaceae). This pathogen was often referred to as *M. penicilliariae* in earlier literature and was previously described as a *Tolyposporium* sp. (Ajrekar & Likhite, 1933; Wilson, 2000). The taxonomic placement of the *Moesziomyces* genus, and *M. bullatus* in particular, have been based on DNA phylogeny, life cycle characteristics, and teliospore characters (Peipenbring et al., 1998; Stoll et al., 2005). *Moesziomyces bullatus* exhibits a narrow host range because *P. glaucum* is the only reported host.

Moesziomyces bullatus produces sori that develop within a single floret of pearl millet (Figure 11.16). Sori combine fungal and plant tissues undergoing hypertrophic and hyperplastic cell expansion. Within the developing sorus, intercellular dikaryotic mycelium converts into masses of dark teliospores that become diploid. In mature sori, diploid teliospores germinate promycelia, and meiosis produces sporidia (Subba Rao & Thakur, 1983) (Figure 11.17). *Moesziomyces bullatus* sporidia are windborne. Thus, infections often appear first on infected fields' outer or border rows.

The development and production of hybrid pearl millet have increased the incidence of *M. bullatus*, which has become a problem, as it was once considered a minor disease on landrace and local cultivar-type varieties (Yadav & Rai, 2013). This smut infects through a floral route or the developing panicle in the boot (Wilson, 1995). Therefore, airborne inoculum and high humidity can cause infections throughout the plant's panicle development and flowering phases (Ajrekar & Likhite, 1933; Vasudeva & Iyengar, 1950).

Because the smut is windborne and takes a boot or floral-infecting route, seed treatments are less effective than soilborne smuts. However, some control could effectively reduce viable teliospores infesting saved seeds that will be used for

Figure 11.16 Pearl millet smut caused by *Moesziomyces bullatus*. (a) Infected (left) and healthy (right) panicles. (b) Mature, black smut sori. "Larry E. Claflin Plant Disease Image Collection," Kansas State University Libraries, Manhattan, Kansas, USA / CC BY 1.0.

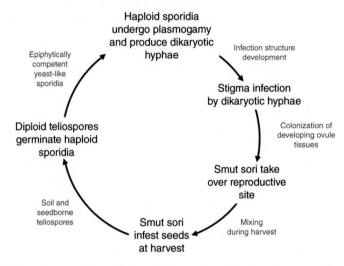

Figure 11.17 Disease cycle of pearl millet smut caused by *Moesziomyces bullatus*. Image by C. R. Little.

planting, although the fungus does not infect seedlings and does not grow systemically in the plant.

Screening for resistance has been successfully undertaken by Thakur et al. (1983). These researchers injected sporidia suspensions into the boot, bagged the inoculated panicles to maintain high relative humidity (>80% R.H.), and scored for severity at 14–21 days post-inoculation. Table 11.8 lists examples of pearl millet genotypes and their reaction to smut infection.

Table 11.8 Some Pearl Millet Entries and Their Reaction to Smut Caused by *Moesziomyces bullatus.*

Plant ID	Reaction	References
B 2301	Resistant	Khanna et al. (2014)
CZP 9802	Resistant	Khanna et al. (2014)
GHB 538	Resistant	Khanna et al. (2014)
GHB 558	Resistant	Khanna et al. (2014)
GHB 719	Resistant	Khanna et al. (2014)
GHB 744	Resistant	Khanna et al. (2014)
GHB 757	Resistant	Khanna et al. (2014)
GHB 732	Resistant	Khanna et al. (2014)
HHB 197	Resistant	Khanna et al. (2014)
ICMA 88006	Resistant	Yadav & Rai (2013), Khanna et al. (2014)
ICMA/B 98006	Resistant	Rai et al. (1998)
ICMB 92777, ICMB 92888	Resistant	Khanna et al. (2014)
ICMH 356	Resistant	Khanna et al. (2014)
ICML 5, ICML 6, ICML 7, ICML 8, ICML 9, ICML 10	Resistant	Thakur & King (1988c)
ICMS 8282, -8283	Resistant	Andrews et al. (1985a)
ICMPS 100-5-1, ICMPS 900-9-3, ICMPS 1600-2-4, ICMPS 2000-5-2	Resistant	Khanna et al. (2014)
ICMV 155	Resistant	Khanna et al. (2014)
ICMV 221	Resistant	Khanna et al. (2014)
ICTP 8203	Resistant	Khanna et al. (2014)
IP 19874	Resistant	Khanna et al. (2014)
JBV 2	Resistant	Khanna et al. (2014)
MH 1317	Resistant	Khanna et al. (2014)
PB 106	Resistant	Khanna et al. (2014)
PUSA 23	Resistant	Khanna et al. (2014)
PUSA 266	Resistant	Khanna et al. (2014)
PUSA 383	Resistant	Khanna et al. (2014)
RAJ 171	Resistant	Khanna et al. (2014)
RHB 121	Resistant	Khanna et al. (2014)
Saburi	Resistant	Khanna et al. (2014)
Shardha	Resistant	Khanna et al. (2014)

Parasitic Plants

Witchweed (*Striga* spp.)

Among the parasitic plants, *Striga* spp. (witchweed; Asterids, Lamiales, Orobanchaceae) are most important on pearl millet. The genus contains numerous recognized species with *S. asiatica* (L.) Kuntze (red or Asiatic witchweed) and *S. hermonthica* (Del.) (purple witchweed) Benth., the most common and severe crop parasitic weeds. Within the individual species, ecotypes (or "morphotypes") may exist that differ based on host pathogenicity, geographic range, and adaptation to climatic or soil edaphic factors (Huang et al., 2012; Jamil et al., 2021; Rodenburg et al., 2016).

The most common species found on pearl millet is *S. hermonthica* (Del.) Benth. (Abbasher et al., 1998). Most *Striga* damage occurs in areas where the environment is favorable for seed germination (warm temperatures, 30–35 °C), soils are light textured (e.g., sandy), and N levels are relatively low. Witchweed damage can be severe in areas with high soil seed populations and poor mineral nutrition. Typically, plants are stunted, appear wilted chlorotic, and may fail to produce sufficient biomass or panicles capable of flowering or setting seed.

The *Striga* plants are visible at the base of the plant and produce colorful flowers (Figure 11.18). Typically, *S. hermonthica* produces a purple flower and thus has been referred to as "purple witchweed."

Striga spp. are hemiparasitic weeds characterized by capsules that produce tiny "dust" seeds, which are no more than the width of a human hair. Individual plants can produce up to half a million seeds during a season (Bebawi et al., 1984). These

Figure 11.18   Example of witchweed (*Striga hermonthica*) growing from the base of prematurely dying pearl millet plants. "Larry E. Claflin Plant Disease Image Collection," Kansas State University Libraries, Manhattan, Kansas, USA / CC BY 1.0.

seeds are long-lived in the soil (10+ years) and germinate in response to strigolactone molecules that are produced by the host plant (Matsutova et al., 2005). Upon germination and successful contact with the host plant, *Striga* produces a holdfast/haustorium that invades the root cortex and eventually establishes a xylem bridge with the host. Because *Striga* is a hemiparasite, it only requires water and minerals from the host. Witchweeds produce chlorophyll and thus are not developmentally dependent on host photosynthates as holoparasites are.

Seed transmission is how *Striga* has spread through many pearl millet production regions in sub-Saharan Africa. Although seeds can be moved by wind, water, and animals, any agricultural operation that moves soil will also move seed. Infested clothing, shoes, tools, and agricultural machinery represent a primary mode of transmission and spread. In addition, the tiny seeds may be present on harvested crop seeds, and long-distance movement of *Striga* can occur via the seed trade.

Control of witchweed has been achieved through several methods, including host resistance, intercropping, herbicides, and biocontrol. Among the control methods, resistance is the most robust approach to deploy. In sorghum, the approach that has been used is to select plants that produce low amounts of the strigolactone *Striga* seed–stimulating root exudate, sorgolactone. A low-stimulation recessive gene for *S. asiatica* seed has been discovered in sorghum lines as well as a major dominant gene for stimulation of *S. hermonthica* seed, and this gene has been mapped (Ezeaku & Gupta, 2004; Haussmann et al., 2001; Hess & Ejeta, 1992; Satish et al., 2011).

Nematodes

More than 50 species of plant-parasitic nematodes across eight countries have been reported as associates of pearl millet, with most reports coming from India (Sharma, 1985; Sharma & McDonald, 1990). Stunt nematodes (*Tylenchorhynchus* spp.), spiral nematodes (*Helicotylenchus* spp.), root-knot nematodes *Meloidogyne incognita* Kofoid & White and *M. javanica* Treub, and the root-lesion nematode *Pratylenchus zea* Graham have the highest reported frequencies and distributions (De Waele et al., 1998; Nayak, 2010; ShanmugaPriya et al., 2019; Sharma & McDonald, 1990). However, pathogenicity has been confirmed in very few cases, and well-documented evidence of damage exists primarily for *M. incognita*. In Georgia, USA, for example, yield increases of 16%–23% have been observed for pearl millet treated with nematicides (Johnson et al., 1995; Timper et al., 2002). Even in the presence of yield loss, the characteristic symptom of root galling was minimal, suggesting some level of tolerance.

The global emphasis on host resistance underscores the importance of root-knot nematodes to pearl millet production. Pearl millet hybrids have been evaluated for resistance to *M. incognita* and *M. arenaria* Neal in Georgia, USA (Timper et al., 2002) and *M. graminicola* Golden & Birchfield in India (Singh et al., 2019). *Meloidogyne incognita* reproduction has also been evaluated on a diverse set of pearl millet cultivars from West and East Africa (Timper & Wilson, 2006). All African cultivars expressed some resistance, with nematode reproduction on most cultivars averaging <10% of that on the susceptible US hybrid HGM-100. Analysis of phenotypic segregation ratios generally supported the hypothesis that one or two dominant genes conferred resistance. In addition to variation in cultivar responses, the assessment of the importance of *M. incognita* on pearl millet is further complicated by the presence of host races. Most US studies utilized race 3, although race 1 has been shown to also reproduce on pearl millet, leading to measurable yield loss (Johnson et al., 1995; Timper et al., 2002; Timper & Wilson, 2006). Surveys outside the United States rarely identify the race(s) present.

Research on nematodes and pearl millet has frequently focused on evaluating pearl millet as a rotation crop. In the southeastern United States, pearl millet has shown potential for managing plant-parasitic nematodes in rotations with cotton, corn, and peanut. Pearl millet hybrids were relatively poor hosts for the sting nematode *Belonolaimus longicaudatus* Rau, the root-lesion nematode *Pratylenchus brachyurus* Godfrey, the stubby-root nematode *Paratrichodorus minor* (Colbran) Siddiqi, and the root-knot nematode *M. javanica* (Timper & Hanna, 2005). Nematode reproduction was generally lower on the *M. incognita* race 3–resistant hybrid TifGrain 102 than on the susceptible hybrid HGM-100. Similarly, several studies in Canada have demonstrated the potential of forage and grain pearl millet hybrids to control *Pratylenchus penetrans* (Cobb) Filip. & Stek. in subsequent potato and tobacco crops (Amankwa et al., 2006; Ball-Coelho et al., 2003; Belair et al., 2004; Dauphinais et al., 2005). Nematode population densities during a pearl millet rotation crop averaged 18%–32% of population densities observed during the standard *P. penetrans*–susceptible rye rotation crop, with corresponding average yield increases of 52% for potato and 54% for tobacco. Surprisingly, *P. penetrans* was the predominant nematode recovered from the roots of pearl millet in Namibia (De Waele et al., 1998), suggesting either intraspecific variation in host adaptation or misattribution of the species designation.

Nematode Life Histories

Plant-parasitic nematodes commonly associated with pearl millet exhibit a range of life cycles and feeding strategies (Table 11.9). The ectoparasitic stunt and spiral nematodes are soil inhabitants throughout their life cycle, which consists of four juvenile stages (J1–J4) and adult egg-laying females with or without accompanying adult males (Anderson & Potter, 1991; Fortuner, 1991). Reproduction is either sexual, in which case males are present, or parthenogenetic, in which case males

Table 11.9   Plant Parasitic Nematode Species Found in Association with Pearl Millet.

Category	Genus	Species	References
Migratory, endoparasitic			
Burrowing nematode	*Radopholus*	*R. similis*	
Lance nematode	*Hoplolaimus*	*H. indicus*	
Root-lesion nematode	*Pratylenchus*	*P. brachyurus,* *P. mulchandi,* *P. penetrans, P. zeae*	Sharma & McDonald (1990), De Waele et al. (1998)
Migratory, ectoparasitic			
Dagger nematode	*Xiphonema*	*X. americanum*	
Ring nematode	*Criconemella*	*C. ornata*	Sharma & McDonald (1990)
Sting nematode	*Belonolaimus*	*B. longicaudatus*	
Stubby root nematode	*Paratrichidorus*	*P. minor*	Sharma & McDonald (1990)
Stunt nematode	*Tylenchorhynchus*	*T. phaseoli,* *T. vulgaris, T. zeae*	Sharma & McDonald (1990)
Sedentary, endoparasitic			
Cyst nematode	*Heterodera*	*H. gambiensis*	
Root-knot nematode	*Meloidogyne*	*M. arenaria,* *M. incognita,* *M. javanica*	Sharma & McDonald (1990)

are rare or absent. Second-stage juveniles (J2) hatch from the eggs, and all subsequent stages are migratory, feeding primarily on root surface tissues (e.g., epidermal cells). However, spiral nematodes may become partially embedded in the root cortex. Damage is minimal except in exceptionally large population densities, and management is rarely required or recommended.

Root-lesion nematodes are migratory endoparasites that burrow through the root cortex, creating lesions that often become necrotic (Bridge & Starr, 2007; Castillo & Volvas, 2007; Khan, 2015). Their life cycles are otherwise similar to the ectoparasitic species discussed above. Due to their global prevalence, broad host range, and potential for substantial disruption of root function, these nematodes are generally considered among the most destructive nematode crop pests. Nevertheless, management is limited to nonhost crop rotation in most cases.

Root-knot nematodes are sedentary endoparasites that establish permanent feeding sites, called giant cells, in root vascular tissues (Bridge & Starr, 2007; Eisenback

& Triantaphyllou, 1991; Khan, 2015). Feeding is typically accompanied by conspicuous root galling caused by the proliferation and hypertrophy of adjacent cells. Hatched J2 invade the roots, initiate feeding sites, and undergo successive molts without feeding to become sedentary swollen adult females or migratory vermiform males. The common agricultural species *M. incognita* is triploid and obligately parthenogenetic, but males may be produced under stress or limited resources. Root-knot nematodes have broad host ranges and are considered the most economically important plant-parasitic nematodes globally. Host resistance is the principal management strategy, with resistant cultivars available for many agronomic crops, including pearl millet (Timper & Hanna, 2005; Timper & Wilson, 2006).

Virus Diseases

A wide array of viruses affects pearl millet, many of which remain unknown or are largely understudied. However, among the viruses, the potyvirus group is the most frequently encountered and is likely to cause the greatest impact on productivity.

Potyviruses

Potyviruses are the most important and widespread group of viruses that infect members of the Poaceae family of grasses. These viruses are characterized by a virion composed of a flexuous rod capsid and a positive-sense (+) single-stranded RNA genome (Inoue-Nagata et al., 2022). Various aphid species can vector them in the field, although mechanical transmission may be used to inoculate potyvirus species under laboratory conditions.

Of the monocot-infecting potyviruses, *Johnsongrass mosaic virus* (JGMV), *Maize dwarf mosaic virus* (MDMV), *Sorghum mosaic virus* (SrMV), and *Sugarcane mosaic virus* (SCMV) infect pearl millet.

Johnsongrass mosaic Virus

JGMV has been found in a range of monocot hosts (Table 11.10). Karan et al. (1992) showed that JGMV could infect all 15 pearl millet accessions tested under glasshouse conditions. Although endemic in Australian Johnsongrass, Laidlaw et al. (2004) indicated that JGMV can infect a wide range of grass hosts, including pearl millet. Further, Siefers et al. (2005) showed that two strains, JGMV-N from Nigeria and JGMV-MDO, could infect pearl millet cultivar C-130.

Maize dwarf mosaic Virus

MDMV is characterized by a monopartite, +ssRNA genome, and a flexuous rod virion structure ($12 \times 750\,nm$). Several MDMV genomes have been characterized

Table 11.10 Monocot Hosts of Pearl Millet Viruses.

Monocot host	Common name	IPCV	JgMV	MDMV	MSV	PMV	SRBsDV	WSMV
Alopecurus brachystachys	Foxtail			x				
Avena sativa	Oats				x			
Axonopus compressus	Carpetgrass				x			
Bromus arvensis	Field brome		x					
B. commutatus	Meadow brome		x					
B. macrostachys	Mediterranean brome		x					
B. racemosus	Bald brome		x					
B. ramosa	Browntop millet					x		
B. sterilis	Poverty brome		x					
B. tectorum	Downy brome			x				
Chrysopogon fulvus	Red false beardgrass			x				
Coix lacyma-jacobi	Jacob's tears		x		x			
Dactyloctenium aegyptium	Egyptian crowfoot grass				x			
Dichelachne micrantha	Plumegrass			x				
Digitaria horizontalis	Jamaican crabgrass				x			
D. sanguinalis	Crabgrass			x		x	x	
Echinochloa colona	Junglerice				x			
E. crus-galli	Barnyardgrass		x	x		x	x	
E. esculentum	Barnyard millet						x	

(Continued)

Table 11.10 (Continued)

Monocot host	Common name	IPCV	JgMV	MDMV	MSV	PMV	SRBsDV	WSMV
Eleusine coracana	Finger millet	x						
E. indica	Goosegrass			x	x		x	
E. stagnina	Burgu millet				x			
Eragrostis trichodes	Sand lovegrass			x				
Gynerium sagittatum	Wildcane			x				
Hordeum vulgare	Barley				x		x	x
Isachne globosa	Bloodgrass						x	
Megathyrsus (Panicum) maximum	Guineagrass		x		x			
Melinis repens	Natal grass				x			
Oplismenus hirtelus	Basketgrass		x					
Oryza glabberima	African rice				x			
O. sativa	Rice			x	x		x	
Panicum bulbosum	Bulb panicgrass		x					
P. capillare	Witchgrass					x		
P. decompositum	Australian millet					x		
P. dichotomiflorum	Fall panicgrass					x		
P. hallii	Hall's panicgrass					x		
P. miliaceum	Proso millet		x	x		x		
P. sumatrense (syn. *P. miliare*)	Little millet			x				
P. turgidum	Bunchgrass					x		

Species	Common name						
P. virgatum	Switchgrass			x			
Paspalum conjugatum	Hilograss				x		x
P. notatum	Bahiagrass				x		x
P. racemosum	—						x
P. scrobiculatum	Kodo millet				x	x	
Pennisetum polystachion	Mission grass				x	x	
P. purpureum	Elephantgrass, Napier grass					x	
Rottboellia cochinchinensis	Itchgrass				x		
Saccharum officinarum	Sugarcane				x	x	x
Setaria barbata	Bristly foxtail				x		
S. italica	Foxtail millet	x		x		x	x
S. verticillata	Hooked bristlegrass			x			x
S. pumila ssp. pumila	Yellow foxtail			x			
Secale cereale	Rye		x				
Sorghastrum nutans	Indiangrass					x	
Sorghum arundinaceum	Wild sorghum						x
S. bicolor	Cultivated sorghum	x				x	
S. halepense	Johnsongrass					x	
Stenotaphrum secondatum	St. Augustine grass			x			x

(Continued)

Table 11.10 (Continued)

Monocot host	Common name	IPCV	JgMV	MDMV	MSV	PMV	SRBsDV	WSMV
Themeda quadrivalvis	Grader grass			x				
Triticum aestivum	Wheat				x		x	x
Urochloa brizantha	Palisade grass		x					
U. decumbens	Spreading liverseed grass		x					
U. deflexa	Deflexed brachiaria		x		x			
U. dictyoneura	–		x					
U. humidicola	Koronivia grass		x					
U. lata	–				x			
U. mosambicensis	African liverseed grass		x	x				
Zea luxurians	Teosinte			x				
Z. mays	Corn, maize		x	x	x	x	x	x

Abbreviations: IPCV, *Indian peanut clump virus;* JgMV, *Johnsongrass mosaic virus;* MDMV, *Maize dwarf mosaic virus;* MSV, *Maize streak virus;* PMV, *Panicum mosaic virus;* SRBsDV, *Southern rice black-streaked dwarf virus;* WSMV, *Wheat streak mosaic virus.*
References: Choi et al. (1989a, 1989b), Jensen et al. (1983), Karavina (2014), Kumar et al. (2018), McGee (1988), Morales et al. (1994), Rishi et al. (1973), Seth et al. (1972), Thouvenel et al. (1976), Wilson (2000).

using next-generation sequencing technology (Wijayasekara & Ali, 2021). Although most work has focused on maize infection by this virus, MDMV can infect 200+ grass species, including pearl millet, sorghum, and Johnsongrass (Achon et al., 2011) (Table 11.10).

MDMV is stylet-borne and nonpersistently vectored by aphids, including *Rhopalosiphum maidis* Fitch. (Thongmeearkom et al., 1976). Aphids acquire MDMV virions as they feed on infected plants. Acquisition occurs within seconds, and viruliferous aphids can transmit MDMV within minutes (Gadhave et al., 2020). An HC-Pro protein forms a bridge between the aphid stylet and a DAG triplet on the MDMV coat protein (Ng & Falk, 2006). This molecular interaction prolongs the retention of virions in the stylet and allows aphids to carry virus particles long distances and continue transmitting, although not indefinitely, as in persistent vectors.

The most prominent symptom of MDMV on pearl millet is mosaics on the upper tier of leaves. Along with mosaic, longitudinal streaking, and mottling that coincide with the veins can be observed, as is also observed in sorghum. Along with foliar symptoms, some pearl millet genotypes express a stunted phenotype.

Jensen et al. (1983) tested 50 pearl millet landraces and 49 breeding lines against isolates of MDMV-A and MDMV-B. Of the landraces tested, 18 (36%) were infected by MDMV-A, and MDMV-B infected 30 (61%). In this case, the geographic origin of the pearl millet landrace did not influence the reaction to MDMV-A or MDMV-B. Eleven (22%) of the breeding lines were infected by MDMV-A, and MDMV-B infected 28 (57%).

Sorghum mosaic Virus

Yahaya et al. (2014) found SrMV in a pearl millet sample obtained from an area near sugarcane production in Nigeria. Other authors have identified SrMV as the dominant virus in mixed infections of sugarcane by SrMV, SCMV, and SCSMV (Guo et al., 2015). SCSMV is a member of the *Potyviridae* but resides in a different genus, *Poacevirus*. This virus did not cause mosaic disease in pearl millet after mechanical inoculation (Mali, 2007).

Sugarcane mosaic Virus

In Texas, USA, Fazli (1971) tested SCMV strain A against pearl millet and found that this strain could significantly reduce head length and plant height in the field. In Australia, Karan et al. (1992) showed that the sabi grass [*Urochloa trichopus* (Hochst.) Stapf.], blue couch grass (*Digitaria didactyla* Willd.), and sugarcane strains of SCMV could infect pearl millet. Wakman et al. (2001) found a mosaic of maize and grasses on the island of Sulawesi, Indonesia. They found that the cause was a virus that exhibited the highest response to two SCMV antisera after inoculation in pearl millet. Yahaya et al. (2014) found SCMV in a pearl millet sample obtained from an area near sugarcane production in Nigeria.

Mastreviruses

Maize streak virus (MSV) is a Mastrevirus that belongs to the family Geminiviridae. As geminiviruses, virions are quasi-icosahedral, geminate, and contain a monopartite single-stranded DNA genome (Muhire et al., 2013; Shepherd et al., 2010). MSV infects a wide range of monocot hosts besides maize, including barley, finger millet, oats, pearl millet, rice, rye, sorghum, sugarcane, wheat, and other grass species (van Antwerpen et al., 2008; Willment et al., 2001) (Table 11.10). MSV is related to several other *African streak virus* species, including *Eragrostis streak virus*, *Sugarcane streak virus*, *Sugarcane streak Réunion virus*, *Urochloa streak virus*, *Sugarcane streak Egypt virus*, and *Panicum streak virus* (Shepherd et al., 2010).

Cicadulina mbila (Naude) (Homoptera, Cicadellidae) is a widespread and persistent leafhopper vector of MSV, although as many as eight other leafhopper species may contribute to virus spread in pearl millet. Further, pearl millet is often used to rear this insect vector (Bosque et al., 1998; Fajinmi et al., 2012). Thus, it is a preferred feeding host for *C. mbila*.

The distribution of MSV likely began as maize was moved from the New World to Africa in the 1500s (Buddenhagen, 1983; Johnson, 1997). Therefore, it is logical that, with a suitable vector and numerous susceptible hosts, the virus would spread throughout many African regions where maize and millets are cropped in proximity and reservoir hosts co-exist near production fields.

MSV in Africa was likely first detected in Nigerian pearl millet in the mid-1990s by Briddon et al. (1996), when they reported an *African streak virus* belonging to the Geminiviridae, vectored by *C. mbila*, and closely related to sugarcane streak virus. As such, this disease has also been observed in Egypt, Malawi, Mauritius, South Africa, and Uganda. MSV has been observed off the African mainland in Madagascar and Réunion, and in India (Wilson, 2000). Seth et al. (1972) observed the disease much earlier in India, described it as "bajra streak," and noted leafhopper transmission.

The symptoms of the virus include light chlorotic streaks, coalescing spots, striping, and banding that can occur on one side of the midrib or the other (Seth et al., 1972; Wilson, 2000). Symptoms may occur depending upon the strain, including panicum, maize, Eleusine, and sugarcane strains. However, these strains have been expanded from MSV-A to MSV-K, with MSV-A the most common strain infecting maize and representative isolates from other strains supposedly infecting pearl millet (Shepherd et al., 2010). According to Nagaraja and Viswanath (1983), pearl millet is susceptible to the Eleusine strain but not the maize strain.

The lessons learned from the control of MSV in maize apply to the control of this virus in pearl millet. Leaf hoppers likely move between established maize plots and grassy weed reservoir hosts. Therefore, controlling leafhoppers and maintaining "clean" fields is essential as a management strategy to reduce virus populations. Autry and Ricaud (1983) described both sources as contributing to epidemics of MSV, although they focused on the strain that caused most disease in maize (presumably MSV-A). Although MSV does not replicate in the leafhopper vector, MSV virions remain circulatory in the hemolymph to the salivary glands and can be transmitted for over a month (Reynaud & Peterschmidt, 1992).

Resistant lines and hybrids represent the best strategy for the management of MSV in pearl millet. Complete resistance, or immunity, in maize occurs through a combination of several minor genes and *msv-1* (Welz et al., 1998). Such tolerance mechanisms and specific genes have not been elucidated in pearl millet.

Panicoviruses

Panicum mosaic virus (PMV) is a Panicovirus in the Tombusviridae, the same family that contains *Tomato bushy stunt virus*. These viruses are characterized by a monopartite, positive-sense single-stranded RNA genome and an icosahedral (T = 3) virion structure. The *Satellite panicum mosaic virus* (SPMV) also possesses a +ssRNA genome and a structurally different icosahedral (T = 1) virion structure. Further, at least two satellite RNAs are associated with PMV, including satC and satS (Pyle et al., 2017).

PMV is responsible for grain losses in pearl millet, proso millet (*Panicum miliaceum* L.), and foxtail millet [*Setaria italica* (L.) P. Beauvois] and infects switchgrass (*Panicum virgatum* L.) (Buzen et al., 1984; Mandadi et al., 2014; Sill & Pickett, 1957) (Table 11.10). Together with SPMV, symptoms such as foliar chlorosis and necrosis, reduced seed development, and stunting can be observed (Scholthof, 1999).

Pearl millet has been used to maintain isolates of PMV (Pyle et al., 2017). However, these authors did not indicate the specific variety used.

Tritimoviruses

Wheat streak mosaic virus (WSMV) is a Tritimovirus belonging to the Potyviridae. A monopartite, +ssRNA genome, and a flexuous rod virion structure characterize the virus. WSMV has been isolated or detected from a range of hosts in the Poaceae, including wheat, oats, foxtail millet, sorghum, and barley, in addition to many weedy monocot hosts (Chalupníková et al., 2017; Dráb et al., 2014; Ito et al., 2012; Siefers et al., 1996) (Table 11.10). Further, pearl millet can support populations of the wheat curl mite vector (*Aceria tosichella* Keifer) (Jeppson et al., 1975).

WSMV can naturally infect pearl millet (Siefers et al., 1996). However, greenhouse experiments using mechanical inoculation showed differences in infection frequency based on isolate origin and the pearl millet variety tested (Siefers et al., 2003).

High plains virus, an *Emaravirus* often associated with WSMV due to transmission by the wheat curl mite and producing WSMV-like symptoms in wheat, was shown not to infect pearl millet. However, other millets were infected, such as proso millet (*P. miliaceum*) (Siefers et al., 1998).

Other Viruses

According to Eyvazi et al. (2021), millets are considered hosts for *Rice black-streaked dwarf virus* (RBSDV) and *Maize rough dwarf virus*, which belong to the genus *Fijivirus* and family *Reoviridae*. RBSDV is persistently transmitted by insects, such as the small brown planthopper (*Laodelphax striatellus* Fallén) (Lv et al., 2016).

Palanga et al. (2022) described *Pennisetum glaucum marfavirus* (*Marfavirus, Tymoviridae*) in pearl millet in Burkina Faso.

The soilborne plasmodiophorids *Polymyxa graminis* Ledingham f.sp. *tropicalis* and *P. graminis* f.sp. *subtropicalis* (Plasmodiophoraceae, Plasmodiophorida) infect pearl millet in India and Africa (Legrève et al., 2002). Neither of these formae speciales causes symptoms in pearl millet, but these organisms can transmit numerous viruses (Dieryck et al., 2011). *Indian peanut clump virus* (IPCV) and *Peanut clump virus* (PCV) are *Pecluvirus* species that can infect pearl millet and be transmitted by these organisms. Dieryck et al. (2011) inoculated pearl millet genotype ICMH-451 with viruliferous strains of *P. graminis* f.sp. *tropicalis* and *P. graminis* f.sp. *subtropicalis* and could achieve 44% PCV infection if the viruliferous zoospores came from PCV-infected sugarcane. In pearl millet, Reddy et al. (1998) could demonstrate 0.9% transmission of IPCV in 'WCC 75' seed. However, seed transmission rates were much higher for foxtail millet (9.7%; 'ISE 147') and finger millet (*E. coracana* Gaertn.) (5.2%; 'Arjuna'). In sorghum ('ICSV 1'), IPCV seed transmission was 0%.

Bacterial Diseases

Bacterial Spot

Pseudomonas syringae Van Hall (Gammaproteobacteria, Pseudomonadales, Pseudomonadaceae) is a pathogenic bacterium found on and in many plant hosts, including pearl millet. Like many plant pathogenic bacteria, it is a bacillus (or rod) shaped bacterium, possesses polar flagella, and is gram negative (Palleroni, 2015) (Table 11.11). Many isolates of *P. syringae* exhibit a strong fluorescence phenotype on King's B medium (Lelliot & Stead, 1987).

Table 11.11 Characteristics of Bacterial Genera Associated with Pearl Millet.

Character	Acidovorax	Klebsiella	Pantoea	Pseudomonas	Xanthomonas
Morphology	Bacillus	Bacillus	Bacillus	Bacillus	Bacillus
Cell size	0.2–1.2 μm w × 0.8–2.0 μm l	0.3–1.0 μm w × 0.6–6.0 μm l	0.5–1.3 μm w × 1.0–3.8 μm l	0.5–1.0 μm w × 1.5–5.0 μm l	0.4–0.6 μm w × 0.8–2.0 μm l
Motility	Single, two or three polar flagella	Non-motile	Peritrichous flagella	Single or several polar flagella	Single polar flagellum
Metabolism	Facultative aerobe	Facultative anaerobe	Facultative anaerobe	Facultative aerobe	Obligate aerobe
Colonies	Non-pigmented	Non-pigmented	Non-pigmented to yellow	Non-pigmented	Yellow
Catalase	Negative	Positive	Positive	Positive	Positive
Oxidase	Positive	Negative	Negative	Positive/ negative	Negative/ weak
G + C content, %	62–70	53–58	52–60	58–69	63–70

Note: Grimont & Grimont (2015), Palleroni (2015), Palmer & Coutinho (2015), Saddler & Bradbury (2015), Willems & Gillis (2015).
Abbreviations: l, length; w, width.

Currently, >50 pathovars of *P. syringae* exist, and the pathovar most commonly found on grasses is *P. syringae* pv. *panici*. An isolate from proso millet has been sequenced (Arnold & Preston, 2019; Gardan et al., 1999; Liu et al., 2012). However, Young and Fletcher (1994) reject the pathovar name "panici." Nevertheless, the prevalent pathovar on pearl millet has not been rigorously established and may vary depending upon geography. Most reports of *P. syringae* on pearl millet have not included a pathovar name (Frederickson et al., 1997; Kendrick, 1926).

Bacterial leaf spot is characterized by watersoaking and circular, elliptical, or elongated lesions and streaking that can become necrotic and be darkly pigmented at the edge (Wilson, 2000). Depending on the host variety and environmental conditions, such lesions can generally be confused with those of other bacterial or fungal diseases. These do not always favor a watersoaking response if the tissue becomes dry.

Frederickson et al. (1999) used Koch's postulates to determine that *P. syringae* was the etiological agent of a leaf spot disease on the pearl millet line '825B' and 'Okashana-1,' among others, in Zimbabwe.

Due to this pathogen's limited impact on pearl millet grain and forage yields, intensive control measures have yet to be established. However, for all foliar leaf diseases, continued surveillance of newly identified germplasm is a good practice to contain the disease (Frederickson et al., 1999). This general consideration should apply to the other bacterial diseases mentioned below.

Bacterial Leaf Streak

Bacterial leaf streak is caused by the gram-negative bacterium *Xanthomonas campestris* (Pammel) Dowson pv. *pennamericanum* sp. nov. (Qhobela & Claflin, 1988) (Gammaproteobacteria, Xanthomonadales, Xanthomonadaceae). However, a recent phylogenomic study by Harrison et al. (2023) moves this pathovar from *X. campestris* to *X. euvesicatoria* as the genome of *X. campestris* pv. *pennamericanum* exhibited 98% identity with the type strain of *X. euvesicatoria* and only 88% with the type strain of *X. campestris*. Regardless, the xanthomonads belong to the class Gammaproteobacteria and the family Xanthamonadaceae (Tang et al., 2021). Like many other plant pathogenic genera, *Xanthomonas* spp. are gram negative and characterized as bacilli. *Xanthomonas* use a single, polar flagellum for motility, averaging 2.1 μm long and 0.7 μm wide (An et al., 2020; Saddler & Bradbury, 2015) (Table 11.11).

Qhobela and Claflin's (1988) original observation of this disease noted that it closely resembled the striping and streaking observed in the bacterial leaf streak and stripe diseases complex observed in sorghum. These symptoms include discolored, elongated lesions trapped within the veins that may extend the entire length of lower to middle-tier leaves. Further, signs such as bacterial exudates and crusting may be observed along the adaxial and abaxial lesion surfaces (Claflin, 2000a, 2000b).

Bacterial Stripe

Bacterial stripe of pearl millet is caused by the gram-negative, rod-shaped bacterium *Acidovorax avenae* ssp. *avenae* (Claflin et al., 1989; Willems & Gillis, 2015) (Betaproteobacteria, Burkholderiales, Comamonadaceae). Giatatis et al. (2002) identified *A. avenae* ssp. *avenae* from pearl millet breeding lines in Georgia, USA. This bacterium produces linear, water-soaked, dark red to brown lesions on the foliage. Some corn strains of *A. avenae* ssp. *avenae* can infect pearl millet and sweet corn (Giatatis et al., 2002).

Acidovorax spp. were previously recognized as species of *Pseudomonas*. *Pseudomonas avenae* was renamed *A. avenae* ssp. *avenae* by Willems et al. (1992). Subsequently, Schaad et al. (2008) suggest four species in place of the previously described subspecies, including *A. avenae*, *A. citrulli*, *A. cattleyae*, and *A. oryzae* sp. nov. At this point, it is unclear which species causes the pearl millet disease. For example, these authors indicated that the strains from finger millet were <70% related to most other isolates used in the study. However, their data showed that the strain from finger millet, FC-500, showed 99.5% and 99.7% internal transcribed spacer similarity with FC-501 and FC-506, isolated from tea plants [*Camellia sinensis* (L.) Kuntze].

Several authors have indicated that *A. avenae* ssp. *avenae* is a seedborne bacterium in graminaceous species, including pearl millet (Giatatis et al., 2002; Giatatis & Walcott, 2007; Song et al., 2004).

Pantoea Leaf Spot and Blight

Pantoea (Gavini et al., 1989) (Gammaproteobacteria, Enterobacterales, Erwiniaceae) is a widespread genus of rod-shaped, gram-negative, and facultatively anaerobic bacteria (Table 11.11). Many species were once referred to as *Enterobacter* or *Erwinia* and may be seen this way in the older literature. The most notable example is *P. stewartii* (Smith et al., 1989; Margert et al., 1993), which was previously referred to as *Erwinia stewartii* and causes Stewart's wilt of sweet corn, among other diseases, depending on bacterial subspecies and host (Abidin et al., 2020; Azizi et al., 2019; Cui et al., 2020; Roper, 2011).

The most important *Pantoea* species in pearl millet are *P. agglomerans*, which causes leaf spot disease, and *P. stewartii* ssp. *indologenes*. *Pantoea agglomerans* was first described as a pearl millet pathogen by Rajagopalan and Rangaswami (1958); at that time, it was known as *Erwinia herbicola*. Subsequently, Frederickson et al. (1997) identified *P. agglomerans* as a species infecting pearl millet in Zimbabwe using Koch's postulates and a range of physiological tests.

Ashajyothi et al. (2021) recovered *P. stewartii* ssp. *indologenes* from pearl millet in Karnataka, India. During these authors' Koch's postulates tests, they reproduced the watersoaking and "typical" leaf blight symptoms that occurred in the field.

Pantoea stewartii ssp. *indologenes* causes leafspots on pearl millet and foxtail millet [*Setaria italica* (L.) P. Beauvois]. Recently, Koirala et al. (2021) recovered leaf spot isolates from onion. A comparison between the Alliaceae and Poaceae isolates revealed that the onion isolates could cause disease in pearl millet, foxtail millet, and corn. However, the Poaceae isolates could not cause disease on *Allium* spp. These authors suggest that the Poaceae and Alliaceae isolates also be categorized at the pathovar level, given that the DNA similarity between isolates from the two host families is relatively high. Therefore, Koirala et al. (2021) propose that the Poaceae isolates be renamed *P. stewartii* ssp. *indologenes* pv. *setariae* and the Alliaceae isolates be renamed *P. stewartii* ssp. *indologenes* pv. *cepaciae*.

Interestingly, *P. stewartii*, considered by Kumar et al. (2021b) to be a "seed endophytic bacterium" in pearl millet, along with *Bacillus subtilis*, *B. tequilensis*, *B. velezensis*, *K. cowanii*, *P. dendritiformis*, and *Pseudomonas aeruginosa*, was shown to be inhibitory to fungal pathogens, including *Alternaria* spp., *Curvularia* spp., *Epicoccum sorghinum*, *E. rostratum*, *Fusarium* spp., and *Rhizoctonia solani* in dual-plate cultures.

Bacterial Stem Rot

In India, bacterial stem rot of pearl millet is caused by *Klebsiella* sp. (Gammaproteobacteria, Enterobacteriales, Enterobacteriaceae) with high sequence identity to *Klebsiella aerogenes* (Malik et al., 2022). Typically, *K. aerogenes* is considered an opportunistic animal pathogen, but it has been isolated from soil and other sources and possesses numerous antibiotic-resistance genes (Mann et al., 2021; Passarelli-Araujo et al., 2019). In some cases, it can function as a spoilage bacterium. Therefore, it is unsurprising that *K. aerogenes*, a facultative anaerobe and mesophile, could colonize the interior stalk tissues of pearl millet, especially if environmental conditions and high levels of readily usable stalk sugars are present. IN *K. aerogenes* pathogenicity studies, Malik et al. (2022) they observed water soaking, leaf streaking, dead leaves, and hollow stems.

Huang et al. (2016) found that a strain (KpC4) of *Klebsiella pneumoniae*, a naturally occurring soilborne bacterial pathogen associated with human nosocomial infections, caused a widespread bacterial top rot disease of maize in Yunnan Province of China. Although this bacterium has not been found in association with pearl millet, it raises questions about the connection between human, animal, and plant pathogens that occasionally occur in modern agroecosystems.

Phytoplasma Diseases

Phytoplasmas, or *Candidatus* Phytoplasma (Mycoplasmatota, Mollcutes, Acholeplasmatales, Acholeplasmataceae), are prokaryotic, bacteria-like organisms (Mollicutes) that persist in insect vectors and the host plant phloem. Phytoplasmas have been found in a wide range of hosts, including pearl millet and many other grasses (Arocha et al., 2005; Hemalatha et al., 2023; Jung et al., 2003; Marcone et al., 2004). Structurally, phytoplasmas do not possess a cell wall, are polymorphic in shape, are small (<1 μm), and have small transposon-rich genomes (Jomantiene & Davis, 2006). The relationship with the host is obligate and cannot be artificially cultured in the laboratory, although many attempts have been made.

Upon their discovery, phytoplasmas were termed mycoplasma-like organisms (Doi et al., 1967). However, species names have recently been divided based on 16Sr groups and subgroups (Bertaccini, 2019; Parte et al., 2020).

Thorat et al. (2017) reported a 16SrII-D phytoplasma associated with millet and millet × Napiergrass hybrids in India. In this study, the authors found that the three isolates tested (BJ01, -02, and -03) grouped into a clade closely related to 'Candidatus P. australasia. A phytoplasma disease of pearl millet caused by *Candidatus* P. aurantifolia has also been observed in India (Hemalatha et al., 2023). Symptoms included phyllody, stunting, and small leaves. The 16S rDNA and secA genes of the pathogen showed that *Ca*. P. aurantifolia belonged to the 16SrII-D group. An earlier study (Kumar et al., 2010) identified another phytoplasma from pearl millet (in the 16SrI-B group) associated with green ear–like symptoms in North India. This pathogen was given the name *Candidatus* P. asteris.

Dedication

This book chapter is dedicated to Dr. Larry E. Claflin, Professor Emeritus of Plant Pathology, Department of Plant Pathology, Kansas State University, who passed away in 2019. Dr. Claflin studied corn, sorghum, and pearl millet diseases, emphasizing bacterial pathogens, for 32 years until his retirement in 2006. During his tenure, he took numerous disease photos from around the world, and these can be found in an extensive collection housed online by Kansas State University Libraries under the "Larry Claflin Plant Disease Image Collection" (https://krex.k-state.edu/handle/2097/19711).

Acknowledgments

The authors are grateful to the anonymous peer reviewers whose comments and edits improved the presentation and completeness of this manuscript. This is Publication No. 24-066-B from the Kansas Agricultural Experiment Station, Manhattan, Kansas, USA.

References

Abbasher, A. A., Hess, D. E., & Sauerborn, J. (1998). Fungal pathogens for biological control of *Striga hermonthica* on sorghum and pearl millet in West Africa. *African Crop Science Journal, 6*, 179–188.

Abidin, N., Ismail, S. I., Vadamalai, G., Yusof, M. T., Hakiman, M., Karam, D. S., Isamail-Suhaimy, N. W., Rohaya, I., & Zulperi, D. (2020). Genetic diversity of *Pantoea stewartii* subspecies *stewartii* causing jackfruit-bronzing disease in Malaysia. *PLoS One, 15*, e0234350.

Abraham, P., Alimpta, R. D., & Bdliya, B. S. (2019). Inheritance of resistance to ergot disease in a diallel cross of pearl millet (*Pennisetum glaucum*) (L.) R. Br.). *Tanzania Journal of Agricultural Sciences, 18*, 50–58.

Achon, M., Alonso-Dueñas, N., & Serrano, L. (2011). Maize dwarf mosaic virus diversity in the Johnsongrass native reservoir and in maize: Evidence of geographical, host and temporal differentiation. *Plant Pathology, 60*, 369–377.

Adeola, O., & Orban, J. I. (2005). Chemical composition and nutrient digestibility of pearl millet (*Pennisetum glaucum* L.) fed to growing pigs. *Journal of Cereal Science, 22,* 174–184.

Ahmed, K. M., & Reddy, C. R. (1993). *A pictorial guide to the identification of seedborne fungi of sorghum, pearl millet, finger millet, chickpea, pigeonpea, and groundnut.* Information Bulletin No. 34. ICRISAT.

Ajrekar, S. L., & Likhite, V. N. (1933). Observations on *Tolyposporium penicillariae* Bref. (the bajra smut fungus). *Current Science, 1,* 215.

Akanmu, A. O., Abiala, M. A., & Odebode, A. C. (2013). Pathogenic effect of soilborne *Fusarium* species on the growth of millet seedlings. *World Journal of Agriculture Sciences, 9,* 60–68.

Amalfitano, C., Pengue, R., Andolfi, A., Vurro, M., Zonno, M. C., & Evidente, A. (2002). HPLC analysis of fusaric acid, 9,10-dehydrofusaric acid and their methyl esters, toxic metabolites from weed pathogenic *Fusarium* species. *Phytochemical Analysis, 13,* 277–282.

Amankwa, G. A., White, A. D., McDowell, T. W., & Van Hooren, D. L. (2006). Pearl millet as a rotation crop with flue-cured tobacco for control of root-lesion nematodes in Ontario. *Canadian Journal of Plant Science, 86,* 1265–1271.

An, S.-Q., Potnis, N., Dow, M., Vorhölter, F.-J., He, Y.-Q., Becker, A., Teper, D., Li, Y., Wang, N., Bleris, L., & Tang, J.-L. (2020). Mechanistic insights into host adaptation, virulence and epidemiology of the phytopathogen *Xanthomonas*. *FEMS Microbiology Reviews, 44,* 1–32.

Anand Kumar, A., Rai, K. N., Andrews, D. J., Talukdar, B. S., Singh, S. D., Rao, A. S., Babu, P. P., & Reddy, B. P. (1995). Registration of ICMP 451 parental line of pearl millet. *Crop Science, 35,* 605.

Anderson, R. V., & Potter, J. W. (1991). Stunt nematodes: *Tylenchorhynchus, Merlinius,* and related genera.In W. In R. Nickle (Ed.), *Manual of agricultural nematology* (pp. 529–586). Marcel Dekker.

Andrews, D. J., King, S. B., Witcombe, J. R., Singh, S. D., Rai, K. N., Thakur, R. P., Talukdar, B. S., Chavan, S. B., & Singh, P. (1985a). Breeding for disease resistance and yield in pearl millet. *Field Crops Research, 11,* 241–258.

Andrews, D. J., Rai, K. N., & Singh, S. D. (1985b). A single dominant gene for rust resistance in pearl millet. *Crop Science, 25,* 565–566.

Arnold, D. L., & Preston, G. M. (2019). *Pseudomonas syringae*: Enterprising epiphyte and stealthy parasite. *Microbiology, 165,* 251–253.

Arocha, Y., López, M., Piñol, B., Fernández, M., Picornell, B., Almeida, R., Palenzuela, I., Wilson, M. R., & Jones, P. (2005). 'Candidatus Phytoplasma graminis' and 'Candidatus Phytoplasma caricae,' two novel phytoplasmas associated with diseases of sugarcane, weeds and papaya in Cuba. *International Journal of Systematic and Evolutionary Microbiology, 55,* 2451–2463.

Ashajyothi, M., Balamurugan, A., Shashikumara, P., Pandey, N., Agarwal, D. K., Tarasatyavati, C. C., Varshney, R. K., & Nayaka, S. C. (2021). First report of pearl millet bacterial leaf blight caused by *Pantoea stewartii* subspecies *indologenes* in India. *Plant Disease, 105,* 3736.

Autrey, L. J. C., & Ricaud, C. (1983). The comparative epidemiology of two diseases of maize caused by leafhopper-borne viruses in Mauritius. In R. T. Plumb & J. M. Thresh (Eds.), *Plant virus epidemiology* (pp. 277–285). Blackwell.

Aycock, R. (1966). *Stem rot and other diseases caused by Sclerotium rolfsii.* North Carolina Agricultural Experiment Station.

Azizi, M. M., Ismail, S. I., Hata, E. M., Zulperi, D., Ina-Salwany, M. Y., & Abdullah, M. A. (2019). First report of *Pantoea stewartii* subsp. *Indologenes* causing leaf blight on rice in Malaysia. *Plant Disease, 103,* 1407.

Bacon, C. W., Porter, J. K., Norred, W. P., & Leslie, J. F. (1996). Production of fusaric acid by *Fusarium* species. *Applied and Environmental Microbiology, 62,* 4039–4043.

Balasubramanian, K. A. (1980). Association of *Drechslera setariae* with downy mildew affected pearl millet. *Current Science, 49,* 233–234.

Ball-Coelho, B., Bruin, A. J., Roy, R. C., & Riga, E. (2003). Forage pearl millet and marigold as rotation crops for biological control of root-lesion nematodes in potato. *Agronomy Journal, 95,* 282–292.

Banks, G. T., Mantle, P. G., & Szczyrbak, C. A. (1974). Large-scale production of clavine alkaloids by *Claviceps fusiformis. Journal of General Microbiology, 82,* 345–361.

Basavaraju, R., Safeeulla, K. M., & Murty, B. R. (1981). Inheritance of resistance to downy mildew in pearl millet. *Indian Journal of Genetics and Plant Breeding, 41,* 144–149.

Bebawi, F. F., Eplee, R. F., & Norris, R. S. (1984). Effects of seed size and weight on witchweed (*Striga asiatica*) seed germination, emergence, and host-parasitization. *Weed Science, 32,* 202–205.

Bélair, G., Dauphinais, N., Fournier, Y., & Dangi, O. P. (2004). Pearl millet for the management of *Pratylenchus penetrans* in flue-cured tobacco in Quebec. *Plant Disease, 88,* 989–992.

Bertaccini, A. (2019). The phytoplasma classification between 'Candidatus species' provisional status and ribosomal grouping system. *Phytopathogenic Mollicutes, 9,* 1–2.

Bhardawaj, N. R., Atri, A., Banyal, D. K., Dhal, A., & Roy, A. K. (2022). Multi-location evaluation of fungicides for managing blast (*Magnaporthe grisea*) disease of forage pearl millet in India. *Crop Protection, 159,* 106019.

Bhowmik, T. P. (1972). *Bipolaris setariae* on two new hosts in India. *Indian Journal of Phytopathology, 25,* 590–591.

Bosio, P., Sicilliano, I., Gilardi, G., Gullino, M. L., & Garibaldi, A. (2017). Verrucarin A and roridin E produced on rocket by *Myrothecium roridum* under different temperatures and CO_2 levels. *World Mycotoxin Journal, 10,* 229–236.

Bosque-Pérez, N. A., Ologjede, S. O., & Buddenhagen, I. W. (1998). Effect of maize streak virus disease on the growth and yield of maize as influenced by varietal resistance levels and plant stage at time of challenge. *Euphytica, 101,* 307–317.

Bottalico, A., & Logrieco, A. (1994). Toxigenic *Alternaria* species of economic importance. In K. K. Shinha & D. Bhatnagar (Eds.), *Mycotoxins in agriculture and food safety* (pp. 65–108). Marcel Dekker.

Breese, W. A., Hash, C. T., Devos, K. M., & Howarth, C. J. (2002). Pearl millet genomics—an overview with respect to breeding for resistance to downy mildew. In J. F. Leslie (Ed.), *Sorghum and millets diseases* (pp. 243–246). Iowa State Press.

Briddon, R. W., Lunness, P., Bedford, I. D., Chamberlin, L. C. L., Mesfin, T., & Markham, P. G. (1996). A streak disease of pearl millet caused by a leafhopper-transmitted geminivirus. *European Journal of Plant Pathology, 102*, 397–400.

Bridge, J., & Starr, J. L. (2007). *Plant nematodes of agricultural importance: A color handbook*. Manson Publishing.

Buddenhagen, I. W. (1983). Crop improvement in relation to virus diseases and their epidemiology. In R. T. Plumb & J. M. Thresh (Eds.), *Plant virus epidemiology* (pp. 25–37). Blackwell.

Burton, G. W., Wallace, A. T., & Rachie, K. O. (1972). Chemical composition and nutritive value of pearl millet (*Pennisetum typhoides* (Burm.) Stapf and E.C. Hubbard) grain. *Crop Science, 12*, 187–188.

Burton, G. W., & Wells, E. D. (1981). Use of near-isogenic host populations to estimate the effect of three foliage diseases on pearl millet forage yield. *Phytopathology, 71*, 331–333.

Butler, E. J. (1907). Some diseases of cereals caused by *Sclerospora graminicola*. *Memoirs of the Department of Agriculture in India. Botanical Series, 2*, 1–24.

Butler, E. J. (1918). *Fungi and diseases in plants*. Thacker Spinck and Company.

Buzen, F. G., Niblett, C. L., Hooper, G. R., Hubbard, J., & Newman, M. A. (1984). Further characterization of Panicum mosaic virus and its associated satellite virus. *Phytopathology, 74*, 313–318.

CABI. (2023). *CABI compendium*. https://www.cabidigitallibrary.org/journal/cabicompendium

Capasso, R., Evidente, A., Cutignano, A., Vurro, M., Zonno, M. C., & Bottalico, A. (1996). Fusaric and 9,10-dehydro-fusaric acids and their methyl esters from *Fusarium nygamai*. *Phytochemistry, 41*, 1035–1039.

Castillo, P., and Volvas, N. 2007. *Pratylenchus* (Nematoda: Pratylenchidae): Diagnosis, biology, pathogenicity and management. Nematology Monographs and Perspectives, Vol. 6. (Hunt, D.J., and Perry, R.N., Series Ed.). Brill Leiden, Boston, USA

Chalal, S. S., Gill, K. S., Phul, P. S., & Singh, N. B. (1981). Effectiveness of recurrent selection for generating ergot resistance in pearl millet. *SABRAO Journal, 13*, 184–186.

Chalupníková, J., Kundu, J. K., Singh, K., Bartaková, P., & Beoni, E. (2017). Wheat streak mosaic virus: Incidence in field crops, potential reservoir within grass species and uptake in winter wheat cultivars. *Journal of Integrative Agriculture, 16*, 523–531.

Chauhan, M. S., Yadav, J. P. S., & Benniwal, J. (1997). Field resistance of cotton to Myrothecium leaf spot and Alternaria leaf spot (*A. macrospora*) diseases. *Journal of Cotton Research and Development, 11*, 196–205.

Chelkowski, J., Zawadzki, M., Zajkowski, P., Logrieco, A., & Bottalico, A. (1990). Moniliformin production by *Fusarium* species. *Mycotoxin Research, 6*, 41–45.

Chelpuri, D., Sharma, R., Durga, K. K., Katiyar, P., Mahendrakar, M. D., Singh, R. B., Yadav, R. S., Gupta, R., & Srivastava, R. K. (2019). Mapping quantitative trait loci (QTLs) associated with resistance to major pathotype-isolates of pearl millet downy mildew pathogen. *European Journal of Plant Pathology, 154*, 983–994.

Chintapalli, R., Wilson, J. P., & Little, C. R. (2006). Using fungal isolation rates from pearl millet caryopses to estimate grain mold resistance. *International Sorghum and Millets Newsletter, 47*, 146–148.

Choi, B. H., Park, K. Y., & Park, R. K. (1989b). Direct and indirect effects of important traits of green fodder yield in crosses of male-sterile and fertile inbred lines of pearl millet (*Pennisetum americanum* L. Leeke). *Korean Journal of Breeding, 21*, 115–121.

Choi, Y. M., Lee, S. M., & Ryu, G. H. (1989a). Studies on the host range of rice black-streaked dwarf virus. *Research Reports of the Rural Development Administration (Suweon), 31*, 14–18.

Claflin, L. (2000a). Bacterial streak. In R. A. Frederiksen & G. N. Odvody (Eds.), *Compendium of sorghum diseases* (2nd ed., p. 6). APS Press.

Claflin, L. E. (2000b). Bacterial stripe. In R. A. Frederiksen & G. N. Odvody (Eds.), *Compendium of sorghum disease* (2nd ed., pp. 5–6). APS Press.

Claflin, L. E., Ramundo, B. A., Leach, J. E., & Erinle, I. D. (1989). *Pseudomonas avenae*, causal agent of bacterial leaf stripe of pearl millet. *Plant Disease, 73*, 1010–1014.

Costa, M. M., Saleh, A. A., Melo, M. P., Guimarães, E. A., Esele, J. P., Zeller, K. A., Summerell, B. A., Pfenning, L. H., & Leslie, J. F. (2022). *Fusarium mirum* sp. nov., intertwining *Fusarium madaense* and *Fusarium andiyazi*, pathogens of tropical grasses. *Fungal Biology, 126*, 250–266.

Cui, D., Huang, M. T., Hu, C. Y., Su, J. B., Lin, L. H., Javed, T., Deng, Z.-H., & Gao, S.-J. (2020). First report of *Pantoea stewartii* subsp. *stewartii* causing bacterial leaf wilt of sugarcane in China. *Plant Disease, 105*, 1190.

Dakshinamoorthy, T., & Sivaprakasam, K. (1988). Relationship between meteorological factors and the incidence of pearl millet ergot disease. *The Madras Agricultural Journal, 75*, 48–50.

Dakshinamoorthy, T., Sivaprakasam, K., & Rangarajan, A. V. (1988). Role of insects in secondary spread of ergot disease of pearl millet. *The Madras Agricultural Journal, 75*, 304–307.

Dangaria, C. J., Dhedhi, K. K., Raghavani, K. L., Sorathia, J. S., Mungra, K. D., & Bunsa, B. D. (2012). A high yielding, early maturing downy mildew resistant pearl millet hybrid: GHB-719. *Journal of Agricultural Research and Technology, 37*, 201–205.

Das, I. K., Kumar, S., Kannababu, N., & Tonapi, V. A. (2019). Pokkah boeng resistance in popular rabi sorghum cultivars and effects of the disease on leaf chlorophyll. *Indian Phytopathology, 72*, 421–426.

Dass, S., Kapoor, R. L., Paroda, R. S., & Jatasra, D. S. (1984). Gene effects for downy mildew (*Sclerospora graminicola*) resistance in pearl millet. *Indian Journal of Genetics, 44*, 280–285.

Dauphinais, N., Bélair, G., Fournier, Y., & Dangi, O. P. (2005). Effect of crop rotation with grain pearl millet on *Pratylenchus penetrans* and subsequent potato yields in Quebec. *Phytoprotection, 86*, 195–199.

de Milliano, W. A. J. (1992). Pearl millet diseases in southern Africa. In W. A. J. de Milliano, R. A. Frederksen, & G. D. Bengston (Eds.), *Sorghum and millets diseases: A second world review* (pp. 115–122). ICRISAT.

De Waele, D., McDonald, A. H., Jordaan, E. M., Orion, D., van den Berg, E., & d Loots, G. C. (1998). Plant-parasitic nematodes associated with maize and pearl millet in Namibia. *African Plant Protection, 4*, 113–117.

Deepak, S. A., Niranjan Raj, S., Umemura, K., Kono, T., & Shetty, H. S. (2003). Cerebroside as an elicitor for induced resistance against the downy mildew pathogen in pearl millet. *Annals of Applied Biology, 143*, 169–173.

Devaiah, S. P., Mahadevappa, G. H., & Shetty, H. S. (2009). Induction of systemic resistance in pearl millet (*Pennisetum glaucum*) against downy mildew (*Sclerospora graminicola*) by *Datura metel* extract. *Crop Protection, 28*, 783–791.

Di Matteo, A., Federici, L., Mattei, B., & Cervone, F. (2003). The crystal structure of polygalacturonase-inhibiting protein (PGIP), a leucine-rich repeat protein involved in plant defense. *Proceedings of the National Academy Sciences – USA, 100*, 10124–10128.

Dieryck, B., Weyns, J., Doucet, D., Bragard, C., & Legrève, A. (2011). Acquisition and transmission of *Peanut clump virus* by *Polymyxa graminis* on cereal species. *Phytopathology, 101*, 1149–1158.

Doi, Y., Teranaka, M., Yora, K., & Asuyama, H. (1967). Mycoplasma or PLT-group-like organisms found in the phloem elements of plants infected with mulberry dwarf, potato witches' broom, aster yellows or paulownia witches' broom. *Annals of the Phytopathological Society of Japan, 33*, 259–266.

Dráb, T., Svobodová, E., Ripl, J., Jarošová, J., Rabenstein, F., Melcher, U., & Kundu, J. K. (2014). SYBR Green I based RT-qPCR assays for the detection of RNA viruses of cereals and grasses. *Crop and Pasture Science, 65*, 1323–1328.

Drew, M. C., & Lynch, J. M. (1980). Soil anaerobiosis, microorganisms, and root function. *Annual Review of Phytopathology, 18*, 37–66.

Eged, S. (2002). Changes in the content of fusaric acid during *Fusarium oxysporum* ontogenesis. *Biologia, 57*, 725–728.

Eisenback, J. D., & Triantaphyllou, H. H. (1991). Root-knot nematodes: Meloidogyne species and races. In W. R. Nickle (Ed.), *Manual of agricultural nematology* (pp. 281–286). Marcel Dekker.

Elisabeth, Z. P., Paco, S., Vibeke, L., Phillipe, S., Irenee, S., & Adama, N. (2008). Importance of seed borne fungi of sorghum and pearl millet in Burkina Faso and their control using plant extracts. *Pakistan Journal of Biological Sciences*, *11*, 321–331.

Eyvazi, A., Massah, A., Soorni, A., & Babaie, G. (2021). Molecular phylogenetic analysis shows that causal agent of maize rough dwarf diseases in Iran is closer to rice black-streaked dwarf virus. *European Journal of Plant Pathology*, *160*, 411–425.

Ezeaku, I. E., & Gupta, S. C. (2004). Development of sorghum populations for resistance to *Striga hermonthica* in the Nigerian Sudan Savanna. *African Journal of Biotechnology*, *3*, 324–329.

Fajinmi, A. A., Dokunmu, A. O., Akheituamen, D. O., & Onanagu, K. A. (2012). Incidence and infection rate of *Maize streak virus* by *Cicadulina triangular* on maize plants and its distribution from lowest diseased leaf under tropical conditions. *Archives of Phytopathology*, *45*, 1591–1598.

FAO. (2023). *Statistics*. https://www.fao.org/statistics/en/

Fawusi, M. O. A., & Agboola, A. A. (1980). Soil moisture requirements for germination of sorghum, millet, tomato, and *Celosia*. *Agronomy Journal*, *72*, 353–357.

Fazli, S.F.I. (1971). *Response of sorghum, millet and corn to different strains of sugarcane mosaic virus and strain A of maize dwarf mosaic virus* [Doctoral Dissertation, Texas A&M University]. https://www.proquest.com/openview/9e19fb220ffe480dd79411c457b57ef5/1.pdf

Ferrar, P. H., & Walker, J. R. L. (1993). O-Diphenol oxidase inhibition – an additional role for oxalic acid in the phytopathogenic arsenal of *Sclerotinia sclerotiorum* and *Sclerotium rolfsii*. *Physiological and Molecular Plant Pathology*, *43*, 415–422.

Fortuner, R. (1991). The Hoplolaiminae. In W. R. Nickle (Ed.), *Manual of agricultural nematology* (pp. 669–719). Marcel Dekker.

Fotso, J., Leslie, J. F., & Smith, J. S. (2002). Production of beauvericin, moniliformin, fusaproliferin, and fumonisins B1, B2 and B3 by ex-type strains of fifteen *Fusarium* species. *Applied and Environmental Microbiology*, *68*, 5195–5197.

Francis, S. M., & Williams, R. J. (1983). *Sclerospora graminicola. Descriptions of Pathogenic Fungi and Bacteria*, *77*.

Frederickson, D. E., Monyo, E. S., King, S. B., & Odvody, G. N. (1997). A disease of pearl millet in Zimbabwe caused by *Pantoea agglomerans*. *Plant Disease*, *81*, 959.

Frederickson, D. E., Monyo, E. S., King, S. B., Odvody, G. N., & Clafilin, L. E. (1999). Presumptive identification of *Pseudomonas syringae*, the cause of foliar leafspots and streaks on pearl millet in Zimbabwe. *Journal of Phytopathology*, *147*, 701–706.

Gadhave, K. R., Gautam, S., Rasmussen, D. A., & Srinivasan, R. (2020). Aphid transmission of *Potyvirus*: The largest plant-infecting RNA virus genus. *Viruses*, *12*, 773.

Gardan, L., Shafik, H., Broch, R., Grimont, F., & Grimont, P. A. D. (1999). DNA relatedness among the pathovars of *Pseudomonas syringae* and description on *Pseudomonas tremae* sp. Nov. and *Pseudomonas cannabina* sp. Nov. (ex Sutic and Dowson 1959). *International Journal of Systematic and Evolutionary Microbiology*, *49*, 469–478.

Gavini, F., Mergaert, J., Beji, A., Mielcarek, C., Izard, D., Kersters, K., & De Ley, J. (1989). Transfer of *Enterobacter agglomerans* (Beijerinck 1888) Ewing and Fife 1972 to *Pantoea* gen. nov. as *Pantoea agglomerans* comb., nov. and description of *Pantoea dispersa* sp. nov. *International Journal of Systematic Bacteriology, 39,* 337–345.

Geetha, H. S., & Shetty, H. S. (2002). Induction of resistant in pearl millet against downy mildew disease caused by *Sclerospora graminicola* using benzothiadizole, calcium chloride and hydrogen peroxide: A comparative evaluation. *Crop Protection, 21,* 601–610.

Giatatis, R., & Walcott, R. (2007). The epidemiology and management of seedborne bacterial diseases. *Annual Review of Phytopathology, 45,* 371–397.

Giatatis, R., Wilson, J., Walcott, R., Sanders, H., & Hanna, W. (2002). Occurrence of bacterial stripe of pearl millet in Georgia. *Plant Disease, 86,* 326.

Girisham, S., Rao, G. V., & Reddy, S. M. (1985). Mycotoxin producing fungi associated with pearl millet. (*Pennisetum typhoides* (Burm. F.) Stapf & C.E. Hubb.). *National Academy of Science Letters, 8,* 333–335.

Goud, T. Y., Sharma, R., Gupta, S. K., Uma Devi, G., Gate, V. L., & Boratkar, M. (2016). Evaluation of designated hybrid seed parents of pearl millet for blast resistance. *Indian Journal of Plant Protection, 44,* 83–87.

Gour, H. N., Nitharwal, P. D., & Agarwal, S. (1992). Partial purification of toxin from *Curvularia lunata* (Wakker) Boedijn. *Biochemie und Physiologie der Pflanzen, 188,* 128–135.

Grimont, P. A. D., & Grimont, F. (2015). *Klebsiella*. In *Bergey's manual of systematics of archaea and bacteria*. Wiley. https://doi.org/10.1002/9781118960608.gbm01150

Guo, J. L., Gao, S. W., Lin, Q. L., Wang, H. B., Que, Y. X., & Xu, L. P. (2015). Transgenic sugarcane resistant to *Sorghum mosaic virus* based on coat protein gene silencing by RNA interference. *Biomed Research International, 2015,* 861907.

Gupta, S. K., Sharma, R., Rai, K. N., & Thakur, R. P. (2012). Inheritance of foliar blast resistance in pearl millet (*Pennisetum glaucum*). *Plant Breeding, 13,* 217–219.

Gupta, S. K., & Singh, D. (1996). Studies on the influence of downy mildew infection on yield and yield-contributing plant characters of pearl millet in India. *International Journal of Pest Management, 42,* 89–93.

Hanna, W. W., & Wells, H. D. (1989). Inheritance of pyricularia leaf spot resistance in pearl millet. *Journal of Heredity, 80,* 145–147.

Hanna, W. W., Wells, H. D., & Burton, G. W. (1982). Transfer of disease immunity from a wild subspecies to pearl millet. In *Agronomy abstracts* (p. 69). ASA.

Harrison, J., Hussain, R. M. F., Aspin, A., Grant, M. R., Vincente, J. G., & Studholme, D. J. (2023). Phylogenomic analysis supports the transfer of 20 pathovars from *Xanthomonas campestris* into *Xanthomonas euvesicatoria*. *Taxonomy, 3,* 29–45.

Hash, C. T., Singh, S. D., Thakur, R. P., & Talukdar, B. S. (1999). Breeding for disease resistance. In I. S. Khairwal, K. N. Rai, D. J. Andrews, & G. Harinrayana (Eds.), *Pearl millet breeding* (337–379). Oxford & IBH Publishing.

Hash, C. T., Witcombe, J. R., Thakur, R. P., Bhatnagar, S. K., Singh, S. D., & Wilson, J. P. (1997). Breeding for pearl millet disease resistance. In *Proceedings of the International Conference on Genetic Improvement of Sorghum and Pearl Millet* (pp. 337–373). International Crop Research Institute.

Haussmann, B. I. G., Hess, D. E., Omanya, G. O., Reddy, B. V. S., Welz, H. G., & Geiger, H. H. (2001). Major and minor genes for stimulation of *Striga hermonthica* seed germination in sorghum, interaction with different *Striga* populations. *Crop Science, 41*, 1507–1512.

Hemalatha, T. M., Shanthipriya, M., Naik, D. V. K., Reddy, B. V. B., Reddy, M. G., Madhavilatha, L., Kumar, M. H., Vajantha, B., Sarala, N. V., & Tagore, K. R. (2023). First report of 'Candidatus *Phytoplasma aurantifolia*' related strains (16SrII-D) associated with stunting, little leaf and phyllody disease of pearl millet from south India. *Plant Disease, 107*, 2209.

Hess, D. E., & Ejeta, G. (1992). Inheritance of resistance to *Striga* in sorghum genotype SRN39. *Plant Breeding, 109*, 233–241.

Hestbjerg, H., Nielsen, K. F., Thrane, U., & Elmholt, S. (2002). Production of trichothecenes and other secondary metabolites by *Fusarium culmorum* and *Fusarium equiseti* on common laboratory media and a soil organic matter agar: An ecological interpretation. *Journal of Agricultural and Food Chemistry, 50*, 7593–7599.

Hill, G. M., & Hanna, W. W. (1990). Nutritive characteristics of pearl-millet grain in beef-cattle diets. *Journal of Animal Science, 68*, 2061–2066.

Horst, R. K. (2013). *Field manual of diseases of grasses and native plants*. Springer Science + Business Media.

Houissa, H., Lasram, S., Sulyok, M., Šarkanj, B., Fontana, A., Strub, C., Krska, R., Schorr-Galindo, S., & Ghorbel, A. (2019). Multimycotoxin LC-MS/MS analysis in pearl millet (*Pennisetum glaucum*) from Tunisia. *Food Control, 106*, 106738.

Huang, K., Whitlock, R., Press, M. C., & Scholes, J. D. (2012). Variation for host range within and among populations of the parasitic plant *Striga hermonthica*. *Heredity, 108*, 96–104.

Huang, M., Lin, L., Wu, Y.-X., Honhing, H., He, P.-F., Li, G.-Z., He, P.-B., Xiong, G.-R., Yuan, Y., & He, Y.-Q. (2016). Pathogenicity of *Klebsiella pneumonia* (KpC4) infecting maize and mice. *Journal of Integrative Agriculture, 15*, 1510–1520.

Hulse, J. H., Liang, E. M., & Pearson, O. E. (1980). *Sorghum and the millets: Their composition and nutritive value*. Academic Press.

Hussein, A., Anwar, S. A., Sahi, G. M., Abbas, Q., & Imran. (2009). Seed borne fungal pathogens associated with pearl millet (*Pennisetum typhoides*) and their impact on seed germination. *Pakistan Journal of Phytopathology, 21*, 55–60.

Indira, S., Xu, X., Iamsupasit, N., Shetty, H. S., Vasanthi, N. S., Singh, S. D., & Bandyopadhyay, R. (2002). Diseases of sorghum and pearl millet in Asia. In J. F. Leslie (Ed.), *Sorghum and millets diseases* (pp. 393–402). Iowa State University Press.

Ingle, R. W., & Raut, J. G. (1994). Effect of fungicidal sprays on incidence of seed-borne fungi and germination of pearl millet. *Seed Research*, *22*, 85–87.

Inoue-Nagata, A. K., Jordan, R., Kreuze, J., Li, F., López-Moya, J. J., Mäkinen, K., Ohshima, K., Wylie, S. J., & ICTV Report Consortium. (2022). ICTV virus taxonomy profile: Potyviridae 2022. *Journal of General Virology*, *103*, 001738.

Ismaiel, A. A., & Papenbrock, J. (2015). Mycotoxins: Producing fungi and mechanisms of phytotoxicity. *Agriculture*, *5*, 492–537.

Ito, D., Miller, Z., Menalled, F., Moffet, M., & Burrows, M. (2012). Relative susceptibility among alternative host species prevalent in the Great Plains to *Wheat streak mosaic virus*. *Plant Disease*, *96*, 1185–1192.

Jamil, M., Kountche, B., & Al-Babili, S. (2021). Current progress in *Striga* management. *Plant Physiology*, *185*, 1339–1352.

Jeger, M. J., Gilijamse, E., Bock, C. H., & Frinking, H. D. (1998). The epidemiology, variability and control of the downy mildews of pearl millet and sorghum, with particular reference to Africa. *Plant Pathology*, *47*, 544–569.

Jensen, S. G., Andrews, D. J., & DeVries, N. E. (1983). Reactions of pearl millet germ plasm from the world collection to maize dwarf mosaic virus strains A and B. *Plant Disease*, *67*, 1105–1108.

Jeppson, L. R., Kiefer, H. H., & Baker, E. W. (1975). *Mites injurious to economic plants*. University of California Press.

Johnson, A. W., Hanna, W. W., & Dowler, C. C. (1995). Effects of irrigation, nitrogen, and a nematicide on pearl millet. *Journal of Nematology*, *27*(4S), 571–574.

Johnson, S. (1997). *Tomatoes, potatoes, corn, and beans: How the foods of the Americas changed eating around the world. Atheneum Books, New York.*

Jomentiene, R., & Davis, R. E. (2006). Clusters of diverse genes existing as multiple, sequence-variable mosaics in a phytoplasma genome. *FEMS Microbiology Letters*, *255*, 59–65.

Jones, E. S., Breese, W. A., Liu, C. J., Singh, S. D., Shaw, D. S., & Witcombe, J. R. (2002). Mapping quantitative trait loci for resistance to downy mildew in pearl millet: Field and glasshouse screens detect the same QTL. *Crop Science*, *42*, 1316–1323.

Jones, E. S., Liu, C. J., Gale, M. D., Hash, C. T., & Witcomb, J. R. (1995). Mapping quantitative trait loci for downy mildew resistance in pearl millet. *Theoretical and Applied Genetics*, *91*, 448–456.

Jung, H.-Y., Sawayanagi, T., Wongkaew, P., Kakizawa, S., Nishigawa, H., Wei, W., Oshima, K., Miyata, S., Ugaki, M., Hibi, T., & Namba, S. (2003). 'Candidatus Phytoplasma oryzae,' a novel phytoplasma taxon associated with rice yellow dwarf disease. *International Journal of Systematic and Evolutionary Microbiology*, *53*, 1925–1929.

Jurjevic, Z., Wilson, D. M., Wilson, J. P., Geiser, D. M., Juba, J. H., Mubatanhema, W., Widstrom, N. W., & Rains, G. C. (2005). *Fusarium* species of the *Gibberella fujikuroi* complex and fumonisin contamination of pearl millet and corn in Georgia, USA. *Mycopathologia*, *159*, 401–406.

Karan, M., Noone, D. F., Teakle, D. S., & Hacker, J. B. (1992). Susceptibility of pearl millet accessions and cultivars to Johnsongrass mosaic and sugarcane mosaic viruses in Queensland. *Australasian Plant Pathology, 21*, 128–130.

Karavina, C. (2014). Maize streak virus: A review of pathogen occurrence, biology and management options for smallholder farmers. *African Journal of Agricultural Research, 9*, 2736–2742.

Kardam, V. K., Meena, A. K., & Godara, S. L. (2023). Status of Bipolaris leaf spot disease (*Bipolaris setariae*) of pearl millet in Rajasthan. *Indian Phytopathology, 76*, 415–427.

Kaurav, A., Pandya, R. K., & Singh, B. (2018). Performance of botanicals and fungicides against blast of pearl millet (*Pennisetum glaucum*). *Annals of Plant and Soil Research, 20*, 258–262.

Kendrick, J. B. (1926). Holcus bacterial spot of species of *Holcus* and *Zea mays*. *Phytopathology, 16*, 236–237.

Kenneth, R., & Kranz, J. (1973). *Plasmopara penniseti* sp. Nov., a downy mildew of pearl millet in Ethiopia. *Transactions of the British Mycological Society, 60*, 590–593.

Kenneth, R. G. (1976). The downy mildews of corn and other Gramineae in Africa and Israel, and the present state of knowledge and research. *Agriculture and Natural Resources, 10*, 148–159.

Khairwal, I. S., Rai, K. N., Diwakar, B., Sharma, Y. K., Rajpurohit, B. S., Nirwan, B., & Bhattacharjee, R. (2007). *Pearl millet: Crop management and seed production manual*. ICRISAT.

Khan, M. R. (2015). Nematode diseases of crops in India. In L. P. Awasthi (Ed.), *Recent advances in the diagnosis and management of plant diseases* (pp. 183–224). Springer.

Khanna, A., Raj, K., and Sangwan, P. 2014. Smut of pearl millet: Current status and future prospects. Pages 123–129 in: Researh Trends in Agriculture Sciences, Vol. 10 (Naresh, R.K., ed.). AkiNik Publications, Delhi, India.

King, G. S., Waight, E. S., Mantle, P. G., & Szczyrbak, C. A. (1977). The structure of clavicipitic acid, and azepinoindole derivative from *Claviceps fusiformis*. *Journal of the Chemical Society, Perkin Transactions, 1*, 2099–2013.

Kogika, M. M., Hagiwara, M. K., & Mirandola, R. M. (1993). Experimental citrinin nephrotoxicosis in dogs: Renal function evaluation. *Veterinary and Human Toxicology, 35*, 136–140.

Koirala, S., Zhao, M., Agarwal, G., Gitaitis, R., Stice, S., Kvitko, B., & Dutta, B. (2021). Identification of two novel pathovars of *Pantoea stewartii* subsp. *indologenes* affecting *Allium* sp. and millets. *Phytopathology, 111*, 1509–1519.

Konde, B. K., Dhage, B. V., & More, B. B. (1980). Seed-borne fungi of some pearl millet cultivars. *Seed Research, 8*, 59–63.

Kritzman, G., Chet, I., & Henis, Y. (1977). The role of oxalic acid in the pathogenic behavior of *Sclerotium rolfsii* Sacc. *Experimental Mycology, 1*, 280–285.

Kulkarni, G. S. (1913). Observations on the downy mildew (*Sclerospora graminicola* Sacc.) (Schroet.) of bajra and jowar. *Memoirs of the Department of Agriculture in India, Botanical Series, 5*, 268–273.

Kulkarni, K. S., Zala, H. N., Bosamia, T. C., Shukla, Y. M., Kumar, S., Fougat, S., Patel, M. S., Narayanan, S., & Joshi, C. G. (2016). De novo transcriptome sequencing to dissect candidate genes associated with pearl millet-downy mildew (*Sclerospora graminicola* Sacc.) interaction. *Frontiers in Plant Science, 7*, 847.

Kumar, A., & Gour, H. N. (1992). Downy mildew of pearl millet: I. Leaf diffusive resistance and transpiration. *Biochemie und Physiologie der Pflanzen, 188*, 39–43.

Kumar, A., Manga, V. K., Gour, H. N., & Purohit, A. K. (2012). Pearl millet downy mildew: Challenges and prospects. *Review of Plant Pathology, 5*, 140–177.

Kumar, A., Mulatu, F., Assefa, Y., Hanson, J., & Jones, C. S. (2018). First report of Johnsongrass mosaic virus infecting *Urochloa mosambicensis* in Ethiopia. *Plant Disease, 102*, 459.

Kumar, A., Sheoran, N., Prakash, G., Ghosh, A., Chikara, S. K., Rajashekara, H., Singh, U. D., Aggarwal, R., & Jain, R. K. (2017). Genome sequence of a unique *Magnaporthe oryzae* RMg-Dl isolate from India that causes blast disease in diverse cereal crops, obtained using PacBio single-molecule and Illumina HiSeq2500 sequencing. *Genome Announcements, 5*, e01570–e01516.

Kumar, K., Pal, G., Verma, A., & Verma, S. K. (2021a). Role of rhizospheric bacteria in disease suppression during seedling formation in millet. In S. K. Dubey & S. K. Verma (Eds.), *Plant, soil and microbes in tropical ecosystems* (pp. 263–274). Springer.

Kumar, K., Verma, A., Pal, G., Anubha, White, J. F., & Verma, S. K. (2021b). Seed endophytic bacteria of pearl millet (*Pennisetum glaucum* L.) promote seedling development and defend against a fungal phytopathogen. *Frontiers in Microbiology, 12*, 774293.

Kumar, P., Gupta, A., Mahato, D. K., Pandhi, S., Pandey, A. K., Kargwal, R., Mishra, S., Suhag, R., Sharma, N., Saurabh, V., Paul, V., Kumar, M., Selvakumar, R., Gamiath, S., Kamle, M., El Enshasy, H. A., Mokhtar, J. A., & Harakeh, S. (2022). Aflatoxins in cereals and cereal-based products: Occurrence, toxicity, impact on human health, and their detoxification and management strategies. *Toxins, 14*, 687.

Kumar, S., Singh, V., & Lakhanpaul, S. (2010). First report of 'Candidatus Phytoplasma asteris' associated with green ear disease of bajra (pearl millet) in India. *New Disease Reports, 22*, 27.

Kumar, S., Thakur, D. P., Duhan, L. C., & Arya, S. (1997). Dispersal of ergot disease in pearl millet field. *Crop Research, 13*, 511–513.

Kumar, V. U., Meera, M. S., Shishupala, S., & Shetty, H. S. (1993a). Quantification of *Sclerospora graminicola* in tissues of *Pennisetum glaucum* using an enzyme-linked immunosorbent assay. *Plant Pathology, 42*, 251–255.

Kumar, V. U., Shishupala, S., Shetty, H. S., & Umesh-Kumar, S. (1993b). Serological evidence for the occurrence of races in *Sclerospora graminicola* and identification of a race-specific surface protein involved in host recognition. *Canadian Journal of Botany, 11*, 1467–1471.

Kumari, P., Godika, S., Ghasolia, R. P., Goyal, S. K., Khan, I., Deora, A., Meena, S., Kumar, S., Kumar, S., & Kumar, L. (2022). Validation of stable resistance in pearl millet hybrids to ergot disease caused by *Claviceps fusiformis*. *Agricultural Mechanization in Asia, 53*, 5967–5974.

Kumudini, B. S., Vasanthi, N. S., & Shetty, H. S. (2005). Hypersensitive response, cell death and histochemical localisation of hydrogen peroxide in host and non-host seedlings infected with the downy mildew pathogen *Sclerospora graminicola*. *Annals of Applied Biology, 139*, 217–225.

Laidlaw, H. K. C., Persley, D. M., Pallaghy, C. K., & Godwin, I. D. (2004). Sequence diversity in the coat protein coding region of the genome RNA of *Johnsongrass mosaic virus* in Australia. *Archives of Virology, 149*, 1633–1641.

Lakose, C. M., Kadvani, D. L., & Dangaria, C. J. (2007). Efficacy of fungicides in controlling blast disease in pearl millet. *Indian Phytopathology, 60*, 68–71.

Lamichhane, J. R., Debaeke, P., Steinberg, C., You, M. P., Barbetti, M. J., & Aubertot, J.-N. (2018). Abiotic and biotic factors affecting crop seed germination and seedling emergence: A conceptual framework. *Plant and Soil, 432*, 1–32.

Lamichhane, J. R., Dürr, C., Schwanck, A. A., Robin, M.-H., Sarthou, J.-P., Cellier, V., Messéan, A., & Aubertot, J.-N. (2017). Integrated management of damping-off diseases: A review. *Agronomy for Sustainable Development, 37*, 10.

Lasram, S., Hamdi, Z., Chenenaoui, S., Mliki, A., & Ghorbel, A. (2016). Comparative study of toxigenic potential of *Aspergillus flavus* and *Aspergillus niger* isolated from barley as affected by temperature, water activity and carbon source. *Journal of Stored Products, 69*, 58–64.

Leakey, C. L. A. (1964). *Dactuliophora*, a new genus of mycelia sterilia from tropical Africa. *Transactions of the British Mycological Society, 47*, 341–350.

Legrève, A., Delfosse, P., & Maraite, H. (2002). Phylogenetic analysis of *Polymyxa* species based on nuclear 5.8S and internal transcribed spacers ribosomal DNA sequences. *Mycological Research, 106*, 138–147.

Lelliot, R. A., & Stead, D. E. (1987). *Methods for the diagnosis of bacterial diseases of plants*. Blackwell Scientific Publications.

Leslie, J. F., & Summerell, B. A. (2006). *The Fusarium laboratory manual*. Blackwell Publishing.

Leslie, J. F., Zeller, K. A., Lamprecht, S. C., Rheeder, J. P., & Marasas, W. F. O. (2005). Toxicity, pathogenicity and genetic differentiation of five species of *Fusarium* for sorghum and millet. *Phytopathology, 95*, 275–283.

Lew, H., Chelkowski, J., Pronczuk, P., & Edinger, W. (1996). Occurrence of the mycotoxin moniliformin in maize (*Zea mays* L.) ears infected by *Fusarium subglutinans* (Wollenweber and Reinking) Nelson et al. *Food Additives and Contaminants, 13*, 321–324.

Li, J., Zhang, R., Li, Y., Long, C., Li, W., Shan, H., Lu, W., & Huang, Y. (2022). Characterization, phylogenetic analyses and pathogenicity of *Bipolaris setariae* associated with brown stripe disease on sugarcane in Yunnan, China. *Plant Pathology, 72*, 89–99.

Liu, C. J., Witcombe, J. R., Pittaway, T. S., Nash, M., Hash, C. T., Busso, C. S., & Gale, M. D. (1994). An RFLP-based genetic map of pearl millet (*Pennisetum glaucum*). *Theoretical and Applied Genetics, 89*, 481–487.

Liu, H., Qiu, H., Zhao, W., Cui, Z., Ibrahim, H., Jin, G., Li, B., Zhu, B., & Xie, G. L. (2012). Genome sequence of the plant pathogen *Pseudomonas syringae* pv. *Panici* LMG 2367. *Journal of Bacteriology, 194*, 5693–5694.

Liu, Y. X., Shi, Y. P., Deng, Y. Y., & Cai, Z. Y. (2016). First report of leaf spot caused by *Bipolaris setariae* on rubber tree in China. *Plant Disease, 100*, 1240.

Lockard, V. G., Phillips, R. D., Hayes, A. W., Berndt, W. O., & O'Neal, R. M. (1980). Citrinin nephrotoxicity in rats: A light and electron microscopic study. *Experimental and Molecular Pathology, 32*, 226–240.

Logrieco, A., Moretti, A., Castellá, G., Kostecki, M., Golinski, P., Ritieni, A., & Chelkowski, J. (1998). Beauvericin production by *Fusarium* species. *Applied and Environmental Microbiology, 64*, 3084–3088.

Loveless, A. R. (1967). *Claviceps fusiformis* sp. Nov., the causal agent of agalactia of sows. *Transactions of the British Mycological Society, 50*, 15–18.

Lubadde, G., Tongoona, P., Derera, J., & Sibiya, J. (2014). Major pearl millet diseases and their effects on-farm grain yield in Uganda. *African Journal of Agricultural Research, 9*, 2911–2918.

Luttrell, E. S. (1954). Diseases of pearl millet in Georgia. *Plant Disease Report, 38*, 507–514.

Luttrell, E. S., Crowder, L. V., & Wells, H. D. (1955). Seed treatment tests with pearl millet, sudan grass, and browntop millet. *Plant Disease Report, 39*, 756–761.

Lv, M., Xie, L., Yang, J., Chen, J., & Zhang, H.-M. (2016). Complete genomic sequence of maize rough dwarf virus, a Fijivirus transmitted by the small brown planthopper. *Genome Announcements, 4*, e01529–e01515.

Mahatma, M. K., Bhatnagar, R., Dhandhukia, P., & Thakkar, V. R. (2009). Variation in metabolites constituent in leaves of downy mildew resistant and susceptible genotypes of pearl millet. *Physiology and Molecular Biology of Plants, 15*, 249–255.

Maher, C. (1936). Elephant grass (*Pennisetum purpureum*) as a cattle fodder in Kenya. *East African Agricultural Journal, 1*, 340–342.

Mahmood, T., & Qureshi, S. H. (1984). Incidence and reaction of pearl millet cultivars to various diseases under rainfed condition. *Pakistan Journal of Agricultural Research, 5*, 32–34.

Makun, H. A., Gbodi, T. A., Tijani, A. S., Abai, A., & Kadiri, G. U. (2007). Toxicologic screening of fungi isolated from millet (*Pennisetum* spp) during the rainy and dry harmattan seasons in Niger state, Nigeria. *African Journal of Biotechnology*, *6*, 34–40.

Mali, V. R. (2007). An immunity (QL3) breaking strain of sorghum mosaic Potyvirus in India. *Plant Disease*, *83*, 877.

Malik, V. K., Sangwan, P., Singh, M., Punia, R., Yadav, D. V., Kumari, P., & Pahuja, S. K. (2022). First report of *Klebsiella aerogenes* inciting stem rot of pearl millet in Haryana, India. *Plant Disease*, *106*, 754.

Manamgoda, D. S., Cai, L., McKenzie, E. H. C., Crous, P. W., Madrid, H., Chukeatirote, E., Shivas, R. G., Tan, Y. P., & Hyde, K. D. (2012). A phylogenetic and taxonomic re-evaluation of the *Bipolaris – Cochliobolus – Curvularia* complex. *Fungal Diversity*, *56*, 131–144.

Manamgoda, D. S., Rossman, A. Y., Castlebury, L. A., Crous, P. W., Madrid, H., Chukeatirote, E., & Hyde, K. D. (2014). The genus *Bipolaris*. *Studies in Mycology*, *79*, 221–288.

Mandadi, K. K., Pyle, J. D., & Scholthof, K.-B. G. (2014). Comparative analysis of antiviral responses in *Brachypodium distachyon* and *Setaria viridis* reveals conserved and unique outcomes among C_3 and C_4 plant defenses. *Molecular Plant-Microbe Interactions*, *27*, 1277–1290.

Manjunatha, G., Deepak, S., Geetha, P. N., Niranjan-Raj, S., Kini, R. K., & Shetty, H. S. (2009). Hypersensitive reaction and P/HRGP accumulation is modulated by nitric oxide through hydrogen peroxide in pearl millet during *Sclerospora graminicola* infection. *Physiological and Molecular Plant Pathology*, *74*, 191–198.

Mann, A., Malik, S., Rana, J. S., & Nehra, K. (2021). Whole genome sequencing data of *Klebsiella aerogenes* isolated from agricultural soil of Haryana, India. *Data in Brief*, *38*, 107311.

Mantle, P. G. (1991). Miscellaneous toxigenic fungi. In J. E. Smith & R. S. Henderson (Eds.), *Mycotoxins and animal foods* (pp. 141–152). CRC Press.

Marasas, W. F. O., Nelson, P. E., & Toussoun, T. A. (1984). *Toxigenic Fusarium species: Identity and mycotoxicology*. The Pennsylvania State University Press.

Marasas, W. F. O., Rabie, C. J., Lübben, A., Nelson, P. E., Toussoun, T. A., & van Wyk, P. S. (1988). *Fusarium nygamai* from millet in Southern Africa. *Mycologia*, *80*, 263–266.

Marasas, W. F. O., Thiel, P. G., Rabie, C. J., Nelson, P. E., & Toussoun, T. E. (1986). Moniliformin production in *Fusarium* section Liseola. *Mycologia*, *78*, 242–247.

Marasas, W. F. O., Thiel, P. G., Sydenham, E. W., Rabie, C. J., Lubben, A., & Nelson, P. E. (1991). Toxicity and moniliformin production by four recently described species of *Fusarium* and two uncertain taxa. *Mycopathologia*, *113*, 191–197.

Marcone, C., Schneider, B., & Seemüller, E. (2004). 'Candidatus Phytoplasma cynodontis,' the phytoplasma associated with Bermuda grass white leaf disease. *International Journal of Systematic and Evolutionary Microbiology*, *54*, 1077–1082.

Markell, S. G., & Francl, L. J. (2003). Fusarium head blight inoculum: Species prevalence and *Gibberella zeae* spore type. *Plant Disease, 87*, 814–820.

Marley, P. S., Diourté, M., Neya, A., Nutsugah, S. K., Sérémé, P., Katilé, S. O., Hess, D. E., Mbaye, D. F., & Ngoko, Z. (2002). Sorghum and pearl millet diseases in West and Central Africa. In J. F. Leslie (Ed.), *Sorghum and millets diseases* (pp. 419–425). Iowa State University Press.

Maryam, A. H. (2015). Evaluation of heterosis in pearl millet (*Pennisetum glaucum* (L.) R. Br.) for agronomic traits and resistance to downy mildew (*Sclerospora graminicola*). *Journal of Agriculture and Crops, 1*, 1–8.

Mathur, S. B., Chalal, S. S., & Thakur, R. P. (1994). *Seed-borne diseases and seed health testing of pearl millet*. Institute of Seed Pathology for Developing Countries.

Mathur, S. K., Nath, R., & Mathur, S. B. (1973). Seed-borne fungi of pearl millet (*Pennisetum typhoides*) and their significance. *Seed Science and Technology, 1*, 811–820.

Matsutova, R., Rani, K., Verstappen, F. W. A., Franssen, M. C. R., Beale, M. H., & Bouwmeester, H. J. (2005). The strigolactone germination stimulants of the plant-parasitic *Striga* and *Orobanche* spp. Are derived from the carotenoid pathway. *Plant Physiology, 139*, 920–934.

Mbwaga, A. M., & Mdolwa, S. I. (1995). Breeding for disease resistance with emphasis on durability. In D. L. Danial (Ed.), *Proceedings of a regional workshop for Eastern, Central and Southern Africa* (pp. 239–243). Wageningen Agricultural University.

McGee, D. C. (1988). *Maize diseases: A reference source for seed technologists*. APS Press.

Mehta, P. R., Singh, B., & Mathur, S. C. (1953). A new leaf spot disease of bajra (*Pennisetum typhoides* Staph and Hubbard) caused by a species of *Piricularia*. *Indian Phytopathology, 5*, 140–143.

Melhus, I. E., van Haltern, F. H., & Bliss, D. E. (1927). A study of the downy mildew *Sclerospora graminicola* (Sacc.) Schroet. *Phytopathology, 17*, 57.

Melhus, I. E., van Haltern, F. H., & Bliss, D. E. (1928). A study of *Sclerospora graminicola* (Sacc.) Schroet. on *Setaria viridis* (L.) Beauv. and *Zea mays* L. *The Review of Applied Mycology, 7*, 712–713.

Mergaert, J., Verdonck, L., & Kersters, K. (1993). Transfer of *Erwinia ananas* (synonym, *Erwinia uredovora*) and *Erwinia stewartii* to the genus *Pantoea* emend. as *Pantoea ananas* (Serrano 1928) comb. nov. and *Pantoea stewartii* (Smith 1898) comb. nov., respectively, and description of *Pantoea stewartii* subsp. indologenes subsp. nov. *International Journal of Systematic and Evolutionary Microbiology, 43*, 162–173.

Miedaner, T., & Gieger, H. H. (2015). Biology, genetics, and management of ergot (*Claviceps* spp.) in rye, sorghum, and pearl millet. *Toxins, 7*, 659–678.

Mishra, B., & Prakish, O. (1975). Alternaria leaf spot of soybean from India. *Indian Journal of Mycology and Plant Pathology, 5*, 95.

Mishra, N. K., & Daradhiyar, S. K. (1991). Mold flora and aflatoxin contamination of stored and cooked samples of pearl millet in the Paharia tribal belt of Santhal Pargana, Bihar, India. *Applied and Environmental Microbiology, 57,* 1223–1226.

Mitter, J. H., & Mitra, A. K. (1941). Occurrence of *Sclerospora graminicola* (Sacc.) Schroet. On *Setaria verticillata* Beauv. In Allahabad. *The Review of Applied Mycology, 20,* 169.

Mitter, J. H., & Tandon, R. N. (1930). A note on *Sclerospora graminicola* (Sacc.) Schroet. In Allahabad. *Journal of the Indian Botanical Society, 9,* 243.

Mofini, M.-T., Diedhiou, A. G., Simonin, M., Dondjou, D. T., Pignoly, S., Ndiaye, C., Min, D., Vigouroux, Y., Laplaze, L., & Kane, A. (2022). Cultivated and wild pearl millet display contrasting patterns of abundance and co-occurrence in their root mycobiome. *Scientific Reports, 12,* 207.

Monson, W. G., Hanna, W. W., & Gaines, T. P. (1986). Effects of rust on yield and quality of pearl millet forage. *Crop Science, 26,* 637–639.

Monyo, E. S. (1998). 15 years of pearl millet improvement in the SADC region. *International Sorghum and Millets Newsletter, 39,* 17–33.

Morales, F. J., Castano, M., Velasco, A. C., & Arroyave, J. (1994). Detection of a potyvirus related to guineagrass mosaic virus infecting *Brachiaria* spp. In South America. *Plant Disease, 78,* 425–428.

Moretti, A., Logrieco, A., Bottalico, A., Ritieni, A., Randazzo, G., & Corda, P. (1995). Beauvericin production by *Fusarium subglutinans* from different geographical areas. *Mycological Research, 99,* 282–286.

Mtisi, E. (1992). Sorghum and pearl millet pathology in Zimbabwe. In W. A. J. Milliano, R. A. Frederiksen, & G. D. Bengston (Eds.), *Sorghum and millets diseases: A second world review* (pp. 3–7). ICRISAT.

Mtisi, E., & de Milliano, W. A. J. (1991). Occurrence and host range of false mildew, caused by *Beniowskia sphaeroidea*, in Zimbabwe. *Plant Disease, 75,* 215.

Mtisi, E., & de Milliano, W. A. J. (1993). False mildew on pearl millet and other hosts in Zimbabwe. *East African Agriculture and Forestry Journal, 59,* 145–153.

Mtisi, E., & McClaren, N. W. (2002). Diseases of sorghum and pearl millet in some southern Africa countries. In *Sorghum and millets diseases* (pp. 427–430). Iowa State University Press (Blackwell).

Muhire, B., Martin, D. P., Brown, J. K., Nava-Castillo, J., Moriones, E., Zerbini, F. M., Rivera-Bustamante, M. V. G., Briddon, R. W., & Varsani, A. (2013). A genome-wide pairwise-identity-based proposal for the classification of viruses in the genus *Mastrevirus* (family *Geminiviridae*). *Archives of Virology, 158,* 1411–1424.

Musser, S. M., & Plattner, R. D. (1997). Fumonisin composition in cultures of *Fusarium moniliforme, Fusarium proliferatum* and *Fusarium nygamai. Journal of Agricultural and Food Chemistry, 45,* 1169–1173.

Nagaraja, A., & Anjaneya Reddy, B. (2009). Foot rot of finger millet: An increasing disease problem in Karnataka. *Crop Research, 38,* 224–225.

Nagaraja, A., & Viswanath, S. (1983). Screening of pearl millet germplasm for resistance to Eleusine strain of maize streak virus. *Indian Journal of Mycology and Plant Pathology, 13*, 221–222.

Nandhini, M., Rajini, S. B., Udayashankar, A. C., Niranjana, S. R., Lund, O. S., Shetty, H. S., & Prakash, H. S. (2018). Diversity, plant growth promoting and downy mildew disease suppression potential of cultivable endophytic fungal communities associated with pearl millet. *Biological Control, 127*, 127–138.

Narayanaswami, R., & Veeraju, V. (1970). A new Cercospora leaf spot of pearl millet. *Indian Phytopathology, 23*, 135–136.

Nattrass, R. M. (1941). Notes on plant diseases. *East African Agricultural Journal, 7*, 56.

Navi, S. S., King, S. B., & Singh, S. D. (1997). A new report of *Bipolaris panici-miliacei* on pearl millet. *International Sorghum and Millets Newsletter, 38*, 124.

Nayak, D. K. (2010). Nematode problems in millets and their management. *International Journal of Plant Sciences, 5*, 693–695.

Nayaka, S. C., Shetty, H. S., Satyavathi, C. T., Yadav, R. S., Kishor, P. B. K., Nagaruju, M., Anoop, T. A., Kumar, M. M., Kurakose, B., Chakravartty, N., Katta, A. V. S. K. M., Lachargari, V. B. R., Singh, O. V., Sahu, P. P., Puranik, S., Kaushal, P., & Srivastava, R. K. (2017). Draft genome sequence of *Sclerospora graminicola*, the pearl millet downy mildew pathogen. *Biotechnology Reports, 16*, 18–20.

N'Diaye, M. (1995). *Rapport annuel d'activités sur l'amélioration du controle du mildiou*. ROCAFREMI/ICRISAT.

Nelson, P. E., Plattner, R. D., Shackelford, D. D., & Desjardins, A. E. (1992). Fumonisin B1 production by *Fusarium* species other than *F. moniliforme* in section *Liseola* and some related species. *Applied and Environmental Microbiology, 58*, 984–989.

Ng, J. C., & Falk, B. W. (2006). Virus-vector interactions mediating nonpersistent and semipersistent transmission of plant viruses. *Annual Review of Phytopathology, 44*, 183–212.

Nitharwal, P. D., Gour, H. N., & Agarwal, S. (1991). Effects of different factors on the production of cellulase by *Curvularia lunata. Folia Microbiologica, 36*, 357–361.

Niu, X.-Q., Yu, F.-Y., Zhu, H., & Qin, W.-Q. (2014). First report of leaf spot disease in coconut seedling caused by *Bipolaris setariae* in China. *Plant Disease, 98*, 1742.

Noor, A., Perumal, R., and Little, C.R. 2023. Identification and molecular characterization of pearl millet seed-associated fungi in Kansas. Phytopathology 113, S3.180.

Odvody, G. N. (2000). Southern sclerotial rot. In R. A. Frederiksen & G. N. Odvody (Eds.), *Compendium of sorghum diseases* (Vol. 2, pp. 37–38). APS Press.

Okabe, I., Morikawa, C., Matsumoto, N., & Yokoyama, K. (1998). Variation in *Sclerotium rolfsii* isolates in Japan. *Mycoscience, 39*, 399–407.

Okunowo, W. O., Osuntoki, A. A., Adekunle, A. A., & Gbenle, G. O. (2013). Occurrence and effectiveness of an indigenous strain of *Myrothecium roridum* Tode: Fries as a bioherbicide for water hyacinth (*Eichhorinia crassipes*) in Nigeria. *Biocontrol Science and Technology, 23,* 1387–1401.

Ong, C. K. (1983). Response to temperature in a stand of pearl millet (*Pennisetum typhoides* S. & H.). *Journal of Experimental Botany, 34,* 322–336.

Onyike, N. B. N., Nelson, P. E., & Marasas, W. F. O. (1991). *Fusarium* species associated with millet grain from Nigeria, Lesotho, and Zimbabwe. *Mycologia, 83,* 708–712.

Pal, K. K., Dey, R., & Tilak, K. (2014). Fungal diseases of groundnut: Control and future challenges. In A. Goyal & C. Manoharachary (Eds.), *Future challenges in crops protection against fungal pathogens* (pp. 1–29). Springer.

Palanga, E., Tibri, E. B., Bangratz, M., Filloux, D., Julian, C., Pinel-Galzi, A., Koala, M., Néva, J. B., Brugidou, C., Tiendrébéogo, F., Roumagnac, P., & Hébrard, E. (2022). Complete genome sequence of a novel marfavirus infecting pearl millet in Burkina Faso. *Archives of Virology, 167,* 245–248.

Palleroni, N. J. (2015). *Pseudomonas.* In *Bergey's manual of systematics of archaea and bacteria.* Wiley. https://doi.org/10.1002/9781118960608.gbm01210

Palmer, M., & Coutinho, T. A. (2015). *Pantoea.* In *Bergey's manual of systematics of archaea and bacteria.* Wiley. https://doi.org/10.1002/9781118960608 .gbm01157.pub2

Parte, A. C., Carbasse, J. S., Meier-Kolthoff, J. P., Reimer, L. C., & Göker, M. (2020). List of prokaryotic names with standing in nomenclature (LPSN) moves to the DSMZ. *International Journal of Systematic and Evolutionary Microbiology, 70,* 5607–5612.

Passarelli-Araujo, H., Palmeiro, J. K., Moharana, K. C., Pedrosa-Silva, F., Dalla-Costa, L. M., & Venancio, T. M. (2019). Genomics analysis unveils important aspects of population structure, virulence, and antimicrobial resistance in *Klebsiella aerogenes. The FEBS Journal, 286,* 3797–3810.

Paterson, R., & Lima, N. (2011). Further mycotoxin effects from climate change. *Food Research International, 44,* 2555–2566.

Pažoutová, S., Kolarik, M., Odvody, G. N., Frederickson, D. E., Olšovská, J., & Man, P. (2008). A new species complex including *Claviceps fusiformis* and *Claviceps hirtella. Fungal Diversity, 31,* 95–110.

Perincherry, L., Lalak-Kańczugowska, J., & Stępierí, Ł. (2019). *Fusarium*-produced mycotoxins in plant-pathogen interactions. *Toxins, 11,* 664.

Pfohl-Leszkowisz, A., & Manderville, R. A. (2007). Ochratoxin A: An overview on toxicity and carcinogenicity in animals and humans. *Molecular Nutrition & Food Research, 51,* 61–99.

Piepenbring, M., Bauer, R., & Oberwinkler, F. (1998). Teliospores of smut fungi: Teliospore connections, appendages, and germ pores studies by electron microscopy; phylogenetic discussion of characteristics of teliospores. *Protoplasma*, *204*, 202–218.

Pitt, J. I., & Miller, J. D. (2017). A concise history of mycotoxin research. *Journal of Agricultural and Food Chemistry*, *65*, 7021–7033.

Prabhu, A. S., Filippi, M. C., & Castro, N. (1992). Pathogenic variation among isolates of *Pyricularia grisea* infecting rice, wheat, and grasses in Brazil. *Tropical Pest Management*, *38*, 367–371.

Prabhu, S. A., Kini, K. R., Raj, S. N., Moerschbacher, B. M., & Shetty, H. S. (2012). Polygalacturonase-inhibitor proteins in pearl millet: Possible involvement in resistance against downy mildew. *Acta Biochemica et Biophysica Sinica*, *44*, 415–423.

Puel, O., Galtier, P., & Oswald, I. P. (2010). Biosynthesis and toxicological effects of patulin. *Toxins*, *2*, 613–631.

Pujar, M., Kumar, S., Sharma, R., Ramu, P., Babu, R., & Gupta, S. K. (2023). Identification of genomic regions linked to blast (*Pyricularia grisea*) resistance in pearl millet. *Plant Breeding*, *142*, 506–517.

Punja, Z. K., Huang, J.-S., & Jenkins, S. F. (1985). Relationship of mycelial growth and production of oxalic acid and cell wall degrading enzymes to virulence in *Sclerotium rolfsii*. *Canadian Journal of Plant Pathology*, *7*, 109–117.

Pyle, J. D., Monis, J., & Scholthof, K.-B. (2017). Complete nucleotide sequences and virion particle association of two satellite RNAs of panicum mosaic virus. *Virus Research*, *240*, 87–93.

Qhobela, M., & Claflin, L. E. (1988). Characterization of *Xanthomonas campestris* pv. *Pennamericanum* pv. Nov., causal agent of bacterial leaf streak of pearl millet. *International Journal of Systematic Bacteriology*, *38*, 362–366.

Rabie, C. J., Lübben, A., Louw, A. I., Rathbone, E. B., Steyn, P. S., & Vleggaar, R. (1978). Moniliformin, a mycotoxin from *Fusarium fusarioides*. *Journal of Agricultural and Food Chemistry*, *26*, 375–379.

Rabie, C. J., Marasas, W. F. O., Thiel, P. G., Lübben, A., & Vleggaar, R. (1982). Moniliformin production and toxicity of different *Fusarium* species from southern Africa. *Applied and Environmental Microbiology*, *43*, 517–521.

Rai, K. N., Talukdar, B. S., Singh, S. D., Rao, A. S., Rao, A. M., & Andrews, D. J. (1994). Registration of ICMP 423 parental line of pearl millet. *Crop Science*, *34*, 1430.

Rai, K. N., & Thakur, R. P. (1995). Ergot reaction of pearl millet hybrids affected by fertility restoration and genetic resistance of parental lines. *Euphytica*, *83*, 225–231.

Rai, K. N., Thakur, R. P., & Rao, A. S. (1998). Registration of smut-resistant pearl millet parental lines ICMA 88006 and ICMB 88006. *Crop Science*, *38*, 575–576.

Raj, C., Sharma, R., Pushpavathi, B., Gupta, S. K., & Radhika, K. (2018). Inheritance and allelic relationship among downy mildew resistance genes in pearl millet. *Plant Disease*, *102*, 1136–1140.

Raj, S. N., Chaluvaraju, G., Amruthesh, K. N., Shetty, H. S., Reddy, M. S., & Kloepper, J. W. (2003). Induction of growth promotion and resistance against downy mildew on pearl millet (*Pennisetum glaucum*) by rhizobacteria. *Plant Disease, 87*, 380–384.

Rajagopalan, C. K. S., & Rangaswami, G. (1958). Bacterial leafspot of *Pennisetum typhoides. Current Science, 27*, 30–31.

Ramakrishnan, T. S. (1941). Top-rot ('twisted top' or 'pokkah boeng') of sugarcane, sorghum, and cumbu. *Current Science, 10*, 406–408.

Ramesh, G. V., Palanna, K. B., Praveen, B., Kumar, H. D. V., Koti, P. S., Sonavane, P., & Tonapi, V. A. (2021). First confirmed report of leaf blight on browntop millet caused by *Bipolaris setariae* in southern peninsular India. *Plant Disease, 105*, 1561.

Rathi, A., Kawatra, A., & Sehgal, S. (2004). Influence of depigmentation of pearl millet (*Pennisetum glaucum* L.) on sensory attributes, nutrient composition, in vitro protein and starch digestibility of pasta. *Food Chemistry, 85*, 275–280.

Reddy, A. S., Hobbs, H. A., Delfosse, P., Murhty, A. K., & Reddy, D. V. R. (1998). Seed transmission of Indian peanut clump virus (IPCV) in peanut and millets. *Plant Disease, 82*, 343–346.

Reddy, B., Mehta, S., Prakash, G., Sheoran, N., & Kumar, A. (2022). Structured framework and genome analysis of *Magnaporthe grisea* Inciting pearl pillet plast disease reveals versatile metabolic pathways, protein families, and virulence factors. *Journal of Fungi, 8*, 614.

Reddy, P. S., Satyavathi, C. T., Khandelwal, V., Patil, H., Gupta, P. C., Sharma, L. D., Mungra, K. D., Singh, S. P., Narasimhulu, R., Bhadarge, H. H., Iyanar, K., Tripathi, M. K., Yadav, D., Bhardwaj, R., Talwar, A. M., Tiwan, V. K., Kachole, U. G., Sravanti, K., Priya, M., & S.,. . .Tonapi, V. A. (2021). Performance and stability of pearl millet varieties for grain yield and micronutrients in arid and semi-arid regions of India. *Frontiers in Plant Science, 12*, 670201.

Reynaud, B., & Peterschmitt, M. (1992). A study of the mode of transmission of MSV by *Cicadulina mbila* using an enzyme-linked immunosorbent assay. *Annals of Applied Biology, 121*, 85–94.

Rishi, N., Bhargava, K. S., & Josh, R. D. (1973). Perpetuation of mosaic virus disease in sugarcane. *Annals of the Phytopathological Society of Japan, 39*, 361–363.

Rodenburg, J., Demont, M., Zwart, S. J., & Bastiaans, L. (2016). Parasitic weed incidence and related economic losses in rice in Africa. *Agriculture, Ecosystems & Environment, 235*, 306–317.

Roper, M. C. (2011). *Pantoea stewartii* subsp. *Stewartii*: Lessons learned from a xylem-dwelling pathogen of sweet corn. *Molecular Plant Pathology, 12*, 628–637.

Saddler, G. S., & Bradbury, J. F. (2015). Xanthomonas. In *Bergey's manual of systematics of archaea and bacteria*. Wiley. https://doi.org/10.1002/9781118960608.gbm01239

Saksena, H. K. (1979). Epidemiology of diseases caused by *Rhizoctonia* species. In Y. L. Nene (Ed.), *Proceedings of the consultants' group discussion on the resistance to soil-borne diseases of legumes* (pp. 59–64). ICRISAT.

Sarma, B. K., Singh, U. P., & Singh, K. P. (2002). Variability in Indian isolates of *Sclerotium rolfsii*. *Mycologia*, *94*, 1051–1058.

Satish, K., Gutema, Z., Grenier, C., Rich, P. J., & Ejeta, G. (2011). Tagging and validation of microsatellite markers linked to low germination stimulant gene (*lgs*) for *Striga* resistance in sorghum [*Sorghum bicolor* (L.) Moench]. *Theoretical and Applied Genetics*, *124*, 989–1003.

Schaad, N. W., Postnikova, E., Sechler, A., Claflin, L. A., Vidaver, A. K., Jones, J. B., Agarkova, I., Ignatov, A., Dickstein, E., & Ramunda, B. A. (2008). Reclassification of subspecies of *Acidovorax* as *A. avenae* (Manns 1905) emend., *A. cattleyae* (Pavarino, 1911) comb. Nov., *A. citrulli* (Schaad et al., 1978) comb. Nov., and proposal of *A. oryzae* sp. Nov. *Systematic and Applied Microbiology*, *31*, 434–446.

Scholthof, K.-B. G. (1999). A synergism induced by satellite panicum mosaic virus. *Molecular Plant-Microbe Interactions*, *12*, 163–166.

Seo, J.-A., Kim, J.-C., & Lee, Y.-W. (1999). N-Acetyl derivatives of type C fumonisins produced by *Fusarium oxysporum*. *Journal of Natural Products*, *62*, 355–357.

Seth, M. L., Raychaudhuri, S. P., & Singh, D. V. (1972). Bajra (pearl millet) streak: A leafhopper borne cereal virus in India. *Plant Disease Report*, *56*, 424–428.

Sewram, V., Mshicileli, N., Shephard, G. S., Vismer, H. F., Rheeder, J. P., Lee, Y.-W., Leslie, J. F., & Marasas, W. F. O. (2005). Production of fumonisin B and C analogs by several *Fusarium* species. *Journal of Agricultural and Food Chemistry*, *53*, 4861–4866.

Sewram, V., Nieuwoudt, T. W., Marasas, W. F. O., Shephard, G. S., & Ritieni, A. (1999). Determination of the mycotoxin moniliformin in cultures of *Fusarium subglutinans* and in naturally contaminated maize by high-performance liquid chromatography-atmospheric pressure chemical ionization mass spectrometry. *Journal of Chromatography A*, *848*, 185–191.

Shailasree, S., Kini, K. R., Deepak, S., Kumudini, B. S., & Shetty, H. S. (2004). Accumulation of hydroxyproline-rich glycoproteins in pearl millet seedlings in response to *Sclerospora graminicola* infection. *Plant Science*, *167*, 1227–1234.

Shanmuga Priya, M., Poornima, K., & Vigila, V. (2019). Distribution of plant parasitic nematodes associated with millets. *International Journal of Current Microbiology and Applied Sciences*, *8*, 795–799.

Sharma, R., Gate, V. L., & Madhavan, S. (2018). Evaluation of fungicides for the management of pearl millet [*Pennisetum glaucum* (L.)] blast caused by *Magnaporthe grisea*. *Crop Protection*, *112*, 209–213.

Sharma, R., Goud, T. Y., Prasad, Y. P., Nimmala, N., Kadvani, D. L., Mathur, A. C., Thakare, C. S., Uma Devi, G., & Naik, M. (2021). Pathogenic variability amongst Indian isolates of *Magnaporthe grisea* causing blast in pearl millet. *Crop Protection*, *139*, 105372.

Sharma, R., Sharma, S., & Gate, V. L. (2020). Tapping *Pennisetum violaceum*, a wild relative of pearl millet (*Pennisetum glaucum*), for resistance to blast (caused by

Magnaporthe grisea) and rust (caused by *Puccinia substriata* var. *indica*). *Plant Disease, 104*, 1487–1491.

Sharma, R., Thakur, R. P., Rai, K. N., Gupta, S. K., Rao, V. P., Rao, A. S., & Kumar, S. (2009). Identification of rust resistance in hybrid parents and advanced breeding lines of pearl millet. *Journal of SAT Agricultural Research, 7*, 1–4.

Sharma, R., Upadhyay, H. D., Manjunatha, S. V., Rai, K. N., Gupta, S. K., & Thakur, R. P. (2013). Pathogenic variation in the pearl millet blast pathogen *Magnaporthe grisea* and identification of resistance to diverse pathotypes. *Plant Disease, 97*, 189–195.

Sharma, S. B. (1985). *A world list of nematodes associated with chickpea, groundnut, pearl millet, pigeonpea, and sorghum* (Pulse Pathology Progress Report 42). ICRISAT.

Sharma, S. B., & McDonald, D. (1990). Global status of nematode problems of groundnut, pigeonpea, chickpea, sorghum and pearl millet, and suggestions for future work. *Crop Protection, 9*, 453–458.

Shephard, G. S., Sewram, V., Nieuwoudt, T. W., Marasas, W. F. O., & Ritieni, A. (1999). Production of the mycotoxins fusaproliferin and beauvericin by South African isolates in the *Fusarium* section *Liseola*. *Journal of Agricultural and Food Chemistry, 47*, 5111–5115.

Shepherd, D. N., Martin, D. P., van der Walt, E., Dent, K., Varsani, A., & Rybicki, E. P. (2010). *Maize streak virus*: An old enemy and complex "emerging" pathogen. *Molecular Plant Pathology, 11*, 1–12.

Shetty, H. S., Mathur, S. B., Neergaard, P., & Safeelua, K. M. (1982). *Drechslera setariae* in Indian pearl millet seeds, its seed-borne nature, transmission and significance. *Transactions of the British Mycological Society, 78*, 170–173.

Shinde, R. B., Patil, F. B., & Sangave, R. A. (1984). Resistance to downy mildew in pearl millet. *Journal of Maharashtra Agricultural University, 9*, 337–338.

Shishupala, S., Kumar, V. U., Shekar, H. S., & Umesh-Kumar, S. (1996). Screening pearl millet cultivars by ELISA for resistance to downy mildew disease. *Plant Pathology, 45*, 978–983.

Shivakumar, P. D., Vasanthi, N. S., Shetty, H. S., & Smedegaard-Petersen, V. (2000). Ribonucleases in the seedlings of pearl millet and their involvement in resistance against downy mildew disease. *European Journal of Plant Pathology, 106*, 825–836.

Shivanna, M. B., & Shetty, H. S. (1986). Myrothecium pod spot of cluster bean and its significance. *Current Science, 55*, 574–576.

Siefers, D. L., French, R. C., Stenger, D. C., & Martin, T. J. (2003). Biological variation among wheat streak mosaic virus isolates. *Phytopathology, 93*, S78.

Siefers, D. L., Haber, S., Ens, W., She, Y.-M., Standing, K. G., & Salomon, R. (2005). Characterization of a distinct Johnsongrass mosaic virus strain isolated from sorghum in Nigeria. *Archives of Virology, 150*, 557–576.

Siefers, D. L., Harvey, T. L., Kofoid, K. D., & Stegmeier, W. D. (1996). Natural infection of pearl millet and sorghum by wheat streak mosaic virus in Kansas. *Plant Disease, 80*, 179–185.

Siefers, D. L., Harvey, T. L., Martin, T. L., & Jensen, S. G. (1998). A partial host range of the High Plains virus of corn and wheat. *Plant Disease, 82*, 875–879.

Sill, W. H., & Pickett, R. C. (1957). A virus diseases switchgrass, *Panicum virgatum* L. *Plant Disease Reporter, 41*, 241–249.

Singh, A., Meena, R. L., & Chattopadhyaya, C. (2008). Location and transmission of *Curvularia penniseti* in pearl millet. *Journal of Mycology and Plant Pathology, 38*, 185–189.

Singh, B. R., & Singh, D. P. (1995). Agronomic and physiological responses of sorghum, maize and pearl millet to irrigation. *Field Crops Research, 42*, 57–67.

Singh, D. S., & Pavgi, M. S. (1997). Perpetuation of *Pyricularia penniseti* causing brown leaf spot of bajra. *Indian Phytopathology, 30*, 242–244.

Singh, G., Kanwar, R. S., & Dev vart. (2019). Studies on resistance in some pearl millet hybrids against *Meloidogyne graminicola*. *Journal of Pharmacognosy and Phytochemistry, 8*, 1513–1515.

Singh, G., Thakur, R. P., & Weltzien, R. E. (1997). Dry stalk rot – A new disease of pearl millet in Rajasthan. *International Sorghum and Millets Newsletter, 38*, 124–128.

Singh, S., Sharma, R., Pushpavathi, B., Gupta, S. K., Durgarani, C. V., & Raj, C. (2018). Inheritance and allelic relationship among gene(s) for blast resistance in pearl millet [*Pennisetum glaucum* (L.) R. Br.]. *Plant Breeding, 137*, 573–584.

Singh, S. D. (1990). Sources of resistance to downy mildew and rust in pearl millet. *Plant Disease, 74*, 871–874.

Singh, S. D. (1995). Downy mildew of pearl millet. *Plant Disease, 79*, 545–550.

Singh, S. D., Alagarswamy, G., Talukdar, B. S., & Hash, T. S. (1994). Registration of ICML 22 photoperiod insensitive, downy mildew resistant pearl millet germplasm. *Crop Science, 34*, 1421.

Singh, S. D., Andrews, D. J., & Rai, K. N. (1987). Registration of ICML 11 rust resistant pearl millet germplasm. *Crop Science, 27*, 367–386.

Singh, S. D., de Milliano, W. A. J., Mtisi, E., & Chingombe, P. (1990a). Brown leaf spot of pearl millet caused by *Bipolaris urochloae* in Zimbabwe. *Plant Disease, 74*, 931–932.

Singh, S. D., & Gopinath, R. (1985). A seedling inoculation technique for detecting downy mildew resistance in pearl millet. *Plant Disease, 69*, 582–584.

Singh, S. D., King, S. B., & Reddy, P. M. (1990b). Registration of five pearl millet germplasm sources with stable resistance to downy mildew. *Crop Science, 30*, 1164.

Singh, S. D., King, S. B., & Werder, J. (1993). *Downy mildew disease of pearl millet.* Information Bulletin No. 37. ICRISAT.

Singh, S. D., Singh, P., Andrews, D. J., Talukdar, B. S., & King, S. B. (1992). Reselection of pearl millet cultivar utilizing residual variability for downy mildew reaction. *Plant Breeding, 109*, 54–59.

Singh, S. D., Singh, P., Rai, K. N., & Andrews, D. J. (1990c). Registration of ICMA 841 and ICMB 841 pearl millet parental lines with A1 cytoplasmic male-sterility system. *Crop Science, 36*, 1378.

Sivanesan, A. (1987). *Graminicolous species of Bipolaris, Curvularia, Drechslera, Exserohilum and their teleomorphs* (Paper 158). Commonwealth Mycological Institute.

Sivanesan, A. (1990). CMI descriptions of fungi and bacteria no. 1006 *Curvularia penniseti. Mycopathologia, 111*, 121–122.

Smith, R. L., Jensen, L. S., Hoveland, C. S., & Hanna, W. W. (1989). Use of pearl millet, sorghum, and triticale grain in broiler diets. *Journal of Production Agriculture, 2*, 78–82.

Song, W. Y., Kim, H. M., Hwang, C. Y., & Schaad, N. W. (2004). Detection of *Acidovorax avenae* ssp. *Avenae* in rice seeds using BIO-PCR. *Journal of Phytopathology, 152*, 667–676.

Srinivasachary, S., Hittalmani, S., Shivayogi, S., Vaishali, M. G., Shashidhar, H. E., & Kumar, G. K. (2002). Genetic analysis of rice blast fungus of southern Karnataka using DNA markers and reaction of popular rice genotypes. *Current Science, 82*, 732–735.

Stoll, M., Begerow, D., & Oberwinkler, F. (2005). Molecular phylogeny of *Ustilago, Sporisorium*, and related taxa based upon combined analysis of rDNA sequences. *Mycological Research, 109*, 342–356.

Subba Rao, K. V., & Thakur, R. P. (1983). *Tolyposporium pennicillariae*, the causal agent of pearl millet smut. *Transactions of the British Mycological Society, 81*, 597–603.

Sudisha, J., Kumar, A., Amruthesh, K. N., Niranjana, S. R., & Shetty, H. S. (2011). Elicitation of resistance and defense enzymes by raw cow milk and amino acids in pearl millet against downy mildew disease caused by *Sclerospora graminicola. Crop Protection, 30*, 794–801.

Suh, J. P., Roh, J. H., Cho, Y. C., Han, S. S., Kim, Y. G., & Jena, K. K. (2009). The Pi40 gene for durable resistance to rice blast and molecular analysis of Pi40-advanced backcross breeding lines. *Phytopathology, 99*, 243–250.

Taber, R. A., Taber, W. A., Pettit, R. E., & Horne, C. W. (1978). Benowskia blight on bristlegrass in Texas. *Plant Disease Report, 62*, 157–160.

Takan, J. P., Chipili, J., Muthumeenakshi, S., Talbot, N. J., Manyasa, E. O., Bandyopadhyay, R., Sere, Y., Nutsugah, S. K., Talhinhas, P., Hossain, M., Brown, A. E., & Sreenivasaprasad, S. (2012). *Magnaporthe oryzae* populations adapted to finger millet and rice exhibit distinctive patterns of genetic diversity, sexuality and host interaction. *Molecular Biotechnology, 50*, 145–158.

Talukdar, B. S., Prakash Babu, P. P., Rao, A. M., Ramakrishna, C., Witcombe, J. R., King, S. B., & Hash, C. T. (1998). Registration of ICMP 85410: Dwarf, downy mildew resistant, restorer parental line of pearl millet. *Crop Science*, *38*, 904–905.

Tang, J., Tang, D.-J., Dubrow, Z. E., Bogdanove, A., & An, S. (2021). *Xanthomonas campestris* pathovars. *Trends in Microbiology*, *29*, 122–123.

Tapsoba, H., & Wilson, J. P. (1999). Increasing complexity of resistance in host populations through intermating to manage rust of pearl millet. *Phytopathology*, *89*, 450–455.

Tchala, A.W. (1989). *Factors influencing grain yield in pearl millet* Pennisetum glaucum *(L.) R. Br.* [Doctoral dissertation, University of Georgia, Athens]. https://www.doc-development-durable.org/file/Culture/Culture-plantes-alimentaires/FICHES_PLANTES/Millet-Mil/FACTORS%20INFLUENCING%20GRAIN%20YIELD%20IN%20PEARL%20MILLET-CS_01864.pdf

Tesso, T., Perumal, R., Little, C. R., Adeyanu, A., Radwan, G. L., Prom, L. K., & Magill, C. W. (2012). Sorghum pathology and biotechnology—a fungal disease perspective: Part II. Anthracnose, stalk rot, and downy mildew. *The European Journal of Plant Science and Biotechnology*, *6*, 31–44.

Thakur, R. P., & King, S. B. (1988a). *Ergot disease of pearl millet*. Information Bulletin No. 24. ICRISAT.

Thakur, R. P., & King, S. B. (1988b). *Smut disease of pearl millet*. Information Bulletin No. 25. ICRISAT.

Thakur, R. P., & King, S. B. (1988c). Registration of six smut resistant germplasms of pearl millet. *Crop Science*, *28*, 382283.

Thakur, R. P., Pushpavathi, B., & Rao, V. P. (1998). Viruence characterization of single-zoospore isolates of *Sclerospora graminicola* from pearl millet. *Plant Disease*, *82*, 747–751.

Thakur, R. P., Rai, K. N., Khairwal, I. S., & Mahala, R. S. (2008). Strategy for downy mildew resistance breeding in pearl millet in India. *Journal of SAT Agricultural Research*, *6*, 1–11.

Thakur, R. P., Rai, K. N., King, S. B., & Rao, V. P. (1993). *Identification and utilization of ergot resistance in pearl millet*. Research Bulletin No. 17. ICRISAT.

Thakur, R. P., Rao, V. P., & Hash, C. T. (1998). A highly virulent pathotype of *Sclerospora graminicola* from Jodhpur, Rajasthan, India. *International Sorghum and Millets Newsletter*, *39*, 140–142.

Thakur, R. P., Rao, V. P., & King, S. B. (1991). Influence of temperature and wetness duration on infection of pearl millet by *Claviceps fusiformis*. *Phytopathology*, *81*, 835–838.

Thakur, R. P., Rao, V. P., & Sharma, R. (2007). Evidence for temporal virulence change in pearl millet downy mildew pathogen populations in India. *Journal of SAT Agricultural Research*, *3*, 1–3.

Thakur, R. P., Rao, V. P., & Sharma, R. (2009). Temporal virulence change and identification of resistance in pearl millet germplasm to diverse pathotypes of *Sclerospora graminicola*. *Journal of Plant Pathology, 91*, 629–636.

Thakur, R. P., Rao, V. P., & Williams, R. J. (1984). The morphology and disease cycle of ergot caused by *Claviceps fusiformis* in pearl millet. *Phytopathology, 74*, 201–205.

Thakur, R. P., Rao, V. P., Williams, R. J., Chahal, S. S., Mathur, S. B., Nafade, S. D., Shetty, H. S., Singh, G., & Bangar, S. G. (1985). Identification of stable resistance to ergot in pearl millet. *Plant Disease, 69*, 982–985.

Thakur, R. P., Rao, V. P., Wu, B. M., Subbarao, K. V., Shetty, H. S., Singh, G., Lukose, C., Panwar, M. S., Sereme, P., Hess, D. E., Gupta, S. C., Dattar, V. V., Panicker, S., Bhangale, G. T., & Panchbhai, S. D. (2004). Host resistance stability to downy mildew in pearl millet and pathogenic variability in *Sclerospora graminicola*. *Crop Protection, 23*, 901–908.

Thakur, R. P., Sharma, R., & Rao, V. P. (2011). *Screening techniques for pearl millet diseases*. Information Bulletin No. 89. ICRISAT.

Thakur, R. P., Subba Rao, K. V., & Williams, R. J. (1983). Evaluation of a new field screening technique for smut resistance in pearl millet. *Phytopathology, 73*, 1255–1258.

Thakur, R. P., & Williams, R. J. (1980). Pollination effects on pearl millet ergot. *Phytopathology, 70*, 80–84.

Thakur, R. P., Williams, R. J., & Rao, V. P. (1982). Development of ergot resistance in pearl millet. *Phytopathology, 72*, 406–408.

Thiel, P. G., Marasas, W. F. O., Sydenham, E. W., Shephard, G. S., Gelderblom, W. C. A., & Nieuwenhuis, J. J. (1991). Survey of fumonisin production by *Fusarium* spp. *Applied and Environmental Microbiology, 57*, 1089–1093.

Thines, M., Göker, M., Oberwinkler, F., & Spring, O. (2007). A revision of *Plasmopara penniseti*, with implications for the host range of the downy mildews with pyriform haustoria. *Mycological Research, 111*, 1377–1385.

Thongmeearkom, P., Ford, R. E., & Jedlinski, H. (1976). Aphid transmission of maize dwarf mosaic virus strains. *Phytopathology, 66*, 332–335.

Thorat, V., Kiran, K., Pramod, T., & Yadav, A. (2017). First report of 16SrII-D phytoplasmas associated with fodder crops in India. *Phytopathogenic Mollicutes, 7*, 106–110.

Thouvenel, J. C., Givord, L., & Pfeiffer, P. (1976). Guinea grass mosaic virus, a new member of the potato virus Y group. *Phytopathology, 66*, 954–957.

Timper, P., & Hanna, W. W. (2005). Reproduction of *Belonolaimus longicaudatus*, *Meloidogyne javanica*, *Paratrichodorus minor*, and *Pratylenchus brachyurus* on pearl millet (*Pennisetum glaucum*). *Journal of Nematology, 37*, 214–219.

Timper, P., & Wilson, J. P. (2006). Root-knot nematode resistance in pearl millet from west and east Africa. *Plant Disease, 90*, 339–344.

Timper, P., Wilson, J. P., Johnson, A. W., & Hanna, W. W. (2002). Evaluation of pearl millet grain hybrids for resistance to *Meloidogyne* spp. and leaf blight caused by *Pyricularia grisea*. *Plant Disease, 86*, 909–914.

Tyagi, P. D. (1985). Identity of dactuliophora leaf spot of pearl millet. *Proceedings of the Indiana Academy of Sciences, 94*, 407–413.

van Antwerpen, T., McFarlane, S. A., Buchanan, G. F., Shepherd, D. N., Martin, D. P., Rybicki, E. P., & Varsani, A. (2008). First report of maize streak virus field infection of sugarcane in South Africa. *Plant Disease, 92*, 982.

van Oosterom, E. J., Weltzien, E., Yadav, O. P., & Bidinger, F. R. (2006). Grain yield components of pearl millet under optimum conditions can be used to identify germplasm with adaptation to arid zones. *Field Crops Research, 96*, 407–421.

Vasudeva, R. S., & Iyengar, M. R. S. (1950). Secondary infection in bajra smut disease caused by *Tolyposporium penicilliariae* Bref. *Current Science, 19*, 123.

Vismer, H. F., Shepherd, G. S., van der Westhuizen, L., Mngqawa, P., Bushula-Njah, B., & Leslie, J. F. (2019). Mycotoxins produced by *Fusarium proliferatum* and *F. pseudonygamai* on maize, sorghum and pearl millet grains in vitro. *International Journal of Food Microbiology, 296*, 31–36.

Voss, K. A., & Riley, R. T. (2013). Fumonisin toxicity and mechanism of action: Overview and current perspectives. *Food Safety, 1*, 2013006.

Wakman, W., Kontong, M. S., Muis, A., Persley, D. M., & Teakle, D. S. (2001). Mosaic disease of maize caused by sugarcane mosaic potyvirus in Sulawesi. *Indonesian Journal of Agricultural Science, 2*, 56–59.

Waterhouse, G. M. (1964). *The genus Sclerospora, diagnoses (or descriptions) from the original papers and a key* (publication no. 17). Commonwealth Mycological Institute.

Weber, G. F. (1963). *Pellicularia* spp. Pathogenic on sorghum and pearl millet in Florida. *Plant Disease Report, 47*, 654–656.

Wells, H. D. (1956). Top rot of pearl millet caused by *Fusarium moniliforme. Plant Disease Report, 40*, 387.

Wells, H. D. (1967). Effects of temperature on pathogenicity of *Helminthosporium setariae* in seedlings of pearl millet, *Pennisetum typhoides. Phytopathology, 57*, 1002.

Wells, H. D., & Hanna, W. W. (1988). Genetics of resistance to *Bipolaris setariae* in pearl millet. *Phytopathology, 78*, 1179–1181.

Wells, H. D., & Winstead, E. E. (1965). Seed-borne fungi in Georgia-grown and Western-grown pearl millet seed on sale in Georgia during 1960. *Plant Disease Report, 49*, 487–489.

Welz, H. G., Schechert, A., Pernet, A., Pixley, K. V., & Geiger, H. H. (1998). A gene for resistance to the maize streak virus in the African CIMMYT maize inbred line CML 202. *Molecular Breeding, 4*, 147–154.

Wenndt, A. J., Sudini, H. K., Mehta, R., Pingali, P., & Nelson, R. (2021). Spatiotemporal assessment of post-harvest mycotoxin contamination in rural North Indian food systems. *Food Control, 126*, 108071.

Wijayasekara, D., & Ali, A. (2021). Evolutionary study of maize dwarf mosaic virus using nearly complete genome sequences acquired by next-generation sequencing. *Scientific Reports, 11*, 18786.

Willems, A., & Gillis, M. (2015). Acidovorax. In *Bergey's manual of systematics of archaea and bacteria*. Wiley. https://doi.org/10.1002/9781118960608.gbm00943

Williams, A., Goor, M., Thielemans, S., Gillis, M., Kersters, K., & De Ley, J. (1992). Transfer of several phytopathogenic *Pseudomonas* species to *Acidovorax* as *Acidovorax avenae* subsp. *avenae* subsp. nov., comb. nov., *Acidovorax avenae* subsp. *citrulli*, *Acidovorax avenae* subsp. *cattleyae*, and *Acidovorax konjaci*. *International Journal of Systematic Bacteriology, 42*, 107–119.

Williams, R. J., & McDonald, D. (1983). Grain molds in the tropics: Problems and importance. *Annual Review of Phytopathology, 21*, 153–178.

Williams, R. J., Singh, S. D., & Pawar, M. N. (1981). An improved field screening technique for downy mildew resistance in pearl millet. *Plant Disease, 65*, 239–241.

Willingale, J., & Mantle, P. G. (1985). Stigma constriction in pearl millet, a factor influencing reproduction and disease. *Annals of Botany, 56*, 109–115.

Willingale, J., Mantle, P. G., & Thakur, R. P. (1986). Postpollination stigmatic constriction, the basis of ergot resistance in selected lines of pearl millet. *Phytopathology, 76*, 536–539.

Willment, J. A., Martin, D. P., & Rybicki, E. P. (2001). Analysis of the diversity of African streak mastreviruses using PCR-generated RFLP's and partial sequence data. *Journal of Virology, 93*, 75–87.

Wilson, J. P., Hanna, W. W., & Gascho, G. J. (1996). Pearl millet grain yield reduction from rust infection. *Journal of Production Agriculture, 9*, 543–545.

Wilson, J. P. (1994). Field and greenhouse evaluations of pearl millet for partial rust resistance to *Puccinia substriata* var. *indica*. *Plant Disease, 78*, 1202–1205.

Wilson, J. P. (1995). Mechanisms associated with the *tr* allele contributing to reduced smut susceptibility of pearl millet. *Phytopathology, 85*, 966–969.

Wilson, J. P. (2000). *Pearl millet diseases: A compilation of information on the known pathogens of pearl millet, Pennisetum glaucum (L.) R. Br.* (Agriculture handbook no. 716). USDA-ARS.

Wilson, J. P. (2002a). Diseases of pearl millet in the Americas. In J. F. Leslie (Ed.), *Sorghum and millets diseases* (pp. 465–469). Iowa State University Press.

Wilson, J. P. (2002b). Fungi associated with the stalk rot complex of pearl millet. *Plant Disease, 86*, 833–839.

Wilson, J. P., & Burton, G. W. (1990). *Phyllosticta pennicillariae* on pearl millet in the United States. *Plant Disease, 74*, 331.

Wilson, J. P., & Burton, G. W. (1991). Registration of Tift #3 and Tift #4 rust resistant pearl millet germplasms. *Crop Science, 31*, 1713.

Wilson, J. P., Burton, G. W., Wells, H. D., Zongo, J. D., & Dicko, I. O. (1989). Leaf spot, rust and smut resistance in pearl millet landraces from central Burkina Faso. *Plant Disease, 73*, 345–349.

Wilson, J. P., Cunfer, B. M., & Phillips, D. V. (1999). Double-cropping and crop rotation effects on diseases and grain yield of pearl millet. *Journal of Production Agriculture*, *12*, 198–202.

Wilson, J. P., & Gates, R. N. (1993). Forage yield losses in hybrid pearl millet due to leaf blight caused primarily by *Pyricularia grisea*. *Phytopathology*, *83*, 739–743.

Wilson, J. P., & Gates, R. N. (1999). Disease resistance and biomass stability of forage pearl millet hybrids with partial rust resistance. *Plant Disease*, *83*, 733–738.

Wilson, J. P., Gates, R. N., & Hanna, W. W. (1991). Effect of rust on yield and digestibility of pearl millet forage. *Phytopathology*, *81*, 233–236.

Wilson, J. P., Gates, R. N., & Panwar, M. S. (2001). Dynamic multiline population approach to resistance gene management. *Phytopathology*, *91*, 255–260.

Wilson, J. P., & Hanna, W. W. (1992). Effects of gene and cytoplasm substitutions in pearl millet on leaf blight epidemics and infection by *Pyricularia grisea*. *Phytopathology*, *82*, 839–842.

Wilson, J. P., Hanna, W. W., Wilson, D. M., Beaver, R. W., & Casper, H. H. (1993). Fungal and mycotoxin contamination of pearl millet grain in response to environmental conditions in Georgia. *Plant Disease*, *77*, 121–124.

Wilson, J. P., Hess, D. E., & Kumar, K. A. (2000). Dactuliophora leaf spot of pearl millet in Niger and Mali. *Plant Disease*, *84*, 201.

Wilson, J. P., & Williamson, W. (1997). Natural infection of eggplant by *Puccinia substriata* var. *indica* in the United States. *Plant Disease*, *81*, 1093.

Witcombe, J. R., Rao, M. N. V. R., Talukdar, B. S., Singh, S. D., & Rao, A. M. (1996). Registration of ICMP 312 pearl millet topcross pollinator germplasm. *Crop Science*, *36*, 471.

Wu, X., Leslie, J. F., Thakur, R. A., & Smith, J. S. (2003). Preparation of a fusaproliferin standard from the culture of *Fusarium subglutinans* E-1583 by high performance liquid chromatography. *Journal of Food and Agricultural Chemistry*, *51*, 383–388.

Xiao, S. Q., Zhang, D., Zhao, J. M., Yuan, M. Y., Wang, J. H., Xu, R. D., Li, G. F., & Xue, C. S. (2019). First report of leaf spot of maize (*Zea mays*) caused by *Bipolaris setariae* in China. *Plant Disease*, *104*, 582.

Yadav, O. P. (1994). Influence of A1 cytoplasm in pearl millet: A review. *Plant Breeding Abstracts*, *64*, 1375–1379.

Yadav, O. P., & Rai, K. N. (2013). Genetic improvement of pearl millet in India. *Agricultural Research*, *2*, 275–292.

Yahaya, A., Dangora, D. B., Khan, A. U., & Zangoma, M. A. (2014). Detection of sugarcane mosaic disease (SCMD) in crops and weeds associated with sugarcane fields in Makarfiand Sabon Gari Local Government Areas of Kaduna State, Nigeria. *International Journal of Current Science*, *11*, 99–104.

Young, G. Y., Lefebvre, C. L., & Johnson, A. G. (1947). *Helminthosporium rostratum* on corn, sorghum, and pearl millet. *Phytopathology, 37*, 180–183.

Young, J. M., & Fletcher, M. J. (1994). *Pseudomonas syringae* pv. *Panici* (Elliot 1923) Young, Dye & Wilkie 1978 is a doubtful name. *Australasian Plant Pathology, 23*, 66–68.

Zarafi, A. B. (1997). *Final report on WCAMRN Project P3 activities conducted in Nigeria during the 1992-1996 cropping seasons.* ROCAFREMI/IAR, Institute for Agricultural Research, Ahmadu Bello University.

Zarafi, A. B., Emechebe, A. M., & Akpa, A. D. (2004). Curvularia midrib spot of pearl millet in the Nigerian savanna. *Nigerian Journal of Plant Protection, 21*, 101–107.

12

Pearl Millet: Pest Management

Rajkumar P. Juneja, Jagdish Jaba, Riyazaddin Mohammed, and Rajan Sharma

Chapter Overview

It is generally believed that pearl millet [*Pennisetum glaucum* (L.) R. Br.] grown as a monocrop or mixed crop or in a relay cropping system has minimal pest problems. However, a perusal of the literature on insect pests of this crop gives quite a different picture. More than 100 insect pests have been reported to be associated with pearl millet–based cropping systems (Kishore & Solomon, 1989), but only a few of them are potential pests of significant economic importance. The status of many of these important millet pests is given as "major pest" in these publications. Also, the economic importance of many other pests has not been defined. The pest status in pearl millet has been classified either as regular (regular occurrence, economic damage, high loss in grain yield) or occasional (occasional occurrence, sporadic attack with economic damage, high/moderate loss in grain yield) (Gahukar & Reddy, 2019). White grub (*Holotrichia consanguinea* Blanch) and shoot fly (*Atherigona approximata* Malloch) are regular pests in India (Biradar & Sajjan, 2018; Choudhary et al., 2018). Millet stem borer (*Acigona ignefusalis* Hampson), millet head miner [*Heiliocheilus albipunctella* (de. Joannis)], and blister beetle (*Psalydolytta fusca* and *Psalydolytta vestita* Olivier) have been recorded as regular pests. Grasshoppers [*Kraussaria angulifera* (Krauss, 1877) and *Oedaleus senegalensis* (Krauss, 1877)] have been recorded as an occasional pest in the Sahel, whereas chinch bug [*Blissus leucopterus leucopterus* (Say)] is a regular pest in the United States (Degri et al., 2014; Goudiaby et al., 2018; Maiga et al., 2008; Wright, 2013; Zethner & Lawrence, 1988).

The pest problem has increased with the introduction of high-yielding varieties. Shoot fly, stem borer, white grubs, earhead worms, and gray weevil (*Cyrtepistomus castaneus*) are the key pests of pearl millet in India and need proper control measures.

Pearl Millet: A Resilient Cereal Crop for Food, Nutrition, and Climate Security, First Edition.
Edited by Ramasamy Perumal, P. V. Vara Prasad, C. Tara Satyavathi, Mahalingam Govindaraj, and Abdou Tenkouano.
© 2025 American Society of Agronomy, Inc., Crop Science Society of America, Inc., Soil Science Society of America, Inc. Published 2025 by John Wiley & Sons, Inc.

Other pests, such as grasshoppers, blister beetles, surface grasshoppers, armyworms, and hairy caterpillars, are of secondary importance and have been reported in certain regions. Juneja et al. (2022) reported shoot fly, stem borer, and earhead worm as major insect pests of pearl millet at Jamnagar during kharif (i.e., monsoon season) 2021. The study conducted at Pearl Millet Research Station, Junagadh Agricultural University, Jamnagar, during kharif 2022 revealed that the losses due to insect pest complexes were 25.32% and 21.21% in grain and fodder yield, respectively (Juneja et al., 2023). During an on-farm survey, Juneja and Parmar (2022) recorded shoot fly (*A. approximata*), stem borers (*Chilo partellus, Coniesta ignefusalis*; *Sesamia inferens*) cotton bollworm (*Helicoverpa armigera*)), fall armyworm (*Spodoptera frugiperda*) (FAW), and gray weevil damaging the crop in India. The major insect pests of pearl millet, with their economic importance, status, nature of damage, and control measures, are discussed in this chapter.

Insect Pests of Pearl Millet

Shoot Fly

Host range: Pearl millet, sorghum, maize, wheat, minor millets, and fodder crops.

   Importance, status, and nature of damage: Many species of shoot flies (Diptera: Muscidae) attack all millets in India (Kalaisekar et al., 2017) and Africa (Nigus & Damte, 2018). The shoot fly has assumed the status of a serious pest in several states in India, especially in Gujarat and Tamil Nadu, followed by Rajasthan. During the early stages of development, shoot fly larvae eat through the leaf sheath, cutting the growing point, which results in the shoot wilting and yellowing, subsequently leading to the death of the seedling, commonly known as "deadheart." Mature larvae feed on decaying material just above the cut. Pest damage from these flies has been found to result in 20%–50% grain yield loss in pearl millet (Kishore, 1996). Mote and Nakat (2001) observed shoot fly incidence in kharif 2000–2001 in Rahuri and Akola tehsils of the Ahmednagar district of Western Maharashtra (deadheart, 20%). The maggots feed on the seedlings and produce deadhearts (Figure 12.1). Sometimes the plant escapes the shoot fly damage because of its vigor. The shoot fly also causes damage to ear heads in the later stages of the crop, and the ear head appears like cat's tail (Figure 12.2). Kishore (1996) reported approximately 23.3%–36.5% grain yield losses caused by shoot fly. The population fluctuation of shoot fly was studied from 1995 to 2006 by Raghvani et al. (2008) at Jamnagar, Gujarat, India, and the study revealed that the incidence ranged between 6.4% and 13.2% during 15–50 days after seedling emergence (DAE).

Figure 12.1 Dead heart symptom at seedling stage.

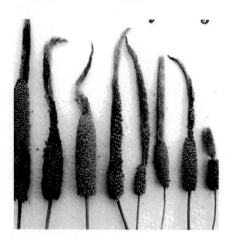

Figure 12.2 Cat's tail symptom at panicle stage.

Control measures: No single method of control is effective against this pest. In case of severe damage or endemic areas, the following control measures can be taken:

- Initiate the plant protection measures at 8% economic threshold level infestation (Raghvani et al., 2005).
- Increase the seed rate to 4–5 kg/ha (i.e., 10% higher than the normal seed rate). Removing and destroying the affected seedlings will also help in controlling the shoot fly population if a regular campaign is organized.
- Dusting of malathion 5% at 25 kg/ha or two sprays of neem oil 0.05% with soap 4 g/L of water is effective.

- Three sprays of neemark 0.3% or triazophos 0.04% at 10, 20, and 30 DAE effectively control the shoot fly.
- Small and marginal farmers who cannot afford the chemical pesticides can apply any one of the available botanicals (e.g., neem seed kernel suspension, neem leaves suspension, or mint leaves suspension) at 5% spray (Juneja et al., 2004).
- Green gram or pigeonpea taken as an intercrop in the ratio of 2:1 can reduce the incidence of shoot fly and to give an equivalent pearl millet yield of 2568 and 2540 kg/ha (incremental cost benefit ratio [ICBR] 1:5.80 and 1:1.30, respectively) as compared to the sole crop (2318 kg/ha).
- Two sprays of profenophos 50 emulsifiable concentrate (EC) at 0.05% or fenobucarb 50 EC at 0.1% at 20 and 40 DAE are effective based on a 3-year study at Jamnagar for the control of shoot fly (Parmar et al., 2015).
- Seed treatment with imidacloprid 600 FS at 8.75 mL/kg seeds at the time of sowing followed by spraying with imidacloprid 17.8 SL at 0.009% at 35 DAE is effective for the management of shoot fly and stem borer.
- The integrated pest management (IPM) module consisting of seed treatment with imidacloprid 600 FS at 8.75 mL/kg at the time of sowing, removal of shoot fly dead hearts, installation of fish meal traps at 10 per hectare (fish meal to be replaced once a week), and spraying of dimethoate 30 EC 0.03% (10 mL per 10 L of water) at 35 DAE was effective and economical for the management of shoot fly (Parmar et al., 2021).
- For the eco-friendly management of shoot fly, treat the seeds with imidacloprid 600 FS (8.75 mL/kg) followed by two sprays at 20 and 40 DAE with *Beauveria bassiana* 1.15 WP (minimum 1×10^8 cfu/g), 0.007% (60 g per 10 L of water), or *Panchgavya* 3% (300 mL per 10 L of water) (Parmar et al., 2022a).

Stem Borer

Stem borer species include *Chilo partellus* (Swinhoe), *Coniesta ignefusalis* Hampson, *Sesamia inferens* Walker, and *Sesamia calamistis* Hampson.

Host range: Pearl millet, sorghum, maize, Sudan grass, Johnsongrass [*Sorghum halepense* (L.) Pers.], and other grasses.

Importance, status, and nature of damage: Stem borer is a major pest and is the most destructive and cosmopolitan species (Vashisth et al., 2022). It occurs in all the pearl millet growing areas in India, predominantly in the Gujarat and Delhi regions. In Rajasthan, particularly in the Jaipur area, the stem borer is emerging as a major pest. Losses of 20%–60% have been reported (Kishore, 1996). The stem borer has been reported to be a major pest of pearl millet in the Sahelian and sub-Saharan regions. According to Sastawa et al. (2002) and Barau et al. (2015), it is the most widespread and damaging insect pest of pearl millet in West Africa.

Depending on the density of the insect population and the genetic nature of the cultivars used, crop yield losses of 24%– 100% have been reported in West Africa (Drame-Yaye et al., 2003; Nwanze & Harris, 1992). At the seedling stage, borer larvae feed in plant whorls, causing parallel small holes in the opening leaves in a specific pattern followed by dead hearts at a later stage (Figure 12.3), whereas at the earhead stage, the silver earhead/empty earhead (chaffy panicles) appears (Figure 12.4). A study carried out for 10 years on population fluctuations of pearl millet stem borer revealed that its incidence is noticed at 15 DAE (4.4%), which gradually increases to its peak (15.1%) at 77 DAE of the crop (Raghvani et al., 2008). During a study on the estimation of the economic threshold level of stem borer at Jamnagar, 5% plant damage recorded the highest net return and ICBR (Parmar et al., 2022b).

Control measures:
- Use of higher seed rate of 5 kg/ha.
- Among chemicals, either quinalphos 0.05% or fenitrothion 0.05%, if applied at 15 DAE, has been found effective in pearl millet.
- Effective control of stem borer can be achieved by spraying fenvalerate 0.01%, cypermethrin 0.01%, or indoxacarb 0.0075% at 20 and 40 DAE of the crop (Raghvani et al., 2010).

Figure 12.3   Stem borer symptom.

Figure 12.4 Silver earhead/empty earhead symptom.

- Two sprays of profenophos 50 EC at 0.05% or fenobucarb 50 EC at 0.1% at 20 and 40 DAE were effective against stem borer based on a 3-year study (2007–2009) at Pearl Millet Research Station, JAU, Jamnagar (Parmar et al., 2015).
- Seed treatment with clothianidin 50 WDG at 7.5 g/kg seed followed by spraying chlorantraniliprole 20 SC at 0.006% at 35 DAE was effective against stem borer (Juneja et al., 2021).
- Farmers growing organic crops are advised to apply two sprays of *Beauveria bassiana* 1.15 WP (2×10^6 cfu/g) 5 g/L at 30 and 60 DAE for effective and economical management of stem borer (Parmar et al., 2019).

White Grub

Major species found on this crop are *Anomala* spp., *Holotrichia consanguinea* Blanchard, *Holotrichia longipennis* Blanchard, and *Holotrichia serrata* Fabricius.

Host range: White grub is a serious pest of groundnut, sorghum, and pearl millet in parts of Gujarat and Rajasthan states of India.

Importance, status, and nature of damage: The pest is prevalent in its larval stage in the fields from March to October. The grubs feed on roots (Figure 12.5). The problem is more serious in lighter soils. The grub rapidly moves from one plant to another under the soil. The damaged plant starts drying up and ultimately dies. The seedlings are attacked by this pest. Larvae cause major devastation, resulting in failure of the crop.

Figure 12.5 White grub.

Control measures:

- Deep summer plowing and growing of Sunn hemp (*Crotalaria juncea*) as a trap crop.
- Collection and destruction of grubs during intercultural operations.
- In endemic areas, collection and destruction of adult beetles can be done in kerosene water immediately after the first rain showers when they visit trees for mating. For this, a campaign should be done in a collaborative manner in the villages.
- Host trees can be sprayed with chlorpyriphos 0.2% within 2–3 days after the rains.
- Seed treatment with imidacloprid 600 FS at 8.75 mL or clothianidin 50 WDG at 7.5 g/kg seed effectively controls the white grubs infesting pearl millet (Anonymous, 2017).
- In a 4-year study, soil drenching of imidacloprid 17.8 SL (300 mL/ha) in standing crop after 21 days of sowing recorded the lowest white grub and termite incidence at harvest and higher grain and fodder yield. Alternatively, mixing chlorpyriphos 20 EC or quinalphos 25 EC at 4.0 L/ha along with irrigation water at 3 weeks after seedling emergence is effective (Kalaisekar et al., 2017). Clothianidin 50 WDG at 150 g ai/ha is effective (ICBR 1:4.01) (Anonymous, 2021).
- For furrow application, chlorpyriphos or quinalphos dust mixed with farmyard manure in a 2:3 ratio and applied at planting is effective. Pesticide applications are recommended in areas with high endemic pressure of white grubs and during pest outbreaks.

Earhead Worms

Cotton bollworm (*Helicoverpa armigera* Hubner), hairy caterpillars (*Estigmene lactinea* Black and *Amsacta moorei* Butl), and Earhead caterpillar (*Eublemma silicula* Swinh) feed on pearl millet earheads; however, *H. armigera* is of significant importance.

Cotton Bollworm

Host range: Pearl millet, sorghum, maize, wheat, minor millets, groundnut, gram, cotton, bean, lucerne, and many other hosts.

Importance, status, and nature of damage: Up to 10%–15% grain yield losses were reported by Juneja and Raghvani (2000) at Jamnagar. Female moths lay eggs on the earheads at emergence, and freshly hatched larvae feed on stigma, leading to poor grain setting. Most of the larvae are dark greenish brown, but they can also be pink, cream, or almost black (Figure 12.6). The larvae can be easily found in the earheads because they do not hide in the soil during the daytime. Pupation takes place in the soil. Studies carried out during kharif 1995–2007 revealed that the incidence starts at earhead emergence, which was found to be higher at 63 DAE and to decrease toward maturity. Its correlation was significantly positive with minimum temperature and significantly negative with number of rainy days (Raghvani et al., 2008).

Control measures:
- Installation of sex pheromone traps (five traps per hectare) at 1 ft above earhead formation is recommended for monitoring of adult male moths (*H. armigera*) (Juneja et al., 2015).
- Because *Helicoverpa* larvae generally appear at the earhead stage, hand collection and destruction of this pest is suggested.
- Green gram or pigeonpea taken as intercrop at a ratio of 2:1 reduces the incidence of *Helicoverpa*, which also gives a higher equivalent yield than sole crop (Anonymous, 1993).

Figure 12.6   *Helicoverpa armigera.*

- Spray *Helicoverpa* nuclear polyhedrosis virus at 450 larval equivalents/ha, *Bacillus thuringiensis* 5% WP at 1.0 kg/ha, or *Beauveria bassiana* 1.15 WP at 2×10^6 cfu/g at 2.0 kg/ha at the appearance of *H. armigera* at earhead stage for effective and economical pest management (Anonymous, 2018).
- Spraying of phosalone 35% EC 750 mL in 500 L water at earhead stage is also effective and economical (Anonymous, 2018).

Hairy Caterpillar

Hairy caterpillar species include *Amsacta* spp., *Estigmene lactinea* Black, and *Amsacta moorei* Butler.

Importance, status, and nature of damage: Hairy caterpillars have been reported as sporadic pests in semi-arid areas. These caterpillars (Figure 12.7) are difficult to kill in the advanced stage; however, advantage can be taken of their habit of pupating in the soil of infested fields. The grain yield losses due to these caterpillars vary from 10% to 60% (Kishore, 1996).

Control measures: Infested fields should be deep plowed immediately after harvest to expose the pupae to the sun and to predators. This operation will considerably reduce the incidence of hairy caterpillars in the next season. The caterpillars in their early larval stage feed extensively on alternate hosts. Hand picking and mechanical destruction of larvae at this stage can control the spread of larvae. The egg masses can also be collected and destroyed. The use of light traps reduces the adult population.

Gray *Eublemma*

Gray Eublemma feed on the maturing grains that are hidden under dome-shaped or elongated structures formed from the dry anthers (Figure 12.8). The caterpillars

Figure 12.7 *Amsacta* spp.

Figure 12.8   *Eublemma silicule.*

feed mostly on the upper part of maturing grains. The excreta of insects lead to fungal attack, deteriorating the quality of grain. The infestation by this pest varies in different varieties and hybrids. The pest remains active from late August to early September in India. The attack is visible immediately with the commencement of grain formation. Kishore (1996) observed two to six larvae per earhead and 6.3%–38.6% infested earheads in different genotypes during kharif 1995 at Delhi. The pest can be managed with two to three sprays of phosalone 750 mL in 500 L water at 20-day intervals.

Leaf Roller/Binder (*Marasmia trapezalis* Guenée)

Leaf roller/binder attacks pearl millet, maize, sorghum, and grasses. The larvae feed on the tips of leaves, causing leaf rolling (Figure 12.9). The larvae also feed by scrapping the green tissues inside leaf folds. The incidence of leaf roller is moderate to

Figure 12.9   Leaf roller/binder.

low in pearl millet growing areas. Although this is an occasional pest, sometimes heavy attacks are observed. Special control measures are seldom necessary, and the control measures applied for major pests like shoot fly and stem borer effectively control the leaf roller.

Millet Head Miner [*Heliocheilus albipunctella* (De Joannis) (Lepidoptera: Noctuidae)]

Importance, status, and nature of damage: Millet head miner (MHM) became a major pest during the droughts of 1972–1974 in the Sahel, and it is considered a deadly problem in the food-insecure region where no other crop is grown except pearl millet (Payne et al., 2011). Damage by MHM is due to the larvae feeding on the panicles and preventing grain formation. Young larvae perforate millet glumes and eat flowers, whereas mature larvae cut the floral peduncles, thus preventing the grain formation, or causing mature grains to spill. Grain yield losses ranging from 16% to 85% have been observed (Gahukar & Ba, 2019; Mohammed et al., 2020). The larvae feed between the rachis and flowers and lift the destroyed flowers or developing grains, leaving a characteristic spiral pattern on the millet head. Because of potentially heavy yield losses, the head miner is considered a major pest of pearl millet in the Sahel region of West and Central Africa (Figure 12.10).

Figure 12.10   Millet head miner.

Control measures: Several measures were undertaken to control MHM, including the planting of pest-tolerant/resistant cultivars, the application of chemical pesticides, and the use of natural enemies that have significant impacts on larval mortality (Gahukar & Ba, 2019). Plowing fields up to 15–25 cm deep in the off-season to expose the pupae to desiccation and predation resulted in 20% pupal mortality (Gahukar, 1990a). This is not practiced in the sandy soils of Sahel because the soils are prone to erosion. The use of chemical pesticides is not realistic for subsistence farmers in Sahel because of high cost, limited availability, lack of trained manpower, and environmental and health risks. Biological control with the release of the larval parasitoid wasp *Habrobracon* (=*Bracon*) *hebetor* Say (Hymenoptera: Braconidae) resulted in parasitism of 80% of the MHM larvae and a 30% increase in grain yield (Ba et al., 2013). Even though biocontrol agents have been proven effective, their large-scale availability is still a challenge (Guerci et al., 2018). Therefore, host plant resistance can be exploited to complement the ongoing biological control efforts as an IPM strategy for MHM (Mohammed et al., 2020). Insecticides, acephate, chlordimeform, chlorpyriphos, and insect growth regulators (diflubenzuron, lufenuron) significantly reduced the pest incidence in Senegal and Niger (Gahukar et al., 1986) with a single application at 75% flowering or with two applications at the beginning of flowering and 5–7 days later. Similarly, cypermethrin was effective in Sudan (Hughes & Rhind, 1988). In Mali, Passerini (1991) contrasted three applications of neem seed kernel extract (3%) at 500 L/ha with an untreated control field; 34.7% infestation was observed in the untreated field, whereas the head infestation rate was lowest (14.1%) in the cypermethrin-treated field, followed by neem seed kernel extract (27.5% infestation). IPM is recommended to control this pest by growing tolerant cultivars, using pheromone traps, timely release of bio-control agents, and usage of short-cycle cultivars.

Blister Beetles

Psalydolytta fusca Olivier., *P. vestita* Duf, and *Mylabris pustulata* Thunberg have been observed on pearl millet earheads.

Host range: Pearl millet, sorghum, maize, pulses, and beans.

Importance, status, and nature of damage: *Psalydolytta fusca* Olivier and *P. vestita* Duf. are common species in the Sahelian countries from Senegal to Chad (Gahukar, 1991), whereas species in the genera *Mylabris* and *Coryna* are widely distributed in Nigeria (Lale & Sastawa, 2000). Heavy infestations of *Mylabris* spp., *Coryna hermaniae* (F.), and *C. chevrolati* (Blair) have occurred on early plantings of pearl millet in Nigeria (Lale & Sastawa, 2000). Adult beetles feed on the flowers or stigma or pollen and sometimes even the milky grain. Because the pest feeds on the flower petals and the pollen grain, the seed setting in the pearl millet spike is affected, and grain yield losses of 4%–48% have been reported in West Africa (Zethner & Lawrence, 1988). Some insects are large and black with yellow stripes across the wings; others are smaller in size (<1 inch in length and thin) and have a brownish or greenish-blue color. They have soft bodies and thin wings. The beetles secrete an acidic substance, which causes a "blister" when it encounters the human body; hence the name "blister beetle."

Control measures:
- The beetles can be caught by hand nets or through light traps and destroyed in kerosene water.
- Spraying flowering spikes with malathion at 500–750 g ai/ha (Zethner & Lawrence, 1988) can prevent infestation.
- Intercropping of pearl millet with sorghum or cowpea in Nigeria resulted in reduced pest infestation, although the highest yield was obtained from millet grown alone (Lale & Sastawa, 2000).
- Use of cultivars with long bristles on the panicle can deter the beetles.

Cotton Grey Weevil (*Myllocerus subfasciatus* Guerin-Meneville)

Host range: Pearl millet, cotton, maize, sugarcane, finger millet, okra, etc.

Importance, status, and nature of damage: This is an occasional serious pest of pearl millet. The adult beetle causes severe damage to the leaves, leaving only the midribs (Figure 12.11). Beetle grubs feed on the roots. In some cases, grub damage at the seedling stage becomes very serious.

Control measures: Dusting of malathion 5% at 20–25 kg/ha is a satisfactory control measure.

Grasshoppers

Different grasshopper species have been found feeding on pearl millet, such as *Hieroglyphus nigrorepletus* Bol. (Phadka grasshopper); *Chrotogonus trachypterus*, Blanchard, (surface grasshopper); *Colemania sphenarioides*, Bol. (Deccan wingless

Figure 12.11 *Myllocerus subfasciatus.*

grasshopper); *Orthacris simulans* Bol. (wingless grasshopper); and *Patanga succincta* Linn. (Bombay locust).

Importance, status, nature of damage, and control measures: These polyphagous insects assume the status of serious pests under favorable conditions. These insects cause severe defoliation of the crop. They prefer to lay eggs on bunds and uncultivated areas of the infested fields. At this stage, scrapping of bunds and plowing of the uncultivated boundaries will help in exposing the egg pods to sun heat. The emergence of nymphs from eggs takes place soon after the first monsoon shower in India. Nymphs remain congregated in their early stage of growth, feeding on wild grasses near cultivated areas. Thus, dusting bunds and grasses immediately after the onset of the rainy season will reduce the population of grasshoppers. Mechanical methods (e.g., digging trenches) can be adopted at this stage to control the pest.

Armyworm [*Mythimna separata* Wlk. (Rice armyworm), *Mythimna loreyi* (Dup.), and *Mythimna unipuncta* Haw]

Importance, status, nature of damage, and control measures: Both in India and Africa, millet plants are regularly attacked by the lepidopterans, including leaf folders (Pyralidae), leaf caterpillars (Noctuidae, Lymantriidae), hairy caterpillars (Arctiidae), and armyworms (Noctuidae) (Gahukar, 1989; Nwanze & Harris, 1992). These caterpillars are nocturnal. They defoliate the crop and badly damage oat, wheat, barley, winter rye, and maize in addition to pearl millet. The larvae mainly feed on leaves, and the older larvae of four to six instars cause the main harm, roughly gnawing out and eating around leaf plate and damaging inflorescences, ears, growth points, and grains in ears. Starting with weeds, the larvae pass on to the cultivated plants. The maximum damage is caused by first-generation larvae. Control measures include weeding, inter-row cultivation, removal of crop residues from fields after harvesting, deep plowing, optimal dates of early sowing, cultivation of resistant varieties, and insecticidal treatments of crops.

Fall Armyworm [*Spodoptera frugiperda* (J.E. Smith) (Lepidoptera: Noctuidae)]

Host range: Maize, sorghum, pearl millet, and other minor millets.

Importance, status, nature of damage, and control measures: Fall armyworm is a destructive pest native to the Americas that has recently been introduced in India and is presently causing economic damage primarily in maize and sorghum crops. If both the hosts are not available, it attacks other crops belonging to Poaceae, such as sugarcane, pearl millet, rice, wheat, finger millet, and fodder grasses (Figure 12.12). The adult moth is a strong flier and can fly over 100 km in

Figure 12.12   Fall armyworm damage on pearl millet.

search of host plants. The male moth has two characteristic markings: a colored spot toward the center and a white patch at the apical margin of the forewing. The forewing of a female is dull with faint markings. A female moth lays over 1000 eggs in single or multiple clusters, and the eggs are covered with hairs. The total life cycle takes 30–38 days as observed under natural/artificial rearing on diet conditions (Deshmukh et al., 2021; Jaba et al., 2020).

A crop may have been infested with FAW if lengthy, papery windows of all sizes are spotted scattered across the leaves in a few nearby plants. This condition is brought on by FAW larvae in their first and second instars, which feed by scraping on leaf surfaces. Early identification of such symptoms is a must for effective management of FAW. The feeding by third instar larvae causes ragged-edged, round to oblong holes on leaves. The size of holes increases with the growth of larvae. Once the larva enters the fifth instar, it feeds voraciously, consuming larger areas of leaves. The sixth instar larva extensively defoliates the plant and produces a large amount of fecal matter.

Cost-effective IPM for FAW

- The very crucial periods to manage this pest or protect crops are up to 50–60 days after emergence. Thereafter, the infestation comes down due to plant phenological structures (the plant whorl space gets reduced for hiding larvae of FAW).

- Seed treatment with FORTENZA DUO (cyantraniliprole 19.8% + thiamethoxam 19.8%)/Benevia (cyantraniliprole-10% OD) (4 mL/kg seed) has been reported to offer protection up to 2–3 weeks after seedling emergence.
- Monitoring/mass trapping: 10 traps per acre.
- Profenophos: 2 mL/L is an effective ovicidal.
- Azadirachtin (5 mL/L) 1500 ppm is an ovipositional deterrent.
- *Nomuraea rileyi* formulation (1×10^8 cfu/g) at 5 g/L is effective for second to third instar larvae.
- Based on the stage of larvae, stepwise/rotation-wise application of insecticide can be followed: emamectin benzoate at 0.5 g/L of water, followed by thiodicarb at 0.5 g/L of water, chlorantraniliprole 18.5% SC at 0.3 mL/L, thiamethoxam 12.6% + lambda cyhalothrin 9.5% at 0.5 mL/L, and spinetoram (Delegate) at 0.4 mL/L.

Poison baiting: Keep the mixture of 10 kg rice bran + 2 kg jaggery with 2–3 L of water for 24 hours to ferment. Add 100 g thiodicarb half an hour before application in the field. The bait should be applied to the whorl of plants. Poison baiting is recommended for the control of late instar larvae of the second window.

All sprays should be directed into whorls (by bottle spray or knapsack spray) either in the early hours of the day or in the evening.

Sugarcane Leafhopper (*Pyrilla perpusilla* Wlk)

The attack of sugarcane leaf hopper is becoming a major constraint in maximizing the production of pearl millet. The intermittent periods of drought during July–September, heavy foliage growth with sugarcane leaf–like color, and high humidity with low temperatures between crop seasons are some of the factors that have resulted in the preference of *Pyrilla* for this crop. Until 1968, the sugarcane leaf hopper was not a pest of pearl millet. An unusually heavy infestation of *Pyrilla* occurred in North India between 1990 and 1991. In addition to pearl millet, sorghum and maize were subjected to heavy infestation by sugarcane leafhopper. Thereafter, these outbreaks occurred every year. Both adults and nymphs suck the sap from tender leaves (Kishore, 1994). The application of chlorpyriphos 20% EC (1500 mL/ha) effectively manages this pest.

Earhead Beetle/Blossom Beetles (Chaffer Beetle)

Anatona stillata Newman, *Oxycetonia versicolor* Fab., and *Chiloloba acuta* Wiedemann. have been found on pearl millet crops.

These insects are principally pollen feeders. They are seen on pearl millet earheads. The larvae develop in organic matter in the soil, and a few infest the roots of the plant as well. These are medium sized to large beetles and are brilliantly colored and dorsally flattened with a large scutellum (Nayar et al., 1976).

Other minor insect pests recorded on this crop in the field are flea beetles, jassids, thrips, aphids, termites, cutworm, and earhead bugs. In addition to these field pests, stored grain pests, such as red rust flour beetle (*Tribolium* spp.), lesser grain borer (*Rhyzopertha dominica* Fab.), and rice weevils (*Sitophilus oryzae* L.), are the principal storage pests of pearl millet grains. The infestation of *S. oryzae* and *R. dominica* commences from the field or threshing floors, where matured grains of pearl millet may get infested. In recent years, storage of pearl millet seeds has become increasingly difficult due to the attack of these stored grain pests. The average damage by stored grain pests to pearl millet has been reported to be 14%–28% (Dhamdhare & Kishore, 1992; Kishore & Sharma, 1984).

Termites or White Ants [*Odontotermes* spp., *Microtermes* spp., *Macrotermes* spp. (Termitidae: Isoptera)]

Host range: Pearl millet, sorghum, finger millet, maize, wheat, sugarcane, upland rice.

Damage symptoms: Termites also attack the roots of maize and sorghum, and the damaged plants topple. They eventually disrupt the movement of nutrients and water through the vascular system, resulting in death of the plant. The dead plants are sheathed with soil.

Control measures: In the areas of regular termite occurrence, the soil should be mixed with chlorpyriphos 5% dust at 35 kg/ha at the time of sowing. When the incidence of pest is noticed in the standing crop, chlorpyriphos 20% EC at 4 L mixed with 50 kg of soil per hectare can be broadcasted evenly, followed by light irrigation.

Status of Host Plant Resistance in Pearl Millet for Major Insect Pests

Insect pests cause biotic stress that needs to be tackled effectively to sustain grain yields and achieve food security for poor households. Reports have indicated the existence of around 500 species globally (Sharma & Davies, 1988). The development of new genotypes with superiority in terms of yield and quality aspects has increased the incidence of pests and the prevalence of minor pests. Although the incidence of pests in pearl millet is less compared with sorghum and other cereals, the polyphagous nature of the pests may pose a threat to the millet cropping system as well. Losses incurred in millets due to insects are lesser in India (10%–20%) (Gahukar & Jotwani, 1980) compared with Ghana (50%) (Tanzubil & Yakubu, 1997).

Different types of host plant resistance mechanisms, such as antibiosis (cv. Zongo) or tolerance (cv. IBV-8004) (Gahukar et al., 1986), have been observed in pearl millet.

Borer infestation of 30%–58% with 15–89 larvae/10 stems in short-cycle cultivars and 27%–42% infestation with 9–17 larvae/10 stems in long-cycle cultivars were observed among 33 cultivars screened (Gahukar, 1990a). Because there were no significant differences, none of these entries was considered resistant or tolerant. Stem infestation varied from 2% to 19% and larval population from 5 to 19 larvae/10 stems in the nine cultivars screened in Senegal (Gahukar, 1990b). Currently, chemical control is the major control measure taken by farmers to regulate the damage caused. However, it is less viable both in terms of economic and environmental sustainability. This demands solutions that offer more effective control measures through host plant resistance. Kishore (1991) evaluated pearl millet germplasm for a wide range of insect pests and identified accessions IP 79, IP 263, IP 1307, and IP 1395 as resistant to leaf hopper (*Pyrilla perpusilla*), whereas 10 entries (IP 57, IP 164, IP 326, IP 1046, IP 1130, IP 1316, IP 1324, IP 1364, IP 1944, and IP 1964) were found resistant to earhead worms *E. silicula* and *Cryptoblabes gnidiella*. The germplasm accessions IP 205, IP 213, IP 225, IP 252, IP 256, IP 314, IP 315, IP 323, IP 375, IP 427, IP-432, IP 467, IP 476, IP 478, IP 501, IP 513, IP 514, IP 835, IP 1450, IP 1538, IP 1546, IP 1158, IP 1365, IP 1411, and IP 1550 were resistant to white grub (Kishore, 1991; Pradhan, 1971). The accessions C-591, Pak-75211, Pak-75212, Pak-75219, Pak-75194, Pak-75227, Pak-75238 Pak-75272, Pak-75276, WCA-78, C-47, Pak-75322, Pak-75323, Pak-75329, Pak-75331, Pak-75334, Pak-75337, Pak-75338, Pak-75339, Pak-75353, and Pak-75359 were found to have the least susceptibility and therefore can be used in the breeding program for green bug resistance (Akhtar et al., 2012). The antibiosis resistance mechanism to millet head miner was observed in the genotypes Gamoji, ICMP 177001, ICMP 177002, ICMV 177003, ICMV IS 90311, LCIC9702, Souna 3, ICMV IS 94206, and PE08043, with appreciable grain yield compared with the susceptible check ICMV IS 92222 (Mohammed et al., 2020).

A study was carried out to screen pearl millet lines against major shoot fly at Jamnagar (Pearl Millet Research Station, Junagadh Agricultural University) in Gujarat state and at Agricultural Research Station, Mandor-Jodhpur in Rajasthan, during kharif 2022 under the All India Coordinated Research Project on Pearl Millet. Thirty-four entries were screened at vegetative and earhead stages. At vegetative stage, none of the entries was found resistant (0.0%), 12 entries [MP 610, MH 2661, MH 2619, AHB 1269, 86M01, HHB 272, ICMV 221, MH 2577, MH 2555, Dhanshakti, MPMH 17, and MH 2564 (R)] were moderately resistant (0.1%–5.0% damage), 22 entries (KBH 108, RHB 177, PB 1705, AHB 1200, MP-7792, MPMH 21, HHB 67 Imp., Pratap, GHB 538, MH 2580, Pusa Comp. 701, NBH 5767, MH 2553 MP 613, MP 612, MH 2606, 86M86, MH 2631, Raj 171, Pusa Comp. 383 JBV 2, and MH 2626) were tolerant (5.1%–10.0%), and the overall incidence ranged between 3.32% and 9.93%. At the earhead stage, three entries (MH 2619, 86M01, and MH 2555) were free from pest damage (Resistant), 30 entries [MP 610, MH 2564 (R), RHB 177, HHB 272, MH 2553, MH 2580, Pratap, MPMH 17, MH 2661, Pusa Comp. 701, MP 613, AHB 1269, MH 2626, MPMH 21, MH 2577,

KBH 108, GHB 538, HHB 67 Imp., Pusa Comp. 383, Dhanshakti, MH 2631, PB 1705, MP 612, AHB 1200, ICMV 221, NBH 5767, JBV 2, MP-7792, 86M86, and Raj 171] were moderately resistant, one entry was tolerant (MH 2606), and the overall incidence ranged between 0.0% and 5.43% (Anonymous, 2023).

The progress in genetic resistance to insect pests in pearl millet is limited, which calls for the evaluation of large genetic resources for the identification of resistance sources. Thus, to monitor pest status, surveys and surveillance for major insect pests should be done on a regular basis. Monitoring of insect pests is also necessary to get information on economic losses caused by different pests and to implement pest avoidance techniques. For the control of major insect pests, host plant resistance/tolerance is the best and most economically viable method. The use of botanicals is also economical, and they are easily available and eco-friendly. There are a good number of pesticides/biopesticides available for seed treatment that are cost effective and that are required in less quantity, and thus less chemical goes into the soil and plant environment. These are helpful in the management of soil insect pests (white grub and termite) and initial foliage insects (shoot fly, thrips, leaf roller, grasshopper, and stem borer). Seed treatment can be combined with one or two foliar sprays to control major insect pests both at the vegetative and earhead stages. The newer chemicals can also be tested for their effectiveness against these pests. Although a high level of resistance may not be available in the breeding lines/cultivars, the inclusion of cultivars with moderate levels of resistance along with chemical and biological control measures forms a good strategy to develop IPM packages for pest management.

References

Akhtar, N., Ahmad, Y., Shakeel, M., Gillani, W. A., Khan, J., Yasmin, T., & Begum, I. (2012). Resistance in pearl millet germplasm to greenbug, *Schizaphis graminum* (Rondani). *Pakistan Journal of Agriculture Research*, *25*(3), 228–232.

Anonymous. (1993). *Annual research report*. Gujarat Agricultural University.

Anonymous. (2017). *Proceedings of the 52nd Annual Group Meeting of AICRP on Pearl Millet*. Indian Council of Agricultural Research.

Anonymous. (2018). *Annual research report*. In *Proceedings of the 14th Meeting of the Plant Protection Sub-Committee of Agricultural Research Council* (pp. 42–51). Junagadh Agricultural University.

Anonymous. (2021). *Proceedings of 56th Annual Group Meeting of AICRP on Pearl Millet* (p. 38). Junagadh Agricultural University.

Anonymous. (2023). Summary of experiments 2022-23. In *Annual Report of AICRP on pearl millet* (pp. 1–79). Indian Council of Agricultural Research.

Ba, M. N., Baoua, I. B., N'Diaye, M., Dabire-Binso, C., Sanon, A., & Tamò, M. (2013). Biological control of the millet head miner *Heliocheilus albipunctella* in the Sahelian region by augmentative releases of the parasitoid wasp *Habrobracon hebetor*: Effectiveness and farmers' perceptions. *Phytoparasitica, 41*, 569–576.

Barau, B. W., Wama, B. E., & Maikeri, T. C. (2015). Incidence of the millet stem borer *Coniesta ignefusalis* (Hampson) in farmers' fields of Lamma and Dingding villages of zing local government area. *International Journal of Bioscience, 10*(1), 107–111.

Biradar, A., & Sajjan, S. (2018). Management of shoot fly in major cereal crops. *International Journal of Pure and Applied Bioscience, 6*, 971–975.

Blanchard, E. (1836). Monographie du genre Ommexecha de la famille des Acridiens. *Annales de la Société Entomologique de France, 5*, 603–624.

Choudhary, S. K., Tandi, B. L., & Singh, S. (2018). Management of white grub, *Holotrichia consanguinea* Blanchard in pearl millet. *Indian Journal of Entomology, 80*, 619–622.

Degri, M. M., Mailaflya, D. M., & Mshelia, J. S. (2014). Effect of intercropping pattern on stem borer infestation in pearl millet (*Pennisetum glaucum* L.) grown in the Nigeria Sudan savanna. *Advanced Entomology, 2*, 81–86.

Deshmukh, S. S., Prasanna, B. M., & Jaba, J. (2021). Fall armyworm (*Spodoptera frugiperda*). In *Polyphagous pests of crops* (pp. 349–372). Springer. https://doi.org/10.1007/978-981-15-8075-8_8

Dhamdhare, S. V., & Kishore, P. (1992). Resistance in pearl millet to khapra beetle, *Trogoderma granrium* everts in storage. *Journal of Entomological Research, 16*(1), 82–84.

Drame-Yaye, A., Youm, O., & Ayertey, J. N. (2003). Assessment of grain yield losses in pearl millet due to the millet stem borer, *Coniesta ignefusalis* (HAMPSON). *Insect Science and its Application, 23*(3), 259–265.

Gahukar, R. T. (1989). Insect pests of millets and their management: A review. *International Journal of Pest Management, 35*, 382–391.

Gahukar, R. T. (1990a). Field screening of pearl millet cultivars in relation to insects and diseases. *International Journal of Tropical Insect Science, 11*, 13–19.

Gahukar, R. T. (1990b). Reaction of locally improved pearl millets to three insect and two diseases in Senegal. *Journal of Economic Entomology, 85*, 2102–2106.

Gahukar, R. T. (1991). Pest status and control of blister beetles in West Africa. *International Journal of Pest Management, 37*, 415–420.

Gahukar, R. T., & Ba, M. N. (2019). An updated review of research on *Heliocheilus albipunctella* (Lepidoptera: Noctuidae), in Sahelian West Africa. *Journal of Integrated Pest Management, 10*, 1–9.

Gahukar, R. T., Bos, W. S., Bhatnagar, V. S., Dieme, E., Bal, A. B., & Fytizas, E. (1986). Acquis recents en entomologie du mil au Senegal. In *Rapport, reunion d'evaluation du programme mil-sorgho du 19-21 mars 1986* (p. 29). Institut Senegalais de Recherches Agronomiques.

Gahukar, R. T., & Jotwani, M. G. (1980). Present status of field pests of sorghum and millets in India. *International Journal of Pest Management, 26*(2), 138–151.

Gahukar, R. T., & Reddy Gadi, V. P. (2019). Management of economically important insect pests of millet. *Journal of Integrated Pest Management, 10*(1), 28. https://doi.org/10.1093/jipm/pmz026

Goudiaby, M. P., Sarr, I., & Sembene, M. (2018). Source of resistance in pearl millet varieties against stem borers and earhead miner. *Journal of Entomology and Zoology Studies, 6*, 1702–1708.

Guerci, M. J., Norton, G. W., Ba, M. N., Baoua, I., Alwang, J., Amadou, L., Moumouni, O., Karimoune, L., & Muniappan, R. (2018). Economic feasibility of an augmentative biological control industry in Niger. *Crop Protection, 110*, 34–40.

Hughes, D., & Rhind, D. (1988). Control of the millet head worm in the western Sudan. *International Journal of Pest Management, 34*, 346–348.

Jaba, J., Sathish, K., & Mishra, S. P. (2020). Biology of fall army worm *S. Frugiperda* (J. E. Smith) on artificial diets. *Indian Journal of Entomology, 82*(3), 543.

Juneja, R. P., & Parmar, G. M. (2022). Survey of insect pests of summer pearl millet in Gujarat. *Insect Environment, 25*(3), 410–413.

Juneja, R. P., Parmar, G. M., Chaudhari, R. J., & Mungra, K. D. (2022). Monitoring of major insect pests, correlation and yield loss in pearl millet. *Insect Environment, 25*(2), 202–206.

Juneja, R. P., Parmar, G. M., Chuadhari, R. J., Parmar, S. K., & Mungra, K. D. (2023). Seasonal *Incidence of major insect pests, correlation with abiotic parameters and losses in yield due to insect pest complex in pearl millet* [Paper presentation]. International Conference on Development and Promotion of Millets and Seed Spices for Livelihood Security ICDPMSSLS 2023 (p. 88). Agriculture University, Jodhpur, Rajasthan.

Juneja, R. P., Parmar, G. M., Ghelani, Y. H., Mungra, K. D., Patel, P. R., & Chaudhari, N. N. (2015). Monitoring of ear head worm *Helicoverpa armigera* (Hubner) through sex pheromone in pearl millet crop. *International Journal of Plant Protection, 8*(2), 245–249.

Juneja, R. P., Parmar, G. M., Mungra, K. D., Detroja, A. C., Kadvani, D. L., & Parmar, S. K. (2021). Development of low cost protection technology for shoot fly and stem borer in pearl millet crop. *Journal of Entomology and Zoology Studies, 9*(1), 1562–1565.

Juneja, R. P., & Raghvani, K. L. (2000). Feeding behaviour of *Helicoverpa armigera* (Hubner) and its damage in pearl millet. *Insect Environment, 6*(3), 141–142.

Juneja, R. P., Raghvani, K. L., Godhani, B. G., & Dangaria, C. J. (2004). Effectiveness of plant origin extracts against pearl millet shoot fly. *Insect Environment, 10*(2), 90–92.

Kalaisekar, A., Padmaja, P. G., Bhagwat, V. R., & Patil, J. V. (2017). *Insect pests of millets: Systematics, bionomics and management* (1st ed.). Elsevier.

Kishore, P. (1991). Sources of resistance amongst world pearl millet, *Pennisetum typhoides* (Burm.) germplasms to important insect pests. *Journal of Entomological Research*, *15*(3), 212–217.

Kishore, P. (1994). Changing pest status of sugarcane leafhopper, *Pyrilla perpusilla* Walker on pearl millet together with identification of sources of resistance amongst bajra germ plasms. *Journal of Entomological Research*, *18*(4), 293–296.

Kishore, P. (1996). Evolving management strategies for pests of millets. *Indian Journal of Entomological Research*, *20*(4), 287–297.

Kishore, P., & Sharma, G. C. (1984). Note on relative susceptibility of seeds of promising bajra (*Pennisetum typhoides*) hybrids and varieties to insect attack in storage. *Seed Research*, *12*(1), 90–92.

Kishore, P., & Solomon, S. (1989). Research needs and future strategy for controlling insect pest problems on bajra based cropping system. *Seeds & Farms*, *15*(7&8), 23–28.

Krauss, H. A. (1877). Orthopteren vom Senegal gesammelt von Dr Franz Steindechner. In *Sitzungsberichte der kaiserlichen Akademie der Wissenschaften, Mathematisch - naturwissenschaftliche Classe, Wien* (pp. 7629–7663).

Lale, N. E. S., & Sastawa, B. M. (2000). Evaluation of host plant resistance, sowing date modification and intercropping as methods for the control of Mylabris and Coryna species (Coleoptera: Meloidae) infesting pearl millet in the Nigerian Sudan savanna. *Journal of Arid Environments*, *46*, 263–280.

Maiga, I. H., Lecoq, M., & Kooyman, C. (2008). Ecology and management of the Senegalese grasshopper, *Oedaleus senegalensis* (Krauss 1877) (Orthoptera: Acrididae) in West Africa: Review and prospects. *Annales de la Societe Entomologique de France*, *44*(3), 271–288.

Mohammed, R., Gangashetty, P. I., & Karimoune, L. (2020). Genetic variation and diversity of pearl millet [*Pennisetum glaucum* (L.)] genotypes assessed for millet head miner, *Heliocheilus albipunctella* resistance, in West Africa. *Euphytica*, *216*(10), 158. https://doi.org/10.1007/s10681-020-02690-y

Mote, U. N., & Nakat, R. V. (2001). Unusual incidence of shoot fly (*Atherigona approximate* mall.) on pearl millet in Western Maharashtra. *Insect Environment*, *7*(2), 76–77.

Nayar, K. K., Anathakrishnan, T. N., & David, B. V. (1976). *General and applied entomology*. McGraw-Hill.

Nigus, C., & Damte, T. (2018). Identification of the tef shoot fly species from tef, *Eragrostis tef* (Zucc.), trotter growing areas of Ethiopia. *African Journal of Insects*, *5*, 181–184.

Nwanze, K. F., & Harris, K. M. (1992). Insect pests of pearl millet in West Africa. *Review of Agriculture Entomology*, *80*(12), 1132–1185.

Parmar, G. M., Juneja, R. P., Chaudhari, R. J., Mungra, K. D., & Parmar, S. K. (2022a). Eco-friendly management strategy for shoot fly and stem borer infesting pearl millet crop. *Frontiers in Crop Improvement, 10,* 364–368.

Parmar, G. M., Juneja, R. P., & Chaudhary, N. N. (2019). Management of major insect pests of pearl millet under organic cultivation. *International Journal of Plant Protection, 12*(1), 62–66.

Parmar, G. M., Juneja, R. P., & Mungra, K. D. (2015). Management of shoot fly and stem borer on pearl millet crop. *International Journal of Plant Protection, 8*(1), 104–107.

Parmar, G. M., Juneja, R. P., Mungra, K. D., Chuadhari, R. J., & Detroja, A. C. (2022b). Determination of economic threshold level for the chemical control of pearl millet stem borer, *Chilo partellus* (Swinhoe). *International Journal of Agricultural Sciences, 14*(4), 11237–11239.

Parmar, G. M., Juneja, R. P., Patel, P. R., Parmar, S. K., Chaudhary, N. N., & Mungra, K. D. (2021). Evaluation of different pest management modules against major insect pests of pearl millet. *Journal of Entomology and Zoology Studies, 9*(3), 390–394.

Passerini, J. (1991). *Field and lab trials in Mali to determine the effects of neem extracts on the millet pests: Heliocheilus albipunctella De Joannis (Lepidoptera: Noctuidae), Coniesta igneusalis Hampson (Lepidoptera: Pyralidae) and Kraussaria angulifera Krauss (Orthoptera: Acrididae).* Department of Entomology, McGill University.

Payne, W., Tapsoba, H., Baoua, I. B., Malick, B. N., N'Diaye, M., & Dabire-Binso, C. (2011). On-farm biological control of the pearl millet head miner: Realization of 35 years of unsteady progress in Mali, Burkina Faso and Niger. *International Journal of Agricultural Sustainability, 9*(1), 186–193.

Pradhan, S. (1971). *Investigations on insect pests of sorghum and millets (1965–1970). Final technical report.* Division of Entomology, Indian Agricultural Research Institute.

Raghvani, K. L., Dangaria, C. J., Juneja, R. P., Parmar, G. M., & Ghelani, Y. H. (2010). Chemical control of stem borer *Chilo partellus* (Swinhoe) in pearl millet. *Indian Journal of Applied Entomology, 24*(1), 85–87.

Raghvani, K. L., Juneja, R. P., & Dangaria, C. J. (2005). *Estimation of economic threshold level of shoot fly Atherigona approximata (Malloch) in pearl millet* [Paper presentation]. National Symposium on Alternative uses of Millets in India, Bikaner, India.

Raghvani, K. L., Juneja, R. P., Ghelani, Y. H., Parmar, G. M., & Dangaria, C. J. (2008). Influence of abiotic factors on population fluctuations of major insect pest of pearl millet. *Indian Journal of Applied Entomology, 22*(1), 48–50.

Sastawa, B. M., Lale, N. E. S., & Ajayi, O. (2002). Evaluating host plant resistance and sowing date modification for the management of the stem borer, *Coniesta ignefusalis* Hampson and the head miner *Heliocheilus albipunctella* de Joannis infesting pearl millet in the Nigerian Sudan Savanna. *Journal of Plant Diseases and Protection*, *109*(5), 530–542. https://www.jstor.org/stable/43215474?seq=1

Sharma, H. C., & Davies, J. C. (1988). *Insect and other animal pests of millets*. ICRISAT.

Tanzubil, P. B., & Yakubu, E. A. (1997). Insect pests of millet in northern Ghana. 1. Farmers' perceptions and damage potential. *International Journal of Pest Management*, *43*(2), 133–136.

Vashisth, S., Jaba, J., Sharma, S. P., & Sharma, H. C. (2022). Biochemical mechanisms of induced resistance to *Chilo partellus* in sorghum. *International Journal of Pest Management*. https://doi.org/10.1080/09670874.2022.2036863

Wright, R. J. (2013). *Chinch bug management*. Institute of Agriculture and Natural Resources, University of Nebraska.

Zethner, O., & Lawrence, A. A. (1988). The economic importance and control of adult blister beetle, *Psalydolytta fusca* Olivier (Coleoptera: Meloidae). *International Journal of Pest Management*, *34*, 407–412.

13

Pearl Millet Biomass for Fodder in West Africa Region

Talla Lo, Aliou Faye, Mamadou T. Diaw, Abdoulaye Dieng, Doohong Min, Augustine Obour, Ramasamy Parumal, and P. V. Vara Prasad

Chapter Overview

In the agro-pastoralism zones of West Africa, native pastures constitute the main feed resources for livestock (Bessin, 1982). However, the quality and quantity of available forage are very variable and depend on the rainfall regime (Boudet et al., 1975; Rivière, 1978), which are affected by quantitative and qualitative seasonal variations exacerbated by losses due to trampling and termites (Lhoste, 1987). Studies have shown that only a third of the natural fodder produced at the end of the rainy season is available for livestock, with decreased quality in the dry season (Boudet, 1983). Another important effect of the decreased available grazing areas is the demographic pressure, the extension of cultivated land, and the reduced density of palatable perennial species (Diouf, 2001). The grazing lands of the region cannot support the maintenance needs of the regions livestock, which leads to transhumant pastoral systems (Aboh et al., 2012), where herds are moved seasonally in search of good quality feeds and fodders (Bationo & Mokwunye, 1991; Melesse et al., 2021).

Introduction

Feeding livestock remains a permanent challenge that contributes to poor livestock performance and animal undernourishment, especially in extensive livestock farming with decreased rainfall and perennial destruction of vegetation due to bushfires and excessive harvesting of trees (FAO, 2004; Tialla, 2011). West Africa (WA) is dominated by agropastoral farming systems with crop and livestock integration, where grain harvested from cereal crops provides food and income to households, while crop residues are used for livestock feeding

Pearl Millet: A Resilient Cereal Crop for Food, Nutrition, and Climate Security, First Edition.
Edited by Ramasamy Perumal, P. V. Vara Prasad, C. Tara Satyavathi, Mahalingam Govindaraj, and Abdou Tenkouano.
© 2025 American Society of Agronomy, Inc., Crop Science Society of America, Inc., Soil Science Society of America, Inc. Published 2025 by John Wiley & Sons, Inc.

(Eeswaran et al., 2022). In return, livestock contribute to soil productivity improvement because they are often the main sources of animal manure for soil fertility improvement because low-income farmers cannot afford inorganic fertilizers. Livestock also provide power for agricultural traction and improve human nutrition through access to animal-source proteins. However, increasing crop yield and livestock products increases demand for inputs and reduces grazing land, leading to frequent conflicts between livestock holders and farmers (Rufino et al., 2011). Cereals and legume crops provide alternatives to human and livestock food constraints; replenishing soil fertility and maintaining the balance of agroecosystems is necessary for sustainable agricultural development and harmonious integration of crop and livestock. Similarly, agro-industrial byproducts like flour mills (brans, meal, and flour), rice mills (bran, cone, and broken flour), sugar refineries (molasses and bagasse), and oil mills (cakes, groundnut shells, and cottonseed) are also used but essentially for complementing animal feeds (Camara, 2007). Crop residue from legumes such as groundnut and cowpea (stems and leaves), after the pods are harvested, was widely used as a supplement in feed rations of ruminants as an important source of protein for animal production (Ditaroh, 1993).

The use of cereal bran in livestock farming is restricted to certain species that cannot make the most of the cellulose (pigs and poultry). On the other hand, tropical cattle are excellent users of bran (Calvet, 1973). Cereal crop residues constitute an important feed resource for livestock and are made up of cereal straw and legume haulms that represent up to 20% of the diet of cattle in sub-Saharan Africa (SSA) (Bayer & Waters-Bayer, 1999). They play a considerable role in animal fodder resources, especially in the dry season (Lynda, 2016; Zoungrana, 2010). Cereal straws, mainly rice, maize, sorghum, and millet, are harvested at maturity when all the principal nutritional components have migrated to the grain. Consequently, the remaining residue fed as fodder is poor in energy, minerals, and vitamins and has highly lignified walls, reducing ingestion and digestibility (Cissé et al., 1996).

The recent discovery of dual-purpose millet by the West Africa National Agricultural Research Systems (WA NARS) in collaboration with the International Crops Research Institute for the Semi-Arid Tropics (ICRISAT) has created high expectations regarding the use of the region as a source of livestock feed because they produced much more grain for human consumption and stay-green biomass. In addition, the dual-purpose pearl millet varieties are well adapted to many regions of WA. Mathur (2012) collected the global georeferenced database of pearl millets (16,855 accesses) and developed a map of pearl millet distribution patterns in Africa, the Middle East, and India (Figure 13.1). The collected data include ICRISAT (13,542 accessions), the United States Department of Agriculture, Agricultural Research Service (472 accessions), and the International Livestock Research Institute (13 accessions). The map shows that most pearl millet occurrences in SSA and India were from ICRISAT. However, biodiversity collections mainly occurred in the WA region.

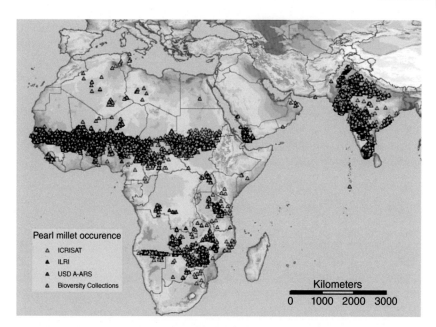

Figure 13.1 Mapping of pearl millet accessions based on the global database of georeference (Mathur et al., 2012 / with permission of Crop Trust).

Dual-purpose pearl millet has an advantage as grains for humans and fodder for livestock (Figure 13.2). The fodder consumed by livestock increases meat and milk production to improve human nutrition. In addition, millet residues left after harvesting can improve soil fertility through in situ carbon addition to

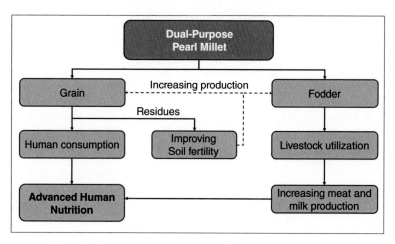

Figure 13.2 Potential uses of grain and fodder of the dual-purpose pearl millet in smallholder farming systems.

the soil. The use of pearl millet residue as livestock feed and fodder has been the subject of many studies, particularly regarding the nutrient value of the stems. However, there is very little information on the quality differences along the plant stem, and studies on nutritional characterization in dual-purpose pearl millet have not been widely evaluated. In this chapter, we present a general use of crop residues for livestock feed in the region and specifically highlight works supported by the Feed the Future Sustainable Intensification Innovation Lab program in Senegal on using pearl millet biomass for livestock feed as an integral part of the crop–livestock farming systems in the drier regions of WA.

Millet Straws for Livestock Feed

Crop residue is the aerial vegetative apparatus of a cereal after harvesting ripe grains (FAO, 2004). Many authors (Allard et al., 1983; Dieng, 1984; Kima, 2008; Ly, 1981; Mongondin & Tacher, 1979) reported the technical standards of their use and stated that, most of the time, crop residues in agropastoral systems are generally wasted because conservation and recovery techniques are not often mastered. The leaves are distributed to animals, and the stems are used to make or renew house fences, while a small portion of the straw is left in the fields for grazing. According to Kane et al. (1982), this fraction can reach 80% in the south of Sine-Saloum in Senegal for animal consumption during the dry season, while a small part (10%–15%) is harvested for domestic use, including confined animals, and 5%–10% is burned before plowing. Millet straws contain on average 85%–90% dry matter (DM), good cellulose quantity (pecto-cellulose walls), and energy from fresh material (0.15–0.35 fresh matter/kg) (ORSTOM, 1984). Using cereal crop residues, including millet by livestock, allows better milk production for 2.5 months after the rainy season (Ickowicz, 2001). Millet stalks are used in animal feed; mincing or grinding millet stalks increase feed intake, their addition to salt or molasses increases digestibility (Cissé et al., 1996; Faye, 2018), and their addition to urea increases cellulose digestibility (Nanema et al., 1998; Rivière 1978). In recent unpublished works, Sow and Faye (2023) demonstrated that pearl millet straw with peanut and cowpea residues–based rations increased milk production of local cows by 2.5 times, while young bull fattening was completed in 3 months instead of the 6 months finishing time with farmers' practice (Figure 13.3).

Opportunities and Challenges of Pearl Millet Production as Fodder for Animals

As population density increases, particularly in areas with good access to urban markets, zero-grazing with improved dairy cattle and cultivated fodder becomes

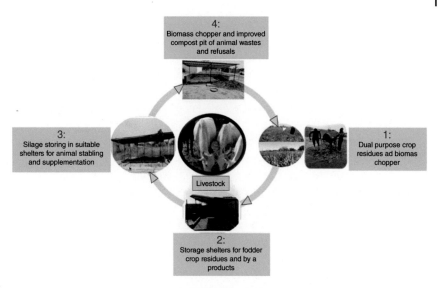

Figure 13.3 Animal production system developed using a millet straw base ration to increase milk and meat production in Senegal (Sow & Faye, 2023).

the predominant form of livestock management (Rufino et al., 2013) because the availability of fodder to support livestock particularly in the dry season is the most limiting factor of this farming system (Bationo & Mokwunye, 1991). Low-quality fodder, bulkiness, difficulty in transporting, challenges in processing and storage, and seasonal variability in fodder availability are the common challenges of livestock feeding in SSA (Amole et al., 2022). Some livestock growers collect crop residues after harvest and store them on rooftops, trees, or sheds until they are fed. This practice, in the case of cereal fodder, results in low digestibility because of high lignin and low crude protein concentrations, which consequently affect animal productivity (Boote et al., 2022). Although the nutritive value of cereal residues is relatively low, feeding crop residues to livestock during dry periods and droughts sustains the survival (Holness, 1999; Masikati, 2011) and body condition of livestock and draught animals used for cropland preparation (Tui et al., 2014). Additionally, access to livestock feed is often challenging due to the high prices (US$14–US$16 per 40-kg bag) and speculation imposed by intermediaries. At the same time, livestock holders do not master fodder production and conservation techniques (CNMDE, 2015), which limits feed availability in the dry season.

Pearl millet is traditionally grown throughout the Sahelo-Sudanian zone of West Africa, where rainfall deficits limit the cultivation of more water-demanding crops, such as maize and sorghum, and could be one of the solutions to livestock feed and nutritional security. Its fodder is an important livestock feed in arid areas (Blümmel et al., 2010) and provides productive pasture for grazing, especially with dwarf varieties, and silage is easily made (Andrews & Kumar, 1992).

Nantoumé and Kouriba (2000) reported crude protein concentrations in pearl millet ranging from 30 to 50 g/kg dry matter. Millet ranks first regarding residue biomass, accounting for about half of the cereal residues used as fodder, followed by sorghum (Abdoulaye & Ly, 2014). Lo (2017) compared new millet varieties (i.e., Thialack 2 and SL169) and a relatively older variety (Souna 3) in Senegal. The results showed that Thialack 2 and SL169 had 44% and 39% more digestible dry matter than Souna 3 (37%).

In recent years, several high-yielding, dual-purpose millet varieties have been developed (ISRA, 2015; Kamara et al., 2017). These varieties offer both the ability to produce nutritious grain for human consumption and high-quality fodder for animal production because they remain green at maturity and thus have the potential to improve forage quality and quantity for regional livestock productivity (Anele et al., 2011), hence their designation as dual-purpose millet varieties. Omanya et al. (2007) indicated that the adoption rates of the "improved" cultivars of pearl millet in Senegal are low, and mostly traditional varieties (i.e., Souna 3) are planted.

In Senegal, dual-purpose millet varieties, which can produce greater nutritious grain yields for human consumption and high-quality fodder for animal feeding, include Thialack 2, SL 28, SL 169, and SL 423 (ISRA, 2015). With appropriate fertilizer application, fodder production from these dual-purpose millet varieties can range from 250 kg/ha (Palé et al., 2009) to 4,100 kg/ha (Faye et al., 2023; Watanabe et al., 2019). In evaluating the performance of millet genotypes in 14 environments in Burkina-Faso, Mali, Niger, and Senegal, Serba et al. (2020) showed that dual purpose millet could produce 4,600 kg/ha fodder.

Dual-purpose pearl millet varieties can be grown as a rainy season crop under rainfed conditions in SSA, in mono-cropping or intercropped systems, or in mixed cropping systems but also in rotation with legumes such as groundnut, cowpea, pigeon pea [*Cajanus cajan* (L.) Millsp.], mung bean [*Vigna radiata* (L.) R. Wilczek], mucuna (*Mucuna* sp.), and soybean [*Glycine max* (L.) Merr.]. In Namibia, Watanabe et al. (2019) reported 2,601 kg/ha grain and 4,396 kg/ha fodder yields in mono-cropping, whereas 2,463 kg/ha grain and 4,074 kg/ha fodder were recorded under the millet–cowpea intercropping system. Typically, pearl millet is planted at low density (<13,000 seeds hill per ha) in traditional farming systems, but the seeding rate can vary depending on the country and the cropping system. In monoculture, the spacing is narrower than the wider rows used in intercropping systems to allow for planting other crops. Previous studies (Siéné et al., 2010) reported that optimum plant density depends on environmental conditions (water supply and soil fertility), beyond which the competition between plants for light, water, and nutrients becomes important and can lead to decreased crop yields. However, Faye et al. (2023) reported that increased sowing density of millet not only increased the grain and biomass (fodder) yield but also improved the system resilience through better soil covering and improved water conservation due to reduced evaporation.

Previous research concluded that inappropriate fertilization was one of the most limiting factors affecting cereal crop yields in SSA (Igue et al., 2016; Saïdou et al., 2018). These result from inappropriate fertilizer application rates, soils with depleted soil organic matter content, low fertilizer use, and poor soil fertility management practices (Stewart et al., 2020). For example, in Senegal, the millet fertilizer recommendation of 150 kg/h N–P–K (15–15–15) as a single rate was developed over 45 years ago and is still used, regardless of soil conditions and climate variability (Diouf, 2008). Such a recommendation does not consider soil types and the specificity of farmers' cropping systems, which can affect both grain and biomass production (Figure 13.4). The use of agricultural byproducts, including biomass for fodder, consolidates the integration between crop and livestock, thus limiting the recurrent conflicts between livestock holders and crop farmers (Kaboré et al., 1999).

Pearl millet straw provides fibrous fodder, which is more assertive as the plant matures and is reflected in the nutritive value of the fodder. In addition to decreased energy and protein, intake and digestibility are reduced. In general, there is a negative relationship between the ingestion of fodder and the proportions of stems, lignified tissues, and membrane constituents, which increase with plant maturity (Mustapha, 1984). In a millet silage and hay study, Bolsen (1985) demonstrated a negative correlation between the age of the plant and the digestibility of the organic nutrients (Table 13.1).

The rations used consisted of 85.5% silage or hay, 4.5% cane molasses, and 10% supplement. These two millet varieties (Souna 3 and Sanio) have low nutrient digestibility (Annet, 1987) (Table 13.2). Except for crude fiber, the digestibility varied little between the two varieties. On the other hand, a very low digestibility of Souna 3 was noted as compared to Sanio.

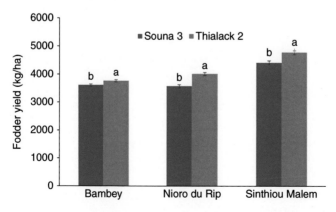

Figure 13.4 Comparison of traditional millet variety (Souna 3) and dual-purpose pearl millet variety (Thialack 2) in Senegal (Faye et al., 2023 / MDPI / CC-BY-4.0.).

Table 13.1   Nutrient Digestibility, VDMI, and ADG Using Silage and Millet Straw at 4 Weeks Growth (Adapted from Bolsen, 1985)

Characteristics	Mid-growth		Grain forming	
	Silage	Fodder	Silage	Fodder
Digestibility				
DMD	60.6	64.6	49.2	53.5
OM	63.8	66.8	49.3	54.2
CF	67.3	68.2	46.7	53.3
VDMI, kg DM/BW$^{0.75}$	1.04	1.3	0.88	0.95
ADG, kg DM	0.08	0.1	0.03	0.02

Abbreviations: ADG, average daily gain; BW, body weight; CF, crude fiber; DMD, dry matter digestibility; OM, organic matter; VDMI voluntary dry matter intake.

Table 13.2   Chemical Composition, Digestibility of the Different Nutrients, and VDMI of Souna 3 and Sanio Varieties for Fodder (Annet, 1987)

Characteristics	Souna 3 fodder	Sanio fodder
Age of the plant, days	55	55
Harvesting stage	Ear appearance	Mid-growth
Chemical composition, %		
DM	96.1	96.0
OM	91.0	89.0
CP	7.8	7.8
CF	37.1	37.5
Digestibility, %		
DMD	53.7	53.9
DMI	56.8	56.2
DCP	38.1	35.8
DCF	59.9	63.7
VDMI, kg DM/BW$^{0.75}$	47.5	69.7

Abbreviations: ADG, average daily gain; BW, body weight; CF, crude fiber; CP, crude protein; DCF, digestible crude fiber; DCP, digestible crude protein; DMD, dry matter digestibility; DMI, dry matter intake; OM, organic matter; VDMI, voluntary dry matter intake.

Biomass of Pearl Millet: Specific Case of the Cropping System in Senegal

As the staple food for the population in Senegal, millet is by far the most important cereal in terms of both sown area and production (ISRA, 2005). According to ANSD (2015), millet occupies 715,996 ha, or 63.6%, of the total cereal production area in the 2015 rainy season, with 408,993 tons, or 32.7%, of national cereal grain production. Two types of millet exist in Senegal: photoperiod-sensitive types called "Sanio," which have longer growth cycles (130–150 days) and are generally grown in the south of the country with >1,000 mm of rainfall. The non-photoperiod sensitive millets, "Souna" varieties, are grown in regions with <1,000 mm annual rainfall (Diourbel, Thiès, Kaolack, and Louga) and produce smaller grain yields. Senegal research programs have focused on the selection of cultivars capable of completing their growth cycles under growing conditions that have become extremely harsh, with shorter rainfall periods and higher temperatures. The characteristics sought are (a) shortened growth cycles and (b) the reduction of the straw/grain ratio. According to ISRA (2005), the first work began in 1931 to improve yields of traditional millet varieties, an early millet variety (Souna), and a late-maturing millet variety (Sanio). The recurrent selection applied by Etasse (1965) on three local populations of millet Souna from 1961 led successively to the creation of the synthetic variety Souna 2 in 1965 and the variety Souna 3 from crossing eight lines taken from PC 28 and PC 32 populations. The latter has a growth cycle of 90 days and can reach a height of 242 cm. In 2006, Thialack 2, a variety of genetic natures from a local Senegalese population, was developed at ISRA in Bambey, Senegal. This variety has a height of 250 cm and a growth cycle of 95 days. This plant breeding research work continues, with a collection of several lines, including improving local populations such as SL28, SL169, and SL423.

Use of Pearl Millet in Animal Production

Ganry and Feller (1977) showed that N fertilization improves forage production. A dose of 90 kg N/ha produces 9,606 kg of DM/ha and 5,365 kg of DM with 30 kg N/ha. This was confirmed by experiments carried out at ENSA's Centre d'Application des Techniques Agricoles in Senegal in 1986, which showed an increase in yield depending on the dose of N. Yields increased from 4,705 kg DM/ha with 50 kg N/ha to 6,229 kg DM/ha with 100 kg N/ha for the Souna 3 variety. Nevertheless, the use of crop residues, in this case millet stalks, is hampered by other uses, such as fuels, building materials (i.e., fences, concessions, etc.), and challenges related to the collection and transportation of large volumes of fodder from farms to residence where the fodder will be fed to livestock (FAO, 2004). On average, new dual-purpose millet varieties (i.e., SL28, SL169, and SL423) had similar nutritive values as other traditional varieties (Table 13.3), and the nutritive values of the lower parts of the stem were generally greater than those in upper parts in pearl millet (Lo, 2017). The SL423 dual-purpose pearl millet variety had the highest concentration of crude protein compared with other pearl millet varieties.

New Pearl Millet Varieties for Fodder Production Capacities and Quality

Evaluation of new dual-purpose pearl millet varieties fodder production in the Animal Production Department at the University of Thies in Senegal showed that SL423 line had the most biomass yield, with 7,403 kg DM/ha, followed by the Thialack 2 variety and the SL169 line, with respective fodder biomass yields of 5,214 and 5,214 kg DM/ha. The SL28 population and the Souna 3 variety produced the least forage biomass yields (305 and 3,045 kg DM/ha, respectively) (Figure 13.5).

Table 13.3 Chemical Composition of The Straw in The Upper and Lower Parts of New Dual-Purpose Pearl Millet Varieties (SL28, SL169, and SL423) as Compared to Souna 3, Thialack 2, and Average per Variety (Lo, 2017)

Varieties	Part of the stem	DM	OM	CF	CP	NDF	DAF	NFE
					–%–			
SL28	Upper	93.8	91.2	40.5	5.4	73.0	48.6	39.1
	Low	93.1	93.6	45.6	4.5	74.1	50.6	36.6
SL169	Upper	93.8	91.7	41.8	6.0	72.4	47.4	37.6
	Low	94.1	94.1	45.7	5.2	76.0	51.3	37.3
SL423	Upper	93.8	93.1	43.3	5.3	77.5	49.5	38.2
	Low	94.0	95.5	48.3	4.0	77.1	54.4	37.1
Souna 3	Upper	93.7	91.4	40.2	5.9	74.4	51.0	39.0
	Low	93.7	94.3	47.5	4.0	77.0	51.5	36.4
Thialack 2	Upper	93.6	93.2	40.2	6.7	76.7	48.1	39.8
	Low	93.6	94.1	45.7	4.6	77.6	53.1	37.4
Average	Upper	93.7	92.1	41.2	5.9	74.8	48.9	38.7
	Low	93.7	94.3	46.6	4.5	76.4	52.2	37.0

Abbreviations: ADF, detergent acid fiber; CF, crude fiber; CP, crude protein; DM, dry matter; NDF, neutral detergent fiber; NNE, non-nitrogenous extractive; OM, organic matter.

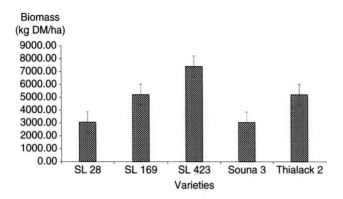

Figure 13.5 Forage biomass yields of two millet varieties (Souna 3 and Thialack 2) compared with those of dual-purpose pearl millet varieties (i.e., SL28, SL169, and SL423) (Lo, 2017).

However, it is worth noting that the fodder yield of the SL423 line was about 2.5 times that of the SL28 line and the Souna 3 variety.

Bastos et al. (2022) assessed the effect of pearl millet varieties (i.e., traditional vs. dual-purpose), plant density, and multiple fertilizer nutrient rates on pearl millet yields. The treatments were management practices including two plant densities (37,000 and 74,000 plants/ha) and 24 different levels of fertilizer nutrient combinations ranging from 0 to 149 kg/ha N (as urea), 0–67 kg/ha phosphorus (P_2O_5, as double super-phosphate), 0–33 kg/ha potassium (K_2O, as potassium chloride), and 0–14,987 kg/ha cow manure.

Pearl millet biomass varied from 137 to 12,414 kg/ha across all site-years and treatments and was affected by management, variety, and the interaction between site-year and variety. Average variety performance ranged from 1,914 to 6,040 kg/ha for Souna 3 and from 1,898 to 5,850 kg/ha for Thialack 2 across site-years (Figure 13.6). Thialack 2 yielded less than Souna 3 in two site-years (11% less), the same as Souna 3 in four site-years, and more than Souna 3 in three site-years (8%–19% more). Grain yield varied from 62 to 6,450 kg/ha across all site-years and treatments and was affected by management, variety, and the interactions between site-year and management and between site-year and variety. Average variety performance ranged from 928 to 2,334 kg/ha for Souna 3 and 1,039 to 2,616 kg/ha for Thialack 2 across different site-years (Figure 13.6). Thialack 2 yielded less than Souna 3 in one site-year (8% less), the same as Souna 3 in four site-years, and more than Souna 3 in four site-years (8%–135% more).

Pearl millet biomass yields was affected by management treatments, which varied from 2,685 (planted at 37,000 plants/ha) to 4,952 kg/ha (planted 74,000 plants/ha, 150 kg N/ha, 67 kg P_2O_5/ha, and 33 kg K_2O/ha). Similarly, grain yield varied from

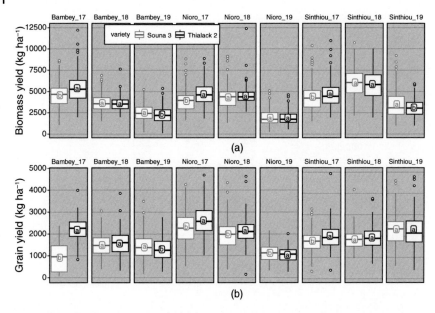

Figure 13.6 Biomass (a) and grain (b) yield for different varieties across nine site years. Boxplots portray the 25th (lower hinge), 50th (solid black line), and 75th (upper hinge) percentiles, largest value no greater than 1.5× interquartile range (lower whisker), smallest value at most 1.5× interquartile range (upper whisker), and outlying observations (points). Variety boxplots within a given site-year followed by the same letter are not significantly different at α = 0.05. BLUE, best linear unbiased estimator; Nioro, Nioro du Rip; Sinthiou, Sinthiou Maleme; 17, 2017; 18, 2018; 19, 2019 (Leonardo et al., 2022 / with permission of Elsevier).

1,183 (planted at 37,000 plants/ha) to 2,117 (planted 74,000 plants/ha, 150 kg N/ha, 67 kg P_2O_5/ha, and 33 kg K_2O/ha) kg/ha (Figure 13.7). Management treatments that optimized biomass yield included planting at 74,000 plants/ha and fertilizer rates ranging from 57 to 149 kg N/ha, 15 to 67 kg P_2O_5/ha, 23 to 33 kg K_2O/ha, and 12,494 kg manure/ha. Furthermore, treatments that optimized grain yield were planted at 74,000 plants/ha and had fertilizer nutrient rates of 23–149 kg N/ha, 15–67 kg P_2O_5/ha, 23–33 kg K_2O/ha, and about 5,000 kg manure/ha. Fertilizer application rates that optimized biomass yield were 102 kg N/ha, 37 kg P_2O_5, and 26 kg K_2O, while grain yield was optimized by applying 91 kg N/ha, 32 kg P_2O_5, and 24 kg K_2O.

Maman et al. (2017) conducted nutrient response functions for the pearl millet sole crop (PMSC) and cowpea–pearl millet intercrop systems. Their results showed that the mean fodder yield of intercrop pearl millet (PMI) was unaffected by N application rates, but PMSC yields increased with N application (Figure 13.8). Application of K had little effect on yields, except for a decline in PMSC fodder yield. Fodder yield for PMSC was unaffected by P application. Grain and fodder yield response to P was greater in Mali with PMI compared with PMSC.

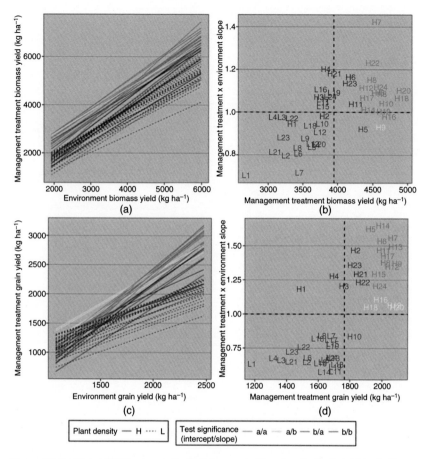

Figure 13.7 Finlay-Wilkinson results for biomass (a, b) and grain (c, d) yield, including management vs. environmental effect line plots (a, c) and management vs. management-environment slope plots (b, d). For all panels, color represents the combination of intercept and slope significance (e.g., a/a is used for treatments with significant greatest intercept and slope). On panels a and c, line type represents plant density level (H, high [74 × 1,000 plants/ha]; L, low [37 × 1,000 plants/ha]). On panels b and d, text labels represent the management (i.e., plant density-nutrient) treatment levels, the vertical black dashed line represents the average yield across all treatments, and the horizontal black dashed line represents the slope of unity (Leonardo et al., 2022 / with permission of Elsevier).

Suzuki et al. (2020) reported that the micro-dosing technique has been disseminated in the Sahel region of West Africa to ensure that farmers use affordable amounts of chemical fertilizer efficiently for cultivating pearl millet. Their findings supported the importance of continuous input of organic matter to keep the residual effect on grain yield and shoot dry weight of pearl millet (Figure 13.9). The positive effect of organic matter input on pearl millet growth and yield

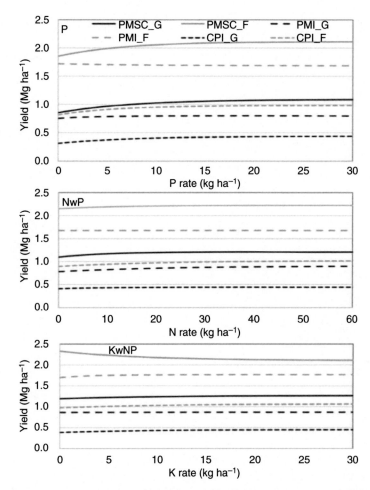

Figure 13.8 Mean grain (_G) and fodder (_F) response of pearl millet sole crop (PMSC) and of intercrop pearl millet (PMI) and cowpea (CPI) to applied P, N with a uniform P application (NwP), and K with uniform N and P applications (KwNP) determined from 12 trials conducted in the Sahel and Sudan Savanna are the same as Mali and Niger (Maman et al., 2017 / John Wiley & Sons).

continued in this 5-year study (Suzuki et al., 2020). Therefore, the continuous application of organic matter, even in small amounts, was a very important management practice to keep consistent levels of grain yield and shoot dry weight each year of the study. Sole fertilizer application by microdosing did not show a positive effect on pearl millet growth and yield compared with control (Figure 13.9). However, the application of organic matter sole and co-application

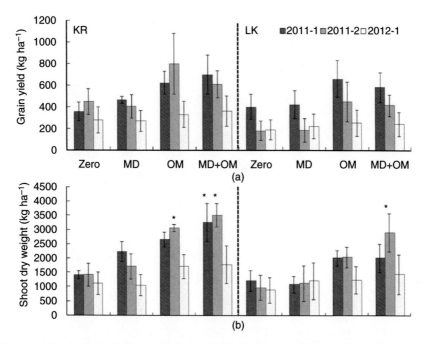

Figure 13.9 Grain yield (a) and shoot dry weight (b) in pearl millet under the different soil management. 2011–2021, the first cultivation in 2011; 2011–2022, the second cultivation from 2011; 2012–2021, the first cultivation in 2012. Asterisks denote significant different at $P < 0.05$ with control (Zero) by LSD. Vertical bar is standard error (Suzuki et al., 2020). JR, Karé village; LK, Lontia Kaina village; MD, micro-dosing of fertilizer; OM, organic matter; MD + OM, co-application of micro-dosing of fertilizer and organic matter; Zero, zero application.

of fertilizer using micro-dosing and organic matter had a high potential to enhance the shoot dry matter and grain yield of pearl millet by improving soil fertility levels and nutrient uptake efficiency.

Lo (2017) compared the digestibility of the lower and upper portion of the stems of three potentially dual-use millet populations (SL28, SL169, and SL423) and those of the Souna 3 and Thialack 2 varieties in Peulh-Peulh sheep. The comparison of the voluntary intake of dry matter of 10 different rations from the two varieties and the three populations shows a highly significant difference between the parts of the stems of millet. The highest voluntary feed intake was given by the SL423 line (42.5 g DM/kg BW$^{0.75}$), followed by the SL169 line (42.00 g DM/kg BW$^{0.75}$). However, even if the upper part of the SL423 line had the highest ingestion, it was the lower portion of this line that had the least ingestion (27.02 g DM/kg BW$^{0.75}$). The lower parts of the three populations had intermediate ingestion values and the Souna 3 variety regardless of the plant part considered (Figure 13.10).

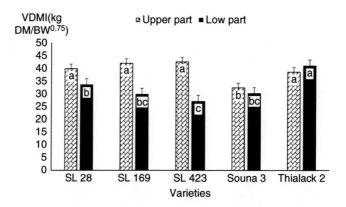

Figure 13.10 Voluntary feed intake of the upper and lower parts of the traditional varieties (Souna 3 and Thialack 2) compared with those of the dual-purpose pearl millet populations (i.e., SL28, SL169, and SL423). The means assigned the same letters are not significantly different at the 5% level.

Regarding the forage bulk values, statistical analysis showed highly significant differences ($P < 0.001$) in the forage bulk values of feed ration prepared from pearl millet varieties and different portions of the stem (Figure 13.11). Contrary to feed intake, the greatest forage bulk values were obtained with the lower parts of the pearl millet stems except those of the Thialack 2 variety, which had intermediate values in all its respective lower and upper parts (1.59 and 1.70, respectively). The highest crowding value was observed with line SL423 (2.94), followed by variety Souna 3 (2.02 and 2.23 for upper and lower parts, respectively) and line

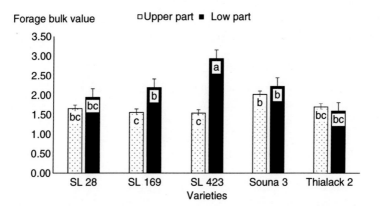

Figure 13.11 Forage bulk values of the upper and lower parts of the millet varieties (i.e., Souna 3 and Thialack 2) compared with those of the three potentially dual-purpose populations (i.e., SL28, SL169, and SL423) (Lo, 2017).

SL169 (2.20). Forage from the upper parts of the SL423 and SL169 populations showed the lowest forage bulk values, 1.54 and 1.56, respectively.

Bhattarai et al. (2019) showed that the nutrient composition of forage pearl millet was comparable to that of forage sorghum and corn harvested for silage, except that corn had greater digestibility than forage sorghum or pearl millet (Table 13.4). The crude protein, neutral detergent fiber, and acid detergent fiber contents were lower in corn than the forage pearl millet and sorghum. In addition, the digestibility of corn was greater than that of forage pearl millet and sorghum. A similar result was reported in the United States that in vitro dry matter digestibility for pearl millet cultivars was comparable with that for maize but higher than sorghum (Ejeta et al., 1987). It was reported that sorghum and pearl millet were likely to replace much of the corn silage crop.

Nutrient Digestibility

Dry Matter Digestibility

In Senegal, Lo (2017) analyzed data on the dry matter digestibility (DMD) of the nutrients of 10 feed rations from pearl millet (Figure 13.12) and showed significant differences ($P = 0.002$). Dry matter of the upper parts shows more digestibility than the lower parts regardless of the pearl millet varieties studied, except for Souna 3 (38.97% against 36.53% for the lower and upper parts, respectively). The best DMD was measured in the upper part of the stem of the pearl millet variety Thialack 2 (44.26% and 41.46% for the respective upper and lower parts), followed by the dual-purpose pearl millet populations like SL423, SL28, and SL169 (43.18%, 40.93%, and 40.89%, respectively). The lowest digestibility was measured in the lower parts of the SL423 population (35.79%). The other pearl millet populations had intermediate values for these same parts considered.

Table 13.4 Nutrient Composition (% of Dry Matter) of Forage Pearl Millet Relative to Forage Sorghum and Corn Harvested for Silage (Bhattarai et al., 2019)

Nutrient composition	Forage pearl millet	Corn	Forage sorghum
		–%–	
NDF	58.0–68.5	54.0	60.5
ADF	31.3–42.5	29.5	37.9
Lignin	3.5–6.3	4.9	8.3
CP	9.0–18.0	8.6	13.3
Digestibility	64.0–69.0	73.0	63.0

Abbreviations: ADF, acid detergent fiber; CP, crude protein; NDF, neutral detergent fiber.

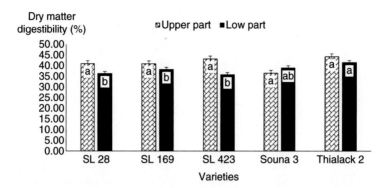

Figure 13.12 Digestibility of the dry matter of fodder consisting of the upper and lower parts of the traditional millet varieties (i.e., Souna 3 and Thialack 2) compared with that of the dual-purpose pearl millet populations (i.e., SL28, SL169, and SL423) (Lo, 2017).

In the same study, the analysis of the results used indicated significant differences ($P = 0.017$) in organic matter digestibility (OMD). Indeed, except for the Souna 3 variety, OMD was better with the upper parts of the plants (Figure 13.13). The highest OMD was obtained with Thialack 2 (47.32% and 44.97% for the upper and lower parts, respectively). The SL28, SL169, and SL423 populations had intermediate OMD values (45.93%, 45.65%, and 43.70%, respectively). However, the lowest OMD value was measured in the upper parts of Souna 3 (38.25%), followed by SL28 and SL423 lines (40.50% and 40.97%, respectively).

Regarding the digestibility of total nitrogenous matter (DTNM), the author found no significant differences ($P = 0.395$) between the pearl millet varieties and the different plant parts studied (Figure 13.14). The DTNM was relatively more important in the upper parts of all the varieties (~35.85% in the upper stem compared with 33.92% in the lower part).

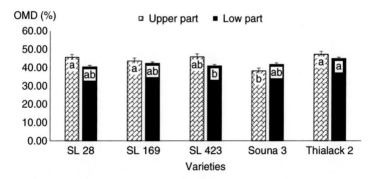

Figure 13.13 Digestibility of organic matter in fodder consisting of the upper and lower parts of traditional pearl millet varieties (i.e., Souna 3 and Thialack 2) compared with that of the dual-purpose pearl millet populations (i.e., SL28, SL169, and SL423) (Lo, 2017).

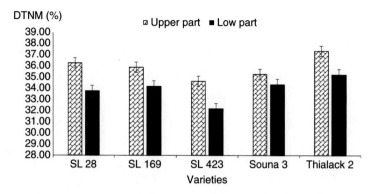

Figure 13.14 Digestibility of total nitrogenous matter in fodder consisting of the upper and lower parts of millet varieties (Souna 3 and Thialack 2) compared with that of the dual-purpose pearl millet populations (SL28, SL169, and SL423) (Lo, 2017).

However, highly significant differences were observed in the digestibility of crude fiber (CF) of different pearl millet feeds ($P = 0.001$) (Figure 13.15). In the upper parts, the digestibility of CF of the SL28 line was higher (48.64%). It was followed by SL423 (43.06%), SL169 (42.17%), and Thialack 2 (40.93%). The Souna 3 variety had the least CF digestibility (CFD) (31.55%). Regarding the lower parts, the SL423 and SL169 populations had similar CFD to that of the Souna 3 and Thialack 2, with an average of 42.58%. The SL28 population had the best CFD for the upper parts as compared to other varieties.

The digestibility of N-free extracts (NFEs) was significantly different ($P = 0.016$) among different the millet feed rations (Figure 13.16). Except for the Souna 3

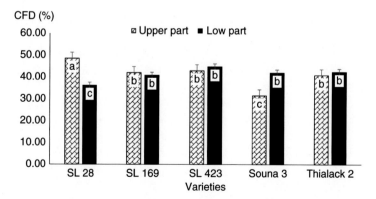

Figure 13.15 Crude fiber digestibility (CFD) of fodder consisting of the upper and lower parts of pearl millet varieties (Souna 3 and Thialack 2) compared with those of the dual-purpose pearl millet populations (SL28, SL169, and SL423) (Lo, 2017).

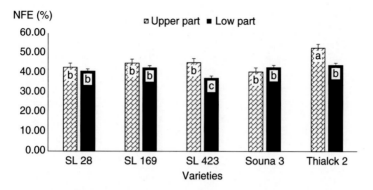

Figure 13.16 Digestibility of non-nitrogenous extracts of fodder consisting of the upper and lower parts of the traditional pearl millet varieties (Souna 3 and Thialack 2) compared with that of the dual-purpose pearl millet populations (SL28, SL169, and SL423) (Lo, 2017).

variety, the NFE of the upper parts was greater than that of the lower parts in all the different rations. The NFE concentration was greater with the Thialack 2 variety (52.64% and 43.85% for the upper and lower parts, respectively) than that of lines SL423 (45.19%), SL169 (44.88%), SL28 (42.73%), and Souna 3 (40.50%). On the lower parts, Souna 3 and the SL169 and Thialack 2 had the highest NFE concentrations (42.73% and 42.49%), while the lowest NFE was observed in the line SL423 (37.24%).

Nutritional Value of Senegalese Pearl Millet Varieties Depending on the Part of the Stem—Energy Values

The energy contents of the upper parts of the Thialack 2 variety and the three dual-purpose populations of pearl millet were greater than their lower parts. For the upper parts of the millet stalks, the Thialack 2 variety had the best milk fodder unit (MFU) (0.53 MFU/kg DM), followed by SL28 and SL423 lines, which had similar energy values of about 0.51 MFU/kg MS (Figure 13.17). The MFU content of SL169 line was 0.48 MFU/kg DM. On the other hand, Souna 3 had greater MFU content in its lower parts (0.43 against 0.41 MFU/kg DM). However, regarding the lower parts, Thialack 2 also offered the best energy content (0.50 MFU/kg DM), followed by the SL169 line (0.46 MFU/kg DM). Lines SL423 and SL28 had similar contents at 0.44 MFU/kg DM. In parallel with the MFU unit, results of the meat fodder units (MtFU) showed similar trends varying from 0.29 to 0.42 MtFU/kg DM and from 0.31 to 0.38 MtFU/kg DM for the upper and lower parts, respectively (Figure 13.18).

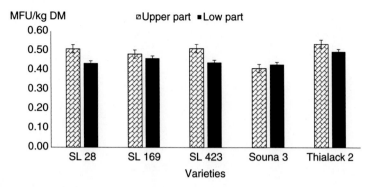

Figure 13.17 Milk fodder unit (MFU) values of the upper and lower parts of the traditional millet varieties (Souna 3 and Thialack 2) compared with those of the dual-purpose pearl millet populations (SL28, SL169, and SL423) (Lo, 2017).

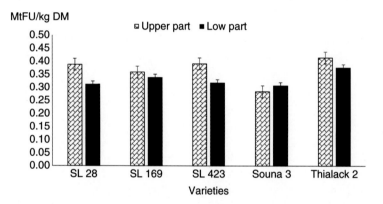

Figure 13.18 Values of the meat fodder units (MtFUs) of the upper and lower parts of the traditional pearl millet varieties (Souna 3 and Thialack 2) compared with those of the dual-purpose pearl millet populations (SL28, SL169, and SL423) (Lo, 2017).

Protein Concentration

The evaluation of protein content was limited to the calculation of digestible nitrogenous matter (DNM) of the millet varieties (Figure 13.19). The general finding was that the upper parts were richer in DNM than the lower parts regardless of pearl millet variety. For the upper parts, the DNM contents were greater with the Thialack 2 variety (25.17 g DNM/kg DM), whereas the SL423 line had the lowest DNM concentration (18.36 g DNM/kg DM). Lines SL169 and SL28 had intermediate concentrations with 21.60 and 19.51 g DNM/kg DM, respectively. In the lower parts, the greatest DNM concentration was noted with the SL169 line (17.77 g DNM/kg DM). The Thialack 2 variety and the SL28 line had intermediate

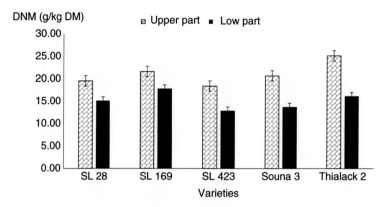

Figure 13.19 Digestible N matter of the upper and lower parts of the traditional pearl millet varieties (Souna 3 and Thialack 2) compared with those of dual-purpose pearl millet populations (SL28, SL169, and SL423) (Lo, 2017).

concentrations with 16.13 and 15.09 g DNM/kg DM, respectively. The lowest DNM contents were obtained with the Souna 3 and SL423 varieties (13.73 and 12.89 g DNM/kg DM, respectively).

Conclusions

This chapter synthesized published and current research data on the biomass production, quality, and trade-offs among biomass uses as well as the opportunities and challenges in pearl millet fodder production in SSA WA. The review showed that estimating the available biomass production of cereals, including pearl millet straw, was difficult because not much research is available on pearl millet for fodder use in WA. It is clear that pearl millet is a climate-resilient crop and has great potential as a good fodder resource for livestock in WA. Current dual-purpose pearl millet varieties with "stay green" traits may improve fodder quality and quantity to feed livestock while improving soil fertility through millet roots left in the field and providing more quality grain production for human consumption. Optimal planting densities must also be determined to maximize the yields of dual-purpose pearl millet varieties in different agroecological zones in SSA. It is expected that wide adoption of current dual-purpose pearl millet varieties in monocropping, in intercropping, or in rotation with legumes and with adequate fertilizer application can significantly enhance biomass production and quality for livestock feeds. More research on soil fertility, plant density, cropping systems, timing, harvesting methods, improvement of nutrient bioavailability, biomass chopping at right size, and storage will be required to optimize their efficient use for livestock feeds. Such studies could significantly contribute to sustainable food/feed and nutrition security for humans and livestock with the benefits of system resilience under climate change and variability in the West African region.

References

Abdoulaye, S. G., & Ly, C. (2014). *Crop residues and agro-industrial byproducts in West Africa situation and way forward for livestock production.* FAO.

Aboh, A. B., Babatounde, S., Oumorou, M., Houinato, M., & Sinsin, B. (2012). Valeur pastorale des parcours naturels en zone soudano-guinéenne et stratégie paysanne d'adaptation aux effets de leur invasion par Chromolaena odorata au Bénin. *International Journal of Biological and Chemical Sciences, 6*(4), 1633–1646.

Allard, S., Berthieau, Y., Drevon, J., Seze, O., & Ganry, F. (1983). Ressources en résidus de récolte et potentialités pour le biogaz au Sénégal. *L'Agronomie Tropicale, 1383*, 213–221.

Amole, T., Augustine, A., Balehegn, M., & Adesogoan, A. T. (2022). Livestock feed resources in the West African Sahel. *Agronomy Journal, 114*, 26–45. http://dx.doi .org/10.1002/agj2.20955

Andrews, D. J., & Kumar, K. A. (1992). Pearl millet for food, feed, and forage. *Advances in Agronomy, 48*, 89–139. http://dx.doi.org/10.1016/ S0065-2113(08)60936-0

Anele, Y. U., Südekum, K. H., Arigbede, M. O., Welp, G., Oni, A. O., Olanite, A. J., & Olubunmi, V. O. (2011). Agronomic performance and nutritive quality of some commercial and improved dual-purpose cowpea (*Vigna unguiculata* L. Walp) varieties on marginal land in Southwest Nigeria. *Grassland Science, 57*, 211–218. http://dx.doi.org/10.1111/j.1744-697X.2011.00229.x

Annet, P. (1987). Etude des divers fourrages tropicaux chez les moutons Peul-peul (Sénégal) - Digestibilité – Ingestibilité – Teneur en lignine. In *Mémoire d'ingénieur de fin d'études non publié.* Supérieure d'Agriculture de Thiès.

ANSD S. (2015). Situation économique et social du Sénégal. Agence Nationale de la Statistique et de la Démographie. https://www.ansd.sn/ressources/ses/SES.

Bastos, L. M., Aliou Faye, Zachary P. Stewart, Tobi Moriaque Akplo, Doohong Min, P. V. Vara Prasad, Ignacio A. Ciampitti (2022). Variety and management selection to optimize pearl millet yield and profit in Senegal. European Journal of Agronomy 139 (2022) 126565. https://url.us.m.mimecastprotect.com/s/ CNdZCVOrRVuEy6kfz9g7l. https://doi.org/10.1016/j.eja.2022.126565

Bationo, A., & Mokwunye, A. U. (1991). Role of manures and crop residue in alleviating soil fertility constraints to crop production: With special reference to the Sahelian and Sudanian zones of West Africa. *Fertilizer Research, 29*(1), 117–125. http://dx.doi.org/10.1007/BF01048993

Bayer W. et Waters-Bayer A., (1999). Gestion et amélioration des fourrages pastoraux dans les régions arides d'Afrique. *SCOONES, I. Les nouvelles orientations du développement pastoral en Afrique.* pp. 111–148. Allemagne, Editions GTZ.

Bessin, R. (1982). *Traitement des pailles et utilisation en alimentation animale: Essai de mise au point d'une ration d'embouche* [Doctoral dissertation]. Ecole Inter-Etats des Sciences et Médicine Vétérinaires.

Bhattarai, B., Singh, S., West, C. P., & Saini, R. (2019). Forage potential of pearl millet and forage Sorghum alternatives to corn under the water-limiting conditions of the Texas High Plains: A review. *Forage & Grazinglands*, *5*(1), 1–12. http://dx.doi.org/10.2134/cftm2019.08.0058

Blümmel, M., Khan, A. A., Vadez, V., Hash, C. T., & Rai, K. N. (2010). Variability in stover quality traits in commercial hybrids of pearl millet (*Pennisetum glaucum* (L.) R. Br.) and grain-stover trait relationships. *Animal Nutrition and Feed Technology*, *10*, 29–38.

Bolsen, A. (1985). *Effect of stage of maturity on feeding value of pearl millet silage or hay.* Kansas State University.

Boote, K. J., Adesogan, A. T., Balehegn, M., Duncan, A., Muir, J. P., Dubeux, J. C., Jr., & Rios, E. F. (2022). Fodder development in sub-Saharan Africa: An introduction. *Agronomy Journal*, *114*(1), 1–7.

Boudet, A. M., Lécussant, R., & Boudet, A. (1975). Mise en évidence et propriétés de deux formes de la 5-déshydroquinate hydro-lyase chez les végétaux supérieurs. *Planta*, *124*, 67–75.

Boudet, G. (1983). Systèmes de production d'élevage au Sénégal, le couvert végétal et le cheptel [Synthèse des comptes rendus de fin d'étude du Groupe d'études et de Recherches pour le Développement de l'Agriculture Tropicale]. (https://agritrop.cirad.fr/370554/)

Calvet, H. (1973). Les aliments actuellement utilisables en embouche au Sénégal. *Revue d'élevage et de médecine vétérinaire des pays tropicaux*, *26*(spécial), 53–56.

Camara, A. (2007). *Diagnostic fourrager pour une amélioration des productions animales dans le Bassin Arachidier du Sénégal: Cas du Niakhar. Mémoire de Fin d'Etudes Approfondies de Productions Animales.*

Cissé, M., Fall, A., Sow, A. M., Gongnet, P., & Korrea, A. (1996). Effet du traitement de la paille de brousse à l'urée et de la complémentation sur la consommation de paille, le poids vif et la note d'état corporel des ovins sahéliens en saison sèche. *Annales de Zootechnie*, *45*, 124–124.

CNMDE. (2015). *Plan d'orientation stratégique 2008–2015 pour un développement durable des exploitations familiales d'élevage provisoire.* Le Conseil National de la Maison des Eleveurs du Sénégal.

Dieng, A. (1984). *Utilisation des sous-produits agricoles et agro-industriels disponibles le long du fleuve Sénégal.* Universite Cheikh Anta Diop de Dakar.

Diouf, A. (2008). *Interaction Société, Nature et climat au Sahel: La rupture socioéconomique et écologique au Centre-Est agrosylvopastoral sénégalais au XXe siècle.* Université de Laval.

Diouf, O. (2001). *La cullture du mil (Pennisetum glaucum (L.) R. Br.) en zone siemi-aride: bases agrophysiologiques justificatives d'une fertilisation azotée. Memoire de titularisation.* ISRA.

Ditaroh, D. (1993). *Valorisation des résidus de récolte et des sous produits agro industriels pour la production de viande au Sénégal* [Doctoral dissertation]. *Ecole Inter-Etats des Sciences et Médicine.* Vétérinaires

Eeswaran, R., Nejadhashemi, A. P., Faye, A., Min, D., Prasad, P. V. V., & Ciampitti, I. A. (2022). Current and future challenges and opportunities for livestock farming in West Africa: Perspectives from the case of Senegal. *Agronomy, 12,* 1818. https://doi.org/10.3390/agronomy12081818

Ejeta, G., Hassen, M. M., & Mertz, E. T. (1987). In vitro digestibility and amino acid composition of pearl millet (*Pennisetum typhoides*) and other cereals. *Proceedings of the National Academy of Sciences of the United States of America, 84,* 6016–6019. http://dx.doi.org/10.1073/pnas.84.17.6016

Etasse, C. (1965). Amélioration du mil pennisetum, au Senegal. *L'agronomie Tropicale, 20,* 976–980.

FAO. (2004). *Conservation du foin et de la paille pour les petits paysans et les pasteurs.* FAO.

Faye, A., Akplo, T. M., Stewart, Z. P., Min, D. H., Obour, A., Assefa, Y., & Vara Prasad, P. V. (2023). Increasing millet planting density with appropriate fertilizer to enhance productivity and system resilience in Senegal. *Sustainability, 15,* 4093. http://dx.doi.org/10.3390/su15054093

Faye, A., Stewart, Z.P., Diome, K., Edward, C.-T., Fall, D., Ganyo, D.K.K., Akplo, T.M., Prasad, P.V.V (2021). Single Application of Biochar Increases Fertilizer Efficiency, C Sequestration, and pH over the Long-Term in Sandy Soils of Senegal. Sustainability 2021, 13, 11817. https://url.us.m.mimecastprotect.com/s/mjr7CR6nNVcpW10i98WWX. https://doi.org/10.3390/su132111817

Ganry, F., & Feller, C. (1977). *Effet de la fertilisation azotée (urée) et de l'amendement organique (compost) sur la productivité du sol et la stabilisation de la matière organique, en monoculture de Mil dans les conditions des zones tropicales semi-arides.* Seminaire Régional Sur Le Recyclage Organique En Agriculture.

Holness, D. H. (1999). *Pigs. The tropical agriculturalist.* Macmillan.

Ickowicz, A. (2001). Potentiel, Sénégal Limites, Forêts soudaniennes et alimentation des bovins au Sénégal. *Bois et Forêts des Tropiques, 270,* 47–61.

Igue, M. A., Oga, A. C., Balogoun, I., Saidou, A., Ezui, G., Youl, S., Kpagbin, G., Mando, A., & Sogbedji, J. M. (2016). Détermination des formules d'engrais minéraux et organiques sur deux types de sols pour une meilleure productivité de maïs (*Zea mays* L.) dans la commune de banikoara (Nord-Est Du Bénin). *European Scientific Journal, 12*(30), 362. http://dx.doi.org/10.19044/esj.2016.v12n30p362

ISRA. (2015). *Fiches variétés Niébé et Sorgho au Sénégal*. Institut Sénégalais de Recherches Agricoles.

ISRA, GRET, CIRAD, & ITA. (2005). *Bilan de la recherche agricole et agroalimentaire au Sénégal*.

Kamara, A. Y., Ewansiha, S., Ajeigbe, H. A., Omoigui, L., Tofa, A. I., & Karim, K. Y. (2017). Agronomic evaluation of cowpea cultivars developed for the West African Savannas. *Legume Research, 40*(4), 669–676. http://dx.doi.org/10.18805/lr.v0i0.8410

KANE, B. et NDIAYE, M. (1982). Utilisation des sous-produits agricoles et agro-industriels dans l'alimentation du bétail au Sénégal. Fao Production et Santé animale (32) P, 81.

Kima, S. A. (2008). Valorisation des gousses depiliostigma thonmngll (schum.) en production animale et étude de l'infestation par des insectes Mémoire, Université Polytechnique de Bobo-Dioulasso/ Institut du Développement Rural. https://beep .ird.fr/collect/upb/index/assoc/IDR-2008-KIM-VAL/IDR-2008-KIM-VAL.pdf (consulté le 16/02/2017).

Lhoste, P. (1987). *Etude de l'élevage dans le développement des zones cotonnières (Burkina Faso, Côte d'Ivoire, Mali). Elevage et relations agriculture-élevage en zone cotonnière. Situation et perspectives*. CIRAD-IEMVT.

Lo, T. (2017). *Etude de la digestibilité des parties inférieures et supérieures de tiges de trois populations de mil potentiellement à double usage (SL28, SL169 et SL423) en comparaison à celles des variétés Souna 3 et Thialack 2 chez le mouton Peulh-Peulh* [Master's thesis]; Ecole Nationale Supérieure d'Agriculrure (ENSA), University Iba Der Thiam, Thies, Senegal.

Ly, C. (1981). *L'utilisation et le potentiel en alimentation animale des résidus et sous produits agricoles au Sine-Saloum (Sénégal) – Essai d'élaboration d'une méthode d'enquêt* [Doctoral dissertation]. EISMV.

Lynda, Y. H. (2016). Le déficit fourrager en zone semi-aride: une contrainte récurrente au développement durable de l'élevage des ruminants. *Revue Agriculture, 1*, 43–51.

Maman, N., Dicko, M., Abdou, G., Kouyate, Z., & Wortmann, C. (2017). Pearl millet and cowpea intercrop response to applied nutrients in West Africa. *Agronomy Journal, 109*, 2333–2342. http://dx.doi.org/10.2134/agronj2017.03.0139

Masikati, P. (2011). *Improving the water productivity of integrated crop-livestock systems in the semi-arid tropics of Zimbabwe: An ex-ante analysis using simulation modeling* [Doctoral dissertation]. *Zentrum für Entwicklungsforschung*.

Mathur, P. N. (2012). *Global strategy for the ex situ conservation of pearl millet*. Global Crop Diversity Trust. https://www.croptrust.org/fileadmin/uploads/croptrust/Documents/Ex_Situ_Crop_Conservation_Strategies/Pearl-Millet-Strategy-FINAL-14May2012.pdf

Melesse, M. B., Tirra, A. N., Ojiewo, C. O., & Hauser, M. (2021). Understanding farmers' trait preferences for dual-purpose crops to improve mixed crop–livestock systems in Zimbabwe. *Sustainability*, *13*(10), 5678. http://dx.doi.org/10.3390/su13105678

Mongondin, B., & Tacher, G. (1979). *Les sous-produits agro-industriels utilisables dans l'alimentation animale au Sénégal*. IEMVT/Ministère de la Coopération.

Mustapha, M. I. (1984). *Intake an utilisation of and millet forages by sheep*. Colorando University.

Nanema, T. C. B., Pokou, K., & Zahonogo, P. (1998). The determinants of household fertility decisions in Burkina Faso. *Research square*. https://doi.org/10.21203/rs.3.rs-4159514/v1

Nantoumé, H., & Kouriba, A. (2000). Mesure de la valeur alimentaire des fourrages et des sous-produits utilisés dans l'a limentation des petits ruminants. *Revue D'élevage et de Médecine Vétérinaire des Pays Tropicaux*, *53*, 279–284.

Omanya, G. O., Weltzien-Rattunde, E., Sogodogo, D., Sanogo, M., Hanssens, N., Guero, Y., & Zangre, R. (2007). Participatory varietal selection with improved pearl millet in West Africa. *Experimental Agriculture*, *43*(1), 5–19. http://dx.doi.org/10.1017/S0014479706004248

ORSTOM. (1984). Rapport de l'inventaire des cultivars de mil en Afrique. Office de Recherches Scientifiques des Territoires d'Outre-Mer. *C.Agr. Pays Chauds 1968, n° 18097/174 code B*

Palé, S., Mason, S. C., & Taonda, S. J. B. (2009). Water and fertilizer influence on yield of grain sorghum varieties produced in Burkina Faso. *South African Journal of Plant and Soil*, *26*(2), 91–97.

Rivière, R. (1978). *Manuel d'alimentation des ruminants domestiques en milieu tropical*. Ministère de la Coopération.

Rufino, M. C., Dury, J., Tittonell, P., Van Wijk, M. T., Herrero, M., Zingore, S., & Giller, K. E. (2011). Competing use of organic resources, village-level interactions between farm types and climate variability in a communal area of NE Zimbabwe. *Agricultural Systems*, *104*(2), 175–190. http://dx.doi.org/10.1016/j.agsy.2010.06.001

Rufino, M. C., Thornton, P. K., Mutie, I., Jones, P. G., Van Wijk, M. T., & Herrero, M. (2013). Transitions in agro-pastoralist systems of East Africa: Impacts on food security and poverty. *Agriculture, Ecosystems & Environment*, *179*, 215–230.

Saïdou, A., Balogoun, I., Ahoton, E. L., Igué, A. M., Youl, S., Ezui, G., & Mando, A. (2018). Fertilizer recommendations for maize production in the South Sudan and Sudano-Guinean zones of Benin. *Nutrient Cycling in Agroecosystems*, *110*(3), 361–373. http://dx.doi.org/10.1007/s10705-017-9902-6

Serba, D. D., Sy, O., Sanogo, M. D., Issaka, A., Ouedraogo, M., Ango, I. K., Drabo, I., & Kanfany, G. M. O. (2020). Performance of dual-purpose pearl millet genotypes in West Africa: Importance of morphology and phenology. *African Crop Science Journal*, *28*(4), 481–498. http://dx.doi.org/10.4314/acsj.v28i4.1

Siene, L. A.C., Muller, B., & Ake, S. (2010). Etude du développement et de la répartition de la biomasse chez deux variétés de mil de longueur de cycle différente sous trois densités de semis. *Journal of Applied Biosciences, 35,* 2260–2278.

Sow, F., & Faye, A. (2023). Concevoir et tester des rations alimentaires à base de ressources localement disponibles pour la production laitière et de viande en saison sèche chez les ruminants; Annual report AICCRA project 2023.

Stewart, Z. P., Pierzynski, G. M., Middendorf, B. J., & Prasad, P. V. V. (2020). Approaches to improve soil fertility in sub-Saharan Africa. *Journal of Experimental Botany, 71,* 632–641. http://dx.doi.org/10.1093/jxb/erz446

Suzuki, K., Okada, K., & Higashimaki, T. (2020). Increasing pearl millet [*Pennisetum glaucum* (L.)] productivity by combining micro-dosing of chemical fertilizer and local organic matters in the Sahel regions. *Niger. Tropical Agriculture and Develpment, 64*(4), 189–200. http://dx.doi.org/10.11248/jsta.64.189

Tialla, D. (2011). *Evaluation des approches d'insémination artificielle sur chaleurs naturelles dans les petits élevages bovins traditionnels de la région de Kaolack au Sénégal* [Doctoral dissertation]. Université Cheikh Anta Diop.

Tui, S. H. K., Onofre Rainde, J., van Rooyen, A., Hauser, M., Siziba, S., Rodriguez, D., & Mazuze, F. (2014). Towards resilient and profitable family farming systems in Central Mozambique. In *RUFORUM Fourth Biennial Conference, Maputo, Mozambique, 19–25 July 2014* (pp. 159–163). Regional Universities Forum for Capacity Building in Agriculture.

Watanabe, Y., Itanna, F., Izumi, Y., Awala, S. K., Fujioka, Y., Tsuchiya, K., & Iijima, M. (2019). Cattle manure and intercropping effects on soil properties and growth and yield of pearl millet and cowpea in Namibia. *Journal of Crop Improvement, 33*(3), 395–409. http://dx.doi.org/10.1080/15427528.2019.1604456

Zoungrana, B. (2010). *Etude de la production de la composition chimique et de la digestibilité de légumineuses fourragères chez les ovins au Burkina Faso.* Université Polytechnique de Bobo-Dioulasso.

14

Pearl Millet: Marketing and Innovation Hubs

*B. Dayakar Rao, Veeresh S. Wali, Shreeja Kulla,
and C. Tara Satyavathi*

Chapter Overview

India is the largest grower and producer of millets in the world, with pearl millet [*Pennisetum glaucum* (L.) R. Br.] being the most important crop in the country. However, there has been a steady decline in millet cultivation in India since 1950, with sorghum being the most affected crop (FAO, 2022). Millets are regionally important in India, and their cultivation is dependent on climatic conditions and food habits, with states like Maharashtra, Karnataka, Rajasthan, Uttar Pradesh, Chhattisgarh, and Madhya Pradesh having significant shares in the production of sorghum, pearl millet, finger millet, and small millets (NAAS, 2022). Millet production in India increased consistently until the 1990s, with a 1.4 times increase in production and a similar increase in productivity; however, millet production witnessed a drastic change after the 1990s, with a decline in area and a 20% decrease in production despite a 73% increase in productivity from 2010 to 2020 compared with 1980–1990.

Millet production has significantly increased due to the adoption of improved cultivars and crop management technologies, but production trends contrast with those of wheat, rice, and maize, possibly because of better technology, extensive cultivation, and policy support (Yadav et al., 2019). India faced a challenge of feeding its fast-rising population in the early 1960s, with a deficit of 10 million t of food grain, necessitating imports. Urgency among stakeholders to enhance food grain production was strongly supported by political will. Technology and policy have influenced the demand and supply of food crops over the last six decades. The increase in maize production is a notable success story achieved through improved productivity and increased cultivation (Swaminathan, 2013, 2016). Despite impressive gains in pearl millet productivity, lack of policy and market-driven demands have led to a decline in cultivation. However, a twofold increase in production was solely technology driven. Sorghum cultivation has decreased due to low productivity and a lack of

Pearl Millet: A Resilient Cereal Crop for Food, Nutrition, and Climate Security, First Edition.
Edited by Ramasamy Perumal, P. V. Vara Prasad, C. Tara Satyavathi, Mahalingam Govindaraj, and Abdou Tenkouano.
© 2025 American Society of Agronomy, Inc., Crop Science Society of America, Inc., Soil Science Society of America, Inc. Published 2025 by John Wiley & Sons, Inc.

demand for kharif produce, resulting in a decline in production and area. However, rabi sorghum is still grown in Maharashtra, Karnataka, and Telangana, and there is a need to strengthen the breeding program for sorghum. The demand for finger millet has remained largely the same since the mid-1960s due to a twofold increase in productivity, whereas the demand for small millets and sorghum has drastically decreased since the 1960s.

Pearl millet is a C4 plant with a high photosynthetic efficiency and the capacity for dry matter production. It is typically produced in the most arid agro-climatic conditions, where other crops such as sorghum and maize fail to generate economically viable yields. Furthermore, because of its short developmental stages and rapid growth rate, pearl millet has a remarkable ability to respond to favorable circumstances, making it an outstanding crop for short growing seasons under enhanced crop management. It is a significant grain in the world, ranking sixth in production after maize, wheat, rice, barley, and sorghum. African countries are the world's largest producers of pearl millet. Pearl millet is grown on over 30 million acres throughout five continents, including Asia, Africa, North America, South America, and Australia. Even though the majority of crop land is in Asia (10 million ha) and Africa (~18 million ha), pearl millet farming is expanding in nontraditional places, with Brazil having the biggest area (~2 million ha). In addition, it is being tested as a grain and forage crop in the United States, Canada, Mexico, West Asia, and North Africa, and Central Asia.

Pearl millet is known as bajra in Hindi and Bengali, bajri in Gujarati and Marathi, bajra in Oriya and Punjabi, sajje in Kannada, sajjalu in Telugu, kambu in Tamil, and kambam in Malayalam. More than 90 million poor people depend on pearl millet for food and income. The crop is well adapted to agricultural areas that are afflicted by severe drought, poor soil fertility, and high temperatures. It also grows well in soils with high salinity or low pH. Because of its tolerance to such challenging environmental conditions, it is often found in the marginal areas where other cereal crops, such as maize or wheat, cannot survive.

After wheat, rice, and maize, pearl millet is India's fourth most extensively farmed food crop. Rajasthan, Maharashtra, Uttar Pradesh, Haryana, and Gujarat account for more than 90% of total pearl millet area and provide 87.7% of total output (Jorben et al., 2020). India's pearl millet growing regions are separated into two primary zones based on agro-climatic conditions: Zone A and Zone B. During 2018–2019, the average production of pearl millet in India was 8.61 million t, with an area of 6.93 million ha (Directorate of Millet Development, 2020). Currently, production of bajra has increased to 9.78 million t (Directorate of Economics and Statistics 2022–2023) from about 7.65 million ha (Press and Information Bureau, 2023). Increased millet production is a crucial strategy for enhancing food security.

Pearl millet is also regarded as one of the crops capable of providing good nutrition and money to small-scale farmers. It has a high energy content (361 Kcal/100 g), comparable to wheat, rice, maize, and sorghum. It is composed of 56%–65% starch. It also has high fiber content (1.2 g/100 g). Pearl millet has a higher protein content than other widely produced grains as well as higher-quality protein (9%–13%), with a larger proportion of important amino acids (Yadav et al., 2017). The flour had a very high lipid content (5%–7%), with a high concentration of linoleic acid (44.8%), oleic acid (23.2%), and palmitic acid (22.3%). Pearl millet and rice grains are gluten-free cereals that are commonly suggested for patients with celiac disease (gluten-sensitive enteropathy). Pearl millet contains vitamins (thiamine, nicotinic acid, and riboflavin) as well as minerals such as Mg, K, Fe, P, Zn, Cu, and Mn. Some genotypes biofortified with Fe and Zn, such as Dhanshakti (ICRISAT), Pusa1201 (IARI, India), and others, have been produced in recent years to compensate for Zn and Fe deficiencies. It also has a high proportion of resistant starch and slowly digested starch (SDS), which helps to explain the low glycemic index (GI). Because of the prevalence of nutrition-related diseases such as diabetes at a young age, foods with low GI and high amounts of resistant starch are in great demand for today's and future generations (Annor et al., 2015).

Understanding Changes in Pearl Millets' Area and Production: The Key Drivers of Market

A structural break analysis was carried out to understand the pattern of declining area under millets in different states. The results of the study are tabulated to understand the change in millets area in the country in identified time periods. It is important to understand the structural break in bajra cultivation in the country after the "green revolution" period. The study was carried out over five time periods, and the results are presented in the Table 14.1. Analysis of decadal growth rate and instability of area under bajra revealed the evolution of states like Madhya Pradesh and West Bengal from negative growth rates in 1960s to positive growth rates in recent years, with the lowest instability in Rajasthan, which is evident from the increasing area under bajra in the recent years.

The compound area growth rate (CAGR) of area and production for the millet crops was analyzed over the years from 1950–1951 to 2020–2021. Bajra has shown a steady decline in production area over the years, with about −0.90 CAGR, while there has been a resistance for declining production over the years with a CAGR of −0.48. All other millets crops except bajra have shown a steady decline in area

Table 14.1   Decadal Growth Rate and Instability of Area Under Bajra

Period I (1966–1975)	Period II (1976–1985)	Period III (1986–1995)	Period IV (1996–2005)	Period V (2006–2017)
Bajra-producing states witnessed stunted growth in the area. Haryana showed a positive growth rate, while Madhya Pradesh showed a negative growth rate at 5% and 1% significance level, respectively. During the decade, high and low instability were witnessed in West Bengal and Haryana, respectively.	States like J&K (2.61%), Rajasthan (2.34%), Odisha (2.1%), Maharashtra (0.19%). The decelerated growth was highest in Punjab (−13.01%). During this decade the instability was highest in West Bengal (49.12%) and lowest in Uttar Pradesh (5.22%).	All states except West Bengal (10.54%), Madhya Pradesh (0.92%), Uttar Pradesh (0.29%), and Gujarat (0.23%) experienced deceleration in the area under bajra. Andhra Pradesh (−10.81%) and Pondicherry (−10.72%) witnessed the highest rate of decline. Instability was highest in Madhya Pradesh (60.14%) and lowest in Andhra Pradesh (4.89%).	Growth in all states (J&K, Madhya Pradesh, Punjab, Rajasthan, and Uttar Pradesh) growth was negative in 1996–2005. The greatest negative growth was observed in Bihar (−15.32%).	West Bengal (8.75%), Delhi (7.45%), Madhya Pradesh (5.24%), and UP (0.92%) witnessed growth. The greatest negative growth was observed in Punjab (−29.09%) followed by Pondicherry (−22.97%). High instability was observed in Punjab (63.01%); lowest instability was observed in Rajasthan (7.75%).

and production over the years (Table 14.2). Further, an attempt was made to study the performance of millets in the country, and the possible increase in the decade was analyzed by considering normal-, best-, and worst-case scenarios.

Worst-Case Scenario

The results of the study indicated that the area under bajra has been declining over the years. The decline was mainly due to an increase in area under cultivation of major cereals like rice and wheat after the green revolution, which increased cultivation of commercial crops like oilseeds, cotton, spices, fruits, and vegetables during recent years due to globalization of agriculture. Despite the decline in area, production of millets increased significantly during the last 80 years. It can be concluded that yield plays a major role.

Table 14.2   Area and Production of Millets Over Different Decades

Year	Bajra		Jowar		Ragi		Small millets	
	A	p	A	p	A	p	A	p
1950–1951	9.30	2.60	15.57	5.50	2.20	1.43	4.61	1.75
1960–1961	10.70	3.28	18.41	9.81	2.52	1.84	4.96	1.91
1970–1971	12.49	8.03	17.37	8.11	2.47	2.16	4.78	1.99
1980–1981	10.58	5.34	15.81	10.43	2.53	2.42	3.98	1.57
1990–1991	10.90	6.65	14.36	11.68	2.17	2.34	2.45	1.19
2000–2001	8.90	6.76	9.86	7.53	1.76	2.73	1.42	0.59
2010–2011	8.90	10.37	7.38	7.00	1.29	2.19	0.80	0.44
2020–2021	7.65	10.86	4.38	4.81	1.16	2.00	0.44	0.35
CAGR	−0.90	−0.48	−0.96	−0.89	−0.93	−0.83	−0.99	−0.98

Note: Data from the Directorate of Economics and Statistics.
Abbreviations: A, area; p, production.

Best- and Normal-Case Scenarios

Increasing use of millets in various ready-to-eat (RTE) food products should be encouraged because it enhances their value and market price. The government should promote blended wheat flour with millet flour (90:10) with incentive to processing units. Diversion toward distilleries could also be explored by the government. There may be an increase in demand for millets due to recent inclusion of millets in the public distribution system with the implementation of a food security bill coupled with increased awareness of people about inclusion of millets in daily meals for healthy living. Millets are candidate crops for tackling climate change. Therefore, there is a need to increase millet production with the increasing demand and to recognize millets as climate change–compliant crops to promote their cultivation and consumption, which makes them an important part of India's food and farming future.

Reality: The Actual Production of Millets

The above factors and increased growth rates in the best-case and normal-case scenarios may lead to an increase in area, production, and yield over the next several years. Bajra production has remained at the normal scenario, indicating the need for interventions in terms of production, processing, and marketing of these grains and their value-added products (VAPs). There is a need for interventions by the government to promote millets, with special emphasis on bajra because it accounts for half of the country's millet production.

Consumption and Storage of Pearl Millet in Ancient and Modern Times

Pearl millet must be cleaned and processed before consumption. These tasks are traditionally done by winnowing and grinding, but modern methods use aspirators and graders. The Indian Council of Agricultural Research–Indian Institute of Millets Research (ICAR-IIMR) has established processing units for pearl millet, which can be used in traditional and modern food products, and further processing can improve shelf life.

Pearl millet was commonly used to make porridge, flatbread, and beer in ancient times and is still used today to make probiotic products like koko and fura. Pearl millet products should be consumed within 1–3 days due to rancidity issues (Jideani et al., 2010; Lei & Jakobsen, 2004). In India, pearl millet is consumed during the winter months for its harmonious pairing with seasonal dishes. It is equally cherished during intense summers, offering a cooling effect to the body and safeguarding against heatstroke. The Egyptians and Harappans made a flatbread similar to pita bread using pearl millet around 2800 BC. In Africa, people made beer from malted pearl millet by 100 BC. Pearl millet grain can be cooked similarly to rice and can be used in various recipes like pongal, lemon rice, pulihora, mint rice, etc. Sometimes, pearl millet is malted and supplemented for use as weaning food along with malted finger millet flour.

Modern processing technology allows for value-added foods to be made from pearl millet, including hot extruded snacks, bakery products, puffs, flakes, and ready-to-cook (RTC) products (Figure 14.1). These products have varying shelf lives, ranging from 1 to 6 months, and can incorporate up to 50% pearl millet flour. Pearl millet grain can also be directly processed to create puffs and flakes using techniques like conditioning and puffing, soaking, roasting, and rolling. These products have a shelf life of about 3–4 months. There are different processing techniques for converting small millet grains into various edible forms like rice, flour, sprouts, salted RTE, flaked, popped, porridge, and fermented products through primary processing techniques.

Pearl millet grain is stored along with neem salt, and granulated salt in earthen pots, bags (cotton, jute, and local grass), and bamboo bins are used to store small quantity of grains. Underground pits, mud houses, silos (mud, thatch, bamboo), and obeh are used to store bulk quantities of grains. Sometimes, pearl millets stems are solarized along with the grain, and small batches of these are hung by a rope to avoid significant losses caused by rats and pests. Nowadays, bulk storage

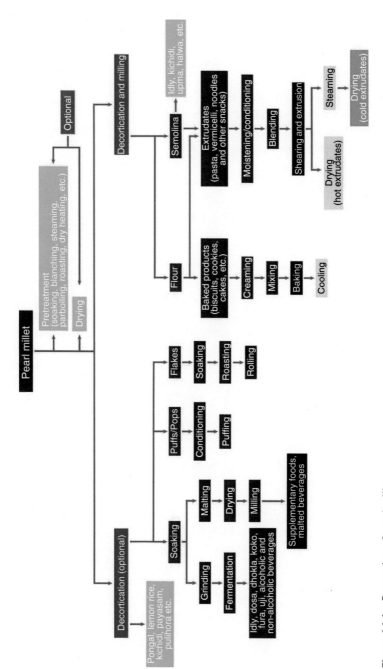

Figure 14.1 Processing of pearl millet.

of pearl millet gains is done in gunny bags and metal silos, whereas the products are stored in high-density polyethylene and metallic pouches.

Pearl Millet Marketing, Value Addition, and Exports

The market segments for millet products can be divided into two broad categories: high-end and low-end. Millet food segments (e.g., instant foods, RTE foods, high-calorie foods, plant proteins, organic foods, healthy foods, immunity foods, and gluten-free) can be grouped into high-end foods. These segments can be addressed by ease, convenience, and easy availability to consumers.

RTC and RTE products of millets are gradually becoming available in Tier 1 and Tier 2 cities. Multi-grain millet flour, flakes of sorghum and pearl millet, finger millet malt, sorghum semolina and pasta, millets-based breakfast cereals, millets-based regional snacks and fast foods, etc., are the commercially available millet products in India both in retail and online.

The market for pearl millet market is projected to grow, with a CAGR of more than 3% over the forecast period. A variety of programs are being launched to improve millet cultivation and consumption to lower health risks associated with diabetes, obesity, and cardiovascular disease. Unified approaches among millet suppliers and food and beverage manufacturing companies to meet millet demand have been critical to the industry's growth. A substantial amount of millet produced is farmed and consumed locally. The remainder of the harvest is used to make beer, infant food, and breakfast items, such as flakes and cereals. Millet beer is prevalent in African culture and is also finding a market in India and the United States.

Millets-based items (e.g., flour, flakes, and cookies) are becoming more prominent in the consumer market. Breakfast cereals such as flakes and local recipes are increasingly popular among African and Asian populations (Figure 14.2). Millets-based breakfast goods produced over US$2 billion in revenue in 2018 and are expected to grow further due to an increase in high-fiber and gluten-free food consumption. There is a need for delectable foods and easy-to-cook products that are enjoyed by all age groups to sustain the promotion of healthy foods like millets. The market functioning can be useful for positioning the millets in the market.

There are 13 countries exporting bajra, of which India is the major exporter. Among 91 importing countries, India exports most of its bajra to United Arab Emirates, Saudi Arabia, Pakistan, Canada, Tunisia, Libya, and the United Kingdom. Of the 732 active bajra suppliers in the world, 688 are located in India. The other major exporters are China and Pakistan. The top three product categories of bajra exports from India are bajra other than seeds (HSN 10082920), bajra for seeds (HSN 10082120), and cereal flour other than wheat (HSN 11029090).

Figure 14.2 Value-added products of pearl millet in markets.

Millets VAPs Scenario in India

Value addition in millets involves primary processing, which includes separating glumes and other foreign materials from the grains to obtain desirable, fine edible grains with optimum quality and consumer acceptability. Secondary processing involves milling the grains to obtain millet flour, which can be further processed into various VAPs such as biscuits, noodles, puff snacks, and beverages. Millet malt, for example, can be used for baking as well as for beverages and has the potential to be a good substitute for gluten-free malt, which is in high demand in the global market.

Studies have shown that while 0.31 million t of millets can be exported as grains, the rest can be exported in value-added form to fetch premium pricing. Diverting about 0.25 million t for secondary and tertiary value addition has the potential to yield up to 0.59 million t of VAPs. The global trade in total malt in 2021 was about US$4 billion, and the global malt-based ingredients market is worth nearly US$30 billion, with most of it being beverages (YES BANK & APEDA, 2022).

A field survey was carried out to understand the extent of value addition in millets. The survey was divided into the major food companies operating in the food sector (FMCG), start-ups in the market, and start-ups supported (and incubated at Nutrihub, ICAR-IIMR). The study revealed that, regarding jowar, out of total production of 4.23 million t, the value addition is only 1.5%–2%. With bajra, the same is noted to be ~3.6% (0.34 million t out of 9.62 million t production). With ragi,

Table 14.3   Share of Value-added Products in Total Millets Production

Millets	Production	Value-added products[a]	Percentage	Major products
		–million t–		
Jowar	4.23	0.064	1.5–2	Rawa, flakes, flour,[b] others[c]
Bajra	9.62	0.349	3.6–4	Flour, flakes, others[c]
Ragi	1.7	0.184	10–11	Flour, biscuits others[c]
Small millets	0.37	0.009	2.30	Rawa, flour, others[c]
Total	**15.92**	**0.606**	**3–4**	

Note: Authors estimation based on survey.
[a] Value-added products from organized (FMCG/startup) and unorganized sectors.
[b] Both single-grain and multigrain flour.
[c] "Others" include biscuits, pops, porridge, vermicelli, pasta, papad, malt-based products, etc.

diversion toward VAPs was higher at 10% (0.18 million t) due to the traditional habit of consuming ragi products mostly in Karnataka and the surrounding states. Small millets production has been falling gradually; however, with more awareness resulting from efforts of Ayurveda-based companies like Patanjali, consumption has improved in the last few years. VAPs in small millets are estimated at 2.3% of total production of ~0.0037 million t (Table 14.3).

A field survey has revealed that, in Jowar VAPs, consumption of rawa and flakes products is higher than other products. Among bajra VAPs, consumption of single- and multi-grain flour and biscuits are higher than other products. Among ragi VAPs, consumption of single- and multi-grain flour, and malt-based biscuits is higher than other products. Among small millets VAPs, consumption of single- and multi-grain flour is higher than other products.

The value addition in millets by the start-ups in the country was studied through surveys carried out by Nutrihub, ICAR-IIMR. It was observed that ragi had the highest share in value addition, with bajra and jowar limited to 9% and 16%, respectively. The start-ups also included other grains such as barley and oats in addition to minor millets such as foxtail millet, kodo millet, barnyard millet, and proso millet. However, it was interesting to note that several start-ups supported by Nutrihub, ICAR-IIMR produced more products (11%) with pearl millet as a major ingredient, which is relatively higher than the market value. The higher share could be due to the support obtained from Nutrihub, ICAR-IIMR, the technology Business Incubator through technology commercialization and incubation (Figures 14.3 and 14.4).

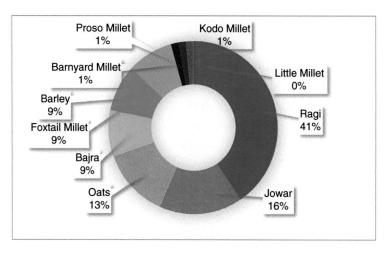

Figure 14.3 Percentage of products by start-ups in the millets' sector.

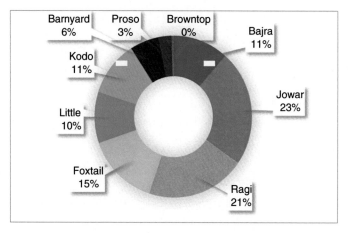

Figure 14.4 Percentage of different millets in products by start-ups supported by Nutrihub, ICAR-IIMR.

Bajra VAP Distribution and Growth Rate

As per the studies carried out by ICAR-IIMR, in the case of bajra, out of total VAPs of 0.375 mm (3.75 lakh ton), the ratio of the organized (FMCG, start-up) sector is estimated at 47% (~1.76 lakh ton), and the ratio of the unorganized sector is estimated at 53% (~1.98 lakh ton) (Dayakar Rao, 2022).

The growth rates of various bajra VAPs in the organized space are depicted in Figure 14.5. Flour and rava were noted to be trading at large volumes. Currently

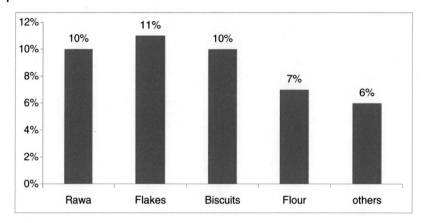

Figure 14.5 Bajra VAP growth rate organized.

rawa, flakes, and biscuits products are growing at 10%, 11%, and 10% per annum, respectively. Bajra flour, both single-grain and multigrain, is growing at 7% per annum. With rising demand for multigrain atta and single-grain 100% atta from Metros and urban households, the production of bajra flour is likely to expand multiple times in the coming year.

The production and value addition in bajra is dependent on several factors, including the minimum support price (MSP) announced. An attempt was made to study the growth in area under bajra vis à vis the MSP announced for bajra. The study revealed that during the last 11 years, area under bajra has declined by 23%, whereas MSPs have risen by 144%, which is lower when compared to jowar. During 2010–2011 until 2017–2018, the increase in MSP in terms of rupees was around INR 545/qtl, and after that until 2020–2021 it was increased by INR 725/qtl. In both periods, acreage declined by 22% for the former and just 1% in the latter (Figure 14.6).

Unlike jowar, acreage under bajra in the last 5 years has been steady, which can be attributed to the increase in procurement activity, especially in state of Haryana, where the state government under its scheme "Meri Pani Meri Virasat" is encouraging farmers to shift from water-guzzling paddy crops to other crops like maize, cotton, bajra, and pulses. In the last 5 years, cultivation of bajra in Haryana has risen by 18.6% at 439,000 ha in 2020–2021. Thus, MSP remains an important factor contributing to increasing area under bajra.

Trends in Millet Exports from India

Exports of sorghum and bajra have gained momentum since 2005; sorghum exports peaked during 2012–2013 and bajra during 2011–2012. Sorghum has witnessed significant decreases in imports from Sudan, Kenya, and Taiwan. The

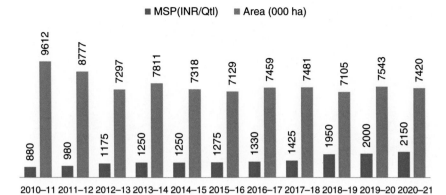

Figure 14.6 Trends in MSP versus area—bajra.

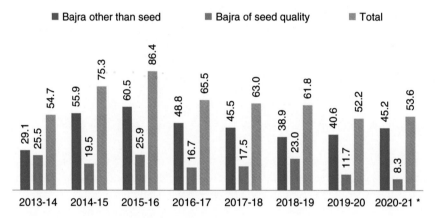

Figure 14.7 Indian Bajra exports in 000' tons.

importing giant of world sorghum market has almost zero sorghum trade with India, which significantly affects the position of India in the international sorghum market. Hence, it is advisable to work on enhancing the exports to major importing countries such as China and Japan to acquire a position in the global market. The volume of sorghum exported has grown at an annual rate of about 10%, and the value of exports grew about 20% annually. The annual growth in international prices of Indian sorghum is surpassing the global inflation rate. Therefore, export prices of sorghum should be amended to achieve competitiveness in the international sorghum market.

Akin to sorghum, the exported value of bajra is growing at a faster rate (21%) than its quantity exported (12%). This may increase exporters' profits; however, this will also affect the competitiveness of Indian bajra in the export market. The quantity exported of bajra from India has declined since 2012, and it has lost its market in Yeman and Pakistan (Figure 14.7).

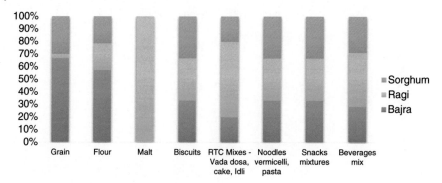

Figure 14.8 Share of different millets in exports from India.

Unlike other major millets, ragi exports have increased, reaching its highest value in 2021. The volume of exports is growing at 22%, and its value is growing at 32% because India is the chief exporter of ragi to the international market and has exported consistently to neighboring countries such Nepal and Sri Lanka. The expanding demand from diversified nations in the past couple of years depicts the wide global market for ragi.

Potential in Bajra Exports—Grains and VAPs in 2030

Bajra exports from India consisting both of seed and "other than seed" have declined during the last 5 years. The majority of the exports of bajra from India can be attributed to the other-than-seed segment. The exports were reported mostly to gulf countries and to the neighboring country of Pakistan. In gulf countries it is mostly used for bird feed. Its consumption as food is limited to south Asia and parts of Africa.

Figure 14.8 shows the percentage of millet exports from India in different forms predicted for the year 2030 as reported by YES BANK and APEDA (2022). The three varieties of millets exported by India are bajra, ragi, and sorghum.

- **Grain:** Bajra is the most exported millet in grain form, accounting for 66.67% of total exports. Ragi and sorghum follow, with 3.33% and 30% of total exports, respectively.
- **Flour:** Bajra is also the most exported millet in flour form, accounting for 57.45% of total exports. Ragi and sorghum follow, with 21.28% and 21.28% of total exports, respectively.
- **Malt:** Sorghum is the only millet exported in malt form, accounting for 100% of total exports.
- **Biscuits:** Bajra, ragi, and sorghum are all exported in biscuit form, with equal shares of 33.33% each.

- **RTC mixes—Vada dosa, cake, Idli:** Bajra and ragi are the only millets exported in RTC mixes, with bajra accounting for 20% of total exports and ragi accounting for 60% of total exports.
- **Pasta:** Bajra, ragi, and sorghum are all exported as pasta (noodles and vermicelli), with equal shares of 33.33% each.
- **Snack mixtures:** Bajra, ragi, and sorghum are all exported as snack mixtures, with equal shares of 33.33% each.
- **Beverage mix:** Bajra and sorghum are the only millets exported in beverage mixes, with bajra accounting for 28.57% of total exports and sorghum accounting for 42.86% of total exports.

Bajra is the most exported millet in all forms, followed by ragi and sorghum and India is a major exporter of millets in a variety of forms and that the demand for millets is growing in both domestic and international markets. In addition to these product categories, research and development may be focused on the development of new-generation products, such as plant-based meat and dairy analogues, probiotics, and prebiotics.

Innovations in Pearl Millet Improvement

Genetic Innovations in Millets

Rapid millet genome sequencing is favored by the new techniques for next-generation sequencing. The multiple omics platforms hold significant promise for improving millets' ability to address micronutrient deficiency and support millets as model systems for long-term crop biofortification. In order to increase the production of nutraceutical substances, essential nutrient accumulation pathways may be genetically manipulated with the help of rate-limiting synthesis step identification. ICRISAT is holding 22,888 pearl millet germplasms, and the NBPGR is holding 7268 (Yadav et al., 2017). Genome editing techniques that target gene expression and gene knockdown may be used to reduce or eliminate antinutritional factors. However, a lack of understanding of the complicated regulatory mechanisms of metabolic pathways limits the ability to increase millet's value. The development of "omics" and "informatics" has allowed millet researchers to simultaneously characterize the function of useful genes in the gene pool and seek valuable genes in the gene pool.

Biofortification: An Innovative Approach to Tackle Malnutrition

In India, pearl millet was the first crop to have a biofortified variety, named "Dhanashakti." Since then, more biofortified pearl millet hybrids, as well as varieties of finger millet and little millet with high Fe and Zn, have been released.

These varieties have the potential to address micronutrient deficiencies in populations that rely heavily on these crops for their diet. Therefore, it is important to promote these biofortified varieties as a means of improving nutrition security.

There is a need for skill development in processing and value addition of millets. Government-funded programs can play a crucial role in promoting the adoption of value addition technologies by providing training and support to farmers, processors, and entrepreneurs. This can help in creating a market for millet products and generating income for rural communities. Government-funded programs should provide training on value addition technologies at the local level. This will help in popularizing several millet food recipes that have been published in books and are available online. The necessary training will help in promoting these recipes and making them more accessible. Skill development in processing and value addition is an important aspect of this strategy, which can help in creating a market for millet products and generating income for rural communities. The availability of millet food recipes in books and online can also help in promoting the consumption of millets. However, it is important to provide necessary training to people to make these recipes and create awareness about the nutritional benefits of millets. This can help in popularizing millets as a healthy and sustainable food option.

Innovation in Millet Processing and Scope for Value Addition in Millets

Increased demand for millet processing has necessitated standardized processing machinery for millets to be developed on a larger scale. This is important to encourage entrepreneurs and increase the efficiency of millet processing. The recovery of grains in primary processing also needs to be addressed, especially in small millets. Efforts are needed to develop more efficient and affordable machines.

The availability of quality raw material is crucial for the food processing industry. To ensure this, it is important to have assured procurement of quality raw material and to use recommended cultivation practices and buyback support. The use of digital buyer-seller platforms can also facilitate networking and final transactions.

Several studies have emphasized the importance of promoting millet production, enhanced value addition, and greater consumption. This includes addressing issues in the value chain through innovations in production, processing, and value addition, as well as mainstreaming millets through policies.

Innovations in Pearl Millet Processing and Value Addition

Processing of grains enhances consumer acceptability and adds to its convenience. A decrease in millet consumption is found to be proportional to an increase in expenditure. Besides, increased income is accompanied by increased consumption of wheat and rice because products made from these cereals are easy to prepare and have better keeping quality. At the same time, people now tend to eat a greater variety of foods. Technological change could perhaps change the scenario for millets, improving their production and utilization. Processing also improves the food value in terms of increasing the variety of products and improving carbohydrate and protein digestibility.

Processing of millets is essential for removing anti-nutritional compounds before consumption to avoid health hazards. Historically, several treatments have been used to make millets suitable for human consumption, such as soaking, heating, roasting, fermentation, and cooking. Nowadays, different unit operations like soaking, dehulling, grinding, roasting, puffing, fermentation, and malting are well established in industries for commercial-scale processing of millets. The selection of appropriate processing operations, equipment, and production scales is based on the targeted output. The adoption of suitable processing technology with appropriate equipment decreases the anti-nutritional constituents and off-flavors from millet, which has led to an increase in the number of consumers of millets and has boosted commercial-scale production of various VAPs of millets. The advancement in the processing technologies of millets opens a new horizon and will help to raise millets to a competitive level among staple foods (Kumar et al., 2021).

Several studies have shown hindrances in the processing and cooking of millets, such as the hard seed coat, antinutritional factors, poor digestibility, and low micronutrient bioavailability. Further decortication can enhance the bioavailability of Ca, Fe, and Zn by 15, 26, and 24 g/100 g, respectively, and enhances the digestibility of protein by reducing the anti-nutritional factors like phytic acid, polyphenols, and tannins (Kumar et al., 2021).

Innovations and Business Opportunities in Millets

Innovation is crucial for startups to attract investments and survive in the ever-changing business world. Incubation facilities provide support and cushion for startups to save on fixed expenses and have a focused approach to their businesses. Nutrihub, an incubator by ICAR-IIMR, offers various services to startups, enabling them to establish a self-sustaining business entity. Investment ventures and the journey of Nutrihub are discussed in detail by Rao and Tonapi (2021). Innovations in research and development and use of technology may be adopted for enhancing millet production and productivity, demand creation, value addition, and entrepreneurship development in millets. Efforts are made to develop high-yielding cultivars of different millets crops, create value addition in millets, and promote entrepreneurship in millets through institutions such as ICAR-IIMR, Hyderabad, and Nutrihub-TBI. Apart from conducting research and development, emphasis may be placed on capacity building and skill development programs and on technology transfer and consultancy services for budding entrepreneurs, existing industrialists, farmers, self-help groups, students, and research scholars in all aspects of millet production and processing, preservation, and value addition and exports. This collectively will lead to India emerging as a millet hub. The researchers need to prioritize sustainable crop substitutes to meet the emerging trends resulting from climate change and the ultimate sustainable development goals. The emphasis must be on quantity, quality, and net benefits that are expected to arise from smart millet cultivation, value addition, and marketing interventions. Therefore, conceptualizing location-specific, demand-driven, and farmer-friendly configurations of millet-based research and development are necessary to achieve this goal.

In the early 1960s, genetic innovations in major food crops such as wheat and rice showed promise under Indian conditions, while the discovery of cytoplasmic male sterility paved the way for developing hybrids in pearl millet and sorghum (Burton, 1965; Crow, 1998; Sharma et al., 2022; Singh et al., 2016; Stephens & Holland, 1954).

Studies have shown that interventions at different stages have resulted in a 51% rise in incremental net income for participating farmers through the technological backstopping of sorghum cultivation with end-product–specific improved cultivars (Rao et al., 2010). The inconvenience in the preparation of sorghum foods has been eliminated through the development of convenient and RTE/RTC foods. The factorization of investment expenditure made per farmer has been worked out to be Rs 356 (equivalent of Rs 890/ha) in a season, resulting in an increase of 51% in incremental net income (average of 58% in kharif sorghum and 44% in rabi sorghum). The study has also revealed that linking up of the entire stakeholders through value addition throughout the value chain system would renew and uplift the diminishing sorghum area and production and its ultimate economic benefits to farmers and other stakeholders in the value chain (Rao et al., 2010). The existing literature on agricultural value chains and their financing mechanisms provides information about strengthening the interface between product and financial markets to enable smallholders to capture the benefits of value addition. Value chain development has created business opportunities, and business model drivers and their goals before delving further into the specifics of each of the prospects are discussed below.

- **Evaluation and quality control:** Fostering innovation in determining grades and standards in millets is an emerging opportunity because they are still in their infancy. People who desire to enter this field have a very bright future.
- **Primary processing:** This stage includes the first steps in cleaning, grading, dehulling, and dehusking. Farmers and farmer organizations can also set up

primary processing, bringing processing to the farm gate. This increases the value of the farmers' products.

- **Secondary processing:** The low usage of millets is mostly due to their flavor and difficulty in preparation. Numerous technologies have been created to produce various value-added millets products, including pasta, vermicelli, atta, noodles, biscuits, puffs, namkeen, and so forth, in order to address these challenges.

- **Sales and marketing consulting:** Marketing plays a significant role in boosting demand for these items, and most companies continue to struggle to receive this assistance. Consequently, this is yet another emerging opportunity for business establishment.

- **Transportation:** Transportation is essential in bringing farmers and consumers together. Many millet farmers continue to produce sparingly and for subsistence. Therefore, a key component of the millets value chain is transportation. At this point, transportation aggregate is still another option.

- **Biofortification:** Because nutrition is a key priority in this COVID era, millet fortification would be a fascinating topic to research. An emerging field in this area would be fortification with other minerals and the creation of products from them.

- **Beverages:** There is still room for improvement in the production of dairy and beer from millets for the Indian market. Among the many alternatives to discover here are millet milk, millet curd, millet ice cream, and millet cool drinks, to name a few.

- **Biofuels:** India's dependence on other nations will significantly decrease if fuel can be produced from millets. Another unexplored field is turning millets, particularly sorghum, into biofuel. Further research can be done on producing jaggery and other products from sorghum.

- **Plant-based protein meat:** Because India is predominantly a vegetarian nation, a new field is opening for developing vegetarian proteins that can replace meat. Millets are a great prospect in this area because they are nutritious.

Successful Millet Value Chain: The Need of the Hour

The green revolution successfully achieved self-reliance in food grain production, but new challenges, such as over-exploitation of groundwater and heavy reliance on chemicals, require high priority on maintenance of soil productivity, integrated nutrient and pest management, efficient irrigation technologies, and diversification of farming systems. Millets can play an important role in diversification and improving livelihood in drylands, but technological innovations and suitable policy and markets are needed for promotion. However, several states in India have shifted from cultivating millets to more profitable crops like maize, cotton, mungbean, and soybean, leading to a significant reduction in millet cultivation. Despite an increase in the MSP for sorghum and pearl millet, the lack of government procurement has affected the economics of millets cultivation.

Millets are favored due to their sustained higher productivity with a short growing season and their ability to withstand dry conditions and high temperatures. The global resolution recognizing 2023 as the International Year of Millets has brought millets into the limelight, and the Government of India has embarked on its comprehensive upscaling the use of millets. Millets can play a leading role in meeting the world's cereal demand and improving the income of small farmers while providing nutritional security and health benefits. The grain is mostly used for food in Rajasthan, compared with only a small percentage in Haryana, which gives a scope for alternative uses of pearl millet grain in the coming years, such as for the cattle and poultry feed industry, breweries, and the starch industry. However, the estimated size of the market for VAPs from pearl millet is very low (Rao et al., 2016a, 2016b). Historical growth rates of per capita consumption in the major pearl millet–producing states for rural and urban regions have indicated that relative shares of different uses of pearl millet grain as cattle feed will increase, whereas food uses will decrease. Malting, blanching, heat treatment, and acid treatment can improve the storability, stability, and digestibility of pearl millet flour (Rao et al., 2016a, 2016b). There is a need to investigate how collaboration among millet supply chain entities leads to innovative practices and how these practices improve the sustainability of the millet value chain. The innovation at each stage (i.e., product, processing technology, packaging, branding, and marketing) enhances collaboration; promotes exploring the economic, environmental, and social benefits of these innovations; and highlights the importance of supply chain innovation for improving the sustainability of the food supply chain.

The value chain model developed by ICAR-IIMR Hyderabad for millets can be replicated in other parts of the world where millets are the main crop. Policymakers and concerned departments should attempt to inspire and encourage millet farmers to strive for commercialization through various approaches, such as inclusion of millets in public distribution systems, subsidies for machinery to process millets, and nation-wide campaigns of millet flagging as a health benefit food (Kumar et al., 2021). The development of processed food (RTE/RTC) using millets is proposed as the way forward, so efforts should be directed toward the development of new products and processing technologies to overcome the constraints and enhance the shelf life of millets and their VAPs. Sustainability and reliability of the model for millet promotion and policy changes on the part of the responsible central/state governments are required. Intensive efforts are needed to create demand for millets through value addition, new farmer producer organizations, and suitable market strategies. There is a need to create public awareness and change the image of millets through emotional and engaging marketing approaches.

Nutrihub: A Classic Case of Holistic Entrepreneurship Development in Millets

The millets business is the new-age opportunity for aspiring entrepreneurs to set up their business and tap the potential of millets at national and international levels. The Government of India is supporting the startup ecosystem with several schemes through the establishment of incubation centers under Department of Science & Technology NIDHI, RKVY-RAFTAAR, BIRAC-BIONEST, etc. The startups can take advantage of the benefits under any of these schemes (Rao & Nune, 2021)

ICAR-IIMR developed over 60 value-added technologies for millets to create demand for the ancient grains, which were declining in consumer diets due to the Green Revolution, and made them available for commercialization but later launched its own brand "Eatrite" to promote the consumption of nutritious millet products (FAO, 1995; Ratnavathi & Patil, 2013). IIMR created its own millet brand, Eatrite, to promote the health benefits of millets and bring VAPs closer to consumers. To further drive demand for millets products, IIMR established Nutrihub, an incubator for millet startups, which provides physical space and infrastructure for up to 25 startups (Ogidi & Abah, 2012; Ratnavathi & Patil, 2013).

Nutrihub-TBISC, the commercial façade of ICAR-IIMR, has a state-of-the-art incubation center and is supporting the millet-based startups in the country under its flagship programs like NEST and NGRAIN by facilitation of the grant-in-aid up to ₹25 lakhs alongside providing incubation and technology licensing. Aspiring entrepreneurs with innovative business ideas may contact Nutrihub, ICAR-IIMR to take advantage of support for scaling up the business and to be directed toward export markets to tap the potential of the national and international markets in the wake of International Year of Millets, 2023.

Nutrihub is the Department of Science & Technology, Government of India supported Technology Business Incubator hosted by the Indian Institute of Millet Research, ICAR-IIMR, Hyderabad. It is the first of its kind to cater to start-ups' needs in the Nutricereals sector in the country. Nutrihub is a focal point where ideas, entrepreneurs, agriprenuers, start-ups, experts, academics, and funding agencies can gravitate toward the creation of a new, knowledge-based economy. Incubation Program at Nutrihub-IIMR is designed to help Nutricereal startups grow in a streamlined fashion by providing them the required technology and business support system. Below are the services provided by Nutrihub Startups in Millets under NEST and NGRAIN to facilitate funding to idea stage and seed stage funding.

Over 400 millet startups have emerged over a 4-year span. This incubation center has contributed much to this figure, while the TBI at ICAR-IIMR has incubated more than 200 millet-specific startups. A compendium of 100 successful millet start-ups, a book is published by ICAR-IIMR (Rao and Tonapi, 2021) Under RKVY-RAFTAAR, Nutrihub has already supported 200 startups, of which 66 took advantage of grants-in-aid up to Rs 7.05 crores. About 170 millet value-added technologies have been transferred to almost 80 start-ups and enterprises.

Challenges to Processing, Value Addition, and Innovation in Pearl Millet

The factors discussed above are broadly responsible for lower yields, reduced shelf life, and low marketability of millets in India. However, there are a few issues pertaining to pearl millet specifically, which are discussed below to highlight the challenges and the possible solutions to overcome these challenges.

Anti-nutritional Factors

Pearl millet is gluten free and has a low GI. Despite these benefits, antinutrients such as phytic acid, polyphenols, and tannins can limit its use as a meal or feed. Anti-nutritional factors are chemicals that limit the availability of nutrients. It is thought that their presence in pearl millet limits protein and starch digestibility, reduces mineral bioavailability, and inhibits proteolytic and amylolytic enzymes.

Studies by Abdalla et al. (1998) showed that pearl millet contains 354–796 mg/g of phytic acid. Phytic acid becomes insoluble at the physiological point of the intestine when it forms a compound with minerals (e.g., Ca, Zn, and Fe). Thus, it decreases the uptake of the aforementioned critical dietary minerals in the human gut. Pelig-Ba (2009) stated that P in this form is not bioavailable to monogastrics because it lacks the digestive enzyme phytase, which is essential when P is separated from the phytate molecule. In addition, the presence of some goitrogenic polyphenols and C-glycosyl flavones, including glucosyl vitexin, glucosyl orientin, and vitexin, may be responsible for some health problems when pearl millet is consumed (Gaitan et al., 1989).

However, by using different processing techniques, the proportions of these antinutritional substances can be lowered. Decortication, malting, blanching, parboiling, acid and heat treatments, and fermentation of pearl millet lowers the antinutritional factors (Table 14.4). Because they are concentrated in the outer layers, they leach out into the soaking medium when they are soaked, fermented, blanched, and parboiled, whereas de-branning removes the outer layers. Egli et al. (2002) reported a 41% reduction in phytic acid content after soaking pearl millet for 72 hours. Sharma and Kapoor (1996) discovered that germination and de-branning, together with autoclaving, decrease phytic acid and polyphenols. Similarly, Rathore et al. (2019) discovered that antinutritional factors can be decreased to a small level by using various processing procedures such as roasting, soaking, boiling, parboiling, fermentation, milling, germination, decortication, and extrusion.

Table 14.4 Effect of Different Treatments on Anti-nutritional Factors of Pearl Millet

Treatment	Process	Result	References
Soaking	24–72 h	Reduced phytic acid content by 41%	Egli et al. (2002)
	12 h	Reduced phytic acid content by 16%	Ocheme & Chimna (2008)
	15 h	Reduced polyphenols by 15% and 11%, respectively	Sihag et al. (2015)
Fermentation	72 h	Significant reduction in the tannin and total phenols	Sade (2009)
	0–36 h (yeast and lactobacilli)	Most of the polyphenols were reduced and observed improvement in IVPD	Elyas et al. (2002)
	24 h	Significant reduction in enzyme inhibitors and phytic acid content	Osman (2011)
	9 h (30–50 °C)	Fermentation for 9 h showed maximum reduction of polyphenols and phytic acid at all temperatures	Dhankher & Chauhan (1987)
Germination	0–96 h	Maximum retardation of phytic acid (0.263%) achieved after 96 h compared with control sample (2.91%)	Badau et al. (2005)
	48 h at 25 °C	Reduction in total phenols (from 3.00 to 0.68 g/100 g) and tannins (from 1.52 to 1.00 g/100 g)	Nithya et al. (2007)
	72 h	Significant reduction in tannin and total phenol content and significant improvement in the IVPD	Pushparaj & Urooj (2011)
	25 h	Enhancement of in vitro Fe availability (from 2.19 to 2.61 mg/100 g) and reduction of polyphenol (from 675.33 to 303.21 mg/100 g)	Bhati et al. (2016)
Hydrothermal treatment	Blanched at 98 °C for 30 s	Significant difference in polyphenols (from 764.45 to 544.45 mg/100 g) and tannins (from 833.42 to 512.10 mg/100 g)	Archana et al. (1998)
Heat treatment	1 h (110 °C)	Substantial reduction in anti-nutritional compounds such as polyphenols (from 3.00 to 2.27 g/100 g), tannin (from 1.52 to 1.30 g/100 g)	Nithya et al. (2007)
	γ-irradiation	Decreased the amount of tannin and phytic acid and gradually increased the IVPD and protein solubility of the grains	Mahmoud et al. (2016)

Abbreviations: IVPD, in vitro protein digestibility.

Shelf Life

Despite having an excellent functional and nutritional profile and improved agronomic features, pearl millet has not attained widespread acceptance when compared to other millets in India and around the world. The main cause for this is pearl millet flour's poor storability and early rancidity. In order to consume pearl millet flour, it must be milled once a week. Products developed from pearl millet last only 1–3 months, which limits the scope of product development. The high concentrations of fats in the germ of seeds and the extremely active lipase enzymes promote hydrolysis of lipid to free fatty acids (FFA), resulting in a short shelf life (4–5 days after milling) (Rani et al., 2018).

Oxidative rancidity produces hydroperoxides and secondary metabolites that cause off-odor. The unpleasant and mousy scent is caused by secondary metabolites such as ketones and aldehydes. The scent is caused by peroxidase working on C-glycosyl flavones, whereas the polyphenol oxidase has been linked to flour browning (Rai et al., 2014). Overall, hydrolytic and oxidative rancidity generate off-odor and a short shelf life in pearl millet flour. Mazumdar et al. (2016) profiled 56 commercial pearl millet lines for enzymatic and oxidative rancidity by extracting fat from their flour. Flour was stored under three storage conditions: refrigerated (4 °C), room temperature (25 °C), and accelerated (35 °C, 70%RH). The acid and peroxide values of their extracted fat measured at regular intervals. Thirteen pearl millet varieties/hybrids least susceptible to rancidity were identified. However, a more comprehensive research effort is required to establish definitive conclusions. Mutations within nonfunctional genes encoding triacylglycerol (TAG) lipases offer potential targets for the cultivation of pearl millet germplasm, aiming to enhance the shelf life of pearl millet flour.

To increase the storability of pearl millet flour, lipase-based hydrolytic rancidity and components responsible for oxidative rancidity must be examined. Dehulling; blanching; heat treatment by dry heating, microwave, and infrared heating; addition of salt and antioxidants; storage at lower temperatures; and different packing techniques are used to limit fat hydrolysis by lipases in pearl millet flour (Rani et al., 2018) (Table 14.5). Decortication of pearl millet has shown less or no lipase activity after up to 50 days of storage (Yadav et al., 2012). FFA content was reduced to 66% when the pearl millet was soaked in 0.1% formaldehyde solution for 6 hours (Baranwal, 2017). Dry heat treatments (e.g., microwave heating, γ-irradiation, oven heating, and extrusion) and hydrothermal treatments have been shown to result in significant reductions in lipase activity, FFA, and fungal incidence (Bhati et al., 2016; Mahmoud et al., 2016; Tiwari et al., 2014; Yadav et al., 2012). Combining these heat treatments has improved shelf life up to 60 days according to Dikkala et al. (2018). Storing of pearl millet at refrigerator temperature has improved storability by reducing rancidity (Mohamed et al., 2011).

Table 14.5 Different Techniques to Improve the Shelf Stability of Pearl Millet and Its Products

S. No.	Processing parameters/ treatment	Results	References
		Grains and flours	
1	Decortication/pearling	Significant decrease in lipid content	Hama et al. (2011)
		No lipase activity up to 50 days of storage	Yadav et al. (2012)
		FFA remained <1% up to sixth day of storage	Tiwari et al. (2014)
2	Fermentation	Lipid content decreased by 97% and 93% for both starter and naturally fermented grains, respectively	Chinenye et al. (2017)
3	Germination/malting	35.4% and 14.8% reduction in the crude fat content	Suma & Urooj (2014)
		FFA stayed <1% up to 8th day of storage when grains were steeped in water containing 0.1% formaldehyde solution for 36 h	Tiwari et al. (2014)
		66.25% reduction of FFA when soaked in 0.1% formaldehyde solution at 25–30 °C for 6 h	Baranwal (2017)
4	Dry heat treatment	63.58% reduction in FFA when treated at 100 °C for 120 min	Bhati et al. (2016)
	Microwave heating	92.9% reduction in LA when treated at 900 W for 100 s	Yadav et al. (2012)
	γ-Irradiation	Significant reduction in the percentage fungal incidence and FFA at a dose level higher than 0.5 kGy	Mahmoud et al. (2016)
	Ohmic heating	Decreased water solubility and absorption indices, harder, and brighter product	Dias-Martins et al. (2019)
	Extrusion	Fat acidity and FFA were found to be the minimum, and the heat-treated pearl millet flour could be stored up to 6 days without undue deterioration in quality at ambient condition	Tiwari et al. (2014)

(continued overleaf)

Table 14.5 (Continued)

S. No.	Processing parameters/ treatment	Results	References
5	Hydrothermal treatment	LA significantly decreased with increase in steaming duration	Nantanga (2007)
		11.5% reduction in FA when grains held in boiling water for 15 min and later tray dried at 60 °C for 2 h	Jalgaonkar et al. (2016)
		53.8% reduction in FFA when blanched for 90 s	Bhati et al. (2016)
		Exhibited no lipase activity and found acceptable for 50 days when stored in polyethylene pouches (75 L) at ambient conditions (15–35 °C)	Yadav et al. (2012)
7	Combined treatments (heat treatment at 150–170 °C for 1.5 min at 300 rpm followed by γ-irradiation)	90.56% reduction in total fungal count when heat treatment combined with irradiation at 2.5 kGy; shelf life improved up to 60 days	Dikkala et al. (2018)
	Value-added products		
8	Pearl millet–based halwa dry mix	6 month stable	Yadav (2011)
9	Pearl millet–based kheer mix (N flushing)	396 days at 8 °C and 288 days at 25 °C	Bunkar et al. (2014)
10	Upma dry mix	No lipase activity after steaming 6 months; stable at ambient conditions (20–35 °C) in polyethylene pouches (75 µm)	Balasubramanian et al. (2014)
11	Extrudates	6-month shelf life	Sumathi et al. (2007)
12	Bread	2–4 days	Nami et al. (2019)
	Other techniques to improve the pearl millet stability		
13	Preservatives	73.97% reduction in FFA at 25–30 °C after 24 h acid treatment with 0.2 N HCl	Bhati et al. (2016)
		BHT and ascorbic acid treatments reduced both hydrolytic and oxidative reactions	Tiwari et al. (2014)

Table 14.5 (Continued)

S. No.	Processing parameters/ treatment	Results	References
14	N flushing	396 days at 8 °C and 288 days at 25 °C	Bunkar et al. (2014)
15	Storing at different temperatures (5, 25, 45 °C)	Storing the millet at lower temperatures (5 °C) has given 15 weeks of shelf life by providing low FFA, AV, and PV, whereas shelf life is only weeks at 45 °C	Selvan et al. (2022)

Abbreviations: AV, p-anisidine value; FFA, free fatty acids; LA, linoleic acid; PV, peroxide value.

Food-processing Technologies

The use of advanced food-processing technologies in pearl millet processing may improve stability and nutritional quality. These methods destroy the enzymes and microorganisms that cause spoilage or rancidity.

High-pressure Processing

High-pressure processing (HPP) is a gentle substitute for thermal pasteurization and sterilization of certain food products. With the development of HPP, consumers now have access to food that has undergone little processing and does not include additives. By successfully retaining bioactive chemicals and inactivating bacteria, yeast, mold, and viruses, HPP increases the shelf life of food products (Nema et al., 2022). Application of HPP in fruits and vegetables has shown significant inactivation of the microorganisms in apple (Vercammen et al., 2012), green beans (Krebbers et al., 2002), mango puree (Gurrero-Beltran et al., 2006), orange juice (Timmermans et al., 2011), and papaya beverage (Chen et al., 2015). By combining HPP with other thermal and nonthermal treatments either sequentially or simultaneously, it is possible to get beyond HPP's inability to inactivate spores. Effective and synergistic results in the inactivation of spores were obtained when combined with other thermal and non-thermal techniques, such as the addition of preservatives (Jofre et al., 2009), ultrafiltration (Zhao et al., 2014), thermosonication (Silva, 2016), irradiation (Paul et al., 1997), and modified atmospheric packaging.

Pulsed Electric Fields

Pulsed electric field (PEF) processing has the potential to create foods with great sensory, nutritional, and shelf-life quality. Foods are placed between two electrodes and processed using a high-intensity PEF. The pulses of high voltage used for this procedure typically range from 20 to 80 kV/cm. Application of a PEF (10–50 kV/cm) has inactivated microorganisms in the processing of apple juice

(Simpson et al., 1995), eggs (Dunn & Pearlman, 1987), green pea juice (Vega-Mercado et al., 1996), and beer (Ulmer et al., 2002).

Pulsed Light Treatment

Pulsed light treatment, a new nonthermal method for disinfecting food surfaces and packaging, uses quick, high-peak pulses of white light across the spectrum. It is seen as an alternative to continuously treating solid and liquid meals with ultraviolet light (Oms-Oliu et al., 2010). In a variety of antimicrobial applications, including the disinfection of water, air, surfaces, and food, exposure to UV short wavelengths between 100 and 280 nm (UV-C) has been used. The inactivation of microorganisms associated with food, such as *Listeria monocytogenes, Escherichia coli, Pseudomonas aeruginosa, Salmonella enteritidis, Bacillus cereus,* and *Staphylococcus aureus,* was 2 and 6 log CFU/mL using low and high UV content, respectively, after 200 pulses.

Cold plasma

Cold plasma is one of the most significant recent advancements in food processing technology. Ionizing a gas either fully or partially produces plasma. In order to sanitize food and/or food contact surfaces without heating or otherwise affecting the treated product, effective plasma systems must have a sufficient concentration of reactive products. Cold plasma is a versatile and effective method for surface sanitization against harmful bacteria, viruses, and parasites.

Innovative Roasting Techniques

Innovative roasting techniques that are more productive and can produce roasted millets of a consistent quality (improved storability) are available for millet processing in addition to traditional roasting methods.

Fluidized Bed Roasting

This method roasts grains quickly and uniformly by suspending grains in a stream of hot air or gas. The grains are cooked for a predetermined period of time and at a predetermined temperature to reach the desired level of roasting (Shingare & Thorat, 2013).

Infrared Roasting

On a conveyor belt, the grains are transported through an infrared radiation chamber where they are heated for a predetermined period of time and at a predetermined temperature (Shingare & Thorat, 2013).

Drum Roasting

This method involves roasting millet grains in a spinning drum. To ensure even roasting, the drum is heated from the outside while the grains are tossed inside.

Vacuum Roasting

This method involves toasting millet grains under low pressure. The grains are roasting in a vacuum chamber for a set amount of time and at a specific temperature. The millets and other grains acquire a unique flavor and aroma as a result of this process (Kaur et al., 2022).

Ultra-filtration

Ultra-filtration is a method of food processing that improves the nutritional value, quality, and safety of food items by using a membrane filtration process to separate food components according to molecular size.

Freezing

Freezing is a popular method of food preservation that results in high-quality, nutrient-dense foods with an extended shelf life. In the process of freezing, heat is removed from the food to lower the temperature below the freezing point, and some of the water changes state to form ice crystals. The recommended storage temperature for all frozen goods is $-18\,°C$, which will give them a longer shelf life. There are different types of novel freezing technologies, such as impingement, hydrofluidization, pressure-assisted freezing, ultrasound-assisted freezing, magnetic resonance–assisted freezing, electrostatic-assisted freezing, microwave-assisted freezing, radiofrequency-assisted freezing, and dehydro freezing (James et al., 2015).

Packaging

Thermal sterilization is a heat-treatment method that eliminates all living microorganisms (yeasts, molds, vegetative bacteria, and spore formers), extending the shelf life of the product. Two types of thermal sterilization are retorting and aseptic processing.

Retorting

Retorting is the process of first wrapping food in a container and then sterilizing it. Foods with pH levels >4.5 require sterilization temperatures $>100\,°C$. It is possible to reach this temperature in batch or continuous retorts. Continuous systems are gradually replacing batch retorts. The most popular continuous systems used in the food industry are hydrostatic retorts and rotary cookers.

Aseptic Packaging

With aseptic packaging, commercially sterilized food is placed in a sterilized package and then sealed in an aseptic setting. Paper and plastic are used in traditional aseptic packaging. Aseptic packing is frequently used to preserve liquids, dairy goods, tomato paste, and fruit slices. It can significantly lengthen food products' shelf lives; for instance, the ultra-high temperature (UHT) pasteurization process can extend liquid milk's shelf life from 19 to 90 days, while UHT processing and aseptic packaging together can extend shelf lives to 6 months or more.

Continuous research on the application of combined techniques may improve the storability of pearl millet grains and its products.

Goitrogens

Goitrogens are dietary compounds that disrupt thyroid metabolism and can exacerbate the impact of insufficient iodine levels. The clinical significance of most goitrogens is generally minimal unless their consumption is elevated and concurrent iodine deficiency exists. Among these, glucosinolates emerge as potent goitrogenic agents. The metabolites derived from glucosinolates engage in a competitive interaction with iodine for uptake by the thyroid gland. Notably, cyanoglucosides represent a distinct class of naturally occurring goitrogens. These compounds undergo metabolic conversion to thiocyanates, which function as anionic competitors with iodine during thyroid hormone synthesis. Elevated concentrations of these substances can impede the active transportation of iodine into the thyroid, diminishing the synthesis of thyroid hormones. Particularly susceptible demographic segments include the developing fetus and neonates, in whom adequate iodine assumes a critical role in facilitating normal thyroid function during the pivotal phase of neurodevelopment.

Studies indicate the presence of goitrogens in pearl millet, predominantly manifested as phenolic flavonoid-type compounds categorized as C-glycosyl flavones, namely vitexin, glucosyl vitexin, and glucosyl orientin. Additional phenolic constituents, such as phloroglucinol, resorcinol, and p-hydroxybenzoic acid, could contribute as well (Taylor, 2004). These compounds seem to impede the conversion of thyroxine hormone to its more bioactive form, triiodothyronine, through inhibition. In vitro investigations involving rats have illustrated 85% inhibition of thyroid peroxidase enzyme by C-glycosylflavones (Gaitan et al., 1989).

Furthermore, studies suggest that the augmented impact of millet on thyroid function in rats, particularly under fermentation, likely stems from the extraction of minerals from millet or potential chemical transformations of the inherent goitrogens within the grain. These goitrogenic elements are predominantly concentrated in the outer layers of the grain and experience substantial reduction upon de-hulling during milling. However, it is essential to exercise caution in attributing excessive nutritional significance to the goitrogens found in pearl millet. Although instances of goiter affliction have been observed among certain rural populations, such as in Sudan, where pearl millet constitutes a dietary staple, it is plausible that these occurrences are linked to dietary limitations leading to iodine deficiency (Taylor, 2004, 2016). Elnour et al. (1998) have also suggested that the amplified thyroid effects of millet in rats due to fermentation may be associated with mineral removal from millet and/or chemical alterations of millet-derived goitrogens. However, evidence from clinical trials on millet consumption must be studied further.

Exploring the Avenues for Tapping the Health Benefits of Millets

Millets have emerged as noteworthy functional foods due to their potent health-promoting properties, being nutrient rich with B vitamins and minerals, aiding digestion due to their fiber content, regulating blood sugar levels, supporting heart health through cholesterol management, promoting weight management, providing antioxidants for combating free radicals, contributing to bone health with Ca and P, potentially reducing the risk of chronic diseases, and offering valuable protein and amino acids. Geetha et al. (2020) found a potential role of millet-based food mix in lowering fasting blood sugar levels and HbA1c. Despite variations in glycemic and insulin responses based on millet type and cooking methods, overall, millets demonstrate a favorable impact on fasting and postprandial blood glucose levels as well as on the plasma–insulin response in both healthy individuals and those with type-2 diabetes (Almaski et al., 2019).

Research has demonstrated that adopting a diet centered around millet can reduce triglycerides and regulate glucose levels among diabetic patients (Vedamanickam et al., 2020). In a meta-analysis conducted by Anitha et al. (2021), it was observed that incorporating millet-based foods into one's diet for varying periods ranging from 21 days to 4 months resulted in a notable decrease in total cholesterol, TAG, very-low-density lipoprotein cholesterol, and/or low-density lipoprotein cholesterol levels. Studies conducted in laboratory settings have suggested that millet phenolics hold promise in impeding the initiation and progression of cancer (Chandrasekara & Shahidi, 2011). Moreover, protein extracts from pearl millet have shown effectiveness in restraining the growth of *Rhizoctonia solani*, *Macrophomina phaseolina*, and *Fusarium oxysporum* (Radhajeyalakshmi et al., 2003). Nonetheless, a more extensive body of research is necessary before conclusive claims regarding the benefits of millet can be established.

Conclusion

The adoption of appropriate processing technology with suitable equipment decreases the anti-nutritional constituents and off-flavors from millets, which has led to an increase in the number of consumers for millets and boosted the commercial scale production of various VAPs of millets.

In the last few years, increased awareness of the health benefits of nutri-cereals has led to an increase in the popularity of VAPs from millets. Now millet-based food products like rawa, flakes, single- and multigrain flour, biscuits, Papad, etc., are gradually finding their place on the plate. Along with the emerging startups, many large FMCG companies (i.e., ITC, TATA Britannia, Kellog's, Parle, Patanjali, HUL, Aachi, etc.) are now operating in this space with various VAPs, which were negligible in previous decades.

Currently, out of total millets production of ~15.92 million t, the ratio of VAPs is ~0.606 million t, equivalent to only 3%–4%. In the next 10 years, there will be substantial efforts to increase consumption of pearl millet, focusing on increasing awareness, branding, and government support. Among the millets, pearl millet has the highest production; therefore, the opportunity for innovation in developing VAPs is quite high compared with other millets. However, there are several challenges that hinder value addition in pearl millet that may be addressed using the right decortication methods, which can considerably improve the acceptability of pearl millet flour and its products. It is imperative that adequate and motorized milling technology be promoted on a commercial basis to improve decortication, milling, and product features. Research says that TAG lipases from low-rancidity pearl millet lines display less TAG breakdown and lipid oxidation after milling. Mutations in these genes may provide targets for the development of pearl millet germplasm with longer flour shelf life.

The Government of India has introduced schemes and subsidies to promote millet cultivation and consumption, such as the National Food Security Mission to enhance production, Rashtriya Krishi Vikas Yojana for financial support, the Prime Minister Formalization of Micro Food Processing Enterprise for food processing, Production Linked Incentive for Millet Based Products, One District One Product, and Pradhan Mantri Krishi Sinchayee Yojana for irrigation. These initiatives, along with various state millet missions, raise awareness about millet's nutrition, ensure fair prices via MSPs, offer seed subsidies, and improve market linkages for millet farmers. The millet sector in the world has been facing several challenges due to policies biased toward finer cereals, lack of investment in infrastructure, lack of effective demand for millet products, lower shelf life for millet products, lack of upscaled technologies, and low funding of agricultural research and extension institutions. There is need for strengthening the capacity of research and development partners for further knowledge transfer, shared learning, innovations, and sustainable impacts. Constraints such as rancidity, bitterness, antinutrients, and dark color hinder the utilization of pearl millet despite its nutritional benefits. Efforts are being made to develop cream/white-colored pearl millet genotypes by several organizations such as ICAR-IIMR, ICAR-IARI, CCSHAU, and ICRISAT. Thus, rancidity in pearl millet flour is the main constraint in enhancing demand. Research and development are needed to tackle this issue and to create a sustainable demand. An integrated value chain approach linking farmers to markets through processing units equipped with machinery can result in additional income for farmers.

References

Abdalla, A. A., El Tinay, A. H., Mohamed, B. E., & Abdalla, A. H. (1998). Proximate composition, starch, phytate and mineral contents of 10 pearl millet genotypes. *Food Chemistry, 63*(2), 243–246.

Almaski, A., Shelly, C. O. E., Lightowler, H., & Thondre, S. (2019). Millet intake and risk factors of type 2 diabetes: A systematic review. *Journal of Food and Nutritional Disorders, 8*(3), 2.

Anitha, S., Botha, R., Kane-Potaka, J., Givens, D. I., Rajendran, A., Tsusaka, T. W., & Bhandari, R. K. (2021). Can millet consumption help manage hyperlipidemia and obesity?: A systematic review and meta-analysis. *Frontiers in Nutrition, 8*, 478.

Annor, G. A., Marcone, M., Corredig, M., Bertoft, E., & Seetharaman, K. (2015). Effects of the amount and type of fatty acids present in millets on their in vitro starch digestibility and expected glycemic index (eGI). *Journal of Cereal Science, 64*, 76–81.

Archana, S., & Kawatra, A. S. H. A. (1998). Reduction of polyphenol and phytic acid content of pearl millet grains by malting and blanching. *Plant Foods for Human Nutrition, 53*, 93–98.

Badau, M. H., Nkama, I., & Jideani, I. A. (2005). Phytic acid content and hydrochloric acid extractability of minerals in pearl millet as affected by germination time and cultivar. *Food Chemistry, 92*(3), 425–435.

Balasubramanian, S., Yadav, D. N., Kaur, J., & Anand, T. (2014). Development and shelf-life evaluation of pearl millet based upma dry mix. *Journal of Food Science and Technology, 51*(6), 1110–1117.

Baranwal, D. (2017). Malting: An indigenous technology used for improving the nutritional quality of grains: A review. *Asian Journal of Dairy and Food Research, 36*, 179–183.

Bhati, D. S., Bhatnagar, V., & Acharya, V. (2016). Effect of pre-milling processing techniques on pearl millet grains with special reference to in-vitro iron availability. *Asian Journal of Dairy and Food Research, 35*(1), 76–80.

Bunkar, D. S., Jha, A., Mahajan, A., & Unnikrishnan, V. S. (2014). Kinetics of changes in shelf life parameters during storage of pearl millet based kheer mix and development of a shelf life prediction model. *Journal of Food Science and Technology, 51*(12), 3740–3748.

Burton, G. W. (1965). Pearl millet Tift 23A released. *Crops & Soils, 17*(5), 19.

Chandrasekara, A., & Shahidi, F. (2011). Antiproliferative potential and DNA scission inhibitory activity of phenolics from whole millet grains. *Journal of Functional Foods, 3*(3), 159–170.

Chen, D., Pang, X., Zhao, J., Gao, L., Liao, X., Wu, J., & Li, Q. (2015). Comparing the effects of high hydrostatic pressure and high temperature short time on papaya beverage. *Innovative Food Science & Emerging Technologies, 32*, 16–28.

Chinenye, O. E., Ayodeji, O. A., & Baba, A. J. (2017). Effect of fermentation (natural and starter) on the physicochemical, anti-nutritional and proximate composition of pearl millet used for flour production. *American Journal of Bioscience and Bioengineering, 5*(1), 12–16.

Crow, J. F. (1998). 90 years ago: The beginning of hybrid maize. *Genetics, 148*(3), 923–928.

Datta Mazumdar, S. K., Gupta, R., Banerjee, S., Gite, P., Durgalla, P. B., & Bagade, P. (2016). Determination of variability in rancidity profile of select commercial pearl millet varieties/hybrids. DC 24. Poster presented in CGIAR Research Program on Dryland Cereals Review Meeting held at Hyderabad, India, 5–6 October 2016. International Crops Research Institute for the Semi-Arid Tropics, Patancheru, Telengana, India.

Dayakar Rao, B. (2022). Jowar & Bajra special- May 2022 Edition. E-Newsletter of Pradhan Mantri Formalisation of Micro Food Processing Enterprises, *2*(5), 1.

Dhankher, N., & Chauhan, B. M. (1987). Effect of temperature and fermentation time on phytic acid and polyphenol content of rabadi—A fermented pearl millet food. *Journal of Food Science, 52*(3), 828–829.

Dias-Martins, A. M., Cappato, L. P., da Costa Mattos, M., Rodrigues, F. N., Pacheco, S., & Carvalho, C. W. P. (2019). Impacts of ohmic heating on decorticated and whole pearl millet grains compared to open-pan cooking. *Journal of Cereal Science, 85*, 120–129.

Dikkala, P. K., Hymavathi, T. V., Roberts, P., & Sujatha, M. (2018). Effect of heat treatment and gamma irradiation on the total bacterial count of selected millet grains (jowar, bajra and foxtail). *International Journal of Current Microbiology and Applied Sciences, 7*(2), 1293–1300.

Directorate of Millets Development. (2020). Department of Agriculture, Co-operation & Farmers Welfare, Ministry of Agriculture & Farmers Welfare, Government of India.

Dunn, J. E., & Pearlman, J. S. (1987). *Methods and apparatus for extending the shelf life of fluid food products* (U.S. Patent No. US 469547). U.S. Patent and Trademark Office.

Egli, I., Davidsson, L., Juillerat, M. A., Barclay, D., & Hurrell, R. F. (2002). The influence of soaking and germination on the phytase activity and phytic acid content of grains and seeds potentially useful for complementary feeding. *Journal of Food Science, 67*(9), 3484–3488.

Elnour, A., Liedén, S. Å., Bourdoux, P., Eltom, M., Khalid, S. A., & Hambraeus, L. (1998). Traditional fermentation increases goitrogenic activity in pearl millet. *Annals of Nutrition and Metabolism, 42*(6), 341–349.

Elyas, S. H., El Tinay, A. H., Yousif, N. E., & Elsheikh, E. A. (2002). Effect of natural fermentation on nutritive value and in vitro protein digestibility of pearl millet. *Food Chemistry, 78*(1), 75–79.

FAO. (1995). Sorghum and millets in human nutrition. *FAO Food and Nutrition Series, 27*, 16–19.

FAO. (2022). *Crops and livestock products.* FAOSTAT. https://www.fao.org/faostat/en/#data/QCL/metadata

Gaitan, E., Lindsay, R. H., Reichert, R. D., Ingbar, S. H., Cooksey, R. C., Legan, J., & Kubota, K. (1989). Antithyroid and goitrogenic effects of millet: Role of C-glycosylflavones. *The Journal of Clinical Endocrinology & Metabolism, 68*(4), 707–714.

Geetha, K., Yankanchi, G. M., Hulamani, S., & Hiremath, N. (2020). Glycemic index of millet based food mix and its effect on pre diabetic subjects. *Journal of Food Science and Technology, 57*, 2732–2738.

Gurrero-Beltran, J. A., Barbosa-Canovas, G. V., Moraga-Ballesteros, G. E. M. M. A., Moraga-Ballesteros, M. J., & Swanson, B. G. (2006). Effect of pH and ascorbic acid on high hydrostatic pressure-processed mango puree. *Journal of Food Processing and Preservation, 30*(5), 582–596.

Hama, F., Icard-Vernière, C., Guyot, J.-P., Picq, C., Diawara, B., & Mouquet-Rivier, C. (2011). Changes in micro-and macronutrient composition of pearl millet and white sorghum during in field versus laboratory decortication. *Journal of Cereal Science, 54*(3), 425–433.

Jalgaonkar, K., Jha, S. K., & Sharma, D. K. (2016). Effect of thermal treatments on the storage life of pearl millet (*Pennisetum glaucum*) flour. *Indian Journal of Agricultural Sciences, 86*(6), 762–767.

James, C., Purnell, G., & James, S. J. (2015). A review of novel and innovative food freezing technologies. *Food and Bioprocess Technology, 8*, 1616–1634.

Jideani, V. A., Oloruntoba, R. H., & Jideani, I. A. (2010). Optimization of fura production using response surface methodology. *International Journal of Food Properties, 13*(2), 272–281.

Jofré, A., Aymerich, T., Grèbol, N., & Garriga, M. (2009). Efficiency of high hydrostatic pressure at 600 MPa against food-borne microorganisms by challenge tests on convenience meat products. *LWT - Food Science and Technology, 42*(5), 924–928.

Jorben, J., Singh, S. P., Satyavathi, C. T., Sankar, S. M., Bhat, J. S., Durgesh, K., & Mallik, M. (2020). Inheritance of fertility restoration of A4 cytoplasm in pearl millet [*Pennisetum glaucum* (L.) R. Br.]. *Indian Journal of Genetics and Plant Breeding, 80*(01), 64–69.

Kaur, R., Kumar, A., Kumar, V., Kumar, S., Saini, R. K., Nayi, P., & Gehlot, R. (2022). Recent advancements and applications of explosion puffing. *Food Chemistry, 403*, 134452.

Krebbers, B., Matser, A. M., Koets, M., & Van den Berg, R. W. (2002). Quality and storage-stability of high-pressure preserved green beans. *Journal of Food Engineering, 54*(1), 27–33.

Kumar, A., Kaur, A., Gupta, K., Gat, Y., & Kumar, V. (2021). Assessment of germination time of finger millet for value addition in functional foods. *Current Science, 120*(2), 406–413.

Lei, V., & Jakobsen, M. (2004). Microbiological characterization and probiotic potential of koko and koko sour water, African spontaneously fermented millet porridge and drink. *Journal of Applied Microbiology, 96*(2), 384–397.

Mahmoud, N. S., Awad, S. H., Madani, R. M. A., Osman, F. A., Elmamoun, K., & Hassan, A. B. (2016). Effect of γ radiation processing on fungal growth and quality characteristics of millet grains. *Food Science & Nutrition, 4*(3), 342–347.

Mohamed, E. A., Ahmed, I. A. M., & Babiker, E. E. (2011). Preservation of millet flour by refrigeration: Changes in total protein and amino acids composition during storage. *International Journal of Nutrition and Food Engineering, 5*(4), 346–349.

NAAS. (2022). *Promoting millet production, value addition and consumption.* National Academy of Agricultural Sciences.

Nami, Y., Gharekhani, M., Aalami, M., & Hejazi, M. A. (2019). Lactobacillus-fermented sourdoughs improve the quality of gluten-free bread made from pearl millet flour. *Journal of Food Science and Technology, 56*(9), 4057–4067.

Nantanga, K. K. M. (2007). *Lipid stabilisation and partial pre-cooking of pearl millet by thermal treatments* [Doctoral dissertation, University of Pretoria]. Lipid Stabilisation and Partial Pre-Cooking of Pearl Millet by Thermal Treatments - ProQuest

Nema, P. K., Sehrawat, R., Ravichandran, C., Kaur, B. P., Kumar, A., & Tarafdar, A. (2022). Inactivating food microbes by high-pressure processing and combined nonthermal and thermal treatment: A review. *Journal of Food Quality, 2022*, 5797843.

Nithya, K. S., Ramachandramurty, B., & Krishnamoorthy, V. V. (2007). Effect of processing methods on nutritional and anti-nutritional qualities of hybrid (COHCU-8) and traditional (CO7) pearl millet varieties of India. *Journal of Biological Sciences, 7*(4), 643–647.

Ocheme, O. B., & Chinma, C. E. (2008). Effects of soaking and germination on some physicochemical properties of millet flour for porridge production. *Journal of Food Technology, 6*(5), 185–188.

Ogidi, A. E., & Abah, D. A. (2012). The impact of sorghum value chain on enterprise development: A holistic diagnosis of some actors in Benue State, Nigeria. *International Journal of Agriculture, 4*(4), 79–92.

Oms-Oliu, G., Martín-Belloso, O., & Soliva-Fortuny, R. (2010). Pulsed light treatments for food preservation: A review. *Food and Bioprocess Technology, 3*, 13–23.

Osman, M. A. (2011). Effect of traditional fermentation process on the nutrient and antinutrient contents of pearl millet during preparation of Lohoh. *Journal of the Saudi Society of Agricultural Sciences, 10*(1), 1–6.

Paul, P., Chawla, S. P., Thomas, P., Kesavan, P. C., Fotedar, R., & Arya, R. N. (1997). Effect of high hydrostatic pressure, gamma-irradiation and combination treatments on the microbiological quality of lamb meat during chilled storage. *Journal of Food Safety, 16*(4), 263–271.

Pelig-Ba, K. B. (2009). Assessment of phytic acid levels in some local cereal grains in two districts in the upper east region of Ghana. *Pakistan Journal of Nutrition, 8*(10), 1540–1547.

Pushparaj, F. S., & Urooj, A. (2011). Influence of processing on dietary fiber, tannin and in vitro protein digestibility of pearl millet. *Food and Nutrition Sciences, 2*(08), 895.

Radhajeyalakshmi, R., Yamunarani, K., Seetharaman, K., & Velazhahan, R. (2003). Existence of thaumatin-like proteins (TLPs) in seeds of cereals. *Acta Phytopathologica et Entomologica Hungarica, 38*(3–4), 251–257.

Rai, S., Kaur, A., & Singh, B. (2014). Quality characteristics of gluten free cookies prepared from different flour combinations. *Journal of Food Science and Technology, 51*, 785–789.

Rani, S., Singh, R., Sehrawat, R., Kaur, B. P., & Upadhyay, A. (2018). Pearl millet processing: A review. *Nutrition & Food Science, 48*(1), 30–44.

Rao, B. D., Malleshi, N. G., Annor, G. A., & Patil, J. V. (2016a). *Millets value chain for nutritional security: A replicable success model from India.* CABI.

Rao, B. D., & Nune, S. D. (2021). Role of Nutrihub incubation for the development of business opportunities in millets: An Indian scenario. In A. Kumar, M. K. Tripathi, D. Joshi, & V. Kumar (Eds.), *Millets and millet technology* (pp. 413–438). Springer.

Rao, B. D., Patil, J. V., Rajendraprasad, M. P., Reddy, K. N., Devi, K., Sriharsha, B., & Kachui, N. (2010). Impact of innovations in value chain on sorghum farmers. *Agricultural Economics Research Review, 23*(347-2016-16943), 419–426.

Rao, B. D., & Tonapi, V. A. (2021). *A compendium of millet start-ups, success stories.* ICAR-IIMR.

Rao, B. D., Vishala, A. D., Christina, G. A., & Tonapi, V. A. (2016b). *Millet recipes–A healthy choice.* Indian Institute of Millets Research.

Rathore, T., Singh, R., Kamble, D. B., Upadhyay, A., & Thangalakshmi, S. (2019). Review on finger millet: Processing and value addition. *The Pharma Innovation Journal, 8*(4), 283–291.

Ratnavathi, C. V., & Patil, J. V. (2013). Sorghum utilization as food. *Nutrition & Food Science, 4*(1), 1–8.

Sade, F. O. (2009). Proximate, antinutritional factors and functional properties of processed pearl millet (*Pennisetum glaucum*). *Journal of Food Technology, 7*(3), 92–97.

Selvan, S. S., Mohapatra, D., Anakkallan, S., Kate, A., Tripathi, M. K., Singh, K., & Kar, A. (2022). Oxidation kinetics and ANN model for shelf life estimation of pearl millet (*Pennisetum glaucum* L.) grains during storage. *Journal of Food Processing and Preservation, 46*(12), e17218.

Sharma, A., & Kapoor, A. C. (1996). Levels of antinutritional factors in pearl millet as affected by processing treatments and various types of fermentation. *Plant Foods for Human Nutrition, 49,* 241–252.

Sharma, T. R., Gupta, S., Roy, A. K., Ram, B., Kar, C. S., Yadav, D. K., Kar, G., Singh, G. P., Kumar, P., Venugopalan, M. V., Singh, R. K., Sundaram, R. M., Mitra, S., Jha, S. K., Satpathy, S., Rakshit, S., Tonapi, V., & Prasad, Y. G. (2022). Achievements in field crops in independent India. In H. Pathak, J. P. Mishra, & T. Mohapatra (Eds.), *Indian agriculture after independence* (p. 426). Indian Council of Agricultural Research.

Shingare, S. P., & Thorat, B. N. (2013). Effect of drying temperature and pretreatment on protein content and color changes during fluidized bed drying of finger millets (ragi, *Eleusine coracana*) sprouts. *Drying Technology, 31*(5), 507–518.

Sihag, M. K., Sharma, V., Goyal, A., Arora, S., & Singh, A. K. (2015). Effect of domestic processing treatments on iron, β-carotene, phytic acid and polyphenols of pearl millet. *Cogent Food & Agriculture, 1*(1), 1109171.

Silva, F. V. (2016). High pressure processing pretreatment enhanced the thermosonication inactivation of *Alicyclobacillus acidoterrestris* spores in orange juice. *Food Control, 62,* 365–372.

Simpson, M. V., Barbosa-Cánovas, G. V., & Swanson, B. G. (1995). The combined inhibitory effect of lysozyme and high voltage pulsed electric fields on the growth of *Bacillus subtilis* spores. In *IFT annual meeting: Book of abstracts* (Vol. 267). Institute of Food Technologists.

Singh, R. B., Khanna-Chopra, R., Singh, A. K., Gopal, K., S., Singh, N. K., Prabhu, K. V., Singh, A. K., Bansal, K. C., & Mahadevappa, M. (2016). Crop sciences. In R. B. Singh (Ed.), *100 years of agricultural sciences in India* (pp. 1–97). *National Academy of Agricultural Sciences.*

Stephens, J. C., & Holland, R. F. (1954). Cytoplasmic male-sterility for hybrid sorghum seed production. *Agronomy Journal, 46,* 20–23.

Suma, P. F., & Urooj, A. (2014). Influence of germination on bioaccessible iron and calcium in pearl millet (*Pennisetum typhoideum*). *Journal of Food Science and Technology, 51*(5), 976–981.

Sumathi, A., Ushakumari, S. R., & Malleshi, N. G. (2007). Physico-chemical characteristics, nutritional quality and shelf-life of pearl millet based extrusion cooked supplementary foods. *International Journal of Food Sciences and Nutrition, 58*(5), 350–362.

Swaminathan, M. (2013). Genesis and growth of the yield revolution in wheat in India: Lessons for shaping our agricultural destiny. *Agricultural Research, 2,* 183–188.

Swaminathan, M. S. (2016). *Combating hunger and achieving food security.* Cambridge University Press.

Taylor, J. R. N. (2004). Pearl. In *Encyclopedia of grain science* (pp. 253–261). Elsevier.

Taylor, J. R. N. (2016). Pearl. In *Encyclopedia of food grains* (pp. 190–198). Elsevier.

Timmermans, R. A. H., Mastwijk, H. C., Knol, J. J., Quataert, M. C. J., Vervoort, L., Van der Plancken, I., & Matser, A. M. (2011). Comparing equivalent thermal, high pressure and pulsed electric field processes for mild pasteurization of orange juice. Part I: Impact on overall quality attributes. *Innovative Food Science & Emerging Technologies, 12*(3), 235–243.

Tiwari, A., Jha, S. K., Pal, R. K., Sethi, S., & Krishan, L. (2014). Effect of pre-milling treatments on storage stability of pearl millet flour. *Journal of Food Processing and Preservation, 38*(3), 1215–1223.

Ulmer, H. M., Heinz, V., Gänzle, M. G., Knorr, D., & Vogel, R. F. (2002). Effects of pulsed electric fields on inactivation and metabolic activity of Lactobacillus plantarum in model beer. *Journal of Applied Microbiology, 93*(2), 326–335.

Vedamanickam, R., Anandan, P., Bupesh, G., & Vasanth, S. (2020). Study of millet and non-millet diet on diabetics and associated metabolic syndrome. *Biomedicine, 40*(1), 55–58.

Vega-Mercado, H., Martin-Belloso, O. L. G. A., Chang, F. J., Barbosa-Canovas, G. V., & Swanson, B. G. (1996). Inactivation of *Escherichia coli* and *Bacillus subtilis* suspended in pea soup using pulsed electric fields. *Journal of Food Processing and Preservation, 20*(6), 501–510.

Vercammen, A., Vanoirbeek, K. G., Lemmens, L., Lurquin, I., Hendrickx, M. E., & Michiels, C. W. (2012). High pressure pasteurization of apple pieces in syrup: Microbiological shelf-life and quality evolution during refrigerated storage. *Innovative Food Science & Emerging Technologies, 16*, 259–266.

Yadav, D. N., Anand, T., Kaur, J., & Singh, A. K. (2012). Improved storage stability of pearl millet flour through microwave treatment. *Agricultural Research, 1*, 399–404.

Yadav, O. P. (2011). Review of pearl millet research. In *Proceedings of all India Coordinated Pearl Millet Improvement Project (AICPMIP) workshop* (pp. 12–14). Indian Council of Agricultural Research.

Yadav, O. P., Singh, D. V., Dhillon, B. S., & Mohapatra, T. (2019). India's evergreen revolution in cereals. *Current Science, 116*(11), 1805–1808.

Yadav, O. P., Upadhyaya, H. D., Reddy, K. N., Jukanti, A. K., Pandey, S., & Tyagi, R. K. (2017). Genetic resources of pearl millet: Status and utilization. *Indian Journal of Plant Genetic Resources, 30*, 31–47.

YES BANK & APEDA. (2022). *Superfood millets: A USD 2 billion export opportunity for India.* Agricultural & Processed Food Products Export Development Authority.

Zhao, L., Wang, Y., Qiu, D., & Liao, X. (2014). Effect of ultrafiltration combined with high-pressure processing on safety and quality features of fresh apple juice. *Food and Bioprocess Technology, 7*, 3246–3258.

15

Pearl Millet: Processing and Value Addition for Gluten-free Markets

Shivaprasad Doddabematti Prakash, Manoj Kumar Pulivarthi, and Kaliramesh Siliveru

Chapter Overview

Pearl millet [*Pennisetum glaucum* (L.) R. Br.] is a nutrient-dense cereal crop that is mostly grown in arid and semi-arid regions of Africa and India. This crop is known to have a deep root system that protects it from drought, heat, and other environmental stresses, making it an important crop for farmers facing agricultural and environmental challenges (Gupta et al., 2017). The absence of gluten proteins in pearl millet further increases its scope in formulating nutritious gluten-free foods for those suffering from celiac disease (CD) or gluten intolerance. It also has high protein and fat content, making it superior to sorghum and other millet grains in terms of total energy and protein quality. However, lysine is a limiting essential amino acid in pearl millet, just like it is in other cereal grains (Dias-Martins et al., 2018; Pawase et al., 2021). Whereas pearl millet is a staple food crop in many African and Indian communities, its presence in North America and Europe is rare, and it remains relatively unfamiliar in some countries (Taylor, 2016). Nonetheless, researchers recognize its potential as a valuable component in gluten-free product production.

The consumption of pearl millet is also linked with several potential health benefits. The high Fe, Ca, Mg, K, Zn, P, polyphenols, dietary fiber, and protein contents distinguish pearl millet from other cereal grains (Habiyaremye et al., 2017). Aside from the increased desire for healthy cereal-based foods, there is a high demand for gluten-free foods, particularly among those suffering from gluten-related diseases such as non-celiac gluten or wheat sensitivity and CD (Moss & McSweeney, 2022). Celiac disease, which affects 1%–2% of the world's population, is a hereditary auto-immune illness characterized by long-term gluten sensitivity. Following a strict gluten-free diet is the only effective treatment for CD (Briani et al., 2008; Doddabematti Prakash et al., 2022; Reilly & Green, 2012). Because of its numerous

Pearl Millet: A Resilient Cereal Crop for Food, Nutrition, and Climate Security, First Edition.
Edited by Ramasamy Perumal, P. V. Vara Prasad, C. Tara Satyavathi, Mahalingam Govindaraj, and Abdou Tenkouano.
© 2025 American Society of Agronomy, Inc., Crop Science Society of America, Inc., Soil Science Society of America, Inc. Published 2025 by John Wiley & Sons, Inc.

health and nutritional benefits, pearl millet has enormous potential for use in the human diet. Pearl millet is naturally gluten free and hence is acceptable for people who are gluten sensitive. It is also a suitable dietary option for patients with type II diabetes or those at risk of developing the disease due to its low (<55) to moderate (56–69) glycemic index (Lemgharbi et al., 2017). Additionally, the antioxidant properties of pearl millet may aid in cell protection, lowering the risk of chronic illnesses, including cancer and heart disease. Studies also show that its consumption assists in reducing inflammation and improving gut health (Dekka et al., 2023; Saleh et al., 2013; Selladurai et al., 2023).

Despite the benefits of consuming pearl millet, it is important to address the potential challenges associated with its use (Figure 15.1). The presence of antinutrients, rapid rancidity, and the development of bitterness and discoloration of flour have hindered the acceptability and use of pearl millet. These grains can be preserved for extended periods of time with minimal quality loss when their kernels are kept intact. However, once the grains are decorticated and ground, the flour's shelf life drops rapidly due to hydrolytic and oxidative changes in the flour/meal lipids. These modifications cause the creation of peroxides and the release of free fatty acids (FFAs), which give the flour an off-flavor and a bitter taste (Kumar et al., 2021). Fortunately, these challenges can be easily addressed with the current advancements in processing techniques. The use of adequate processing methods can pave the way to the production of gluten-free foods with an extended shelf life

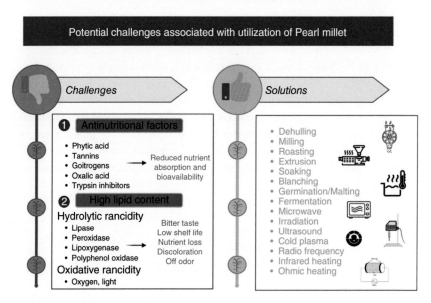

Figure 15.1 Challenges associated with the utilization of pearl millet and solutions.

using pearl millet, enabling its market expansion to unconventional millet consumers. This chapter provides a thorough analysis of pearl millet's nutritional and health advantages as well as a number of processing methods (e.g., dehulling, milling, roasting, and extrusion) and other treatments that improve the value of pearl millet as a gluten-free substitute in a variety of food applications. Additionally, this chapter explores different modern and traditional value-added products made from pearl millet in various parts of the world.

Nutritional Framework

The composition of pearl millet has been extensively studied over the past decade (Table 15.1). Due to its abundant protein content, improved amino acid balance, and high concentration of vitamin A, pearl millet has superior nutritional benefits compared with other grains (Krishnan & Meera, 2018; Sankar et al., 2008). Because it contains more oil than maize, it is frequently referred to as a "high-energy" grain (Bello et al., 2019). Pearl millet has a protein level that varies from 8% to 19%, which is analogous to wheat (11.8 g/100 g) and is greater than other cereal grains, such as maize (4.7 g/100 g), rice (6.8 g/100 g), and sorghum (10.4 g/100 g) (Nambiar et al., 2011). Albumins and globulins make up 22%–28% of the protein in pearl millet, followed by prolamin and prolamin-like compounds (22%–35%) and glutelin and glutelin-like compounds (28%–32%) (Krishnan & Meera, 2018). Additionally, compared with sorghum, the amino acid balance of pearl millet is known to be superior. Methionine and lysine contents in pearl millet are 40% higher than in maize. Compared with corn and sorghum, lysine content in pearl millet is higher by 21% and 36%, respectively (Léder, 2004). In terms of fat content, pearl millet is superior to rice, maize, wheat, and sorghum (5–7 g/100 g). The digestibility of pearl millet fat is greater in comparison to other cereal grains (Pei et al., 2022). The dried grain contains about 70% carbohydrates, mostly in the form of starch (56%–65%), of which 20%–22% is amylose. Pearl millet has a lower carbohydrate content (1.2%–2.6%) than rice (75%–80%) (Pei et al., 2022; Verma & Srivastav, 2017). Similarly, pearl millet has greater total dietary fiber (18.6%) than sorghum (14.2%), wheat (17.2%), and rice (8.3%) (Kamath & Belavady, 1980; Pei et al., 2022). Further, pearl millet is known to have a greater ash percentage (2.3%) than wheat (1.42%), rice (0.56%), or maize (1.16%) (Abdulrahman & Omoniyi, 2016). In terms of minerals, pearl millet is an abundant source of Mn, Cu, Zn, and P. In contrast to their deficiencies in vitamins B and C, mature kernels are high in vitamin A (Krishnan & Meera, 2018). Due to the relatively high Fe content, individuals who consume pearl millet can fulfill their Fe requirement and do not have to suffer from anemia, a Fe deficiency condition (Sabuz et al., 2023).

Table 15.1   Nutritional Composition of Pearl Millet

Composition	Value
Protein (%)	11.8
Fat (%)	4.8
Carbohydrate (%)	67.0
Crude fiber (%)	2.3
Minerals (mg)	
Ca	16
Fe	6
Mg	228
P	570
K	390
Na	10
Zn	3.4
Cu	1.5
Mg	3.3
Vitamins (μg)	
K	1.8
E	100
Riboflavin	580
Thiamine	842
B6	768
Niacin	9.4
Pantothenic acid	170
Foliate	170
Amino acids (g/100 g of protein)	
Glutamic acid	23.0
Leucine	10.7
Isoleucine	4.4
Alanine	8.7
Proline	5.8
Valin	4.9
Arginine	4.6
Threonine	4.0
Lysine	3.1
Cysteine	1.5
Tryptophan	1.4
Methionine	1.1

Note: Adapted from Nambiar et al. (2011) and Pei et al. (2022).

Antinutritional Properties

Similar to other grains, pearl millet has antinutrients such as tannins, polyphenols, and phytic acid. The main form of stored P and minerals, phytate is the salt of phytic acid, myo-inositol-1,2,3,4,5,6 hexakisphosphate. It is found throughout the plant kingdom and makes up 75% of the total P in the kernel (Boncompagni et al., 2018; Raboy, 2002). The strong chelating properties of phytic acid allow it to easily form complexes with monovalent and multivalent cations like P, Ca, Fe, Zn, Mn, and others, lowering their bio-availability and impairing absorption. According to Sandberg (2001), inositol triphosphate and inositol tetraphosphate led to decreased Fe absorption in processed foods containing a mixture of inositol phosphates, which could be related to Fe interaction with highly phosphorylated inositol phosphates. Similarly, Jha et al. (2015) reported that endosperm has less phytic acid than the aleurone layer of pearl millet. According to their result, a complete phytic acid degradation results in a fivefold increase in mineral absorption; however, a 90% degradation (1%–0.1%) caused a twofold increase in absorption, indicating that complete phytic acid degradation may not always be attainable.

In addition to phytic acid, some C-glycosylflavones (C-GFs), including glucosyl vitexin, glucosyl orientin, and vitexin, are goitrogenic polyphenols that may contribute to health issues. In plants and as dietary components, C-GFs are physiologically active and resistant to hydrolysis (Boncompagni et al., 2018). Epidemiologic studies suggest that the consumption of millet as a primary dietary component, commonly observed in rural villages across Asia and Africa, might be linked to the occurrence of endemic goiter in these areas. Several studies have indicated that increased consumption of millet can lead to increased occurrence of goiter. Studies by Eltom et al. (1984), Osman (2012), and Osman and Fatah (1981) reported that goiter was found to be more common in rural villages of Sudan whose dietary energy consisted of 74% of millets.

Polyphenols are antinutritional factors that are widely found in pearl millets (307–714 mg/100 g) (Boncompagni et al., 2018). Jha et al. (2015) reported that the endosperm (0.45 g/100 g) portion of pearl millet has a lower content of polyphenols in comparison to the bran (1.44 g/100 g). Several studies have also indicated that polyphenols can greatly interfere with human enzymatic systems that regulate thyroid hormone synthesis (Gonçalves et al., 2017). This may result in the formation of a goiter, which is a swelling of the larynx or neck brought on by an enlarged thyroid gland. Similarly, some polyphenols have the capability to form complexes with Fe and make them unavailable for absorption. According to Brune et al. (1991), the extent of blockage of Fe absorption roughly corresponds to the quantity of Fe-binding phenolic galloyl groups in meals. Further,

Brune et al. (1991) asserted that the primary inhibitors of Fe absorption were Fe-binding galloyl groups, as opposed to phenolic catechol groups, which had only a minor impact on Fe absorption. Hurrell (2004) suggests that most food polyphenols have the potential to actively restrict the absorption of non-haem Fe in our diet. Therefore, it seems more suitable to correlate Fe bio-accessibility with the specific phenolic group rather than the overall polyphenol amount. However, recent evidence has shown that polyphenols, such as phenolic acids and tannins, are thought to have antioxidant properties that in turn help to strengthen the body's immune system (Bello et al., 2022; Jena et al., 2023). Flavonoids, such as flavanols, chalcones, flavones, anthocyanidins, flavanones, and aminophenolics, are found in sufficient quantity in pearl millet and exhibit antioxidant properties (Hassan et al., 2021). These antioxidants are well-known substances that lessen the body's exposure to free radicals and provide anti-inflammatory properties (Lobo et al., 2010). Studies by Călinoiu and Vodnar (2018) and Hassan et al. (2021) have shown that the most prevalent form of hydroxycinnamic acid, which has strong antioxidant effects, is ferulic acid in pearl millet. They exist in bound form; therefore, their release into the colon can occur without the need for microbial action during digestion (Călinoiu & Vodnar, 2018; Hassan et al., 2021). Similarly, Chandrasekara and Shahidi (2011) reported that ferulic acid and p-coumaric acid, generally found in whole pearl millet, have the capacity to reduce the number of tumor cells. In a recent study by Krishnan et al. (2022), pearl millet phenolics were found to inhibit carbolytic enzymes, control glucose transport, and exhibit anti-diabetic activities. In a similar study, administration of whole-grain pearl millet flour (PMF) or its ethanolic extract to obese rats fed a high-fat diet resulted in significant decreases in fasting glucose, insulins and homeostatic model assessment-insulin resistant levels, supporting the hypoglycemic effect of pearl millet (Alzahrani et al., 2022).

Processing Techniques for Pearl Millet

Pearl millet is processed to enhance palatability, nutritional bioavailability, and digestibility (Table 15.2). This results in increased utility in developing nutrient-rich, gluten-free foods with good sensory profiles. Processing technologies like dehulling, milling, extrusion, roasting, germination, blanching, soaking, and other novel techniques, such as microwave treatment and ohmic heating, raise the quality of the grains by transforming them into various edible forms (e.g., rice, roasted, popped, sprouted, fermented, extruded, and salted) and porridges (Jaybhaye et al., 2014).

Table 15.2   Processing Technologies for Pearl Millet and Their Advantages

Processing method	Advantages of processing	References
Decortication	Increased Fe and Zn bioaccessibility	Krishnan & Meera (2022)
Milling	Increased flour shelf-life up to 3–4 months	Rani et al. (2018)
	Improved in vitro protein digestibility	Pushparaj & Urooj (2011)
Roasting	Increased oil absorption and water solubility index of flour	Obadina et al. (2016)
	Reduced antinutritional factors and improved shelf-life	Jalgaonkar et al. (2016)
Extrusion	Reduced antinutritional factors and enhanced product shelf-life	Patil & Kaur (2018)
	Higher yield	Moussa et al. (2022)
Blanching	Increased flour shelf-life and increased availability of minerals	Pawase et al. (2021)
	High in vitro Fe availability and reduced free fatty acid content	Bhati et al. (2016)
Germination/ malting	Increased bioaccessible polyphenols	Hithamani & Srinivasan (2014)

Dehulling/Decortication

Dehulling is the primary processing step for pearl millet grains before applying any other value-addition techniques. This process involves removing the husk or hull of the kernel using mechanical decorticators. Due to the relatively smaller size of pearl millet kernels compared with larger grains such as rice, wheat, oats, spelt, barley, and quinoa, dehulling can be challenging. Traditionally, the hull/ coarse outer layer (~10%) of pearl millet and other millets are removed by hand pounding. However, this process is tedious and inefficient because some of the husks remain in the grain, which can affect the quality of the flour produced. The steps involved in the mechanical dehulling of pearl millet are grading, destoning, dehulling, and aspiration. During this process, a small amount of flour is also generated, which can be further used in snacks and baked products. A typical de-huller consists of a hopper, abrasive stones, and hull and grain outlets. Some systems might also have an aspirator for the removal of the husk for efficient separation of the hull from grains. The rice huller with a polisher is also being used for dehulling purposes. The Rural Industries Innovation Center in Kane, Botswana, has developed a de-huller that can process 400–600 kg/h and is suitable for both sorghum and pearl millet decortication (Singh et al., 2022). The study by Krishnan and Meera (2022) found that decortication of pearl millet by 10%–15% increased the bio-accessibility of Fe and Zn.

Milling

Pearl millet is milled to produce traditional foods in Africa and Asia. The traditional milling process, which involves grinding grains using stones or other manual tools, can be laborious and time consuming. Additionally, this process often results in a coarse and uneven texture in the flour, which can affect its end-use performance in various applications, such as baking and cooking. Efficient and modern milling technologies can help enhance the nutritional value and marketability of food products made from pearl millet. The main objective of the milling process is to efficiently segregate endosperm, bran, and germ components from it so that they can be further processed to produce fine flour. Several difficulties arise during the milling of pearl millet because of its tiny kernel size, rigidly set germ, and firm endosperm (Pawase et al., 2019). Pearl millet is de-hulled and de-branned before milling to obtain refined flour with good storage stability. The application of pretreatments such as moist heat followed by drying grains to a moisture content of 10%–12% and dehulling has been effective in increasing the shelf life of flour produced from milling. This method was successfully industrialized by Central Food Technology Research Institute, India (Rani et al., 2018). The most common mills used are roller, plate, and hammer mills. Roller milling can be an economical option for separating the bran and germ portions efficiently and generating fine flour from pearl millet without the need for additional de-branning or polishing steps. Hammer mills generate flour with larger particle sizes, making it less suitable for producing thin and stiff porridge with a rough texture as well as baked and steamed food products with a smooth texture. Conversely, fine flour obtained from roller mills is well suited for creating such food products (Rai et al., 2008).

The production of whole-grain PMF is common, but its shelf life is low compared with refined flour due to the presence of bran and germ, which contain oils that are susceptible to rancidity over time. Furthermore, bitterness can occur due to the presence of tannins in certain varieties of pearl millet. The shelf life of PMF can be increased for a couple of months by storing it in an airtight, cool, and dry environment. Additionally, to combat the degradation of PMF, there are several pretreatment methods, such as soaking, steaming, blanching, roasting, microwaving, and fermentation, that reduce the bitterness caused by tannins and enhance the digestibility of starch and protein. Suitable pretreatment methods should be applied based on the end application of the flour, and these treatment efficiencies may also vary based on the pearl millet cultivar.

In a study by Pushparaj and Urooj (2011), two pearl millet cultivars (Kalukombu and Maharashtra Rabi Bajra) were examined under different milling processes, including whole flour, the bran-rich fraction, and semi-refined flour. The results indicated that the bran-rich fraction exhibited higher in vitro protein digestibility compared with the other fractions. These findings suggest that factors beyond tannin content, such as protein–protein interactions and non-protein components, can affect protein digestibility, emphasizing the complexity of the digestion process.

Roasting

Roasting is a simple method that can enhance the bioavailability of micronutrients in pearl millet by reducing its antinutritional properties. In a study conducted by Singh and Raghuvanshi (2012), it was shown that composite flour made from roasted finger millet, pearl millet, lentils, and sorghum had higher net protein utilization compared with composite flour obtained through other treatment methods such as dehulling, boiling, baking, and malting processes. Roasting improves the bioavailability of micronutrients and the functional properties of PMF. Obadina et al. (2016) investigated the outcome of roasting on whole pearl millet, which is further ground to obtain flour using a hammer mill. The study found that nutrient loss increased with higher roasting temperatures. Interestingly, roasting resulted in an increased water solubility index and oil absorption capacity of the flour samples, including its proximate composition (specifically carbohydrates, fat, ash, and energy). In a similar study, Jalgaonkar et al. (2016) examined the effect of roasting and hydrothermal treatments on the shelf life of PMF. The study found that roasting reduced phytic acid contents by 10%, fat acidity by 8%, and trypsin inhibitor activity by 9% in PMF. These findings suggest that roasting can be a great solution for decreasing the levels of antinutritional factors and improving the flavor profile, quality, and shelf life of PMF.

In a study conducted by Hithamani and Srinivasan (2014), the effects of different processing methods on the polyphenolic content and bioaccessibility of finger millet and pearl millet were investigated. Roasting resulted in a threefold increase in the total concentration of phenolic compounds and a 12% increase in polyphenol content. Furthermore, there was a significant improvement in bioaccessible flavonoid content and a 20% increase in bioaccessible polyphenols by sprouting and roasting. Overall, the bioaccessibility of phenolic compounds increased by 40% by sprouting, 85% by roasting, and 37% by pressure cooking (Hithamani & Srinivasan, 2014). These findings highlight the potential of roasting to improve the availability of beneficial phenolic compounds in pearl millet.

Extrusion

Extrusion cooking is a high-temperature and brief process that transforms starchy and protein-rich biomaterials into ready-to-eat (RTE) or ready-to-cook (RTC) food products with improved functionality. In this process, the flour is uniformly mixed with water to form a dough, which is then subjected to high shear and temperatures in a pressurized environment while being driven through an outlet with a narrow opening attached to a die. These processing conditions induce some key physicochemical changes, such as the degradation of lipids, denaturation of protein, and starch gelatinization (Yang et al., 2022). The resulting product (extrudate) is suitable for various applications in food processing, feed production, and the manufacturing of plant-based meats. It also exhibits improved digestibility, texture, and functionality, adding more value to the product. In addition to these advantages, extrusion cooking can help in the reduction of anti-nutritional factors, deactivation of lipases, and enhancement of product shelf life in pearl millet (Patil & Kaur, 2018; Sumathi et al., 2007).

A study conducted by Sumathi et al. (2007) shows that extrudates made from pearl millet grits blended with defatted soy and legumes had an extended shelf life of 6 months under ambient storage conditions. Similarly, Oliveira et al. (2021) explored the potential of combining pearl millet with maize to create healthier and more affordable extruded snacks. The study found that a 1:1 (w/w) ratio of pearl millet and maize produced snacks with good sensory characteristics. These snacks also had high fiber and protein contents, indicating their potential for commercialization compared with existing commercial maize extrudates. Additionally, Sobowale et al. (2021) conducted a study on the production of gluten-free snacks using a twin-screw extruder, where they substituted pearl millet with small amounts of walnut flour and corn starch in an 80:10:10 ratio. The findings revealed that this formulation improved the sensory characteristics of the snacks while maintaining high protein, fiber, and fat content.

Another notable study by Moussa et al. (2022) used extrusion cooking to produce traditional West African pearl millet couscous, which is known for its laborious and time-consuming preparation process. The method developed involves the use of dried and milled low-moisture extrudates in the preparation of couscous. This novel method produced a sevenfold increase in yield while maintaining the final quality of the product, which is closely comparable to couscous made from the traditional method.

Fermentation

Fermentation is a biochemical process during which diverse microorganisms, including bacteria and fungi, convert intricate substrates into simpler compounds. This transformation brings about numerous biochemical modifications, ultimately influencing the balance between nutritional and anti-nutritional components in grains. Consequently, these changes have the potential to affect the grains' storage characteristics (Singh et al., 2015). Fermentation increases the digestibility of pearl millet grain. For instance, a study conducted by Onyango et al. (2013) on the effect of malting and fermentation on pearl millet and sorghum flours has shown desirable effects. The digestibility of PMFs was significantly increased from 21.5% to 97.4%–98.3% upon fermentation at room temperature (25 °C) for 48 hours. This enhancement in digestibility was attributed to a natural fermentation process facilitated by environmental microflora, leading to lactic acid fermentation. Fermentation has also resulted in a significant decrease in levels of tannin and phytate content in both sorghum and PMF.

Fermentation plays a crucial role in enhancing the nutritional composition and preservation of food. This method not only contributes to the preservation of food products but also leads to improvements in flavor, color, and ultimately the nutritional content of the raw materials (Chinenye et al., 2017; Saleh et al., 2013). Similarly, research by Hassan et al. (2006) highlighted that fermentation of pearl millet leads to a reduction in antinutritional content in food grains while increasing protein availability, in vitro protein digestibility, starch digestibility, and overall nutritional composition of the grains. Fermentation also induces significant physicochemical changes in PMFs, potentially leading to improved functionality. A study conducted by Olamiti et al. (2020) showed that fermentation positively influenced color, thermal properties, crystallinity level, and functional groups of PMFs. Moreover, 24-hour fermentation of pearl millet demonstrated that protein and lipid contents remained relatively unchanged. However, there was a notable decrease in carbohydrate content accompanied by a corresponding increase in soluble sugars. Furthermore, amino acid analysis indicated significant reductions in glycine, lysine, and arginine contents due to fermentation. The process also led to a substantial reduction in trypsin and amylase inhibitor activities as well as phytic acid content. Interestingly, tannin content exhibited a significant increase after the fermentation process (Osman, 2012).

Hydrothermal Treatment

Since pearl millet has been used in gluten-free applications as PMF, many processing techniques have been used as pretreatments to improve the quality of PMF, with an emphasis on extending shelf life and increasing mineral availability. Bhati et al. (2016) reported the effects of various processing pretreatments, including blanching, acid treatment, dry heat treatment, and malting treatment, on in vitro Fe availability. Although the total Fe content decreased with processing, all the processing methods showed higher in vitro Fe availability. Among the methods, dry heat treatment for 120 min yielded the highest Fe availability (3.58 mg/100 g), whereas acid treatment for 24 h produced the lowest (2.39 mg/100 g). Additionally, it was established that 90 s of hot water blanching greatly decreased FFA concentration while increasing Fe availability.

Pawase et al. (2021) also evaluated the blanching effect on the shelf life and nutritional quality of different pearl millet cultivars. According to their study, unblanched pearl millet cultivars' lipid acidity increased noticeably over the course of 30 days, reaching a concentration of 210.4–370.8 mg KOH/100 g from 16.6–22.4 mg KOH/100 g. However, the blanched samples showed a much smaller increase, ranging from 11.23–18.6 to 22.4–33.6 mg KOH/100 g from Day 0 to Day 30. The study also revealed a similar trend in polyphenol contents and acid value in the blanched samples during the 30-day storage period. These findings suggest that blanching helps to preserve the quality and extend the shelf life of PMF by reducing lipid oxidation and increasing the availability of minerals such as Ca and P.

Vinutha et al. (2022) subjected a hydro-pretreated PMF to heat treatments, specifically thermal near-infrared (thNIR) and hydrothermal (HTh) treatments. In comparison to their respective individual treatments, the combined thNIR and HTh treatment decreased the enzymatic activity of polyphenol oxidase (100%), lipoxygenase (84.8%), lipase (47.8%), and peroxidase (98.1%). The FFA content (67.84%) and peroxide content (66.4%) significantly decreased in the thNIR-Hth–treated flour samples after 90 days of storage.

Other Processing Technologies

In addition to the aforementioned widely adopted processing methods, pearl millet has also been processed using various processing technologies such as irradiation, popping, malting/germination, toasting, etc. These methods seek to increase the use of pearl millet by lowering its anti-nutritional qualities and regulating rancidity.

ElShazali et al. (2011) explored the impact of radiation and various processing methods on PMF, considering antinutritional properties, sensory attributes, and protein digestibility. Dehulling alone reduced tannin and phytate concentration by over 50% and improved protein digestibility, but it also reduced the quality characteristics of both millet cultivars. Cooking significantly reduced antinutrients in both whole and dehulled flours. Radiation alone had no impact on antinutrients or the digestibility of proteins, but, when combined with cooking, it led to a considerable decrease in antinutritional qualities and an improvement in wheat quality attributes. However, boiling after radiation reduced the ability of proteins to be digested.

Malting is also a promising technique used to elevate the nutritional value of cereal grains. It involves controlled germination, which generates enzymes that initiate the breakdown of starch and other constituents within the grain. Obadina et al. (2017) studied the effect of varying malting durations on the physicochemical and nutritional attributes of pearl millet. The malting process resulted in increased levels of protein, crude fiber, lysine, and aspartic acid in PMF. The water solubility index and oil absorption capacity of the flour were also increased due to malting. On the other hand, the total phenolic, crabohydrate, ash, and fat content of the flour decreased as the malting period was extended. Significant changes were not observed in the phenylalanine, serine, and leucine content of the flours.

Popping is another simple and economic method known to improve the nutritional and sensory attributes of processed kernels. The process involves dry heat application to produce popped pearl millet, which is convenient for consumption as a RTE snack or specialty food. Garud et al. (2022) investigated various millet cultivars to determine their popping characteristics. Notably, the AIMP 92901 millet cultivar exhibited the most favorable popping attributes, characterized by the highest volume expansion ratio and popping yield. Scanning electron microscopy revealed a connection between pericarp thickness and popping yield, suggesting that a thicker pericarp contributes to improved popped pearl millet quality. The study concluded that a temperature of 260 °C yielded the best results for popping pearl millet, resulting in the highest sensory qualities.

Novel processing technologies, including microwave (Yadav et al., 2012), radiofrequency (Yarrakula et al., 2022), infrared (Vinutha et al., 2022), and ohmic heating (Dias-Martins et al., 2019; Mishra et al., 2021), as well as non-thermal methods such as pulsed electric field, cold plasma (Lokeswari et al., 2021), irradiation (ElShazali et al., 2011), high-pressure processing, ozonation, and ultrasound treatment (Vidhyalakshmi & Meera, 2023), are being explored in pearl millet along with other millet types (Joshi et al., 2023). These technologies have gained significant attention due to advantages such as a shorter processing time, less nutrient loss, and lower energy consumption. There are further research opportunities to be explored to enhance the utilization of pearl millet in gluten-free applications.

Value Addition of Pearl Millet

Despite their exceptional nutritional profile, millets consumption remains low among rural and poor populations. To ensure nutritional security and combat the escalating effects of climate change and widespread diseases, diversifying our food resources is imperative, and incorporating common millets is a key step. Pearl millet, with its gluten-free nature, abundance of bioactive substances possessing therapeutic benefits, and high density of micronutrients, presents an ideal choice for producing a range of functional and value-added food products. There is already a broad variety of value-added millets goods on the market, including biscuits, cakes, pasta, and baby food, captivating the attention of both financially prosperous individuals and health-conscious consumers.

Gluten-Free Application of Pearl Millet

Celiac disease, also referred to as gluten-sensitive enteropathy, is a persistent auto-immune disorder that affects people with genetic susceptibilities. It arises due to the consumption of gluten-containing foods, leading to inflammation of the mucosal lining, villous atrophy, and crypt hyperplasia (Nasr et al., 2016). The distinct feature of CD is an exaggerated immune response toward prolamins (grain proteins such as gliadin, avenin, hordein, and secalin), which are prevalent in cereal crops such as wheat, oats, barley, and rye, setting it apart from other auto-immune reactions (Niro et al., 2019). The predominance of CD, which affects about 1% of people worldwide, frequently goes undiagnosed (Rubio-Tapia et al., 2009). People with CD can only be treated by eating solely gluten-free foods throughout their lives.

Gluten-free pseudocereals and minor cereals, such as millet, are regarded as viable substitutes for traditional foods commonly consumed by individuals with CD. These crops, which are native to specific regions, have received limited documentation and are seldom utilized by the food industry. In fact, many of them fall under the category of "orphan crops" or "underutilized crops" (Cheng et al., 2017; Niro et al., 2019). These underutilized crops play a crucial role in numerous developing countries by ensuring food security and generating income for farmers who face resource constraints. Moreover, their inherent drought tolerance surpasses that of most major cereals (Cheng, 2018).

Pearl millet is the most important millet variety for human consumption. It accounts for 40% of global production and has a consistency similar to rice flour. Historically, around 4000 years ago, the West African Sahel region saw the domestication of pearl millet. From there, it spread to East Africa and India (Satyavathi et al., 2021; Sharma et al., 2021). Currently, >30 million ha of land are used to grow pearl millet, with the majority of that area being in Africa (>18 million ha) and Asia (>10 million ha) (Raheem et al., 2021). Pearl millet offers a number of benefits, including high nutritional and functional qualities. It is recognized as one of the best sources of nutrients and is frequently suggested to eradicate nutritional deficiencies in both children and geriatric groups, including the population that is gluten sensitive (Selladurai et al., 2023). People with issues related to CD and gluten intolerance are advised to switch to a diet that includes pearl millet because it is gluten free (Adiamo et al., 2018).

Although the use of these millets has somewhat increased recently, there has not been much attention paid to processing and adding value to these grains because of a lack of appropriate processing technology owing to difficulty in processing. The creation of modern processing techniques geared toward enhancing the production of value-added products is required to promote and popularize millets on a global scale. Only by expanding consumer demand and product acceptance on the worldwide market will this be possible. The dietary needs of the gluten-sensitive population have recently drawn attention to millets-based gluten-free goods. Further, the market for gluten-free goods generated over $4.63 billion in sales in 2017 and reached $6.47 billion in 2023, with a compound annual growth rate of 7.6% (Woomer & Adedeji, 2021). Because of this increasing demand, pearl millet–based products have garnered major attention in the current market scenario.

Bakery Products

Because of their convenient, RTE nature, long shelf life, and appealing sensory properties, bakery items are popular in both urban and rural areas. Products from bakeries are bought and consumed daily (Kale & Vairagar, 2018). Due to their potential as good carriers for bioactive chemicals, bakery products aid in increasing the intake of particular nutrients. Wheat flour or refined wheat flour is the essential ingredient in the preparation of baked goods; however, because of their low fiber content, millets are being used to replace a small percentage of wheat flour in an effort to provide a substitute, lessen reliance on wheat, and produce gluten-free bakery products (Saraswathi & Hameed, 2022).

Several studies have examined the production of bakery products such as bread biscuits and cookies by fully or partially substituting with PMF. Irondi et al. (2022) reported bread preparation by mixing PMF in different ratios with sweet detar flour (SDF) or sodium carboxymethyl cellulose (SCMC). According to their results, PMF and SDF blends (98.5% and 1.5%, respectively) had greater anti-amylase and anti-glucosidase activity than the PMF-SCMC blends. Additionally, breads made with PMF-SDF and PMF-SCMC had similar sensory qualities. As a result, SDF might work well in place of SCMC in the production of gluten-free bread based on PMF, which has improved starch-digesting enzyme inhibitory properties and is aimed at individuals with type II diabetes and gluten sensitivity. Maktouf et al. (2016) also report that adding 5% PMF to dough containing 95% wheat flour produces bread with enhanced rheological characteristics (e.g., dough strength, ratio of elasticity to extensibility, loaf volume, and texture). Fermentation has a substantial impact on the texture and quality of bread produced (Gobbetti et al., 2018). Nami et al. (2019) observed that the utilization of sourdough in the manufacturing process of pearl millet bread leads to enhanced product quality with extended shelf life.

Biscuits and cookies are some of the most popular baked goods consumed all over the world (Nehra et al., 2021). Usually produced with a combination of wheat flour, sugar, and oil, biscuits are a RTE food (Goubgou et al., 2021). Due to their widespread acceptance, low cost, and long shelf life, biscuits and cookies are among the most important RTE foods (Asna Urooj, 2014). Singh et al. (2019) prepared a biscuit with a combination of pearl millet and chickpea flour (50:50). The product had a higher level of protein, fiber, and Fe with improved sensory qualities. Kulkarni et al. (2021) also replaced wheat flour with PMF (40:60). The resulting product had superior textural and sensory properties. Similarly, Asna Urooj (2014) partially replaced refined wheat flour with PMF, and the result indicated no change in the physical characteristics compared with the control, but the overall nutritional and sensory attributes of the pearl millet cookies were higher. Thorat and Lande (2017) replaced refined wheat (40%) with PMF, and the results indicated improved physical and textural characteristics of the cookies. Adebiyi et al. (2016) investigated fermentation and malting's impact on flour and biscuits made from PMF. According to their findings, the physical, chemical, and nutritional attributes of the PMF and resulting biscuits were improved by fermentation and malting.

Puffed/Popped Products

Cereals have been puffed/popped for consumption as snacks for hundreds of years. Millets are popped to create a porous product with a good texture, distinct flavor, and low bulk density (Singh & Sehgal, 2008). Millet and cereal grains are primarily made up of starch. A variety of processing steps, particularly heating grains, causes the starch–protein matrix to undergo physicochemical and structural changes, which causes the grains to expand and become puffy (Deshpande & Nishad, 2021). Salt and sand roasting, oil frying, microwave popping, and gun puffing are the dry heat techniques that are used for popping and puffing in general (Jaybhaye et al., 2014; Mishra et al., 2014); however, puffing in hot oil (200–220 °C) or sand (~250 °C) are the most popular methods in India (Karel Hoke et al., 2005). Puffed/popped pearl millet has been incorporated into the manufacturing of nutritionally rich RTE snacks. Singh and Sehgal (2008) in their research prepared ladoo using puffed pearl millet. According to their results, ladoo prepared from 100% popped millet exhibited improved in vitro digestibility with higher cellulose and lignin content. Similarly, Singh et al. (2021) prepared a RTE snack bar using popped pearl millet. The study revealed that the popping process greatly increased nutrients like protein, Fe, Zn, and Ca, and the prepared bar was very palatable and simple to digest.

Extruded Products

One of the modern food processing techniques used to provide a variety of low-cost snacks, specialized foods, and supplemental nutrients is extrusion cooking

(Lakshmi Devi et al., 2014). It lessens the variables that are anti-nutritional, makes the food microbially safe, and increases consumer appeal (Mukhopadhyay & Bandyopadhyay, 2001). Additionally, it enhances the digestion of protein and carbohydrates and lessens nutritional degradation. Millets have been widely studied for their extrusion capabilities, and over the last decade scientists have evaluated the extrusion processing characteristics of some of the important millets. Kharat et al. (2019) evaluated the effects of the extrusion processing parameters such as moisture screw speed (300, 350, and 400 rpm), barrel temperature (110, 120, and 130 °C), and moisture (15%, 17.5%, and 20%). The result suggested that extrudates of pearl millet displayed an expansion ratio of 3.78 and a water absorption index of 2.69 g/g. In another study, Adebanjo et al. (2020) prepared an extruded product with PMF and carrot flour to compensate for the lack of β-carotene in PMF. According to the study, pearl millet and carrot composite flour are good sources of macronutrients such as Mn, K, Ca, Fe, and Zn. The flour blends are therefore appropriate for the bakery and confectionery sectors' creation of nutrient-dense, flour-based goods. In addition to having more total carotenoids, the sample made with carrot flour also created a flake with a pleasing look, which contributes to its acceptability. The inference is that millet and carrot can be combined to create a flake that is more nutrient dense than commercial cornflakes. Pasta is another popular ready-to-cook (RTC) extruded product. Rathi et al. (2004) compared the sensory attributes of pasta T1 (raw PMF + semolina) and T2 (depigmented PMF + semolina). According to their research, pasta prepared from depigmented PMF pasta (T2) showed enhanced sensory qualities, particularly the color. Similarly, Pushpa Devi et al. (2013) noticed that the blended mixture of PMF and finger millet flour in RTC extruded products resulted in satisfactory levels of nutritional profile, texture, color, and sensory characteristics.

Traditional Products

Koko

Koko is a popular porridge from West Africa (Amadou et al., 2011). The process involves soaking pearl millet overnight to produce koko, followed by wet milling of koko with spices (pepper, chili pepper, clove, and ginger) along with water to form a thick mixture. This mixture is then strained, fermented, and left to settle for 2–3 hours. The upper liquid layer is separated, heated for 1–2 hours, and combined with the settled sedimented layer until the desired consistency is achieved (Lei & Jakobsen, 2004). The millet grain is often steeped in the evening, and the entire process is completed by midday the next day (Amadou et al., 2011). *Lactobacillus fermentum*, *Weisella confuse*, and *Lactobacillus salivarius* are the main microbes in koko (Adimpong et al., 2012; Lei & Jakobsen, 2004).

Fura

Fura is another popular porridge or semi-solid dumpling from Africa (Amadou et al., 2011; Jideani et al., 2001). Amadou et al. (2011) described the detailed process of making fura. Using a locally made disc attrition mill, pearl millet is ground after being mildly moistened with water. The grain is sieved and processed in a hammer mill after being sun-dried, removing the hull from the grain. Black pepper, powdered ginger, and water (95 °C) are combined with PMF in a mortar, and the blend is thoroughly mashed with a pestle to form a smooth dough. The dough is further kneaded to form small balls by hand and cooked for 30 min at atmospheric pressure inside a cooking pan filled with hot water. While still hot, the balls are kneaded to create a smooth, slightly elastic mass. Following that, fura balls are formed from the dough. A porridge-like consistency is achieved by reconstituting the created stiff dough with sour milk.

Ben-saalga

Ben-saalga is a fermented gruel popular among young children in Burkina Faso of West Africa (Guyot et al., 2003). Pearl millet is processed primarily by washing and soaking, followed by grinding, kneading, sifting, settling, and cooking. The addition of aromatic substances is based on individual preferences. Ingredients like pepper, ginger, and mint are typically added in small quantities before pulverizing. The ground sample is allowed to rest, and the supernatant is boiled for 40 min. The paste is then added to the supernatant and cooked for 7 min (Amadou et al., 2011).

Flatbread

Flatbread, also called roti, chapati, etc., is the oldest type of bread consumed today (Neela & Fanta, 2020). Flatbread is customarily eaten in Europe, North and South Africa, China, the Middle East, Central America, and the Indian subcontinent (Boukid, 2022). PMF has been used in place of wheat flour to make flatbread. It provides additional benefits because it is gluten-free and has an increased content of micronutrients required for humans. It is prepared by mixing around 50 g of flour with 45 mL (about 1.52 oz) of warm water in order to make one roti. To create a cohesive dough, the flour–water combination is kneaded on a wooden board. The dough is formed into a ball and pushed with a wooden rod into a thin, circular sheet that is between 1.3 and 3 mm thick. This sheet is then baked for about 1 min on an iron or earthenware skillet (Nehra et al., 2021).

Dalia

Dalia is a traditional porridge made with milk and cereal commonly consumed in the Indian subcontinent (Jha & Patel, 2014). It has a short shelf life and is frequently made using wheat, milk, and sugar. Sorghum, maize, and pearl millet are examples of coarse grains that are more nutrient-dense than wheat and rice (Jha et al., 2013). In an attempt to prepare the nutritionally rich millet-based dalia, Singh et al. (2013) developed and optimized a standard process for instant multigrain dalia mix. They used pearl millet, maize, and sorghum for preparing the dalia mix. The produced mix offered an increased shelf life (>71 weeks) at 10 °C. In a similar study, Mridula et al. (2015) prepared a multigrain dalia mix using sprouted wheat, pearl millet, barley, and sorghum. The product was nutritionally rich (high Ca, Fe, and fiber content) with better sensory attributes.

Kunu

Kunu or kunu-zaki, a popular fermented non-alcoholic beverage from northern Nigeria, is primarily crafted from pearl millet, boasting a delightful combination of low viscosity, a sweet-sour flavor, and a creamy milky appearance (Adelekan et al., 2013). The manufacturing process entails steeping millet grains; wet milling alongside the inclusion of spices such as ginger, cloves, and pepper; partially gelatinizing the resulting paste; blending it with ungelatinized paste; fermenting the mixture; sieving it; blending it with sugar; and bottling it (Jideani et al., 2021). Sugar may be added to the mixture to improve its acceptability (Ndulaka et al., 2014). Fermentation and hydrolytic enzymes help to break complex carbohydrates into simple sugars.

Ontaku

Oshikundu, also called ontaku or oshiwambo, is a low-alcohol fermented beverage from the northern and northeastern regions of Namibia (Embashu et al., 2012). Pearl millet bran, PMF, malted sorghum flour, and water are the important ingredients for the manufacturing of the beverage (Embashu et al., 2013). Embashu et al. (2013) described the traditional method of processing ontaku. The manufacturing procedure includes the blending of mahangu flour and malted sorghum flour with boiling water. The mixture is cooled, and the pearl millet bran is added. Following this, the mixture is fermented, and the natural lactic acid bacteria facilitate the breaking down of complex starch into simple sugar, producing lactic acid. The decreased pH of the environment facilitates the growth of the yeast that ferments sugar to produce alcohol and CO_2.

Omalodu

Omalodu, also called omalovu or omalovugiilya, is a traditional, unpasteurized hazy beer from Namibia (Embashu et al., 2019). Regional variations aside, the customary stages engaged in the traditional brewing of African opaque/hazy beer encompass malting, drying, milling, souring (through lactic acid fermentation), boiling (to achieve starch gelatinization), mashing (for thinning and conversion of gelatinized starch into sugar), straining (to separate the grains), and, finally, alcoholic fermentation (Taylor, 2003).

Conclusion

Processing technologies offer a promising avenue for enhancing the bioavailability of micronutrients and elevating the overall dietary quality of pearl millet. Through effective processing and value addition, the consumption of pearl millet can be extended from rural to urban areas, strengthening its significance in diets across various demographics. The well-balanced nutritional composition and gluten-free nature of pearl millet make it a viable alternative for formulating a diverse range of value-added food products.

References

Abdulrahman, W. F., & Omoniyi, A. O. (2016). Proximate analysis and mineral compositions of different cereals available in Gwagwalada market, F.C.T, Abuja, Nigeria. *Journal of Advances in Food Science & Technology*, *3*(2), 50–55.

Adebanjo, L. A., Olatunde, G. O., Adegunwa, M. O., Dada, O. C., & Alamu, E. O. (2020). Extruded flakes from pearl millet (*Pennisetum glaucum*) – carrot (*Daucus carota*) blended flours- production, nutritional and sensory attributes. *Cogent Food and Agriculture*, *6*(1), 1733332. https://doi.org/10.1080/23311932 .2020.1733332

Adebiyi, J. A., Obadina, A. O., Mulaba-Bafubiandi, A. F., Adebo, O. A., & Kayitesi, E. (2016). Effect of fermentation and malting on the microstructure and selected physicochemical properties of pearl millet (*Pennisetum glaucum*) flour and biscuit. *Journal of Cereal Science*, *70*, 132–139. https://doi.org/10.1016/ j.jcs.2016.05.026

Adelekan, A. O., Alamu, A. E., Arisa, N. U., Adebayo, Y. O., & Dosa, A. S. (2013). Nutritional, microbiological and sensory characteristics of malted soy-Kunu Zaki: An improved traditional beverage. *Advances in Microbiology*, *03*(04), 389–397. https://doi.org/10.4236/aim.2013.34053

Adiamo, O. Q., Fawale, O. S., & Olawoye, B. (2018). Recent trends in the formulation of gluten-free sorghum products. *Journal of Culinary Science and Technology*, *16*(4), 311–325. https://doi.org/10.1080/15428052.2017.1388896

Adimpong, D. B., Nielsen, D. S., Sørensen, K. I., Derkx, P. M. F., & Jespersen, L. (2012). Genotypic characterization and safety assessment of lactic acid bacteria from indigenous African fermented food products. *BMC Microbiology*, *12*, 75. https://doi.org/10.1186/1471-2180-12-75

Alzahrani, N. S., Alshammari, G. M., El-ansary, A., & Yagoub, A. E. A. (2022). Anti-hyperlipidemia, hypoglycemic, and hepatoprotective impacts of pearl millet (*Pennisetum glaucum* L.) grains and their ethanol extract on rats fed a high-fat diet. *Nutrients*, *14*(1791), 1–21.

Amadou, I., Gbadamosi, O. S., & Le, G. W. (2011). Millet-based traditional processed foods and beverages: A review. *Cereal Foods World, 56*(3), 115–121. https://doi.org/10.1094/CFW-56-3-0115

Asna Urooj, F. S. P. (2014). Sensory, physical and nutritional qualities of cookies prepared from pearl millet (*Pennisetum typhoideum*). *Journal of Food Processing & Technology, 5*(10), 1000377. https://doi.org/10.4172/2157-7110.1000377

Bello, O. M., Ogbesejana, A. B., Balkisu, A., Osibemhe, M., Musa, B., & Oguntoye, S. O. (2022). Polyphenolic fractions from three millet types (fonio, finger millet, and pearl millet): Their characterization and biological importance. *Clinical Complementary Medicine and Pharmacology, 2*(1), 100020. https://doi.org/10.1016/j.ccmp.2022.100020

Bello, Z. A., Walker, S., & Tesfuhuney, W. (2019). Water relations and productivity of two lines of pearl millet grown on lysimeter with two different soil types. *Agricultural Water Management, 221*, 528–537. https://doi.org/10.1016/j.agwat.2019.05.024

Bhati, D., Bhatnagar, V., & Acharya, V. (2016). Effect of pre-milling processing techniques on pearl millet grains with special reference to iron availability. *Asian Journal of Dairy and Food Research, 35*(1), 76–80. https://doi.org/10.18805/ajdfr.v35i1.9256

Boncompagni, E., Orozco-Arroyo, G., Cominelli, E., Gangashetty, P. I., Grando, S., Tenutse Kwaku Zu, T., Daminati, M. G., Nielsen, E., & Sparvoli, F. (2018). Antinutritional factors in pearl millet grains: Phytate and goitrogens content variability and molecular characterization of genes involved in their pathways. *PloS One, 13*(6), e0198394. https://doi.org/10.1371/journal.pone.0198394

Boukid, F. (2022). Flatbread – a canvas for innovation: A review. *Applied Food Research, 2*(1), 100071. https://doi.org/10.1016/j.afres.2022.100071

Briani, C., Samaroo, D., & Alaedini, A. (2008). Celiac disease: From gluten to autoimmunity. *Autoimmunity Reviews, 7*(8), 644–650. https://doi.org/10.1016/j.autrev.2008.05.006

Brune, M., Hallberg, L., & Sdnberg, A.-B. (1991). Determination of iron-binding phenolic groups in foods. *Journal of Food Science, 56*(1), 128–131.

Călinoiu, L. F., & Vodnar, D. C. (2018). Whole grains and phenolic acids: A review on bioactivity, functionality, health benefits and bioavailability. *Nutrients, 10*(11), 1615. https://doi.org/10.3390/nu10111615

Chandrasekara, A., & Shahidi, F. (2011). Determination of antioxidant activity in free and hydrolyzed fractions of millet grains and characterization of their phenolic profiles by HPLC-DAD-ESI-MSn. *Journal of Functional Foods, 3*(3), 144–158. https://doi.org/10.1016/j.jff.2011.03.007

Cheng, A. (2018). Review: Shaping a sustainable food future by rediscovering long-forgotten ancient grains. *Plant Science, 269*, 136–142. https://doi.org/10.1016/j.plantsci.2018.01.018

Cheng, A., Mayes, S., Dalle, G., Demissew, S., & Massawe, F. (2017). Diversifying crops for food and nutrition security: A case of teff. *Biological Reviews, 92*(1), 188–198. https://doi.org/10.1111/brv.12225

Chinenye, O. E., Ayodeji, O. A., & Baba, A. J. (2017). Effect of fermentation (natural and starter) on the physicochemical, anti-nutritional and proximate composition of pearl millet used for flour production. *American Journal of Bioscience and Bioengineering, 5*(1), 12–16.

Dekka, S., Paul, A., Vidyalakshmi, R., & Mahendran, R. (2023). Potential processing technologies for utilization of millets: An updated comprehensive review. *Journal of Food Process Engineering, 46*, e14279. https://doi.org/10.1111/jfpe.14279

Deshpande, S. D., & Nishad, P. K. (2021). Technology for millet value-added products. In A. Kumar, M. K. Tripathi, D. Joshi, & V. Kumar (Eds.), *Millets and millet technology* (pp. 293–303). Springer. https://doi.org/10.1007/978-981-16-0676-2_14

Dias-Martins, A. M., Cappato, L. P., da Costa Mattos, M., Rodrigues, F. N., Pacheco, S., & Carvalho, C. W. P. (2019). Impacts of ohmic heating on decorticated and whole pearl millet grains compared to open-pan cooking. *Journal of Cereal Science, 85*, 120–129. https://doi.org/10.1016/j.jcs.2018.11.007

Dias-Martins, A. M., Pessanha, K. L. F., Pacheco, S., Rodrigues, J. A. S., & Carvalho, C. W. P. (2018). Potential use of pearl millet (*Pennisetum glaucum* (L.) R. Br.) in Brazil: Food security, processing, health benefits and nutritional products. *Food Research International, 109*, 175–186. https://doi.org/10.1016/j.foodres.2018.04.023

Doddabematti Prakash, S., Nkurikiye, E., Rajpurohit, B., Li, Y., & Siliveru, K. (2022). Significance of different milling methods on white proso millet flour physicochemical, rheological, and baking properties. *Journal of Texture Studies, 54*, 92–104. https://doi.org/10.1111/jtxs.12717

ElShazali, A. M., Nahid, A. A., Salma, H. A., & Elfadil, E. B. (2011). Effect of radiation process on antinutrients, protein digestibility and sensory quality of pearl millet flour during processing and storage. *International Food Research Journal, 18*(4), 1401–1407.

Eltom, M., Hofvander, Y., Torelm, I., & Fellström, B. (1984). Endemic goitre in the Darfur region (Sudan): Epidemiology and setiology. *Acta Medica Scandinavica, 215*(5), 467–475. https://doi.org/10.1111/j.0954-6820.1984.tb17680.x

Embashu, W., Cheikhyoussef, A., & Kahaka, G. (2012). *Survey on indigenous knowledge and household processing methods of Oshikundu; a cereal-based fermented beverage from Oshana, Oshikoto, Ohangwena and Omusati Regions in Namibia.* University of Namibia.

Embashu, W., Cheikhyoussef, A., Kahaka, G., Lendelvo, S., & Cheikyoussef, A. (2013). Processing methods of Oshikundu, a traditional beverage from sub-tribes within Aawambo culture in the Northern Namibia. *Journal for Studies in Humanities and Social Sciences, 2*(1), 117–127.

Embashu, W., Iileka, O., & Nantanga, K. K. M. (2019). Namibian opaque beer: a review. *Journal of the Institute of Brewing, 125*(1), 4–9. https://doi.org/10.1002/jib.533

Garud, S. R., Lamdande, A. G., Tavanandi, H. A., Mohite, N. K., & Nidoni, U. (2022). Effect of physicochemical properties on popping characteristics of selected pearl millet varieties. *Journal of the Science of Food and Agriculture*, *102*(15), 7370–7378.

Gobbetti, M., Pontonio, E., Filannino, P., Rizzello, C. G., De Angelis, M., & Di Cagno, R. (2018). How to improve the gluten-free diet: The state of the art from a food science perspective. *Food Research International*, *110*, 22–32. https://doi.org/10.1016/j.foodres.2017.04.010

Gonçalves, C. F. L., De Freitas, M. L., & Ferreira, A. C. F. (2017). Flavonoids, thyroid iodide uptake and thyroid cancer—A review. *International Journal of Molecular Sciences*, *18*(6), 1247. https://doi.org/10.3390/ijms18061247

Goubgou, M., Songré-Ouattara, L. T., Bationo, F., Lingani-Sawadogo, H., Traoré, Y., & Savadogo, A. (2021). Biscuits: A systematic review and meta-analysis of improving the nutritional quality and health benefits. *Food Production, Processing and Nutrition*, *3*(1), 26. https://doi.org/10.1186/s43014-021-00071-z

Gupta, S. M., Arora, S., Mirza, N., Pande, A., Lata, C., Puranik, S., Kumar, J., & Kumar, A. (2017). Finger millet: A "certain" crop for an "uncertain" future and a solution to food insecurity and hidden hunger under stressful environments. *Frontiers in Plant Science*, *8*(April), 1–11. https://doi.org/10.3389/fpls.2017.00643

Guyot, J.-P., Mouquet, C., Tou, E. H., Counil, E., Traore, A. S., & Treche, S. (2003). Study of the processing of pearl millet (*Pennisetum glaucum*) into ben-saalga, a fermented gruel from Burkina Faso. *Food-Based Approaches for a Healthy Nutrition*, *23-28*(11), 437–444.

Habiyaremye, C., Matanguihan, J. B., D'Alpoim Guedes, J., Ganjyal, G. M., Whiteman, M. R., Kidwell, K. K., & Murphy, K. M. (2017). Proso millet (*Panicum miliaceum* L.) and its potential for cultivation in the Pacific northwest, U.S.: A review. *Frontiers in Plant Science*, *7*, 1961. https://doi.org/10.3389/fpls.2016.01961

Hassan, A. B., Ahmed, I. A. M., Osman, N. M., Eltayeb, M. M., Osman, G. A., & Babiker, E. E. (2006). Effect of processing treatments followed by fermentation on protein content and digestibility of pearl millet (*Pennisetum typhoideum*) cultivars. *Pakistan Journal of Nutrition*, *5*(1), 86–89.

Hassan, Z. M., Sebola, N. A., & Mabelebele, M. (2021). The nutritional use of millet grain for food and feed: A review. *Agriculture and Food Security*, *10*(1), 1–14. https://doi.org/10.1186/s40066-020-00282-6

Hithamani, G., & Srinivasan, K. (2014). Effect of domestic processing on the polyphenol content and bioaccessibility in finger millet (*Eleusine coracana*) and pearl millet (*Pennisetum glaucum*). *Food Chemistry*, *164*, 55–62. https://doi.org/10.1016/j.foodchem.2014.04.107

Hoke, K., Housova, J., & Houska, M. (2005). Optimum conditions of rice puffing. *Czech Journal of Food Science*, *23*(1), 1–11.

Hurrell, R. F. (2004). Phytic acid degradation as a means of improving iron absorption. *International Journal of Vitamins and Nutrition Research*, *74*(6), 445–452.

Irondi, E. A., Imam, Y. T., & Ajani, E. O. (2022). Physicochemical, antioxidant and starch-digesting enzymes inhibitory properties of pearl millet and sweet detar gluten-free flour blends, and sensory qualities of their breads. *Frontiers in Food Science and Technology, 2.* https://doi.org/10.3389/frfst.2022.974588

Jalgaonkar, K., Jha, S. K., & Sharma, D. K. (2016). Effect of thermal treatments on the storage life of pearl millet (*Pennisetum glaucum*) flour. *Indian Journal of Agricultural Sciences, 86*(6), 762–767.

Jaybhaye, R. V., Pardeshi, I. L., Vengaiah, P. C., & Srivastav, P. P. (2014). Processing and technology for millet based food products: A review. *Journal of Ready to Eat Food, 1*(2), 32–48. http://www.jakraya.com/journal/jref

Jena, A., Sharma, V., & Dutta, U. (2023). Millets as superfoods: Let thy cereal be thy medicine. *Indian Journal of Gastroenterology, 42*(3), 304–307. https://doi.org/10.1007/s12664-023-01377-1

Jha, A., & Patel, A. A. (2014). Kinetics of HMF formation during storage of instant kheer mix powder and development of a shelf-life prediction model. *Journal of Food Processing and Preservation, 38*(1), 125–135. https://doi.org/10.1111/j.1745-4549.2012.00754.x

Jha, A., Tripathi, A. D., Alam, T., & Yadav, R. (2013). Process optimization for manufacture of pearl millet-based dairy dessert by using response surface methodology (RSM). *Journal of Food Science and Technology, 50*(2), 367–373. https://doi.org/10.1007/s13197-011-0347-7

Jha, N., Krishnan, R., & Meera, M. S. (2015). Effect of different soaking conditions on inhibitory factors and bioaccessibility of iron and zinc in pearl millet. *Journal of Cereal Science, 66,* 46–52. https://doi.org/10.1016/j.jcs.2015.10.002

Jideani, V. A., Nkama, I., Agbo, E. B., & Jideani, I. A. (2001). Survey of fura production in some northern states of Nigeria. *Plant Foods for Human Nutrition, 56,* 23–36.

Jideani, V. A., Ratau, M. A., & Okudoh, V. I. (2021). Non-alcoholic pearl millet beverage innovation with own bioburden: Leuconostocmesenteroides, pediococcuspentosaceus and enterococcus gallinarum. *Food, 10*(7), 1447. https://doi.org/10.3390/foods10071447

Joshi, T. J., Singh, S. M., & Rao, P. S. (2023). Novel thermal and non-thermal millet processing technologies: Advances and research trends. *European Food Research and Technology, 249,* 1149–1160. https://doi.org/10.1007/s00217-023-04227-8

Kale, P. G., & Vairagar, P. R. (2018). Application of pearl millet in functional food. *International Journal of Agricultural Engineering, 11*(1), 90–94.

Kamath, M. V., & Belavady, B. (1980). Unavailable carbohydrates of commonly consumed Indian foods. *Journal of the Science of Food and Agriculture, 31,* 192–202.

Kharat, S., Medina-Meza, I. G., Kowalski, R. J., Hosamani, A., Ramachandra, C. T., Hiregoudar, S., & Ganjyal, G. M. (2019). Extrusion processing characteristics of whole grain flours of select major millets (foxtail, finger, and pearl). *Food and Bioproducts Processing, 114*, 60–71. https://doi.org/10.1016/j.fbp.2018.07.002

Krishnan, R., & Meera, M. S. (2018). Pearl millet minerals: Effect of processing on bioaccessibility. *Journal of Food Science and Technology, 55*(9), 3362–3372. https://doi.org/10.1007/s13197-018-3305-9

Krishnan, R., & Meera, M. S. (2022). Monitoring bioaccessibility of iron and zinc in pearl millet grain after sequential milling. *Journal of Food Science and Technology, 59*(2), 784–795. https://doi.org/10.1007/s13197-021-05072-x

Krishnan, V., Verma, P., Saha, S., Singh, B., Vinutha, T., Kumar, R. R., Kulshreshta, A., Singh, S. P., Sathyavathi, T., Sachdev, A., & Praveen, S. (2022). Polyphenol-enriched extract from pearl millet (*Pennisetum glaucum*) inhibits key enzymes involved in post prandial hyper glycemia (α-amylase, α-glucosidase) and regulates hepatic glucose uptake. *Biocatalysis and Agricultural Biotechnology, 43*, 102411. https://doi.org/10.1016/j.bcab.2022.102411

Kulkarni, D. B., Sakhale, B. K., & Chavan, R. F. (2021). Studies on development of low gluten cookies from pearl millet and wheat flour. *Food Research, 5*(4), 114–119. https://doi.org/10.26656/fr.2017.5(4).028

Kumar, R. R., Bhargava, D. V., Pandit, K., Goswami, S., Mukesh Shankar, S., Singh, S. P., Rai, G. K., Tara Satyavathi, C., & Praveen, S. (2021). Lipase – The fascinating dynamics of enzyme in seed storage and germination – A real challenge to pearl millet. *Food Chemistry, 361*, 130031. https://doi.org/10.1016/j.foodchem.2021.130031

Lakshmi Devi, N., Shobha, S., Alavi, S., Kalpana, K., & Soumya, M. (2014). Utilization of extrusion technology for the development of millet based complementary foods. *Journal of Food Science and Technology, 51*(10), 2845–2850. https://doi.org/10.1007/s13197-012-0789-6

Léder, I. (2004). Sorghum and millets. In G. Fuleky (Ed.), *Cultivated plants, primarily as food sources* (Vol. 1, pp. 66–83). EOLSS.

Lei, V., & Jakobsen, M. (2004). Microbiological characterization and probiotic potential of koko and koko sour water, African spontaneously fermented millet porridge and drink. *Journal of Applied Microbiology, 96*(2), 384–397. https://doi.org/10.1046/j.1365-2672.2004.02162.x

Lemgharbi, M., Souilah, R., Belhadi, B., Terbag, L., Djabali, D., & Nadjemi, B. (2017). Starch digestion in pearl millet (*Pennisetum glaucum* (L.) R. Br.) flour from arid area of Algeria. *Journal of Applied Botany and Food Quality, 90*(January), 126–131. https://doi.org/10.5073/JABFQ.2017.090.015

Lobo, V., Patil, A., Phatak, A., & Chandra, N. (2010). Free radicals, antioxidants and functional foods: Impact on human health. *Pharmacognosy Reviews, 4*(8), 118–126. https://doi.org/10.4103/0973-7847.70902

Lokeswari, R., Sharanyakanth, P. S., Jaspin, S., & Mahendran, R. (2021). Cold plasma effects on changes in physical, nutritional, hydration, and pasting properties of pearl millet (*Pennisetum glaucum*). *IEEE Transactions on Plasma Science, 49*(5), 1745–1751. https://doi.org/10.1109/TPS.2021.3074441

Maktouf, S., Jeddou, K. B., Moulis, C., Hajji, H., Remaud-Simeon, M., & Ellouz-Ghorbel, R. (2016). Evaluation of dough rheological properties and bread texture of pearl millet-wheat flour mix. *Journal of Food Science and Technology, 53*(4), 2061–2066. https://doi.org/10.1007/s13197-015-2065-z

Mishra, G., Joshi, D. C., & Kumar Panda, B. (2014). Popping and puffing of cereal grains: A review. *Journal of Grain Processing and Storage, 1*(2), 34–46. http://www.jakraya.com/journal/jgps

Mishra, N., Jyoti, Y., & Jain, S. (2021). Role of ohmic heating in stabilization of pearl millet: An overview. *The Pharma Innovation Journal, 10*(10), 232–234.

Moss, R., & McSweeney, M. B. (2022). Effect of quinoa, chia and millet addition on consumer acceptability of gluten-free bread. *International Journal of Food Science and Technology, 57*(2), 1248–1258. https://doi.org/10.1111/ijfs.15509

Moussa, M., Ponrajan, A., Campanella, O. H., Okos, M. R., Martinez, M. M., & Hamaker, B. R. (2022). Novel pearl millet couscous process for west African markets using a low-cost single-screw extruder. *International Journal of Food Science and Technology, 57*(7), 4594–4601. https://doi.org/10.1111/ijfs.15797

Mridula, D., Sharma, M., & Gupta, R. K. (2015). Development of quick cooking multi-grain dalia utilizing sprouted grains. *Journal of Food Science and Technology, 52*(9), 5826–5833. https://doi.org/10.1007/s13197-014-1634-x

Mukhopadhyay, N., & Bandyopadhyay, S. (2001). Extrusion cooking technology employed to reduce the anti-nutritional factor tannin in sesame (*Sesamum indicum*) meal. *Journal of Food Engineering, 56*, 201–201. http://www.elsevier.com/locate/jfoodeng

Nambiar, V. S., Sareen, N., Shahu, T., Desai, R., Dhaduk, J. J., & Nambiar, S. (2011). Potential functional implications of pearl millet (*Pennisetum glaucum*) in health and disease. *Journal of Applied Pharmaceutical Science, 1*(10), 62–67.

Nami, Y., Gharekhani, M., Aalami, M., & Hejazi, M. A. (2019). Lactobacillus-fermented sourdoughs improve the quality of gluten-free bread made from pearl millet flour. *Journal of Food Science and Technology, 56*(9), 4057–4067. https://doi.org/10.1007/s13197-019-03874-8

Nasr, I., Nasr, I., Al Shekeili, L., Al Wahshi, H. A., Nasr, M. H., & Ciclitira, P. J. (2016). Celiac disease, wheat allergy and non celiac gluten sensitivity. *Integrative Food, Nutrition and Metabolism, 3*(4). https://doi.org/10.15761/ifnm.1000155

Ndulaka, J. C., Obasi, N. E., & Omeire, G. C. (2014). Production and evaluation of reconstitutable kunun-zaki. *Nigerian Food Journal, 32*(2), 66–72.

Neela, S., & Fanta, S. W. (2020). Injera (an ethnic, traditional staple food of Ethiopia): A review on traditional practice to scientific developments. *Journal of Ethnic Foods, 7*(1), 32. https://doi.org/10.1186/s42779-020-00069-x

Nehra, M., Siroha, A. K., Punia, S., & Kumar, S. (2021). Process standardization for bread preparation using composite blend of wheat and pearl millet: Nutritional, antioxidant and sensory approach. *Current Research in Nutrition and Food Science*, *9*(2), 511–520. https://doi.org/10.12944/CRNFSJ.9.2.14

Niro, S., D'Agostino, A., Fratianni, A., Cinquanta, L., & Panfili, G. (2019). Gluten-free alternative grains: Nutritional evaluation and bioactive compounds. *Food*, *8*(6), 208. https://doi.org/10.3390/foods8060208

Obadina, A. O., Arogbokun, C. A., Soares, A. O., de Carvalho, C. W. P., Barboza, H. T., & Adekoya, I. O. (2017). Changes in nutritional and physico-chemical properties of pearl millet (*Pennisetum glaucum*) ex-borno variety flour as a result of malting. *Journal of Food Science and Technology*, *54*, 4442–4451.

Obadina, A. O., Ishola, I. O., Adekoya, I. O., Soares, A. G., de Carvalho, C. W. P., & Barboza, H. T. (2016). Nutritional and physico-chemical properties of flour from native and roasted whole grain pearl millet (*Pennisetum glaucum* [L.]R. Br.). *Journal of Cereal Science*, *70*, 247–252. https://doi.org/10.1016/j.jcs.2016.06.005

Olamiti, G., Takalani, T. K., Beswa, D., & Jideani, A. I. O. (2020). Effect of malting and fermentation on colour, thermal properties, functional groups and crystallinity level of flours from pearl millet (*Pennisetum glaucum*) and sorghum (*Sorghum bicolor*). *Heliyon*, *6*(12), e05467.

Oliveira, D. P. L., Soares Júnior, M. S., Bento, J. A. C., dos Santos, I. G., & de Castro Ferreira, T. A. P. (2021). Influence of extrusion conditions on the physical and nutritional properties of snacks from maize and pearl millet. *Journal of Food Processing and Preservation*, *45*(3), e15215. https://doi.org/10.1111/jfpp.15215

Onyango, C. A., Ochanda, S. O., Mwasaru, M. A., Ochieng, J. K., Mathooko, F. M., & Kinyuru, J. N. (2013). Effects of malting and fermentation on anti-nutrient reduction and protein digestibility of red sorghum, white sorghum and pearl millet. *Journal of Food Research*, *2*(1), 41.

Osman, A. K. (2012). Bulrush millet (*Pennisetum typhoides*) a contributory factor to the endemicity of goitre in western Sudan. *Ecology of Food and Nutrition*, *11*(2), 121–128.

Osman, A. K., & Fatah, A. A. (1981). Factors other than iodine deficiency contributing to the endemicity of goitre in Darfur province (Sudan). *International Journal of Food Sciences and Nutrition*, *35*(4), 302–309. https://doi.org/10.3109/09637488109143057

Patil, S. S., & Kaur, C. (2018). Current trends in extrusion: Development of functional foods and novel ingredients. *Food Science and Technology Research*, *24*(1), 23–34. https://doi.org/10.3136/fstr.24.23

Pawase, P., Chavan, U., & Lande, S. (2019). Pearl millet processing and its effect on antinational factors? *International Journal of Food Sciences and Nutrition*, *4*(6), 10–18.

Pawase, P. A., Chavan, U. D., & Kotecha, P. M. (2021). Effect of blanching on nutritional quality of different pearl millet cultivars. *Biological Forum–An International Journal*, *13*(1), 650–655.

Pei, J. J., Umapathy, V. R., Vengadassalapathy, S., Hussain, S. F. J., Rajagopal, P., Jayaraman, S., Veeraraghavan, V. P., Palanisamy, C. P., & Gopinath, K. (2022).

A review of the potential consequences of pearl millet (*Pennisetum glaucum*) for diabetes mellitus and other biomedical applications. *Nutrients, 14*(14), 2932. https://doi.org/10.3390/nu14142932

Pushpa Devi, M., Sangeetha, N., & Scholar, R. (2013). Extraction and dehydration of millet milk powder for formulation of extruded product. *IOSR Journal of Environmental Science, 7*(1), 63–70.

Pushparaj, F. S., & Urooj, A. (2011). Influence of processing on dietary fiber, tannin and in vitro protein digestibility of pearl millet. *Food and Nutrition Sciences, 2*(8), 895–900.

Raboy, V. (2002). Progress in breeding low phytate crops. *The Journal of Nutrition, 132*, 503S–505S.

Raheem, D., Dayoub, M., Birech, R., & Nakiyemba, A. (2021). The contribution of cereal grains to food security and sustainability in Africa: Potential application of UAV in Ghana, Nigeria, Uganda, and Namibia. *Urban Science, 5*(1), 8. https://doi.org/10.3390/urbansci5010008

Rai, K. N., Gowda, C. L. L., Reddy, B. V. S., & Sehgal, S. (2008). Adaptation and potential uses of sorghum and pearl millet in alternative and health foods. *Comprehensive Reviews in Food Science and Food Safety, 7*(4), 320–396.

Rani, S., Singh, R., Sehrawat, R., Kaur, B. P., & Upadhyay, A. (2018). Pearl millet processing: A review. *Nutrition and Food Science, 48*(1), 30–44. https://doi.org/10.1108/NFS-04-2017-0070

Rathi, A., Kawatra, A., & Sehgal, S. (2004). Influence of depigmentation of pearl millet (*Pennisetum glaucum* L.) on sensory attributes, nutrient composition, in vitro protein and starch digestibility of pasta. *Food Chemistry, 85*(2), 275–280. https://doi.org/10.1016/j.foodchem.2003.06.021

Reilly, N. R., & Green, P. H. R. (2012). Epidemiology and clinical presentations of celiac disease. *Seminars in Immunopathology, 34*(4), 473–478. https://doi.org/10.1007/s00281-012-0311-2

Rubio-Tapia, A., Kyle, R. A., Kaplan, E. L., Johnson, D. R., Page, W., Erdtmann, F., Brantner, T. L., Kim, W. R., Phelps, T. K., Lahr, B. D., Zinsmeister, A. R., Melton, L. J., & Murray, J. A. (2009). Increased prevalence and mortality in undiagnosed celiac disease. *Gastroenterology, 137*(1), 88–93. https://doi.org/10.1053/j.gastro.2009.03.059

Sabuz, A. A., Rana, M. R., Ahmed, T., Molla, M. M., Islam, N., Khan, H. H., Chowdhury, G. F., Zhao, Q., & Shen, Q. (2023). Health-promoting potential of millet: A review. *Separations, 10*(2), 80. https://doi.org/10.3390/separations10020080

Saleh, A. S. M., Zhang, Q., Chen, J., & Shen, Q. (2013). Millet grains: Nutritional quality, processing, and potential health benefits. *Comprehensive Reviews in Food Science and Food Safety, 12*(3), 281–295. https://doi.org/10.1111/1541-4337.12012

Sandberg, A.-S. (2001). In vitro and in vivo degradation of phytate. In N. R. Reddy & S. K. Sathe (Eds.), *Food phytates* (Vol. 1, pp. 139–156). CRC Press.

Sankar, R., Pandav, C. S., & Sripathy, G. (2008). Ethics and public health policy: Lessons from salt iodization program in India. *Comprehensive Reviews in Food Science and Food Safety*, *7*(4), 386–389. https://doi.org/10.1111/j.1541-4337.2008.00049.x

Saraswathi, R., & Hameed, R. S. (2022). Value addition influenced in millet products: A review. *Journal of Nutrition and Food Sciences*, *10*(1), 1161.

Satyavathi, C. T., Ambawat, S., Khandelwal, V., & Srivastava, R. K. (2021). Pearl millet: A climate-resilient Nutricereal for mitigating hidden hunger and provide nutritional security. In *Frontiers in Plant Science* (Vol. *12*, pp. 659938). https://doi.org/10.3389/fpls.2021.659938

Selladurai, M., Pulivarthi, M. K., Raj, A. S., Iftikhar, M., Prasad, P. V. V., & Siliveru, K. (2023). Considerations for gluten free foods - pearl and finger millet processing and market demand. *Grain and Oil Science and Technology*, *6*(2), 59–70. https://doi.org/10.1016/j.gaost.2022.11.003

Sharma, S., Sharma, R., Govindaraj, M., Mahala, R. S., Satyavathi, C. T., Srivastava, R. K., Gumma, M. K., & Kilian, B. (2021). Harnessing wild relatives of pearl millet for germplasm enhancement: Challenges and opportunities. *Crop Science*, *61*(1), 177–200. https://doi.org/10.1002/csc2.20343

Singh, A. K., Rehal, J., Kaur, A., & Jyot, G. (2015). Enhancement of attributes of cereals by germination and fermentation: A review. *Critical Reviews in Food Science and Nutrition*, *55*(11), 1575–1589.

Singh, B. P., Jha, A., Sharma, N., & Rasane, P. (2013). Optimization of a process and development of a shelf life prediction model for instant multigrain dalia mix. *Journal of Food Process Engineering*, *36*(6), 811–823. https://doi.org/10.1111/jfpe.12050

Singh, G., & Sehgal, S. (2008). Nutritional evaluation of ladoo prepared from popped pearl millet. *Nutrition & Food Science*, *38*(4), 310–315. https://doi.org/10.1108/00346650810891360

Singh, P. R., & Raghuvanshi, R. S. (2012). Finger millet for food and nutritional security. *African Journal of Food Science*, *6*(4). https://doi.org/10.5897/ajfsx10.010

Singh, R., Nain, M. S., & Manju, M. (2019). Nutrient analysis and acceptability of different ratio pearl millet (*Pennisetum glaucum*) based biscuits. *Indian Journal of Agricultural Science*, *90*(2), 428–430.

Singh, R., Singh, K., & Singh Nain, M. (2021). Nutritional evaluation and storage stability of popped pearl millet bar. *Current Science*, *120*(8), 1374–1381.

Singh, S., Singh, A. K., Singh, K., & Singh, B. (2022). Millets processing, nutritional quality & fermented product: A review. *The Pharma Innovation Journal*, *11*(5), 304–308.

Sobowale, S. S., Kewuyemi, Y. O., & Olayanju, A. T. (2021). Process optimization of extrusion variables and effects on some quality and sensory characteristics of extruded snacks from whole pearl millet-based flour. *SN Applied Sciences*, *3*(10), 824. https://doi.org/10.1007/s42452-021-04808-w

Sumathi, A., Ushakumari, S. R., & Malleshi, N. G. (2007). Physico-chemical characteristics, nutritional quality and shelf-life of pearl millet based extrusion cooked supplementary foods. *International Journal of Food Sciences and Nutrition*, *58*(5), 350–362. https://doi.org/10.1080/09637480701252187S.S

Taylor, J. R. N. (2003). Fermented foods: Beverages from sorghum and millet. In *Encyclopedia of food sciences and nutrition* (Vol. 1, pp. 2352–2359). Academic Press.

Taylor, J. R. N. (2016). Millet Pearl: Overview. In *Encyclopedia of food grains* (Vol. 1, pp. 190–198). Academic Press.

Thorat, A. A. K., & Lande, S. B. (2017). Evaluation of physical and textural properties of cookies prepared from pearl millet flour. *International Journal of Current Microbiology and Applied Sciences*, *6*(4), 692–701. https://doi.org/10.20546/ijcmas.2017.604.085

Verma, D. K., & Srivastav, P. P. (2017). Proximate composition, mineral content and fatty acids analyses of aromatic and non-aromatic Indian rice. *Rice Science*, *24*(1), 21–31. https://doi.org/10.1016/j.rsci.2016.05.005

Vidhyalakshmi, R., & Meera, M. S. (2023). Dry heat and ultrasonication treatment of pearl millet flour: Effect on thermal, structural, and in-vitro digestibility properties of starch. *Journal of Food Measurement and Characterization*, *17*, 2858–2868. https://doi.org/10.1007/s11694-023-01832-9

Vinutha, T., Kumar, D., Bansal, N., Krishnan, V., Goswami, S., Kumar, R. R., Kundu, A., Poondia, V., Rudra, S. G., Muthusamy, V., Rama Prashat, G., Venkatesh, P., Kumari, S., Jaiswal, P., Singh, A., Sachdev, A., Singh, S. P., Satyavathi, T., Ramesh, S. V., & Praveen, S. (2022). Thermal treatments reduce rancidity and modulate structural and digestive properties of starch in pearl millet flour. *International Journal of Biological Macromolecules*, *195*, 207–216. https://doi.org/10.1016/j.ijbiomac.2021.12.011

Woomer, J. S., & Adedeji, A. A. (2021). Current applications of gluten-free grains: A review. *Critical Reviews in Food Science and Nutrition*, *61*(1), 14–24. https://doi.org10.1080/10408398.2020.1713724

Yadav, D. N., Anand, T., Kaur, J., & Singh, A. K. (2012). Improved storage stability of pearl millet flour through microwave treatment. *Agricultural Research*, *1*(4), 399–404. https://doi.org10.1007/s40003-012-0040-8

Yang, T., Ma, S., Liu, J., Sun, B., & Wang, X. (2022). Influences of four processing methods on main nutritional components of foxtail millet: A review. *Grain and Oil Science and Technology*, *5*(3), 156–165. https://doi.org10.1016/j.gaost.2022.06.005

Yarrakula, S., Suraj, A. B., Rehaman, A., & Saravanan, S. (2022). Effect of hot air assisted radio frequency technology on physical and functional properties of pearl millet. *The Pharma Innovation*, *11*(March), 633–637.

Printed and bound by CPI Group (UK) Ltd, Croydon, CR0 4YY

17/04/2025

14658857-0001